SB189
5
S43
1983

MANTU GUPTA

AF725329

DISCARDED

SEED PROTEINS

ADVANCES IN AGRICULTURAL BIOTECHNOLOGY

Already published in this series

Marcelle R., Clijsters H. and Van Poucke M, eds: Effects of Stress on Photosynthesis. 1983 ISBN 90-247-2799-5

Gottschalk W. and Müller H.P., eds: Seed Proteins: Biochemistry, Genetics, Nutritive Value 1983. ISBN 90-247-2789-8

In preparation

Akazawa T., ed: The New Frontiers in Biotechnology

Seed Proteins

Biochemistry, Genetics, Nutritive Value

edited by

WERNER GOTTSCHALK
HERMANN P. MÜLLER

Institute of Genetics
University of Bonn

1983

MARTINUS NIJHOFF / DR W. JUNK PUBLISHERS
THE HAGUE / BOSTON / LONDON

Distributors:

for the United States and Canada
Kluwer Boston, Inc.
190 Old Derby Street
Hingham, MA 02043
USA

for all other countries
Kluwer Academic Publishers Group
Distribution Center
P.O. Box 322
3300 AH Dordrecht
The Netherlands

Library of Congress Cataloging in Publication Data

Main entry under title:

Seed proteins.

(Advances in agricultural biotechnology)
1. Grain--Breeding. 2. Legumes--Breeding. 3. Grain
--Seed--Composition. 4. Legumes--Seed--Composition.
5. Plant proteins. 6. Grain--Genetics. 7. Legumes--
Genetics. I. Gottschalk, Werner. II. Müller,
Hermann P. III. Series.
SB189.5.S43 1983 633.1'042 82-22377
ISBN 90-247-2789-8

ISBN 90-247-2789-8 (this volume)
ISBN 90-247-2790-1 (series)

Copyright © 1983 by Martinus Nijhoff/Dr W. Junk Publishers, The Hague.

All rights reserved. No part of this publication may be reproduced, stored in a retrieval system, or transmitted in any form or by any means, mechanical, photocopying, recording, or otherwise, without the prior written permission of the publishers, Martinus Nijhoff/Dr W. Junk Publishers, P.O. Box 566, 2501 CN The Hague, The Netherlands.

PRINTED IN THE NETHERLANDS

Contents

List of Contributors

BEAMES, R.M., Department of Animal Science, University of British Columbia, Vancouver, Canada.

BHATIA, C.R., Biology and Agriculture Division, Bhabha Atomic Research Centre, Bombay, India.

BROWN, J.W.S., Agrigenetics Research Park, Madison, Wisconsin, U.S.A.

CASEY, R., John Innes Institute, Norwich, England.

DENIĆ, M., Maize Research Institute 'Zemun Polje', Belgrade, Yugoslavia.

DOLL, H., Risø National Laboratory, Roskilde, Denmark.

EGGUM, B.O., Animal Physiology and Chemistry, National Institute of Animal Science, Copenhagen, Denmark.

ERSLAND, D.R., Agrigenetics Research Park, Madison, Wisconsin, U.S.A.

GOTTSCHALK, W., Institute of Genetics, University of Bonn, Bonn, F.R.Germany.

HALL, T.C., Agrigenetics Research Park, Madison, Wisconsin, U.S.A.

KOSHIYAMA, I., Central Research Laboratory of Kikkoman Shoyu Co., Noda-Shi, Chiba-Ken, Japan.

LADIZINSKY, G., Faculty of Agriculture, The Hebrew University, Rehovot, Israel.

LAFIANDRA, D., Germplasm Institute, CNR, Bari, Italy.

MICKE, A., Plant Breeding and Genetics Section, Joint FAO/IAEA Division, International Atomic Energy Agency, Vienna, Austria.

MIFLIN, B.J., Biochemistry Department, Rothamstead Experimental Station, Harpenden, Herts., England.

MÜLLER, H.P., Institute of Genetics, University of Bonn, Bonn, F.R.Germany.

PIETSCH, G., Bundes-Sortenamt, Hannover, F.R.Germany.

PORCEDDU, E., Germplasm Institute, CNR, Bari, Italy.

SCARASCIA-MUGNOZZA, G.T., Plant Biology Institute, Tuscia University, Viterbo, Italy.

SHEWRY, P.R., Biochemistry Department, Rothamstead Experimental Station, Harpenden, Herts., England.

STEGEMANN, H., Institut für Biochemie, Biologische Bundesanstalt, Braunschweig, F.R.Germany.

SWAMINATHAN, M.S., International Rice Research Institute, Los Baños, Laguna, The Philippines.

TANAKA, S., Institute of Radiation Breeding, Ohmiya-Machi, Ibaraki-Ken, Japan.
WILSON, C.M., U.S. Department of Agriculture, Science and Education Administration, Agricultural Research; Department of Agronomy, University of Illinois, Urbana, U.S.A.
WOLFF, G., Institute of Genetics, University of Bonn, Bonn, F.R.Germany.

Preface

Investigations on seed proteins have been intensively carried out during the past two decades. This is valid with regard to both their chemical composition as well as their nutritive value. The development of new biochemical and physical methods has resulted in obtaining deep insights into the structures of seed proteins and their mutual interactions. Intensive exchange of information between the scientists participating in national and international research programmes has given strong impulses for intensifying the research in this field. For the quantitative and qualitative investigations of seed proteins, not only some model plants were used; on the contrary, they were carried out on a large number of different crops important for different regions of the earth. In this way, a level of knowledge has been reached which could not be expected in this diversity within such a short period. This holds not only true for biochemical but also for physiological characters of the species studied. With regard to nutritional aspects, the problem of the limiting amino acids was of special interest, but also seed proteins acting as antinutritional factors were analysed in detail.

Based on the knowledge of seed protein structures, it was possible to perform investigations on the genetic basis of their synthesis. This was done under two different aspects: The basic knowledge on the genes involved should be widened; moreover, it should be tried to improve the seed proteins quantitatively and qualitatively under the influence of mutant genes. The optimistic expectations in this field have been fulfilled only in a few cases, but plenty of new information has been obtained being of considerable importance for understanding more details of gene action in higher plants.

The aspects just mentioned were decisive for the selection of the contributions compiled in the present book. It contains preferentially review articles supplemented by some original papers. In some mainly theoretical chapters, the relevance of seed protein improvement of our crops as well as the extent of international cooperation in this field are discussed. In the more empirical chapters, methodological as well as biochemical, physiological, genetical, evolutionary and nutritional aspects are considered. Moreover, those cultivated plants, which are of particular importance for human nutrition, are discussed in detail. In this way, a balanced survey on the present state of knowledge on seed proteins is given.

Bonn, September 1982

WERNER GOTTSCHALK
HERMANN P. MÜLLER

1. Relevance of Protein Improvement in Plant Breeding

M.S. SWAMINATHAN

At the 1974 World Food Conference in Rome, it was unanimously resolved that all governments should strive to ensure that by 1984 'no child, woman, or man goes to bed hungry and that no human being's physical or mental potential is stunted by malnutrition.' Unfortunately, available statistics show that the number of people going to bed hungry in 1981 was higher than in 1974.

Why is this target, which was considered to be realistic in 1974, still proving to be only a piece of rhetoric? We should analyse carefully the factors that govern our inability to providing everyone born in this world his or her daily bread, despite the position of food as first among the hierarchical needs of man.

Factors Limiting Food Output Prospects

Lester Brown [6] has drawn attention to five major groups of constraints which can reduce food production in the future:

Conversion of cropland to non-farm use
  5 million ha in USA, 1967 to 1977
Soil erosion
  30% of the world's cropland being degraded
Deterioration and loss of irrigated land
  salinity, lack of maintenance
Declining rate of yield increases
  1960's: 2.2%/year
  1970's: 1.6%/year
Likely higher real costs of energy.

In addition, there could be considerable instability in production arising from weather fluctuations. For example, North American agriculture is essentially

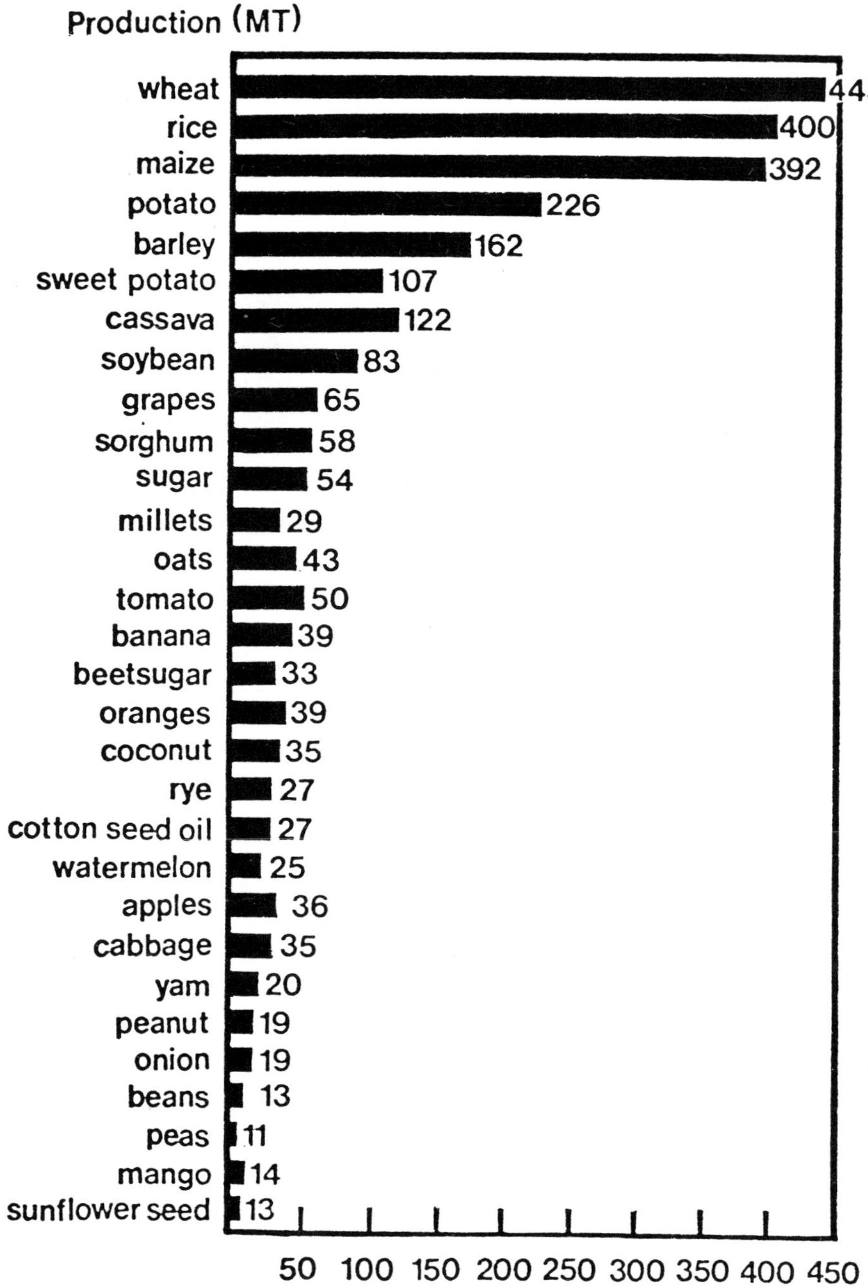

Figure 1. Annual production of the world's major food crops (1980)
Source: FAO Production Year Book, 1980

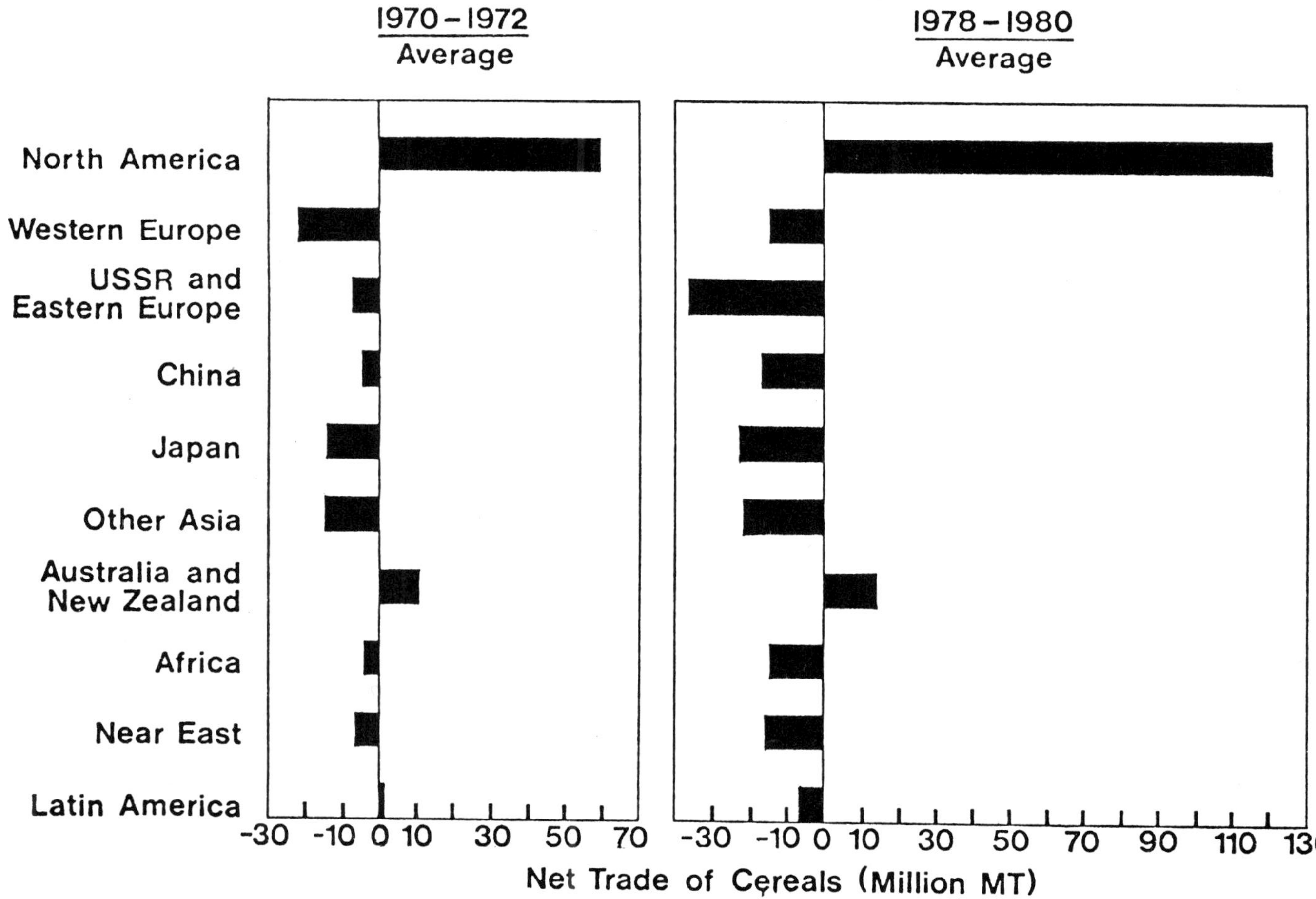

Figure 2. World's increasing dependence of the grain exports of a few countries; USA and Canada supply most of the grain (USDA).

rainfed. Production shortfalls more than 15% below the trend have occurred at fairly regular intervals in the past. Shortfalls in North American production combined with rising world consumption and low reserves could well trigger a major crisis. The vulnerability of the world food situation to the vagaries of the weather arises from dependence on too few countries and too few crops to balance the global food budget (Figures 1 and 2). Even with the relatively recent development of the European Economic Community (EEC) as a net wheat exporter (about 10 million tons in 1980–81), North America is expected this year to supply 68% of all the wheat and 78% of the coarse grains that enter international trade. The U.S. alone will account for over 50% of the wheat, 70% of the coarse grains and about 25% of the rice world exports in 1981–82. The International Wheat Council, at its last meeting in late 1981, came to this conclusion:

> 'It must be recognized that the ratio of carryover stocks to utilization is unlikely to show any significant increase in the near future. Stocks in the five major exporting countries will not approach again the levels of the 1960s, when they sometimes exceeded the volume of world trade in the following year. Rising storage costs and high interest rates mean that carryover stocks will tend to be limited to the amounts required to meet domestic and export needs until the arrival of the new crop. Some exporting countries may even introduce measures to cut down production if stocks became too burdensome.
>
> Clearly, this means that year-to-year production will be all-important to meet requirements. It is therefore vital that importing countries should hold stocks, if they wish to ensure some margin of food security and avoid any interruption in the supply and distribution chain. More than even, the world wheat supply and demand balance will depend on favourable weather conditions in the main producing areas in 1982 and future years'.

Thanks to the development of high yielding varieties of wheat, rice, maize, and other crops during the last 15 years, there has been an improvement in the overall global food situation. However, the fact remains that the food import needs of many developing countries have increased dramatically in recent years. The cereal imports of developing countries as a whole have doubled in the past decade and are now close to 100 million tons a year.

Of considerable concern is the situation that has developed in low-income countries. According to an analysis made by FAO, as many as 37 low-income countries (with 1980 per capita incomes of $730 or less) recorded negative growth rates in per capita production of cereals during the seventies; of these, 19 experienced a decline in total cereal output. Thus, far from achieving greater self-sufficiency, these countries are faced with a widening food gap (Figure 3). The amount of food aid received by these countries has stagnated in recent years. Low-

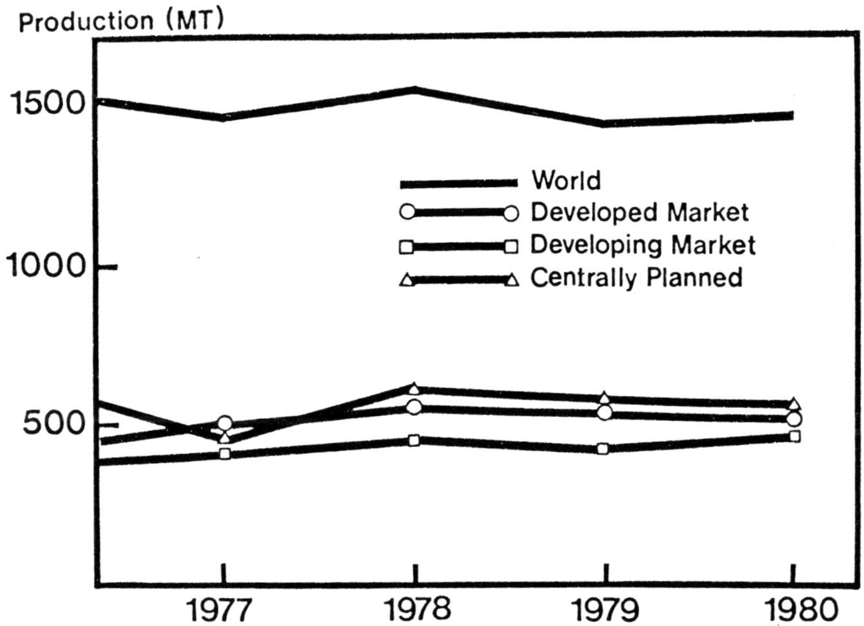

Figure 3. Food production trends in developed, developing and centrally planned countries.
Source: FAO Production Year Books, 1977 to 1980

income countries as a whole now spend over $7 billion a year to cover the value of commercial imports of cereals.

World food security for the medium term is uncertain (Figure 4). While international market prices of cereals have declined during 1982, following some years of rise, costs have continued to rise. This has led to concern in exporting countries that farmers' incentives will weaken and that this may have implications for growth in 1982 and even beyond. The U.S. Government has already decided to limit 1982 production of wheat, coarse grains, and rice through voluntary set aside programmes. According to the USDA, initial projections for 1982–83 U.S. crops point to nearly 4 million tons of wheat and 16 million tons of coarse grains less production than in 1981–82.

NUTRITION STUDY

To develop scientifically sound and economically feasible nutrition strategy, it will be necessary to understand the nature of the nutrition problem, not only within a country but also, in large countries like India, in different parts of a country. The general nutrition problems of the population as well as the specific nutrition problems of pre-school children and pregnant and nursing mothers will have to be

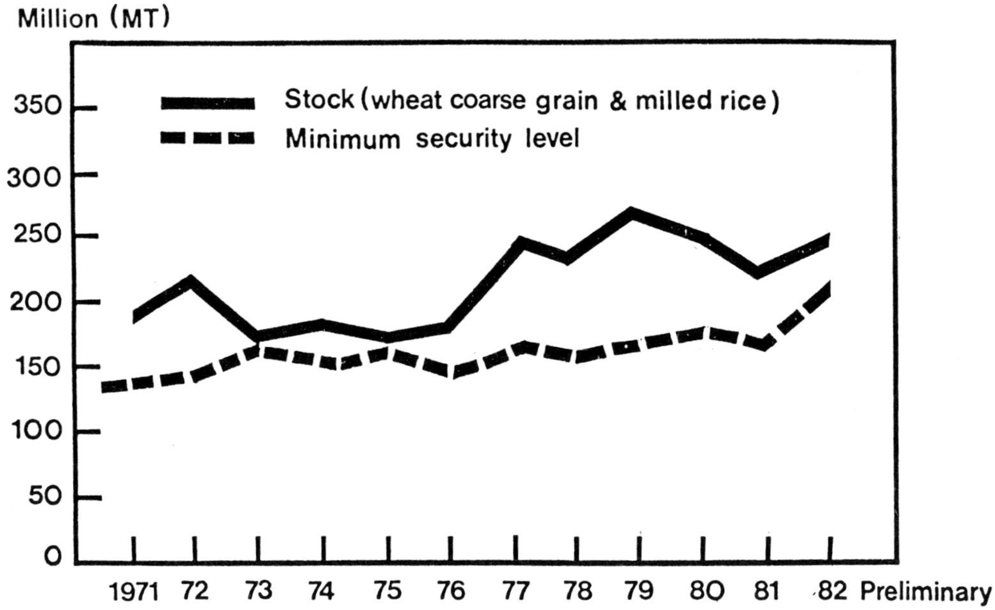

Figure 4. World carry-over stocks and minimum security levels of cereals (including CHINA and USSR)
(Based on FAO data)

clearly understood. For example, in the Indian context, where a cereal like wheat or rice is the staple grain, [29] has shown that inadequacy of caloric intake is the major cause of malnutrition. Protein deficiency may also be a problem in areas where a tuber crop like cassava is the staple.

An important consideration, which was not fully taken into account in assessments of the extent of malnutrition in the past, is the finding that there is no absolute energy requirement for any day or period. The individual is in homeostasis and his requirement is controlled by a regulated system. It is known that there are seasonal changes in incidence of malnutrition in rural areas. Studies in some parts of India have shown that the time spent in breast feeding came down markedly during a period when the post-harvest workload on the mothers was very high. Such seasonal variations in food intake are reflected in similar variations in the incidence of *kwashiorkor*. It is necessary that in-depth studies be conducted on the seasonal incidence of low birth weight, maternal nutrient deficiencies, and kwashiorkor. Such high variabilities in both the qualitative and quantitative dimensions of the nutrition problem underline the wisdom of examining possible solutions in the context of agricultural and rural development.

Because agriculture is the backbone of the rural economy, agricultural advancement and rural development are in most cases concurrent and interrelated events. They are both highly location-specific in reference to the packages of technology, services, and public policies needed to sustain and stimulate them. Integration of nutritional considerations into agricultural and rural development will probably result in more speedy and reliable results than the many intervention and fire fighting operations undertaken thus far. This is where plant breeding for improved nutritive quality assumes significance.

ENERGY IMPLICATIONS

Since the escalation in cost of energy sources based on fossil fuels, it has become important that the energy implications of every aspect of crop improvement be carefully studied. Agriculture is predominantly a solar energy harvesting enterprise. With the modernisation of agriculture, the quantity and variety of cultural energy inputs have increased considerably. In the energy input-output model shown in Figure 5, several leakages exist.

Three important sources of energy loss are related to physiological, pathological, and utilisation factors. The pathological leaks arise from damage caused by the triple alliance of pests, pathogens, and weeds. These can be controlled to a great extent by breeding for multiple resistance coupled with the use of appropriate chemicals. Work done at the International Rice Research Institute in breeding for

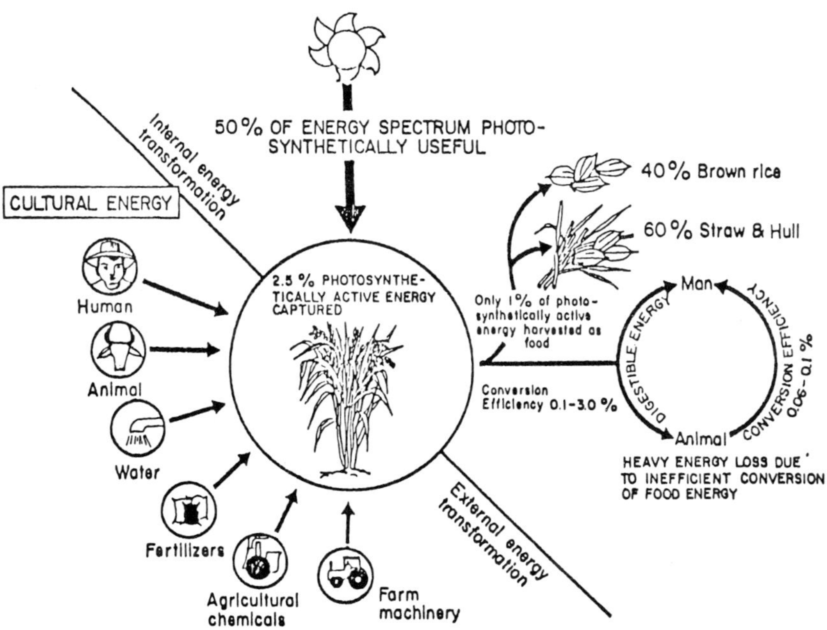

Figure 5. Solar and cultural energy input and output cycle in rice.

Table 1. Disease and insect ratings of IR varieties

Variety	Growth duration (days)	Diseases				Insects					
		BL	BLB	TG	GS	BPH biotypes			GLH	SB	GM
						1	2	3			
IR8	130	S	S	S	S	S	S	S	MR	S	S
IR20	130	MR	R	MR	S	S	S	S	R	MR	S
IR24	125	S	S	S	S	S	S	S	R	S	S
IR26	130	R	R	MR	S	R	S	R	R	MR	S
IR28	105	R	R	R	R	R	S	R	R	MR	S
IR34	125	MR	R	R	R	R	S	R	R	MR	S
IR36	110	MR	R	R	R	R	R	S	R	MR	R
IR42	135	MR	R	R	R	R	R	S	MR	MR	R
IR46	140	R	R	MR	S	R	S	R	MR	MR	S
IR50	105	MR	R	R	R	R	R	S	R	MR	NA
IR52	115	MR	R	R	R	R	R	S	R	S	NA
IR54	120	MR	R	R	R	R	R	S	R	MR	NA
IR56	105	R	R	R	R	R	R	R	R	MR	NA

S = susceptible, MR = moderately resistant, R = resistant, NA = not available, BL = blast, BLB = bacterial blight, TG = tungro, GS = grassy stunt, BPH = brown planthopper, GLH = green leat-hopper, SB = stem borer, GM = gall midge.

multiple resistance to pests and diseases in rice is an illustration (Table 1). Physiological leaks arising from photorespiration also can be minimised. However, the utilisation pattern determines the actual quantities of food grains required to feed a population. For example, nearly two-thirds of the grain in developed countries is used for feeding livestock. Data on the world use of grain, as well as the use of grain for livestock feed, are given in Tables 2 and 3. Improvements in the quality of the feed grain would greatly help the plant-animal-man food chain. Thus, quality improvement of grains has to take into consideration the needs of both man and farm animals.

MEETING THE CHALLENGES

The challenge of producing adequate food of the desired quantity and quality can be met only through a multi-pronged strategy. The major components of such an integrated strategy should be:

a) Breeding varieties characterised by both high yield potential and high performance stability.
b) Breeding varieties which can improve productivity per unit of time, land, water, and energy by taking into consideration the emerging farming systems.

Table 2. World Use of Grains (million tons) 1/

	For human consumption	For livestock feeding
1966	468	354
1967	475	368
1968	484	390
1969	490	421
1970	502	429
1971	506	455
1972	515	476
1973	537	488
1974	553	461
1975	555	460
1976	588	481
1977	594	504
1978	615	536
1979	631	553
1980	642	548

1/ Excluding rice
Source: FAO estimates

Table 3. Use of Grain for Livestock Feeding as a Proportion of Total Domestic Availabilities 1/ 2/ (in percentages)

	Developed	East Europe and USSR	Developing	
			All	Low-income
1966	67.2	42.5	15.5	3.2
1967	67.7	45.3	15.4	3.1
1968	68.7	46.7	15.9	3.3
1969	69.9	49.3	16.1	3.4
1970	69.6	50.4	16.1	3.2
1971	70.8	52.8	16.5	3.4
1972	70.7	54.7	17.3	3.3
1973	70.7	53.3	16.8	3.2
1974	67.4	55.8	16.8	3.3
1975	67.7	54.6	18.0	3.8
1976	66.8	54.9	18.1	3.6
1977	67.7	56.7	19.0	3.3
1978	68.8	57.6	18.2	3.6
1979	68.7	59.7	19.3	3.7
1980	67.1	58.7	20.3	3.8

1/ Excluding rice
2/ Figures do not include China PR

Source: FAO estimates

c) Breeding specifically for high protein content and good amino acid balance in food, feed, and fodder crops.

Before taking up plant breeding approaches in relation to (b) and (c), I would like to refer briefly to emerging farming systems and the need for reorienting land use patterns on the basis of nutritional criteria.

EMERGING FARMING SYSTEMS IN THE TROPICS AND SUB-TROPICS

Multiple Cropping Systems in Irrigated Areas

Various two-, three-, and even four-crop sequences are now being followed. In promoting multiple-cropping systems, attention should be paid to ensuring that grain and fodder legumes find a place in the rotation. Also, crops susceptible to the same pests and diseases should not be grown in succession. Introduction of grain and fodder legumes in the rotation will improve human nutrition as well as soil fertility. A mungbean-rice-wheat rotation is a good method of combining cereals and legumes in north-west India. Short-duration varieties of pigeonpea (*Cajanus cajan*) have made pigeonpea-wheat rotations possible. Jute-rice-wheat rotation is becoming popular in eastern India and Bangladesh.

The introduction into crops of relative insensitivity to photoperiod and temperature through breeding has been responsible for the development of period-fixed rather than season-bound varieties. To breed varieties for multiple cropping, per-day yield has to be used as a selection criterion in segregating generations. Other factors such as seed dormancy will also need attention, because the grain will sprout if there is rainfall at harvest time in a crop that ripens before the monsoon rains have ceased.

Rainfed Farming

Production possibilities in high-rainfall areas are similar to those in irrigated areas. However, in the unirrigated semi-arid areas commonly referred to as dry-farming areas, considerable production risks exist. Grain legumes, sorghum, millets, and oilseed crops are mostly grown in such areas. A wide variety of fruit trees also can be grown. Research thrusts in semi-arid areas should stress water and soil conservation and land-use planning based on precipitation, evapotranspiration, and the moisture holding capacity of the soil. Contingency plans should be developed and introduced to minimise the risk of total crop loss during aberrant weather. It is also necessary to find more profitable crops for some of the semi-arid areas. Many under-exploited plants have potential economic value.

Plant breeders should develop varieties which can be grown in flood-free seasons in chronically flood-prone areas and drought-escaping varieties in drought-prone areas. For this to happen, collaboration between plant breeders and agro-meteorologists is necessary.

Mixed Cropping and Intercropping

Farmers use various crop combinations, particularly in unirrigated areas. But not all are scientifically sound. Therefore, intercropping systems based on complementarity between the companion crops must be developed. Among the major components of complementarity are:

effcient interception of sunlight;
ability to tap nutrients and moisture from different soil profile depths;
non-overlapping susceptibility to pests and diseases;
introduction of legumes to promote biological N fixation and increase protein availability.

Multi-level or Three Dimensional Cropping

In garden lands where a wide variety of plantation crops, fruit trees and other perennial crops are grown, it is possible to design a crop canopy within which the vertical space is utilised more efficiently. Plant architects will have to take into account the effective use of both horizontal and vertical spaces when breeding varieties for use in three-dimensional crop canopies. Efficiency in such a cropping system will again be based on the extent of the complementarity generated among crops in the system. For example, studies in Malaysia and India have shown that coconut, cocoa, and pineapple form a good combination. It intercepts sunlight efficiently in a combined canopy and also extracts nutrients and moisture from different depths in the soil profile. Studies of the root system of companion crops are of particular importance. The introduction of grain and fodder legumes into these 3-dimensional crop canopies will provide opportunities for animal husbandry.

A careful study of all the major garden land cropping systems, based on the extent of symbiosis and syndergy among the system components, will be useful in developing specifications for plant breeders to use in developing ideotypes (i.e. conceptual plant types) for efficient performance in 3-dimensional crop canopies.

Kitchen Gardening and Home Fish Gardening

Kitchen gardening can be one of the most efficient systems of farming from the point of view of solar and cultural energy conversion. Vegetables rich in beta carotene and iron need to be developed and popularised. Winged bean provides a good source of protein. If planned intelligently and scientifically, backyard gardens, roof gardens, and other methods of growing vegetables and fruits in whatever space is available around huts and houses can make a substantial contribution to improved nutrition. Where ponds are available in large numbers, home fish gardening can be an excellent method of supplementing food and income.

Forestry and Agro-forestry

The importance of improving the productivity of forest canopies cannot be overemphasised. Agro-forestry has been defined as a sustainable management system for land which increases overall production, combines agricultural crops, tree crops, forest plants, or animals simultaneously or sequentially. Sylvi-pastoral, sylvi-horticulture, sylvi-agriculture, and other combined land-use systems are extremely important for the food, feed, fuel, and fertiliser needs of people in many hilly regions.

Plant breeders have yet to give attention to breeding varieties suitable for such systems of sylvi-culture. Shrubs and trees suitable for raising energy plantations in villages and initiating gasoline agriculture need to be identified and improved.

Mixed Farming

Mixed farming systems may involve crop-livestock, crop-fish, and crop-livestock-fish production programmes. In South East Asia, fishing in rice fields is common. The minimal use of pesticides will be important in order not to create problems of fish mortality and transfer of toxic residues through the food chain. This will involve maximum use of genetic resistance and the development of integrated pest-management systems. Fodder and grain legumes assume significance in crop-livestock integrated production systems.

Sea Farming

There are considerable opportunities for the spread of scientific sea farming practices involving an appropriate blend of capture and culture fisheries. The rate of growth of oysters, mussels, prawns, lobsters, eels, and a wide variety of marine plants and animals is high in tropical seas. If, along with such integrated sea farming practices, the cultivation of suitable economic trees like casuarina, cashewnut, and coconut can be popularised along the coasts, thriving coastal agriculture-cum-mariculture systems can be developed. In addition to improving income and nutrition, such farming systems can help arrest coastal erosion.

Land Use Boards

In order to help restructure land use patterns on scientific lines based on the results of plant breeding for yield and quality, it would be desirable to organise Land

Use Boards with interdisciplinary expertise. Each Board can cover a specific agro-ecological area. Such Land Use Boards should assist farmers in optimising the economic benefits from land and water through attention to major ingredients of scientific land use:

Ecology. Land use based on ecological considerations will help maximise the economic benefits from a given environment and minimise damage through man-made as well as natural processes of desertification. The aim should be to prevent the destruction or diminution of the biological potential of land. Agro-meteorological research data will have to be integrated into crop breeding and planning models so that contingency plans suited to different weather probabilities can be prepared.

Economics. To reorient land and water use on the basis of sound principles of economics, it is essential that production, storage, processing, and marketing be viewed as a total system. Equal emphasis will have to be paid to both production and post-harvest technologies. The prevailing mismatch between these two areas of the production-consumption chain in many developing countries is harming both producers and consumers. To bridge the gap between potential and actual farm yields, it will be necessary to identify the precise constraints operating in each area and remove them. When post-harvest technology is neglected, opportunities for the preparation of value-added products are lost. For example, food production statistics simply state that during 1978–79 India produced about 131 million tons of foodgrains. This ignores the fact that the plants represented in this statistics produced over 400 million tons of biomass, out of which grains constituted about 131 million tons. If the entire biomass is viewed as an asset and is utilised effectively, new avenues of income generation can be opened up. A part of it is currently used for feeding animals or as fuel. But by looking at the dry matter yield part by part and by introducing techniques of preparing value-added material, rural incomes can be enhanced.

Energy. The energy needs of agriculture will have to be carefully worked out and an integrated energy supply system involving a suitable blend of renewable and non-renewable forms of energy will have to be introduced in each block. So far, the pathway to productivity improvement adopted in both developed and developing countries has tended to rely heavily on a growing consumption of non-renewable forms of energy. We will have to reverse this process through the promotion of organic recycling techniques and through the wide-spread use of biological sources of fertilisers, such as azolla, blue green algae, and symbiotic and non-symbiotic forms of nitrogen fixing organisms. Also, the current tendency to cultivate such energy rich crops as grain legumes and oilseeds under conditions of energy deprivation has to be corrected. Phosphorous conservation and recycling will be particularly important, since phosphorous is a non-renewable resource.

Employment. As early as 1862, Col. Baird Smith, who investigated the causes of the famine of 1860–61 in the North-West Province of India, mentioned that Indian famines are more 'famines of work than of food, since when work can be had and paid for, food is always forthcoming'. In the area of nutrition, the situation today is one of providing the wherewithal to purchase food rather than the availability of food in the market. All estimates of employment potential show that a majority of the people in India and many developing countries will have to depend upon agriculture, agro-industries, and small scale industries as the major source of income until the end of this century.

Nutrition. Every Land Use Board should have as a member a nutrition scientist who could help suggest suitable crops and crop varieties to meet these needs:

Enhancing protein availability;
Preparing homemade weaning foods;
Providing some critical missing nutrients in the diet, such as Vitamin A, Vitamin C, iron, etc.;
Developing a cereal-grain-legume combination so that all the essential amino acids can be provided in the diet;
Introducing appropriate fodder legumes and shrubs which could provide the needed calories and proteins to farm animals, thereby enabling the introduction of larger quantities of such animal food as milk and milk products, eggs, etc., in the diet;
Developing suitable agriculture-cum-aquaculture techniques which could help promote such dietary combinations as rice-fish, potato-fish, etc.; and
Promoting agro-forestry systems of land management in which appropriate botanical remedies to the specific nutritional maladies of the region can be incorporated.

Post-harvest Technology

In addition to detailed attention to scientific land and water use planning, it will be essential to promote relevant research in the area of post-harvest technology. Post-harvest technology research should, on the one hand, aim at providing clean and healthy food to the consumer and, on the other, at preparing value-added products in the village itself. The production technology should be such that it does not cause problems, such as pesticide residues in the edible plant parts. With the introduction of photo-insensitive varieties of crops resulting in crops coming to harvest when rainfall has not stopped, problems of mycotoxicosis arising from inadequate grain drying are becoming widespread. Therefore, the whole area of grain drying and storage requires greater attention. It is a common experience among the poor, who are already undernourished, to purchase grains, fruits, and vegetables infected with *Aspergillus flavus* because of the tendency of the traders to sell them at cheaper prices. Thus the poor are doubly hit, first by caloric inadequacy and second by the consumption of infected grains. Severe liver disorders arising from the consumption of grains infected with *Aspergillus flavus* or contaminated with weed seeds carrying liver toxins have been reported in recent years.

In countries like India, a majority of the farmers cultivate less than one hectare of land. They have to meet their own food requirements first, and should have some surplus produce for sale. An important method for obtaining supplementary income in such cases is the integration of animals into the farming system. However, where even the human population suffers from inadequacies of calories, calorie deprivation is even greater for animals. The yield of milk and meat becomes very low under such conditions. Plant breeding for higher protein content will be valuable in the production of high yielding-cum-high quality feed grains. Also, it is necessary that suitable technologies be developed for preparing fortified feed material from cellulosic wastes and from agricultural raw material. Fortification of straw with molasses and urea as well as microbiological enrichment of starchy materials, such as cassavas, will have to receive much wider adoption.

Pathways of breeding

Production potential. As was mentioned earlier, considerable improvement has been made in the last few decades in yields of cereals such as wheat, rice, maize, sorghum, and other crops through the breeding of plant types which can respond to good management, particularly to the application of mineral fertiliser. Maximum yields reported in temperate and tropical regions are shown in Table 4. These yields are still far below theoretical yield projected [9, 22]. The question remains as to what would be the upper yield limit that a plant breeder could attempt to achieve. Buringh et al [7] analysed absolute production potential in terms of grain equivalency based on soil, climate, precipitation, and radiation in different parts of the world. They concluded that the absolute global production potential was 49,830 million tons of grain equivalents per year. Using a similar methodology, [27] but also considering the duration for which total interception of light is possible, the absolute production potential for India was calculated as 4,572 million tons grain equivalents per year. [28] De Wit et al [10] defined the production potential of a crop as the growth rate of closed green crop surface, optimally supplied with water and nutrients, in a disease and weed free environment under prevailing weather conditions. Accordingly, a crop capable of retaining a closed crop canopy for 100 days could produce 35,000 kg dry matter/ha^{-1}. A 50% harvest index from such a crop would give 17.5 t/ha^{-1}, dry or 20 t/ha^{-1} 12% moisture grain. It is clear that there is considerable scope for improvement in situations with high input agronomy.

Recognising that the potential yield of a crop is determined not only by variety but by the environment in which it grows and the management it receives, FAO has

Table 4. Total dry biomass, grain yield, their energy content and harvest index

Crop	Total biomass g/m²	Grain yield g/m²	Harvest index	Total biomass energy K cal/m²	Grain energy K cal/m²	Harvest index
Wheat (Kalyansona)	1558	583	37.4	6104	2387	39.1
Wheat (cv Moti)	1375	596	43.3	5410	2441	45.1
Triticale (DTS-141)	1520	484	31.8	6062	1997	33.0
Barley (cv Ratna)	1867	445	23.8	6986	1798	25.7
Chickpea (cv JG.62)	1027	327	30.5	3992	1412	35.3
Pigeon pea (cv Prabhat)	1008	310	30.7	3994	1379	34.5
Brassica Campestris (Yellow)	1160	377	29.0	5055	2249	44.4
B. Campestris (Brown)	1380	421	26.0	5828	2636	45.2
B. juncea (Mustard)	1820	486	26.0	7608	2959	38.9

Source: [29]

started delineating areas of Africa and Asia most suitable for different crops. It would be appropriate that breeding objectives are fixed in relation to the expected potential of different crops based on that FAO exercise.

Energy value. Yields of different crops are generally compared on the basis of weight [9, 31]. Weight is also the basis for calculating harvest index [11, 12, 18, 19]. While this is a simpler method of comparing yield and harvest index, it is not necessarily the best. Because the compositions of grains differ, energy contents for the same weight differ for different crops and sometimes even for cultivars (Table 4). On the basis of energy content, it is obvious that 0.6 kg of an oil-containing seed would be equal to 1.0 kg of a cereal grain. Therefore, it would not be realistic to expect similar yields in terms of weight of an oilseed crop such as groundnut and a cereal such as wheat (Table 5). Furthermore, on the basis of energy, even the harvest indices of some oilseed crops are as good as those of high yielding cereals (Table 4).

Consequently, if a change in composition of grains in favour of proteins with a particular amino acid profile or in favour of lipid content, while retaining the same weight is an objective, that would require additional energy [3, 24] Sinclair and de Wit [25] calculated the requirements of photosynthates if nitrogen assimilation is enhanced for protein content. Most of these estimates are based on glucose utilisa-

Table 5. Relative grain yields (weight) of different crops on the basis of their composition. Values are expressed as 1.0 Kg wheat equivalent (FAO Production Yearbook 1980).

	Composition (% of dry weight)				Seed g/g photo-synthate	Wheat equivalent
	Carbo-hydrate	Protein	Oil/ Lipid	Ash		
Cereals						
Wheat	82	14	2	2	0.71	1.0
Rice	88	8	2	2	0.75	1.05
Sorghum	82	12	4	2	0.70	0.98
Pulses						
Chickpea	68	23	5	4	0.64	0.90
Pigeon pea	69	25	2	4	0.66	0.93
Mung bean	69	26	1	4	0.66	0.93
Oil-seeds						
Groundnut	25	27	45	3	0.43	0.60
Sesame	19	20	54	7	0.42	0.59
Rape and mustard	25	23	48	4	0.43	0.60

tion efficiency through different metabolic pathways [23]. In these estimates, it is expected that all biochemical reactions would occur in synchrony, which often does not happen. Consequently, glucose utilisation efficiency in natural systems could be poorer than expected. Nonetheless, these studies are useful to a plant breeder because he is made aware of the biochemical constraints imposed by the system. He either has to seek a compromise between yield and quality or has to be prepared to improve photosynthate availability simultaneously. If this factor is not recognised, improvement in calorie yield may invariably result in a protein or lipid penalty and vice versa.

Breeding for yield improvement. Breeding for yield is one of the principal objectives for different crops in all parts of the world. Development of a new concept or a new technique usually has been responsible for significant advances in yield. Among these, the phenomenon of heterosis and the introduction of dwarfing genes could be considered important landmarks. However, after the quantum jump in yield associated with the exploitation of hybrid vigour and dwarf and fertiliser responsive plant types, progress usually slows down. It could be that at a given time, a plant breeder selected a variety/hybrid which gave a higher yield because a particular character, such as disease resistance, gave an advantage. Evidence was presented to show that genes for yield per se did not exist in barley. [17] Therefore, the question remains: in case, there are no specific genes for yield, how does this character express itself at the phenotypic level? A clear understanding of the basis of yield is obviously essential for launching a successful breeding programme.

Physiological basis of yield. Crop yield is the result of a large number of processes – germination, growth, differentiation development, and senescence. Identification of the particular process that limits yield in a given environment becomes the primary consideration. Nonetheless, an important fact is that a crop must produce an adequate amount of dry matter and appropriate partitioning of the phytomass for it to give good yield. This brings in the following questions:

Has there been any improvement in phytomass (dry matter) production in the course of the improvement of various crops?

What is the relation between dry matter and yield?

Can dry matter production limiting yield be improved?

Recent studies using old and new varieties of wheat released in India and Britain and peanuts released in the United States suggest that there has been no significant change in dry matter production [13, 27, 1]. A major change in the partitioning of dry matter was mainly responsible for improved yields. In peanuts, this change was due to a change in growth pattern. These studies lead to two important conclusions:

In those crops where no major improvement has occurred, as in many leguminous and oileferous crops, improvement in yield in the short-term could be achieved through changing the pattern of growth and the partitioning of phytomass. Efforts to improve dry matter production could be a long term objective.

In crops in which significant improvement has already occurred, such as wheat, rice, sorghum, maize, soybean, etc., improvement in dry matter production per unit area without disturbing the partitioning already obtained could be an important objective.

These conclusions lead to these questions:

How is dry matter obtained and how could it be improved?

What determines the partitioning of dry matter and when is it determined?

What is the effect of the preceding phase on partitioning?

Components of processes leading to dry matter production were described [30]. These include leaf area development, leaf area index, rate of photosynthesis, rates of respiration and photorespiration, nutrient uptake, nitrate assimilation, and water use. Genetic variability in many of these characters, both at the species and the varietal level, has been established.

New parameters. Some of the new parameters for plant breeding of immediate as well as of long term interest are:

Increasing availability of food containing desirable nutritional value and organoleptic properties.

Improvement in the efficiency of food production for each unit of cultural and solar energy invested.

Stability of crop production through resilience to weather behaviour and resistance to the triple alliance of weeds, pests and pathogens.

Identification and improvement of plants suitable as sources of biomass and renewable energy.

Identification of more efficient plant types for crop-livestock and agriculture-aquaculture production systems.

Identification of plant types suitable for multiple, relay and intercropping and for 3-dimensional crop canopies where annuals and perennials are grown in a cooperative farming system.

Identification of genotypes for optimum response to high, low and zero input conditions.

Nutritional goals for plant breeders. Concerted efforts by plant breeders are required to increase the protein quantity and quality in cereals low in protein and lysine content. One pathway for improving protein content in cereals is to break the negative correlation between protein content and grain yield. Grain filling has been the major constraint in the high lysine genotypes so far identified in cereals such as maize, barley, and sorghum. By rigorous breeding and selection, possibilities do exist for recovering plump grains combined with high protein and/or high lysine, as was shown by [19] in wheat and [2] in barley.

The problem in cereals is producing genetically higher levels of protein without sacrificing yield. In recent years considerable progress has been made in wheat where variation concomitant with both increased protein content and increased grain yield could be found. Several wheat genotypes have been identified that have higher than normal levels of yield equivalent to those of parental varieties with lower protein content [20, 26].

The advances being made in identifying specific proteins affecting breadmaking quality in wheat are also an important development which will be exploited in wheat breeding in the near future [21]. The ability to identify useful products biochemically as well as the genes that produce them must increase the efficiency of selection. This ability also has the potential to open up the hitherto untapped reservoir that exists among the wild relatives of wheat, such as the species of genus *Aegilops* which have been considered to be a valuable initial material for developing new wheat forms producing seed of high protein content and high technological and baking qualities [4]. But the use of wild species may pose problems of linkage with undesirable characters.

No major gene for lysine increase has so far been detected in wheat, probably because of its polypoid nature. Such genes may exist in the wild relatives of wheat or, indeed, in bread wheat itself, but the screening necessary to identify them has not been carried out. The number of protein changes required for significant increases in lysine is likely to be large and this could be a major restriction to exploitation of these proteins in breeding. If plant transformation becomes possible, then the transfer of genes from other species having proteins with high lysine levels to wheat could be considered. Alternatively, directed changes at the DNA level may be able to alter the number of lysine residues in a particular protein [21].

Yield can most effectively be expressed as biologically available energy per hectare. In effect, yield depends on the balance in composition of protein and its digestability and the availability of starch, fat, and other nutritive components. A panel to CIMMYT [8] recommended that the amount of protein present in grains should not fall below the best levels now achievable in present crops, but if possible, the composition of the protein should be improved, first with respect to lysine.

Energy is another nutritional factor most likely to be limiting in the diet of persons. Where energy intake is low, the diet is apt to be low in essential nutrients, including protein. Dietary energy supplied by carbohydrates is low in energy density. If dietary energy is taken care of by fat, which is high in energy density, the utilisation of the protein present in cereals and legumes can be improved. If legume protein consumption were to fall appreciably below the present level because of low production and high cost or if a significant proportion of dietary energy were to be derived from cassava or other low protein sources, improvement in the level of essential amino acids in cereal grains becomes an urgent consideration.

Protein improvement in combination with high grain yield is the breeding goal in cereals, whereas in food legumes high yields are the goal. Studies on the production physiology of grain legumes led to knowledge of some of the factors, such as flower shedding, leaf senescence, etc., that are responsible for low yields in legumes. Restructuring the plant type of grain legumes may help obtain reasonable yields. The genetics and relevant selection strategy need to be better understood if optimum yield levels in legume crops are to be realised.

Removal of antinutritional factors which influence the digestibility of legume proteins is another area of research needed to make grain legumes more nutritious and acceptable to larger segments of the world population. It has been shown that field beans (*Vicia faba*) with white flowers do not have the tannins that limit the digestibility of the testas to ruminants [5].

To utilise oilseed proteins for direct human comsumption, breeding efforts need to be directed towards improving oilseed crops with regard to flavour, colour, antinutritional factors, and processing technology. The highly effective breeding work that resulted in the development of varieties of rape seed (*Brassica campestris* and *B. napus*) with very low levels of erucic acid in the storage lipids of the seed is being followed up by attempts to increase linoleic acid and to reduce the linolenic acid content.

The starchy root and tuber crops which form a stable diet in limited areas of some countries need special attention. Research is needed to determine whether breeding can alter protein ratios to achieve improved overall nutritive quality of tuber crops.

Concurrent attention needs to be paid to the biochemical, bioenergetic, and production physiology aspects of plant breeding if efforts to concurrently improve protein and calorie yield are to succeed. It is not enough that the partitioning of the total phytomass between grain and other plant parts be studied (i.e., conventional harvest index studies). Nitrogen and energy harvest indices require equal attention.

CONCLUSION

Data on the annual growth of population and food production are shown in Table 6. This data, as well as the data on per capita daily food supply in terms of protein (Table 7), emphasises the urgency of stepping up both food production and protein supply in developing countries. The plant breeding approach has the great advantage in that it helps farmers to produce better quality food at little extra cost.

Table 6. Percentage annual growth of population and food production.

Region	Population		Food Production			
			Total		Per Capita	
	1966-70	1971-80	1966-70	1971-80	1966-70	1971-80
Developed (including Europe & USSR)	0.9	0.8	1.8	1.8	0.9	1.1
Developing (including Asian CPE)	2.3	2.2	3.6	3.3	1.2	1.1
MSA[a], Total	2.5	2.5	4.8	2.6	2.3	0.1
World	1.9	1.9	2.5	2.4	0.6	0.6

Source: FAO

[a]MSA: most seriously affected countries

Table 7. Per capita daily food supply in terms of protein (grams)

Region	Total Protein 1961-65	Total Protein 1975-77	Animal Protein 1961-65	Animal Protein 1975-77
Developed Countries	91.3	98.5	44.9	55.2
Developing Countries	53.6	56.9	10.2	11.7
World	65.4	68.4	21.1	23.7

Source: FAO, Food Balance Sheets

While conventional breeding approaches should be pursued with vigour, it is important that the modern genetic engineering tools now becoming available are also harnessed fully for developing varieties characterised by high yield potential, good stability of performance, and good nutritive quality, including protein content. It may be possible to engineer seeds that overproduce storage proteins. This kind of technology has been perfected in bacteria. The quality of storage proteins produced in grains can be changed by site-directed mutagenesis technology such that all the essential nutritional requirements of humans can be met by the grain proteins only. It should also be possible to integrate genes of specific animal proteins in place of the storage protein genes.

An appropriate and scientific integration of conventional and emerging techniques of plant breeding can help us successfully face the challenge of overcoming problems of under- and mal-nutrition.

REFERENCES

1. Austin, RB, J Bingham, RD Blackwell, LT Evans, MA Ford, CL Morgan and M Taylor (1980): Genetic improvements in winter wheat yields since 1900 and associated physiological changes. J. Agric. Sci. 94: 675–689.
2. Bansal, HC, RP Singh, S Bhaskaran, JM Santha, and BR Murty (1980): Hybridization and selection for improving seed protein in barley. Theor. Appl. Genet. 58: 129–136.
3. Bhatia, CR and R Rabson (1976): Bioenergetic considerations in cereal breeding for protein improvement. Science 194, 1418–1419.
4. Bochev, B and V Doncheva (1980): Possibilities of utilizing genus *Aegilops* for obtaining high quality wheat forms. Proc. Third Int. Wheat Conf. Madrid: 757–772.
5. Bond, DA (1976): *In vitro* digestibility of the testa in tannin-free field beans (*Vicia faba* L.). J. Agri. Sci. 86: 561–566.
6. Brown, LR (1981): World population growth, soil erosion and food security. Science 214, 995–1002.
7. Buringh, P, HD Van Heemst, and GJ Staring (1975): Computation of absolute maximum food production of the world. Dept. of Tropical Science, Agricultural University, Wageningen, The Netherlands; 46 pp.
8. CIMMYT 1979 – Quinquennial Review Report, CIMMYT, Mexico.
9. De Vries, CA, JD Ferwerda and M Flach (1967): Choice of food crops in relation to actual and potential production in the tropics. Neth. J. Agric. Sci. 15: 241–248.
10. De Wit, CT, HH van Laar and H van Keulen (1979): Physiological potential of crop production. In: Sheep, J, and AJT Hendriksen (eds.) Plant Breeding Perspectives: 47–82.
11. Donald, CM 1968. The breeding of crop ideotypes. Euphytica 17, 385–403.
12. Donald, CM, and Hamblin, J 1976. The biological yield and harvest index of cereals as agronomic and plant breeding criteria. Adv. Agronomy 28, 361–405.
13. Duncan, WG, DE McCloud, RL McGraw, and KJ Boots (1978): Physiological aspects of peanut yield improvement. Crop Sci. 18: 1015–1020.
14. FAO (1971): Agricultural Commodity Projection. 1970–1980. Vol I & II; FAO Rome.
15. FAO/WHO (1973): Energy and protein requirements. FAO Nutrition Meetings Rep. Ser. 52; FAO Rome.
16. Goel, A, SC Bhargava, and SK Sinha, 1981. Energy as the basis of harvest index. Ind. J. Exptl. Biol. (In Press).

17. Grafius, JE (1959): Heterosis in barley. Agron. J. 51: 551–554.
18. Jain, HK, NC Singh, and A Austin (1976): Breeding for higher protein yields in bread wheat: Experimental approach and phenotypic marker. Z. Pflanzenzüchtg. 77: 100–111.
19. Jain, HK and VP Kulshrestha, 1976. Dwarfing genes and breeding for yield in bread wheat. Z. Pflanzenzüchtg. 27: 102–112.
20. Kuhr, SL and VA Johnson (1980): Performance of high protein lines in international trials. Proc. Third Int. Wheat Conf. Madrid: 781–796.
21. Law, CN 1981. Genetical aspects of breeding for improved grain protein content and type in wheat. Third Seminar of FAO/SIDA/SAREC Project on Improvement in Nutritional Quality of Barley and Spring Wheat. May 10–21, Turkey.
22. Loomis, RS and WA Williams (1963): Maximum crop productivity: An estimate. Crop Sci. 3: 67–72.
23. Penning de Vries, FWT, AHM Brunsting, and HH van Lear (1974): Products, requirements and efficiency of biosynthesis; quantitative approach. J. Theor. Biol. 45: 339–377.
24. Rabson, R, CR Bhatia, and RK Mitra (1978): Crop productivity, grain protein and energy: Inputs, subsidies and limitations. Seed Protein Improvement by Nucreal Techniques: 3–20; IAEA Vienna.
25. Sinclair, TR and CT de Wit (1975): Photosynthate and nitrogen requirements for seed production by various crops. Science 189: 565–567.
26. Singhal and Jain, HK 1981 (under publication).
27. Sinha, SK and R Khanna (1975): Physiological, biochemical and genetic basis of heterosis. Adv. Agron. 27: 123–174.
28. Sinha, SK and MS Swaminathan (1979): The absolute maximum food production potential in India. An estimate. Curr. Sci. 48: 425–429.
29. Sukhatme, PV (1975): Human protein needs and the relative role of energy and protein in meeting them. In: Steele, F, and A Bourne (eds.): The Man-Food Equation: 53–75; Academic Press, London.
30. Swaminathan, MS (1979): Opportunities and problems in the developing countries. In: Sneep, J, and AJJ Hendriksen (eds.): Plant Breeding Perspectives: 1–47.
31. Wittwer, SH (1981): Global aspects of food production. Research and technology needs for the 21st century. In: Swaminathan, MS, and SK Sinha (eds.): Global Aspects of Food Production; Academic Press, London.

2. International Research Programmes for the Genetic Improvement of Grain Proteins

A. MICKE

THE PROTEIN GAP

In 1967, the United Nations Advisory Committee on the Application of Science and Technology to Development presented a report to the UN Economic and Social Council entitled 'International action to avert the impending protein crisis' [2]. The report was subject to a resolution by the 22nd Session of the UN General Assembly [3]. In the preface to the report it is stated that 'World food production is falling behind population growth despite all current national, bilateral and international efforts to reverse this trend'. Then attention is called to the fact that even providing enough food for all in terms of calorie needs would not be sufficient: 'Adequate protein is also required for the normal maintenance of body tissue and functions, and additionally for growth, maturation, pregnancy, lactation and recovery from injury and disease'. And the statement continues: 'Today there are over 300 million children, who, for lack of sufficient protein and calories suffer grossly retarded physical growth and development, and for many of these, mental development, learning and behaviour may be impaired as well. Protein-calorie deficiencies also directly affect the health and economic productivity of adult populations'.

The protein-calorie malnutrition problem has been singled out by the UN-Advisory Committee on the Application of Science and Technology to Development as the first one for concerted attack. Identification of other problems was to follow, to which the application of science and technology could make particular contributions for the benefit of man in developing countries. These efforts years later culminated in the United Nations Conference on Science and Technology for Development held 1979 in Vienna. However, in the meantime, the signal given by the United Nations with regard to an 'impending protein crisis' was heard and had a world-wide response. It stimulated an enormous amount of international and national efforts to attack and eventually solve the identified problem.

The mentioned report [2] gave policy directions which, for about 10 years, were seriously followed. Besides calling attention to the needed increase of plant and animal production, to the improvement of fisheries operation and to the prevention of food losses, the Advisory Group strongly recommended the production and use of seed and leaf protein concentrates, of fish-protein concentrates, of synthetic amino acids and of single-cell protein, realizing that there cannot be a single or simple solution to a problem of this magnitude.

The United Nations action on protein-calorie malnutrition found a large part of the scientific community alert and many organizations prepared to react. Besides various national institutions which had already studied the problem before and had contributed data to its identification, there existed already the Protein Advisory Group of WHO, FAO and UNICEF, which had been concerned with action against protein-calorie malnutrition at an international level. This Group was set up already in 1961, originally being composed of medical nutritionists and pediatricians, with the task to advise on the safety and suitability for human consumption of new protein foods. Later also production and processing of protein food became items of concern [2, 12].

The mentioned recommendations of the UN Advisory Committee stress the importance of increasing the production of nutritious food but also point out the need for research training, food testing, consumer education and adaption of appropriate governmental policies for production, processing and marketing high quality food items. The financial investment required for the UN system to meet the world food crisis and to close the protein gap was estimated at US$40 million per year [2].

Many research organizations and institutions followed these recommendations and contributed solutions to specific aspects of the protein-calorie malnutrition problem. In the context of this book, however, only international programmes on genetic improvement of crop plants for higher grain protein production will be considered.

Precise identification of objectives is the most essential prerequisite for the success of a plant breeding programme. When asked what the so-called protein gap amounts to in actual figures, most of the documents available in 1968 failed to give an answer and it is likely that the experts themselves were somewhat uncertain. Most people were satisfied with statements like the one made by AUTRET that 'the deficit is so great that all provisional objectives of production of protein foods for the next decades have no chance, under any circumstances, of being surpassed' [27, 63]. In another paper AUTRET et al. had specified that in 43 out of 88 countries the mean protein consumption was below the requirement, these 43 countries encompassing 900 million people [28]. With further increasing populations, the production of protein foods would have to be increased by 90% between 1970 and 1985. THIELEBEIN, however, attempted some calculations based upon available data and prevailing assumptions in order to answer the plant breeder's question and came to the conclusion, that the world requirement of protein (specified as 'reference protein') was around 77 million tons against a supply at that time of about 83 million and a production exceeding 100 million tons [84]. So he concluded that on a global scale, no protein gap existed. FAO's statistics, however, showed a striking disparity in protein production: With a world average of 43 kg per capita, the highest production was in North America with 156 kg, the lowest in the Far East with 24 kg [53]. The protein gap existed obviously between advanced and developing countries, between the rich and the poor of the same country, and particularly in situations when protein supply of the lowest income class is restricted to cereals only. There was (and still is) a close relationship between income level and protein supply [28]. Of course, any international development programme must be directed towards helping the poor.

It was predicted in 1968 that despite the promise of new technologies in the development of unconventional sources of protein food and feed, crops and livestock would continue to be the major sources of protein for human comsumption [4]. Meat being out of reach for many of the poor and its production a rather wasteful protein conversion process, the plant breeders were convinced that increasing protein quantity and quality of grain crops must be the primary objective [62]. Cereals, the staple food of the poorest, also supply about half of the proteins globally available as food. Logically, an increase in the protein content of cereal grain from, e.g., 10 to 11% would mean 5 million tons of protein more per year. Would one be able ro raise the protein utilization value from 60 to 70% through more appropriate amino acid composition, one could gain another 7–8 million tons. Coupled with an increase in over-all food production as exemplified by the 'Green Revolution' in such countries as India and Pakistan, these improvements could mean raising health standards, vitality and activity of millions of people to take their future in their own hands. This was a challenge for the plant breeders and it appears that they have taken the lead and were the most consistent in converting the UN appeal into action. However, the United Nations World Food Conference in 1974 could not draw such an optimistic picture [10] and six years later the number of the hungry and malnourished had further increased rather than decreased [82].

Protein requirements

One would presume that the identification of a so called 'protein gap' was based upon the difference between supply and demand, supply being identical with production minus losses and demand identified by nutritionists as physiologically required intake. The difference between the two should indicate whether there was enough of protein or not [26]. The problem, however, turned out to be more complicated.

Attempts to determine human protein requirements go back as far as 1935 [1, 52]. The status of knowledge in the late sixties seemed to be that an average adult in a developing country would have to consume between 51 and 83 g of locally available protein (depending upon the main source) as part of a diet that would satisfy also the energy needs [28]. This figure was not reached by large parts of the poorer population in developing countries, while the richer part and most people in developed countries consumed much more than the required amount of protein [82].

The recommendations by a FAO/WHO Expert Group in 1965 on protein requirements [1], which formed an essential basis for the international and national action 'to avert an impending protein crisis', were superseded in 1973 by recommendations of another expert group [8]. A number of crucial assumptions were made regarding protein requirements and these led to the recommendation of much reduced levels of protein in the diet than before. The ca. 61 g of local protein intake previously assumed to be required for an average man in a developing country were reduced by about 35% and as a consequence the 'protein crisis' had disappeared 'with the stroke of a pen' [59]. There remained, however, a substantial calorie deficit in developing countries.

Subsequently, it became a popular and much repeated statement that all efforts should be devoted to develop the technology to increase the quantity and not the quality of food [10, 13, 59, 88]. Soon articles appeared blaming the UN and policy makers for having wasted valuable resources on a fictive problem [87], even calling the whole affair the 'protein fiasco' [63].

Policy makers must indeed have been extremely irritated by facing such consequences of differences in opinion among nutritional scientists, but probably are even more disturbed by the continuing controversy in scientific circles about the recommendations of 1973 and its various subsequent modifications and interpretations by other expert committees [43, 44, 51, 72].

More than irritated, however, must have been the plant breeders. They require 10 or more years and an investment of about 1 Million US$ to produce an improved crop variety. At the end of the sixties many of them had embarked on one of the most difficult but seemingly also most urgent tasks of improving the nutritional quality of cereals by genetic means (in other words to breed high protein/high lysine cultivars with competitive yields), only to be told in the very midst of the first breeding cycle that it was not at all urgent to do so and would be better to go back to developing varieties producing bulky yields [66, 76, 77].

It seems obvious, however, that during the coming years the science concerned with human nutrition – stimulated by the 'protein gap fiasco' and motivated by continuing problems of malnutrition – will make substantial progress in understanding the complex interactions of food, function, growth and health [43, 60, 78]. Then, other expert committees might have other thoughts with regard to requirements of food, taking into account all nutrients, the digestibility, the problems created by processing, and eventually come up with a more solid guidance for setting priorities in crop research and production [51, 62, 66]. It can also be hoped that in the near future nutritional education improves, for the poor that won't have enough to eat, as well as for the rich, that eat too much [49, 73].

70% of the edible protein comes from plants. Of this, nearly half is supplied by cereals like wheat, rice and maize. However, there are wide differences in the contribution of major food crops to the diet of people, reflecting ecological conditions and economic status [28, 53, 82]. In order to determine the target for attacking a 'Protein Gap', the situation would have to be assessed country by country. Food balance sheets however provide only estimates of averages at the national level. They must be supplemented by surveys to assess the food intake and diet composition of different socio-economic groups and even members of a family, in order to identify the target population and the most vulnerable groups.

In connection with the 15th session of the FAO Conference 1969, obstacles contributing to protein malnutrition have been listed as follows [5]:

1. Lack of money among large population groups to purchase more protein food at present prices.
2. Lack of foreign exchange to pay for imports of protein foods from countries able to produce them at relatively low cost.
3. Difficulties in producing adequate quantities of protein foods domestically at prices within the purchasing power of the population.
4. Inadequate processing, handling and marketing.
5. Families do not eat protein foods that are financially within their means because of their expenditure preferences, food habits, ignorance of nutritive value and of methods of cooking, etc.

This statement underlines the complexity of the problem as a whole, and that its solution was not simply to provide more nitrogen fertilizers, to give flour enriched with synthetic amino acids to the bakeries, to replace in farmer's fields cereals by grain legumes, or to set up the manufacture of protein concentrates from oil seeds, leaves or single cell organisms. FAO started its action by adding to existing projects in developing countries the concern about adequate protein supply. In this way, initiatives were taken quickly without too much of the usual bureaucratic programming exercises and lengthy budgetary planning [5].

Progress, however, was slow, partly due to ignorance [49], partly due to the difficulties for governments of developing countries to assign a high priority to food protein improvement in competition with other development targets such as building roads, setting up schools, eradicating pests, establishing industries, increasing power supply, etc. Incentives to farmers for producing high quality food were to be linked with measures to assure such quality during marketing, with nutritional education and with subsidies for the poor to enable them to purchase such high quality food. Such integrated organizational systems are not easy to establish, but financial implications were the overriding restrictive factor. Therefore, in 1971 the

UN again published a document, analysing the reasons for slow progress in protein food matters and suggesting effective strategies to avert the protein problem confronting developing countries [7]. Among these were (a) efforts to balance negative effects of the spread of higher yielding cereal varieties by expanding the 'green revolution' to pulses and oil seed crops, (b) strengthening the breeding programmes for better protein quantity and quality in all important cereals.

GENETIC IMPROVEMENT OF NUTRITIONAL QUALITY OF CROPS

Predominant attention has been paid by plant breeders in the past to the selection of genotypes with higher productivity, also responsive to increasing inputs of nutrients and other growth requirements [76]. Quality was only exceptionally looked at, and if so mostly in relation to better suitability for processing or marketing. Little if any attention was given to attempts to improve (or at least to maintain) the nutritional quality of crop products.

Genes affecting the amino acid composition in maize were known for almost 40 years and maintained in collections for merely academic reasons but were never attempted to be used till the mid sixties [64, 69], when the public was made more aware of the importance of the nutritional value of crops. Two of these genes are now known worldwide as '*opaque-2*' and '*floury-2*'. In 1966, scientists were optimistic that the use of these genes could bring a breakthrough in improving the quality of corn protein without significantly affecting yield or other important agronomic characteristics [67]. Time has shown that this was too much of optimism. It seems that there is a price to be paid whenever higher quality is to be achieved [13].

The Joint FAO/IAEA Division of Atomic Energy in Food and Agriculture, noting that the spontaneous endosperm mutants identified in maize should have parallel genetic variation in other cereals, envisaged mutation induction as a tool of practical significance to supplement existing germ plasm collections, if genotypes with desired endosperm composition could not be identified [79]. A Panel of Experts, convened by FAO and IAEA in 1968 at Röstanga (Sweden) discussed this concept and recommended:

a) to assess systematically the existing genetic variation in protein content and protein composition of crop plants;
b) to study environmental influences and fertilizer effects on protein quantity and quality;
c) to study synthesis and transport of proteins and amino acids during grain formation and the genetic system directing these processes;
d) to develop rapid mass screening techniques for grain protein and amino acids, using biological, chemical and physical approaches;
e) to carry out research on identification and modification or elimination of

deleterious compounds, that interfere with the utilization of high value food and feed;

f) to utilize induced mutations as an approach for plant protein improvement;
g) to alert national governments and international agencies to the relatively simple solution of the protein crisis by plant breeding.

Following the recommendations of this Panel of Experts, FAO and IAEA established in 1968 an International Coordinated Research Programme on the Use of Nuclear Techniques for Seed Protein Improvement.

The lead in breeding maize varieties with improved grain quality using the opaque-2 gene was taken by CIMMYT, the International Centre for Maize and Wheat Improvement in Mexico. The International Rice Research Institute (IRRI) in the Philippines developed a strong programme on rice protein improvement based upon its huge germ plasm collection. Meanwhile national institutes, such as the Swedish Seed Association at Svalöv and the University of Nebraska undertook the screening of available germ plasm collections for genotypes nutritionally more valuable than present day's high yielding varieties [41, 46, 56] and developed their own protein breeding programmes. The Swedish programme later on became associated with FAO in the form of a FAO/SIDA/SAREC Project on Improvement in Nutritional Quality of Barley and Spring Wheat [21, 23, 47].

CIMMYT AND THE OPAQUE-2 MAIZE

The International Centre for Maize and Wheat Improvement, CIMMYT, is an outgrowth of over twenty years cooperative efforts between the Government of Mexico and the Rockefeller Foundation, to improve the basic food crops of Mexico. The Centre was set up in 1963 when the Government of Mexico and the Rockefeller Foundation decided to disseminate to other parts of the world some of the improved varieties of maize and wheat. In order to better fulfill the world wide task, the Centre was reorganized in 1966 and established as an autonomous international research and training institute with financial support from other foundations, national and international sources [9, 18].

Following the discovery by Mertz and co-workers at Purdue University [64] that the mutant gene '*opaque-2*' modifies the endosperm protein of maize, increasing particularly the quantities of lysine and tryptophan, feeding trials confirmed that animals could gain weight up to three and a half times faster when fed with *opaque-2* maize instead of normal maize grain. Children in Colombia suffering from severe protein deficiency were brought back to normal health in 2–3 months on a diet of *opaque-2* maize [25, 67].

One of the recommendations of the United Nations Advisory Committee on the Application of Science and Technology to Development [2] was to 'develop and support projects for the prompt introduction of improved varieties such as corn with higher lysine and tryptophan contents'. This recommendation followed a statement in the report by an *ad hoc* Panel of Experts to the Committee, that 'at present dramatically improved varieties of corn, rice and cotton seed are available for immediate introduction in developing countries' creating erroneously the impression, that the plant breeder's job for these species was already completed.

In August 1969, CIMMYT requested support from UNDP for the purpose of assisting people in developing countries in improving the protein content of their diet through (a) intensifying research on raising the quantity and quality of protein in maize in combination with high yield, and (b) training agronomists to carry out national programmes of nutritive maize production [9]. In March 1970, US$ 1.6 million were approved for activities focussing on Central and South America. When the work started, it was realized soon that a whole series of obstacles had to be removed before *opaque-2* maize could have an impact in actually improving the diet of the people in Latin-America. Among them were:

a) the original mutant stocks were adapted ecologically to the US 'corn belt' but not to the sub-tropical and tropical areas of Latin-America;
b) the grain type of the mutant lines was not acceptable to the majority of farmers and consumers;
c) the soft grain was lighter in weight and thus yields were reduced by 10–15%. Although its lysine and tryptophan content was higher, its protein content tended to be lower;
d) the grain of the mutant lines appeared to be more susceptible to attack by pathogens and insect pests, and to mechanical damage;
e) the analytical techniques for assessing protein and amino acids were complicated, too slow and too expensive and required too much seed material.

After 4 years of work, CIMMYT reviewed the status of maize protein improvement as follows [9]:

'The *opaque-2* gene exerts a pleiotropic effect in the sense that high lysine and tryptophan levels are always associated with opaque, dull, chalky and lustreless grain, lower kernel density, lower grain yield, decreased protein percent and fading of color. In order to make improvements, the pleiotropic effects have to be modified by numerous other genes and this is a difficult and time consuming process. Through a wide array of crosses with corn originating in different countries, considerable progress has been made in modifying the pleiotropic effects of the *opaque-2* gene and in placing it into genotypes of different geographical adaptation. However, the programme is still in the stage of experimenting with improved populations and has not reached the level of competitive varieties to be recommended to farmers'. Further UNDP support for this project was approved for 1973–1976 in order to produce maize populations with wide adaptability, high yield, quality protein, pest resistance and acceptable grain types for use in various regions of the world. At the end of this phase in 1976, a CIMMYT report reviewed the achievements as follows:

'A number of populations and experimental varieties of high quality protein type are ready to undergo testing in countries around the world. Original problems (inacceptability of soft grain, more prevalent ear rots, more insect damage) have been diminished but are to some degree still present. Some experimental varieties are producing good quality grain with high specific gravity and yields that are not significantly different from the normal high-yielding maize. Progress has been impeded by the necessity to accumulate a large number of modifier genes'. In spite of concentrated research no alternative had been found for the identification of high lysine and tryptophan lines other than careful chemical analysis. Animal feeding trials had been discontinued due to difficulties with the chosen meadow vole as test animal. In tests carried out elsewhere, however, it could be confirmed that new hard endosperm forms of quality protein maize were as nutritious as the original soft grain opaque-2 types [18].

A third phase was approved for support by UNDP for 1977–1980. Results of 10 years of work were summarized and evaluated by VASAL *et al.* [85] and VILLEGAS *et al.* [86]. By that time in Colombia, Brazil and the USA, some *opaque-2* hybrids had appeared on the market, but could not compete, even with incentives and premium payment. Straightforward backcrossing programmes failed to improve the kernel weight. The attempt to incorporate a number of modifier genes was making progress more difficult, as softness of grain could not be used anymore as simple marker for the high lysine selection. Precise chemical analysis was required in each generation. Only small gains were made per cycle and these were difficult to transfer because of great environmental influences. Nevertheless, several hard grain *opaque-2* pools with tropical, temperate or highland adaptation were developed. Modified *opaque-2* types almost reached the grain yield of normal maize. In an international trial with 60 locations, the best modified *opaque-2* lines were comparable to normal ones in the field at about half of the test sites. However, no composite variety has been released for commercial production in any country [88].

The continuous difficulties encountered in developing high yielding/high protein quality maize can also be seen from a statement made at a Symposium on Production, Processing and Utilization of Maize, organized by the United Nations Economic Commission for Europe in Belgrade 1980 [37]: '*Opaque-2* maize has not yet been widely accepted by farmers (in the USA) despite a strong selling effort by some seedsmen and much favourable press, probably because the farmers would not accept the combination in *opaque-2* maize of 10–15% less yield, poor endosperm hardness characteristics and increased susceptibility to ear rots'. Nevertheless, breeders and nutritionists are not giving up and are studying favourable combinations of different mutant genes, such as *sugary-2* plus *opaque-2* [45, 61]. At the same time, search for other mutant genes altering endosperm composition continues [68, 74, 89].

BREEDING FOR HIGH PROTEIN CONTENT AND QUALITY IN WHEAT
(The USDA-ARS/University of Nebraska programme)

The Agricultural Research Service, USDA, cooperating with the University of Nebraska, has conducted research on the quantity and quality of wheat protein since 1954 recognizing that more than a billion people depend upon wheat as their basic food [56]. Its broad adaptation to many climates and environments makes wheat unique among the cereals. Unique also is the wheat protein which allows preparation of leavened baked products. Improvement of its protein in terms of nutritional value therefore would have significant and far reaching consequences by providing additional protein for about 400 million people. The value of wheat as a food however, encompasses yield, protein content and nutritional value of the protein. Improvement of productivity at the expense of protein content and quality has dubious value. Equally questionable would be the improvement of protein quantity or quality at the expense of productivity or technological value. However, these characters were found to be interdependent to a large extent.

The USDA-ARS/University of Nebraska cooperative project has since 1966 screened a world collection of ca. 13000 common wheats, 3400 durum wheats and more than 600 spelts for protein content and quality. Identified genetic resources for high protein among the common wheats were Atlas 66 and Atlas 50 (both derived from the Argentinian variety Frondoso), Anniversario and Nap Hal. Nap Hal and CI 13449 were identified furthermore as sources of higher lysine content. Experimental lines developed from crosses with these protein donors achieved much higher values than ordinary varieties. The 100 best lines had protein values increased by ca. 40% and lysine values increased by about 20%. Cooperative relationships with developing countries and international organizations were established particularly through the International Winter Wheat Performance Nurseries organized with assistance from AID and in cooperation with CIMMYT. Seeds of improved lines were made available. In 1975, the variety 'Lancota' (NE701132) combining improved endosperm quality with high yield, disease resistance and excellent agronomic performance was certified and released for the US [57, 88].

Crosses of Atlas 66, Nap Hal and CI 13449 led to transgressive segregation, indicating different and additive genes operating in these genotypes. It had been observed already in 1968, that the higher protein content of Atlas 66 and its genetic derivatives is not the result of more nitrogen uptake but of better nitrogen translocation. This allows the assumption that the benefits of protein improvement resulting from this programme can be realized without extra cost for additional nitrogen fertilization. Calculations were made to demonstrate, that wheats derived from this programme with about 17% protein provide more of all the essential amino acids, including lysine, than opaque-2 maize with 10% protein. Furthermore, one has estimated that one acre of such wheat would provide the protein needs of one man for 877 days, whereas the protein from one acre of rice would only be sufficient for 654 days. This, in simplified terms, underlines the value of a successful protein improvement programme in wheat [55, 57].

THE USE OF NUCLEAR TECHNIQUES FOR GRAIN PROTEIN IMPROVEMENT (The Joint FAO/IAEA Division's Programme)

This programme may serve as another example of the multiple international activities following the alarming resolution by the United Nations. The programme of this FAO Division carried out jointly with IAEA consisted in 1968 of research-based applications of nuclear techniques towards improving fertilizer use efficiency, metabolism and health of farm animals, pest control through the sterile male technique, genetic plant variety improvement through induced mutations and the preservation of food through irradiation. All these activities could contribute to augmenting the quantity of nutritionally valuable food for the people in need. However, it appeared that something more could be done. Judging the potential impact of all approaches, it seemed that developing crop plant varieties with improved quality characteristics would be the most adequate and effective way to provide more protein for direct human consumption as well as for animal feed. Only plant breeding was offering the possibility to adjust the quality of the plant protein to the nutritional requirements of man and of monogastric animals. Evidence for the potential of mutations in individual genes to contribute substantially to the objectives was coming from the re-discovery of recessive alleles of the *opaque-2* and *floury-2* genes in maize which exerted a major impact upon the endosperm composition and raised drastically the level of limiting amino acids. Providing crop plant varieties with improved nutritional value was considered to have the advantage of (a) a long-term return after the initial investment for improvement, in contrast to the continuing cost of diet supplements and (b) adaptability to local cropping and eating habits, unlike new high protein food products and dietary additives. An ad hoc Panel of Experts [4] convened in June 1968 advised on details of a coordinated international research programme that became established still in the same year. The work planned for this programme encompassed the following subjects:

1. Induction of mutations, as by treatment with ionizing radiation, to expand the natural variation for high protein content and higher content of specific amino acids, assumed to be on the low side because conscious selection for these traits had rarely been practiced. Mutation induction was expected to improve the grain protein characters without affecting other desirable characteristics present in a particular good variety.
2. Development of rapid, relatively inexpensive screening techniques for protein and for specific, essential amino acids such as lysine, including the use of

radioisotopes where feasible, order to increase the rate, decrease the cost and improve the reliability of analyses.
3. Development of methods for more rapid and less expensive nutritional tests with micro-organisms and test animals, allowing the estimate of digestibility and utilization of grain protein by human beings.

The question whether it is worthwhile to consider artificially induced genetic variation in protein, when some variation has been found to exist naturally was raised. SIGURBJÖRNSSON reasoned that natural genetic variability in protein quantity and quality may not be so great or so easily available as it is in many other plant characteristics of value to food production [79]. Although possible kinds of mutations causing changes in protein content have most likely already occurred during the eons of evolution, it appeared unreasonable to him to expect the relative quantities of amino acids that meet the nutritional requirements of man, woman and children to have been selected during natural evolution, the aims and forces of which are plant's survival and reproduction. It would also be unrealistic to assume that selection for the nutritionally appropriate composition of amino acids in grains has been much practiced during early phases of intelligent plant breeding. 'Induction of mutations is therefore a suitable means to reproduce the evolutionary potential of crop plants, and stringent selection is likely to lead to the discovery of "new", hitherto unknown mutants containing levels of limiting amino acids and total protein desired by man' [79]. Another most important reason for resorting to mutation induction was certainly that mutations can be induced in any fixed plant variety, including the very high yielding and most adapted ones, thus, eventually accommodating an improved quality in an otherwise probably unchanged, agronomically valuable genotype.

The programme required inter-disciplinary cooperation of geneticists, plant breeders, analytical chemists, biochemists, physicists and nutritionists. It was fortunate that the International Atomic Energy Agency could make available to the programme the services and facilities of its own Seibersdorf Laboratory. The programme gained momentum with the FAO/IAEA International Symposium on Plant Protein Resources: Their Improvement through the Application of Nuclear Techniques organized by FAO and IAEA in Vienna in 1970 [6]. Essential was the generous financial support provided by the Federal Republic of Germany from 1970 till 1982 and by the Swedish International Development Authority from 1980 till 1982.

The work performed and the results achieved within the frame of this programme have been summarized [34, 80] at the end of phase II of the programme which was marked by an International Symposium on Seed Protein Improvement in Cereals and Grain Legumes, organized by FAO and IAEA in cooperation with the German Society for Radiation and Environmental Research at Neuherberg in 1978 [20].

In the subject area of plant breeding, success has been predominantly in barley, likely because of its diploidy and easy handling in large scale experiments [4, 47]. DOLL selected from mutagen treated Danish spring barley about 20 mutants with high lysine properties, thus supplementing the till then unique genetic resource, a promising high lysine line collected in Ethiopia. As these selections were made already in an early stage of the programme, their utilization in cross breeding advanced more than in any other project [54, 80]. A comparable high lysine sorghum mutant has only been reported in 1975 [29, 30], and its value in cross breeding remains still to be seen. Out of a total of 19 institutions 9 searched for protein mutants in wheat, 9 in rice, 4 in barley and one in millets. Despite the more intensive efforts in wheat and rice, no high lysine mutant comparable to the ones in barley or sorghum has been detected [35, 50]. The reason for this lack of success is certainly not the effectiveness of mutagen treatment or the methodology of screening. It must have to do with the inherent biological and genetical characteristics of these plant species. AXTELL has put forward the hypothesis that major gene effects for lysine increase can only be achieved via a reduction of the lysine poor prolamins [29]. Since rice contains already very little of prolamin in its endosperm protein (only 5–10%, compared with 30–40% in barley and 50–60% in corn), its further reduction may face unsurmountable barriers of biochemical and genetic nature [77]. Improvement of protein quantity while maintaining the amino acid composition was the more appropriate objective for rice. Wheat on the other hand, has similar levels of prolamins as barley, however, its allohexaploid genome leads to buffering of many single gene effects, by which mechanism any mutation towards higher lysine may express only to such a small extent, that it remains almost unnoticeable in the 'noise' of environment conditioned variation. Through repeated selection over many generations, the plant breeders of the IAEA Laboratory were able to identify a mutant line of wheat with a noticeable lysine increase [50, 80].

Protein quantity increases were reported more frequently, however, it took a number of generations before the frequent artifacts resulting from reduced starch deposition were separated from the rare valuable mutants having more protein amount per grain, per plant, or per unit of cultivation area. A high protein wheat mutant was released as variety in Chile under the name "Carolina" in 1981 [71].

Details of any of the individual mutation breeding projects can be found in various IAEA publications, particularly in the symposium proceedings of 1979 [20].

The development of suitable analytical procedures for mass screening, a precondition of mutation breeding was followed up through the programme [11, 42, 58, 66], narrowing down the suitable methods to only very few [70].

ASSESSMENT OF NUTRITIONAL VALUE OF CEREAL GRAINS

One of the bottlenecks in breeding programmes for protein improvement was the slowness and the high costs of chemical screening tests [11, 58]. Another problem became more obvious only with advancing progress in grain protein improvement programmes: the unsatisfactory correlation between chemical assays and the true nutritional value for man or monogastric animals [14, 39, 40].

Certain high lysine mutants of barley and derived cross breeding product were found to have poorer protein digestibility than the original varieties. This is due to the fact that the highly digestible prolamin fraction is reduced, while the added lysine appears in a less digestible protein fraction. So, a 20% increase in lysine in a mutant no. 1508 was in fact reduced to only 10%, taking into account the digestibility [38].

It was therefore not without reason that in 1976, 13 years after the re-discovery of high lysine mutants of corn, a group of nutritional experts gathered upon invitation by FAO and IAEA, to critically assess available techniques for nutritional value testing, to discuss prospects for developing more simple but still reliable nutritional evaluation techniques for early screening of large populations and to make corresponding recommendations for lines of research [14]. Chemical, biological, microbiological, enzymatic assay techniques were reviewed. The recommendations supplement the PAG Guideline No. 16 on protein methods for cereal breeders as related to human nutritional requirements [12].

Particular attention was called by the expert group to anti-nutritional factors such as phenolic compounds (tannins, resorcinols), phytates and inhibitors of amylases and protease. Tannins were found to be present in higher amounts in sorghum and barley, and inhibit protein digestibility [39, 81]. Specific trypsin inhibitors have been found mainly in rye and triticale, but are also present in wheat. They not only reduce protein digestibility but may also accentuate the deficiency of sulfur-containing amino acids in many plant proteins. Alpha-amylase inhibitors are present in significant amounts in wheat, barley and rye. They affect not only the availability of starch but also of protein. A particular problem with these anti-nutritional factors is that they are rather stable in the usual cereal processing.

Being aware of these complications, the plant breeder may perhaps be discouraged to continue efforts for grain protein improvement. The solution must be the cooperation with institutes specializing on nutritional assays and the development of suitable mass screening methods to be applied before the conventional animal feeding trials.

GRAIN LEGUMES AS PART OF A BALANCED DIET

Grain legumes on the average contain twice as much protein as cereals. In addition, when added to a mixed diet, they may supplement favourably the amount of essential amino acids insofar as they are richer in lysine and tryptophan than cereals, but poorer in methionine which is easily provided from cereals. Thus, leguminous seeds not only increase the protein quantity, they also improve substantially the utilization of the cereal protein already available in the diet [33]. The problem is that yields of grain legumes are lower and prices therefore tend to be much higher than for cereals, thus making it dependent on the income, whether a family can afford to improve its diet by supplementing it with beans or other pulses [51]. That cultivation of pulses was becoming less and less profitable for farmers in developing countries, resulting in decreasing supply and further increasing prices, is partially due to the success of international cereal breeding programmes (mainly on wheat, rice and maize) [32].

Therefore, first emphasis in legume improvement for better nutrition should be on increasing and stabilizing yields and on reducing costs of production, while protein content and protein quality only need to be maintained at current levels [16, 31]. This consideration was followed in most of the international research programmes aiming directly at grain legume improvement (such as by FAO, IAEA, ICRISAT, IITA and CIAT). However, methods for evaluating the food quality of legume seeds are available [52].

The Protein Advisory Group of the United Nations convened a symposium at the FAO Headquarters Rome in 1972 bringing together agronomists, plant breeders, geneticists, nutritionists, biochemists and food scientists to review nutritional and food use deficiencies, plant physiology, pathology, and production technology problems which needed to be resolved in order to increase the supply of this valuable food item [65]. Six food legumes (dry bean, pigeon pea, cow pea, chickpea, broad bean, pea) and two leguminous oil seeds (peanut, soybean) were identified as priority targets for international and national research efforts. The PAG Statement No. 22 [65] summarizes the recommendations for action agreed upon during the course of this symposium.

In 1975 FAO and IAEA jointly organized a seminar in Sri Lanka to specify the needs for grain legume improvement in South-East Asia (an area where many vegetarians rely upon pulses as their main protein source) and to consider the contribution that could be expected from widening genetic variability through mutation induction [17]. It was recognized that many of the currently cultivated legume species still possess a number of 'wild type' characteristics that were beneficial during past natural evolution, but hinder the upgrading of grain production under more intensive farming conditions. Plant architecture and the occurrence of toxic substances are the main wild type characteristics. Since domestication of crop plants appears to be based largely upon monogenic or oligogenic traits, mutation induction was looked at as a promising means to obtain desired characteristics missing in current legume germ plasm collections [31].

Grain legume crops consist of a number of species specializing for the various climatic conditions [19, 65, 83]. Since they have the ability for symbiotic nitrogen fixation and therefore require little or no fertilization, they are often grown on marginal lands and with sub-optimal management. Genetic improvement for higher production therefore is a complex task, requiring diversified efforts with specific objectives for each species and the particular farming conditions [75, 83].

Substantial progress has been made in most major legume crops by international institutes, such as ICRISAT, IITA, AVRDC and CIAT but also in national research programmes. Induced mutation programmes, supported through IAEA as part of the FAO/IAEA/GSF Programme on the Use of Nuclear Techniques for Seed Protein Improvement, the FAO/IAEA/SIDA Programme on Induced Mutations for Disease Resistance in Legumes and in several other ways, were successful in contributing to improved varieties of groundnut, dry bean, chickpea, broad bean and pea [15, 20, 22, 31, 75]. In general, however, it must be admitted that genetic improvement did not yet match the expectations that can be derived from the advances made in soybeans within this century. Slow progress must partly be attributed to the limits imposed by the symbiotic nitrogen fixation of existing plant genotypes under current cultural conditions and to lack of knowledge as to the optimal plant architecture for various cultivation systems. The increasing costs of nitrogen fertilizers, however, have strengthened the interest in grain legume improvement and further progress is certainly to be expected from world wide national and international efforts [7, 16, 19, 48, 52, 65, 75, 77].

CONCLUSIONS

International cooperation to help alleviating problems faced by individual countries or parts of their population is a new and positive human phenomenon in this century, which otherwise is blamed so often for its cruelty and inhumanity. The international action initiated by the United Nations in 1967, to avert what looked like an impending protein crisis with perhaps catastrophic consequences, can be seen as an impressive example of international concern, cooperation and effort to solve problems that are beyond the capacity of individual countries. It must be taken as a good sign of hope that it was possible to mobilize scientists, policy makers, funding organizations and politicians for a concerted action. Critics have pointed out shortcomings, limitations and failures, but those actively involved in the work are convinced that an enormous amount of knowledge and experience has been generated that will bear fruits in the continuing struggle against malnutrition of further rising populations.

It is certainly easier to set objectives for a plant breeding programme, if the target is to produce adequate quality feed for raising pigs and poultry, than to supply food which fulfills the requirements for good physical and mental development of children and for motivated performance of adults. But nutritional science has been stimulated, plant geneticists have learned how to deal with grain endosperm traits, chemists have been told to reconsider their analytical techniques, multi-disciplinary research has been established and politicians have been made more aware of the complexity of development. These are the outcomes of international programmes for grain protein improvement that will probably have even a longer lasting effect than improved crop cultivars released to farmers.

REFERENCES

1. Anonymous (1965): Protein Requirements. Joint FAO/WHO Expert Group. FAO Nutrition Meetings Report Series No. 37.
2. Anonymous (1968): International action to avert the impending protein crisis. United Nations.
3. Anonymous (1968): Resolution adopted by the UN General Assembly on the increase in the production and use of edible protein. A/RES/2416(XXIII).
4. Anonymous (1969): New Approaches to Breeding for Improved Plant Protein. IAEA Vienna.
5. Anonymous (1969): Filling the protein gap. FAO Rome.
6. Anonymous (1970): Improving Plant Protein by Nuclear Techniques. IAEA Vienna.
7. Anonymous (1971): Strategy statement on action to avert the protein crisis in the developing countries. Report of the Panel of Experts on the Protein Problem Confronting Developing Countries. UN New York.
8. Anonymous (1973): Energy and Protein Requirements. FAO Nutr. Meetings Rept. Ser. No. 52. WHO Tech. Rept. Ser. No. 522.
9. Anonymous (1974): Research and training in the development of high lysine maize. A United Nations Development Programme Global Project. UNDP New York.
10. Anonymous (1974): Preliminary assessment of the world food situation present and future. World Food Conference.
11. Anonymous (1975): Recommendations of an ad hoc panel on analytical screening methods for seed protein content and quality. Annex III in 'Breeding for Seed Protein Improvement Using Nuclear Techniques'. IAEA Vienna.
12. Anonymous (1975): PAG Guideline *16* on Protein Methods for Cereal Breeders as related to Human Nutritional Requirements. AG Bull. *5* (2): 22–48.
13. Anonymous (1976): Genetic Improvement of Seed Proteins. Nat. Acad. Sci., Washington, USA.
14. Anonymous (1977): Nutritional Evaluation of Cereal Mutants. IAEA Vienna.
15. Anonymous (1977): Induced Mutations against Plant Diseases. IAEA Vienna.
16. Anonymous (1977): Food legume crops: Improvement and production. FAO Plant Production and Protection Paper No. 9.
17. Anonymous (1977): Induced Mutations for the Improvement of Grain Legumes in South East Asia (1975). IAEA Vienna.
18. Anonymous (1978): Research and training in the development of high lysine maize (Phase II). DP/GLO/Final Report 4. UNDP New York.
19. Anonymous (1979): Tropical legumes: resources for the future.Nat. Acad. Sci. Washington, USA.
20. Anonymous (1979): Seed Protein Improvement in Cereals and Grain Legumes I, II. IAEA Vienna.
21. Anonymous (1979): FAO/SIDA/SAREC Project on Improvement in Nutritional Quality of Barley and Spring Wheat. FAO Rome.
22. Anonymous (1980): Induced Mutations for Improvement of Grain Legume Production. IAEA Vienna.
23. Anonymous (1981): Third Seminar of the FAO/SIDA/SAREC Project on Improvement in Nutritional Quality of Barley and Spring Wheat. FAO Rome.
24. Anonymous (1982): Induced Mutations for Improvement of Grain Legume Production II. IAEA Vienna.
25. Alexander, DE, RJ Lambert, and JW Dudley (1969): Breeding problems and potentials of modified protein maize. New Approaches to Breeding for Improved Plant Protein: 55–65; IAEA Vienna.
26. Altschul, AM (1976): The protein-calorie trade-off. In: Genetic Improvement of Seed Proteins. Nat. Acad. Sci., Washington: 5–17.
27. Autret, M (1970): Besoins et disponibilités futurs de proteines. Improving Plant Protein by Nuclear Techniques: 33–34; IAEA Vienna.
28. Autret, M, J Perisse, F Sizaret, and M Cresta (1968): Protein value of different types of diet in the world: Their appropriate supplementation. FAO Nutrition Newsl. *6*, No. 4: 1–29.

29. Axtell, JD (1976): Naturally occurring and induced genotypes of high lysine sorghum. Evaluation of Seed Protein Alterations by Mutation Breeding: 45–53; IAEA Vienna.
30. Axtell, JD, SW van Scoyoc, PJ Christensen, and G Ejeta (1979): Current status of protein quality improvement in grain sorghum. Seed Protein Improvement in Cereals and Grain Legumes II: 357–365; IAEA Vienna.
31. Bahl, PN, SP Singh, Hayat Ram, DB Raju, and HK Jain (1979): Breeding for improved plant architecture and high protein yields. Seed Protein Improvement in Cereals and Grain Legumes I: 297–307; IAEA Vienna.
32. Borlaug, NE (1972): Building a protein revolution on grain legumes. In: Milner, M (ed.): Nutritional Improvement of Food Legumes by Breeding: 7–11; Wiley, New York.
33. Bressani, R, and LG Elias (1979): The world protein and nutritional situation. Seed Protein Improvement in Cereals and Grain Legumes I: 3–23, IAEA Vienna.
34. Brock, RD (1978): Seed protein improvement in cereals and grain legumes. Atomic Energy Rev. *16*, 3: 561–565.
35. Coffman, WR, and BO Juliano (1979): Seed protein improvement in rice. Seed Protein Improvement in Cereals and Grain Legumes II: 261–277; IAEA Vienna.
36. Doll, H, B Kφie, and BO Eggum (1974): Induced high lysine mutants in barley. Radiat. Bot. *14*: 73–80.
37. Duvick, DN (1980): Recent advances in maize breeding with a view to raising yield and quality. Paper presented at the Symposium on Production, Processing and Utilization of Maize organized by the UN-Economic Commission for Europe at Belgrade (Yugoslavia) AGRI/SEM.12/R.16.
38. Eggum, BO (1977): Nutritional aspects of cereal proteins. In: Genetic Diversity in Plants: 349–369; Plenum Press, New York.
39. Eggum, BO, and LD Campbell (1979): Nutritional and anti-nutritional assay. Seed Protein Improvement in Cereals and Grain I: 353–368; IAEA Vienna.
40. Eggum, BO, and KD Christensen (1975): Influence of tannin on protein utilization in feedstuffs with special reference to barley. Breeding for Seed Protein Improvement using Nuclear Techniques: 135–143; IAEA Vienna.
41. Frey, KJ (1977): Protein of oats. Z. Pflanzenzüchtg. *78*: 185–215.
42. Georgi, B, EG Niemann, RD Brock, and H Axmann (1979): Comparison of analytical techniques for seed protein and amino acid analysis: An inter-laboratory comparison. Seed Protein Improvement in Cereals and Grain Legumes I: 311–341; IAEA Vienna.
43. Graham, GG (1976): The relative importance of protein and energy in the diets of infants and children. In: Improving the nutrient quality of cereals II: 320–327; AID, Washington.
44. Graham, GG (1977): Factors affecting the human nutritional value of cereal grains. Nutritional Evaluation of Cereal Mutants: 1–12; IAEA Vienna.
45. Graham, GG, DV Glover, GL de Romana, E Moralis, and WC Maclean (1980): Nutritional value of normal, opaque-2 and sugary-2/opaque-2 maize hybrids for infants and children. I. Digestibility and utilization. J. Nutrition *110*: 1061–1069.
46. Hagberg, A, and KE Karlsson (1969): Breeding for high protein content and quality in barley. New Approaches to Breeding for Improved Plant Protein: 17–21; IAEA Vienna.
47. Hagberg, A, G Persson, R Ekman, KE Karlsson, AM Tallberg, V Stoy, NO Bertholdsson, M Mounla, and H Johansson (1979): The Svalöv protein quality breeding programme. Seed Protein Improvement in Cereals and Grain Legumes II: 303–313; IAEA Vienna.
48. Hardy, RWF, UD Havelka, and B Quebedeaux (1976): Opportunities for improved seed yield and protein production: N_2 fixation, CO_2 fixation and O_2 control of reproductive growth. In: Genetic Improvement of Seed Proteins: 196–230; Nat. Acad. Sci. Washington.
49. Heinz, HJ (1974): Nutritional illiteracy: a view from an industrialized nation. Impact of Science on Society *24*, (2): 179–180.
50. Hermelin, T, and G Adam (1978): Selection for increased variability of seed protein and lysine contents in wheat. Seed Protein Improvement by Nuclear Techniques: 293–300; IAEA Vienna.

51. Hulse, JH, and OE Pearson (1980): How nutrition priorities can be integrated into crop improvement programmes. Food and Nutrition *2*, (1): 7–10.
52. Hulse, JH, KO Rachie, and LW Billingsley (1977): Nutritional Standards and Methods of Evaluation for Food Legume Breeders. IDRC Ottawa, Canada.
53. Jalil, ME, and WM Tahir (1970): Review of the world's plant protein resources. Improving Plant Protein by Nuclear Techniques: 21–32; IAEA Vienna.
54. Johansson, H (1978): Improvement of nutritional quality in barley and spring wheat: A FAO/SIDA/SAREC project. Seed Protein Improvement by Nuclear Techniques: 85–90; IAEA Vienna.
55. Johnson, VA, PJ Mattern, and SL Kuhr (1979): Genetic improvement of wheat protein. Seed Protein Improvement in Cereals and Grain Legumes II: 165–181; IAEA Vienna.
56. Johnson, VA, PJ Mattern, DA Whited, and JW Schmidt (1969): Breeding for high protein content and quality in wheat. New Approaches to Breeding for Improved Plant Protein: 29–40; IAEA Vienna.
57. Johnson, VA, PJ Mattern, KD Wilhelmi, and SL Kuhr (1978): Seed protein improvement in common wheat (*Triticum aestivum* L.): Opportunities and constraints. Seed Protein Improvement by Nuclear Techniques: 23–32; IAEA Vienna.
58. Kaul, AK (1973): Mutation breeding and crop improvement. Nuclear Techniques for Seed Protein Improvement: 1–106; IAEA Vienna.
59. Kracht, U (1974): The nutritional aspects of food production and distribution. Impact of Science on Society *24*, (2): 157–169.
60. Lewin, R (1974): The poverty of under-nourished brains. New Scientist *64*, 268–271.
61. Lovato, MB, and WJ Da Silva (1980): Dry matter, nitrogen and zein accumulation in normal and sugary opaque-2 kernels in segregating ears. Maydica *25*: 167–172.
62. Mauron, J (1979): Die Proteinwertigkeit der Nahrung: Aufbesserung oder Verminderung durch industriemässige Herstellung. Z. Ernährungswiss. *23*: 10–39.
63. McLaren, DS (1974): The greet protein fiasco. The Lancet *11*: 93–96.
64. Mertz, ET, LS Bates, and OE Nelson (1964): Mutant gene that changes protein composition and increases lysine content of maize endosperm. Science *145*: 279–280.
65. Milner, M (Ed.) (1973): Nutritional Improvement of Food Legumes by Breeding. Protein Advisory Group; UN New York.
66. Munck, L (1979): Prospects for the future development of food and feed materials. Seed Protein Improvement in Cereals and Grain Legumes II: 413–422; IAEA Vienna.
67. Nelson, OE (1969): The modification by mutation of protein quality in maize. New Approaches to Breeding for Improved Plant Protein: 41–54; IAEA Vienna.
68. Nelson, OE (1979): Inheritance of amino acid content in cereals. Seed Protein Improvement in Cereals and Grain Legumes I: 79–88; IAEA Vienna.
69. Nelson, OE, ET Mertz, and LS Bates (1965): Second mutant gene affecting the amino acid pattern of maize endosperm proteins. Science *150*: 1469–1470.
70. Niemann, EG, D Christoffers, B Georgi, HW Kaestner, M Mohyuddin, and TR Sharma (1978): Screening methods for protein, amino acids and nutritive value in crops and processed food. Seed Protein Improvement by Nuclear Techniques: 331–345; IAEA Vienna.
71. Parodi, PC, and IM Nebreda (1978): Mutation breeding to increase protein content in wheat. Seed Protein Improvement by Nuclear Techniques: 33–39; IAEA Vienna.
72. Passmore, R, BM Nicol, MN Rao, GH Beaton, and EM de Mayer (1974): Handbook of human nutritional requirements. FAO Nutritional Studies No. 28, WHO Monograph Series No. 61. FAO Rome.
73. Pellet, PL (1981): Malnutrition, wealth and development. Food and Nutrition Bull. *3* (1): 17–19.
74. Phillips, RL, PR Morris, F Wold, and BG Gengenbach (1981): Seedling screening for lysine-plus-threonine resistant maize. Crop Sci. *21*: 601–607.

75. Rao, CH, JL Tickoo, Ram Hayat, and HK Jain (1975): Improvement of pulse crops through induced mutations: Reconstruction of plant type. Breeding for Seed Protein Improvement Using Nuclear Techniques: 125–131; IAEA Vienna.
76. Röbbelen, G (1977): Possibilities and limitations of breeding for nutritional improvement of cereals. Nutritional Evaluation of Cereal Mutants: 47–57; IAEA Vienna.
77. Röbbelen, G (1979): The challenge of breeding for improved protein crops. Seed Protein Improvement in Cereals and Grain Legumes I: 27–41; IAEA Vienna.
78. Scrimshaw, NW, and R Lockwood (1980): Interpretation of data on human food availability and nutrient consumption. Food and Nutrition Bull. *2* (1): 29–37.
79. Sigurbjörnsson, B (1970): Role of plant breeding in bridging the protein gap. Improving Plant Protein by Nuclear Techniques: 13–17; IAEA Vienna.
80. Sigurbjörnsson, B, RD Brock, and T Hermelin (1979): A joint FAO/IAEA/GSF programme on grain protein improvement. Seed Protein Improvement in Cereals and Grain Legumes I: 387–421; IAEA Vienna.
81. Silano, V (1977): Factors affecting digestibility and availability of proteins in cereals. Nutritional Evaluation of Cereal Mutants: 13–46; IAEA Vienna.
82. Swaminathan, MS (1980): Introducing nutritional considerations into agricultural and rural development. Food and Nutrition Bull. *3* (3): 30–36.
83. Swaminathan, MS, and HK Jain (1973): Food legumes in Indian agriculture. In: Nutritional Improvement of Food Legumes by Breeding: 69–82; P.A.G. New York.
84. Thielebein, M (1969): The world's protein situation and crop improvement. New Approaches to Breeding for Improved Plant Protein: 3–6; IAEA Vienna.
85. Vasal, SK, K Villegas, and R Bauer (1979): Present status of breeding quality protein maize. Seed Protein Improvement in Cereals and Grain Legumes II: 127–148; IAEA Vienna.
86. Villegas, EM, BO Eggum, SK Vasal, and MM Kohli (1980): Progress in nutritional improvement of maize and triticale. Food and Nutrition Bull. *2* (1): 17–24.
87. Waterlow, JC, and RR Payne (1975): The protein gap. Nature *258*: 113–117.
88. Wilcke, HL (Ed.) (1976): Improving the nutrient quality of cereals II. AID, Washington.
89. Zorilla, HL, and PL Crane (1979): Direct application of ninhydrin reagent solution to exposed maize endosperm to screen for protein mutants. Crop Sci. *19*: 659–662.

3. Methods for Characterization of the Seed Proteins in Cereals and Legumes

H. STEGEMANN and G. PIETSCH

List of Abbreviations

(P)AA	= (Poly)Acrylamide
Bis	= N,N′-Methylene-bis-acrylamide
PAGE	= Polyacrylamide-Gel-Electrophoresis
PoroPAGE	= Porosity gradient-PAGE
PAGIF	= Polyacrylamide Gel Isoelectric Focusing
Mapping	= Separation by PAGIF, followed by PAGE etc. in the perpendicular direction
SGE	= Starch Gel Electrophoresis
kD	= Kilo-Dalton, unit $\times$ 10^3 of apparent MW
MW	= Molecular Weight (apparent)
ME	= Mercaptoethan-2-ol
Tris	= Tris (hydroxymethyl) aminomethane
EDTA	= Ethylenediaminetetraacetic acid. Di-sodium salt
SDS	= Sodium dodecyl-sulfate

The nutritional value, processing quality, genetic aspects, physiological and phytopathological behavior are involved when seed proteins in crops are studied. Proteins mirror some important properties of the plant or its product. For this reason we will describe recent developments of their characterization, e.g. by solubility or by other physical parameters such as size, shape and charge. No methods are covered which use costly equipment or have been discussed in recent reviews, such as ultracentrifugation, sequence analysis, isotachophoresis, HPLC (High Performance-Liquid-Chromatography) [50, 94], affinity chromatography [123], immuno assays with enzymes [33, 114, 168] or isotopes [170], characterization of lectins [92] and proteinases/inhibitors [130]. The excellent surveys on seed proteins of cereals and legumes [88a], of cereals [14, 111, 176a] and legumes [21] should be consulted. For amino acid composition see [45, 88a] and the book 'Seed Protein Improvement in Cereals and Grain Legumes', published in 1979 by the Internat. Atomic Energy Agency in Vienna (Austria).

Our personal experience within the last ten years is limited to proteins in seeds of maize, wheat, rice, barley, different grasses, faba and phaseolus beans and some less important cereals or legumes. The term 'seed' rather than kernel etc. is used to mean the whole grain, bean, and so on. We have included experience with seeds beyond the scope of this book if necessary to show the general applicability of a certain method. Preference is given to citations of papers whose results have been confirmed by us. The history of the sample and the preparation of the extract are exhaustively treated or at least cited since handling strongly influences the quantity, quality and distribution of proteins. There are as many extraction procedures as researchers. Apparently it is not always considered that proteins outside their natural environment are delicate substances and targets for secondary alterations within hours. Prolonged handling is rarely checked for a possible impact on protein composition. Moreover it is not always appreciated that the intact structure of many proteins depends on their primary hydration layer, which should not be removed, e.g. by lyophilization. Protein analysis advanced mainly in the medical field and was later applied to plant proteins which are more difficult to handle due to accompanying phenols, abundant polysaccharides and restricted solubility. Our emphasis is placed on discussion of electrophoretic procedures, since they can accomodate quite a large amount of 'impurities', require little prepurification and permit fast handling of samples. Electrophoretic methods have – at the moment – the highest inherent capacity for resolving proteins; they are very promising in biochemical taxonomy and genetics. 'It became possible, for the first time, to identify large numbers of particular gene loci ..... and a simple matter to ascertain the proportion that was heterozygous in single individuals. Electrophoretic data also are an important source of single gene markers' [58a, b].

The major steps in the development of electrophoretic methods since 1955 can be summarized as follows:

1955: starch gels; separation mainly by charge [143]
1959: polyacrylamide gels; separation by size and charge [127]
1962: ampholytes; separation by charge, 'isoelectric focusing' [166, 171]
1967: SDS; formation of protomers with similar shape and charge, separation by the apparent MW [137]
1968: porosity gradients; separation by the MW of the native proteins [103]
1968: mapping, that is the combination of focusing with regular electrophoresis [36, 96a, b] or the combination
1973: with SDS-electrophoresis [160].

The evolution of methods is reviewed is general [1b] and for those using SDS [153]. Since then these techniques advanced and we will discuss them within the framework of our topic.

The importance of other methods for analytical protein separation has decreased. Ultracentrifugation has lost significance except with density gradients. Separation by dialysis and ultrafiltration has many shortcomings, e.g. secondary alteration (oxidation, proteolysis [151a, 163] within a few hours) or unpredictable adsorption of some proteins to the membrane [54]. Size exclusion chromatography or gel-filtration has the same drawbacks; usually it takes even longer and the separation power is poor, although for preparative applications and a short running time it may be recommended. Either crosslinked dextran (e.g. Sephadex) or polyacrylamide (e.g. Biogel) can be employed. Ion-exchange columns, salting-out procedures, solvent precipitations and all other time-consuming preparations include the risk of unwanted changes in protein structure and composition. Gel-electrophoretic methods are not as critical since molecules which tend to interact are separated immediately after the start and undesired reactions are further hampered in the gel's network. Fractions from a preparative electrophoresis for instance contained proteins not digestible by trypsin, which is quite a safe indication of their native structure [156]. After heating these proteins could be split by the enzyme.

Some newer methods have not yet found wide applications in protein analysis. Chromatofocusing [142] yielded results similar to focusing in a spiral [97, personal communications by G.F. Domagk, Göttingen], however, we consider chromatofocusing a convenient method. High Performance Liquid Chromatography (HPLC) certainly has a promising future also for large molecules [94] and its speed is unsurpassed, but it will surely not match the high resolution of mapping.

COLLECTION OF SAMPLES

Samples should be harvested when mature seeds have developed, since any other state of ripening is a transient state and not necessarily suitable for genetic work. It may be valuable to collect samples at suitable intervals before maturity and to follow the change of protein distribution until it reaches the dormant state.

Unusual climatic conditions must be considered. Hot and dry weather prior to harvest may lead to prematurity, and the synthesis of 'storage proteins' may have been stopped too early. Extremes in plant nutrition, e.g. sulfur deficiency [109] or N-fertilization at some stage of growth [125] may lead to a small aberration, particularly among the 'storage' proteins, with otherwise constant protein patterns governed only by the genetic background. Our findings since 1974 point to a very small influence on water/buffer soluble proteins, whereas alcohol-soluble (storage) proteins may show a certain shift in some years. This has been supported by the results of other groups [11, 21, 120, 125, 176]. Healthy samples are a prerequisite. Fungal or bacterial infections, sometimes not visible at first glance, have to be excluded. Samples should be kept dry, dark and cool. They can be frozen but never freeze-dried unless they are prepared for SDS-extraction and SDS-PAGE. Seed meals should not be stored without prior defatting and gentle drying.

MILLING, CLASSIFICATION AND DEFATTING

The disintegration of seeds for protein research should be done from samples as dry as possible, or even from seeds pre-extracted with solvents to reduce the lipids

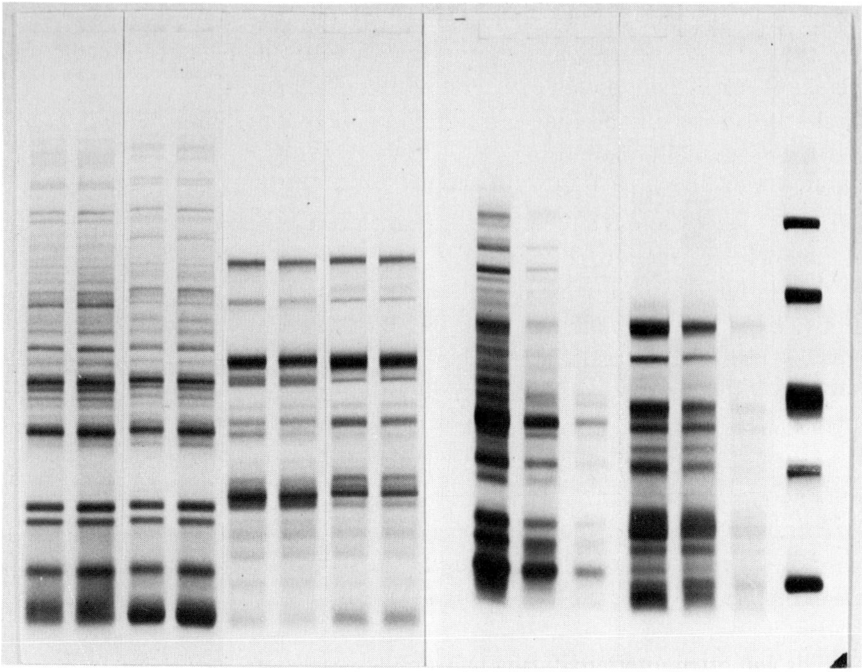

Figure 1. Left: *Influence of the reducing agent* ME (slot 1, 2, 5, 6) and dithiothreitol (3, 4, 7, 8) on maize proteins in the water (1–4) and buffer (5–8) extract. *No difference* is seen between fresh extracts (1, 3, 5, 7) and extracts *after storage* without SDS at −20° for 5 weeks (2, 4, 6, 8). Right: *Influence of particle size* on protein yield and the individual band intensity of maize proteins in PAGE with SDS-Tris-borate buffer pH 7.1 in water (slot 9–11) and buffer (12–14) extracts. Particles below 45 μm slot 9 and 12; from 45 to 90 μm slot 10 and 13; from 90 to 180 μm slot 11 and 14; MW-markers (slot 15) as in 'Comments on the Figures'.

before milling. Mills of the Hammer, vibro, mixer, planetary and centrifuge type, or mortar and pestle, are used and the resulting flour, meal, bran or hulls are extracted. Some seeds, e.g. green coffee beans, are hard to grind. They should be worked up in a centrifuge mill which – by comparison – proved most suited with respect to fineness of the meal and minimum alteration of proteins. Although the seeds are crushed within a second, the catapulting of the particles by the fast rotating blades through the sieve generates enough heat to change some of the components. For maize and coffee beans protomers with higher MW are somewhat enhanced and [personal communication by H.G. Meier, Braunschweig] phenolic substances are altered if the mill is not cooled. When 1 g maize (or 5 kernels) was milled with 10 g of dry ice in a precooled centrifuge mill (type ZM 1, Retsch, D-5657 Haan) using the 250μm sieve, the yield of meal was 70%, and more than 50% of the meal particles were below 25μm. With 5 g maize the yield was 90%. The TECATOR CYCLOTEC mill yielded coarser material.

Whereas in practice the objective of the milling is to separate endosperm from bran and germ subsequentially, the aim in research is sometimes to separate the different tissues as far as possible before milling and then to disintegrate the material completely to a size suitable for even extraction. It has been shown [18] that the bran did not contain additional proteins.

Sieving or air classification is essential for any analysis unless total protein is determined, since there is a difference in protein content and protein composition for any particle fraction (Figure 1). The defatted material should be taken, otherwise proteins do not separate well in later steps. Particle size distribution is a sensitive measure of the milling process [62]. The powdered endosperm is enriched in the fines [118a, b] in usual mills. The fraction below 90 μm (170 mesh) from a centrifuge mill is used by us for protein work since it was found to reflect a good average and allows for a fast and relatively complete extraction. The particle fractionation is done with the dry powder (if particle size is above 25 μm) or with a suspension of the powder in dry acetone (dried over potassium carbonate) if size distribution of very fine material has to be determined. Sieves can be tightly connected by ribbons from fluoroelastomers to give a pile for screening. Stainless steel equipment is preferred to sieves made from bronze to avoid contamination with copper.

Lipids will often interfere during hydrophilic extraction procedures and reduce stability during storage. They should therefore be removed or diminished by extraction with different solvents. The effect on wheat flour has been reported [32]. Short chain hydrocarbons or butanol are quite popular. Hydrocarbons will remove most of the lipids and no (or little) other components, including salts. A mixture of chloroform/benzene was proposed [124]. Cold acetone is used less frequently; however, in the last ten years we have found it to be the most recommendable lipid extractant for protein work. The loss of proteins is negligible. Acetone (dried over potassium carbonate) removes salts and residual water to a considerable extent form the seed powder, which makes the consecutive 'Osborne-extraction' more reliable. Acetone is conveniently evaporated in vacuo. On rare occasions, especially for beans, acetone had to be complemented by ethylacetate to remove a high content of wax before the subsequent protein extraction. The extraction is done in an ultrasonic bath to save time and to intensify the defatting.

PREPARATION OF EXTRACTS

We shall discuss a limited number of extraction methods with respect to frequency of use and general applicability. Osborne's scheme is still the basis for characterizing seed proteins even when there are no alcohol-soluble proteins present. The history of extractions is reviewed [88a]. Osborne's scheme includes consecutive extractions by water, buffer, alcohol, and solvents with extreme pH. For cereals it is synonymous with extraction of albumins – globulins – prolamins – glutenins. (Common names are used for globulins, e.g. legumin, and for prolamins, e.g. gliadin, hordein, zein.) The alcohols used today are ethanol [30], chloroethanol [7, 71] or propanol (the -1 of -2-isomer) [69, 135] with or without reducing agents. Only the last step of the Osborne fractionation has been replaced by using neutral solutions of detergents or urea. Surfactants such as alkyl-benzene sulfonates [53] or SDS in combination with propan-2-ol and reductants [135, 84] for extraction of cereals, legumes and leaves are gaining popularity. The presence of glycerol [65], sucrose [64] or formamide/dimethylformamide [98, 30] in the last step could increase the solubility and stability of the proteins. Occasionally inhibitors of serine proteases and organic mercury compounds are added [44]. Solutions with different pH-values, salts, and urea concentrations are applied. For extractions with respect to a special crop see [117, 173] for maize and sorghum, [83, 85, 175] for maize, [14, 18, 31, 52, 68, 71, 131, 140a] for wheat or rye, [46, 86, 101, 132, 140] for barley, [55] for millet and [37] for Italian millet, [73, 115, 116] for rice, [9, 25] for French beans, [10, 159] for broad beans, [13] for lentils, [39] for peas. Our standardized extraction procedure for cereals, legumes and other seeds is summarized later. See also Figure 1, Figure 2, Figure 4, Figure 6. A different way to sequential extraction is the direct treatment of the meal with one extractant or ions of heavy metals to dissolve a group of proteins with one special feature [18, 106]. With 2M urea [89] and with chloroform/methanol mainly gliadins are dissolved. A recent review covers well the glutenin extraction [15]. Tetramethylurea does extract mainly one part of the zein fraction [158]. Instead of stepwise extraction one may dissolve as much protein as possible, e.g. by SDS, and re-fractionate the extract [59]. However, we found it extremely difficult to remove the SDS in spite of many hints in the literature. We were successful only in one case when a mixture of urea/tetramethylurea/ampholytes helped to free the proteins from SDS in an electric field [43] provided the SDS-concentration was diminished first by precipitation as K-salt [178]. An alternative to SDS is the use of urea ampholytes [77], and possibly ME, giving good results for seed [28].

Some of the working-up conditions have attracted more attention. After water and buffer extraction Landry and Moureaux [84, 85b] used ME for reducing conditions and the anionic detergent SDS, well known for detaching and solubilizing viral [153] and cereal [135] proteins. An effective agent for breaking hydrophobic bonds is Na-dodecanoate- and -hexadecanoate [52, 68, 78] which extracts glutenin from flour without prior reduction, and the combination of urea and tetramethylurea [40]. For legumes the method of Scholz *et al.* [134] is frequently used.

It is self-evident that no extraction procedure will yield a single type of proteins. It is most important to describe and follow any extraction procedure in every detail starting with milling conditions/size distribution and ending with the adequate kind of storage to get reproducible results. No matter how the extract was obtained it is more interesting to analyze the mixture by electrophoresis and to compare the distribution of the individual proteins among the various fractions.

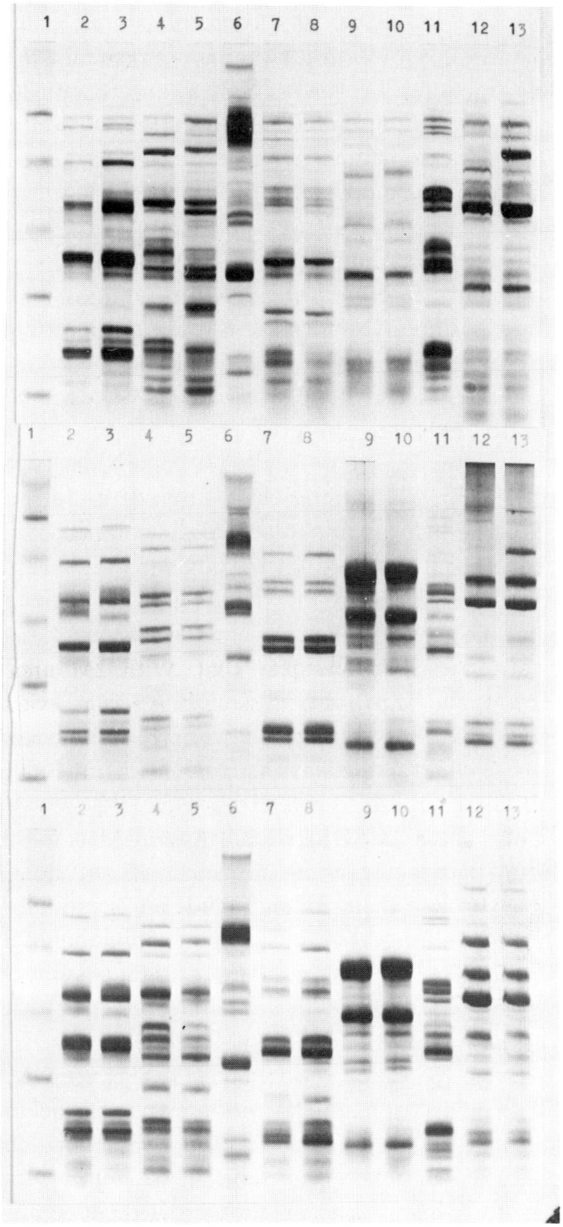

Figure 2. *Performance of a continuous buffer for* separation of seed proteins from *very different legumes*. Those buffers are *more reliable* and the electrophoresis *less tedious* compared to a discontinuous buffer system. PAGE is done in 6% PAA and Tris-borate buffer pH 7.1 with 0.1% SDS. Shown are the proteins extracted by water (top), by buffer after 3 extractions by water (middle) and directly by 4% SDS and 2% ME in water (below). MW-markers (1) as in 'Comments on the Figures'.

Vicia faba: cv Diana (2), cv Giza 1 (3); Pisum sat.: cv Kleine Rheinländerin (4), cv Wunder v. Kelvendon (5); Phaseolus vulg.: cv Dopp. Holl. Prinzeß (6); Cicer arietinum: cv Giza 2 (7), cv Family 2 (8); Lupinus albus (9); Lupinus termis (10); Lens esculenta: cv Giza 9 (11); Trigonella foenum graecum: cv Giza 2 (12), cv Giza 30 (13).

Keeping this in mind and being fully aware of the merits of some published procedures we tried to apply uniform conditions to many different seeds in order to obtain comparable fractions. The more common extractants have been checked with respect to ease in handling, economy and good results after electrophoretic separation [69]. The extraction time at 0° could be shortened from 1 h to 15 min by using an ultrasonic bath and small particles ($<$ 90 μm) which represent the whole kernel. The following sequence proved to be generally applicable: water – Tris/borate buffer pH 8.9 according to [93, 160] – propan-2-ol/water(buffer) – SDS- or urea-containing buffer. Two parallel extractions under reducing conditions in presence and absence of urea give information about buried and unburied disulfide groups. For seeds of wheat [70], rice [116], maize [151a, 163], grass [149], phaseolus [61] and faba beans [159], peas [61], lentils [61], fenugreek [61], jojoba [136], coffee [8] or tubers e.g. potato [160] and oca/sorrel [162] the proposed sequence worked perfectly (Figures 2, 6).

Many proteins have intra- and inter-chain disulfide bridges. These cross-links lower the solubility and increase apparent molecular weights. Therefore it is advisable to reduce the disulfide bonds for maximum yield during extractions. For protomer formation in the presence of SDS the disulfide bonds *must* be reduced. The difference in yield between the reduced and the unreduced extract points to the amount of 'open' disulfides. The addition of urea brings about the further reduction of 'buried' disulfides. The Tris/borate buffer pH 8.9 with ME and SDS is very efficient in extraction [69, 70] also with propanol for the prolamins. It was found for barley that this borate buffer should have a pH not below 8.6 for complete action of ME on disulfide bridges [46]. ME is commonly used or the more expensive dithiothreitol, which does not lead to an equilibrium. 3-Mercaptopropan-1, 2-diol, frequently applied in microbial research, is a substitute for the smelly ME [100].

Undesired autoxidation of phenolic compounds can be prevented by the addition of reducing agents like sulfite, but may also lead to reduction of disulfides [167]. If low concentrations of sulfite are applied, cystines in proteins are not attacked. In the absence of urea, 99% of the disulfides in gliadins did not react with sulfite; however, in the presence of 2%, 4% and 8% urea, 60, 90 and 100% resp. of the disulfides were reduced [164]. Urea concentration should be below 5M urea to avoid swelling of starch. To inhibit formation of pholyphenols our routine is to add 100 mg Na_2SO_3 and 75 mg $Na_2S_2O_5$ to 100 ml cold extract [160] for material with a tendency to become colored, and less or none if applicable. The SH-groups could be stabilized by alkylation with 4-vinylpyridine to yield S-pyridylethylated proteins, with acrylonitrile, N-ethylmale-imide or iodoacetamide as usual in protein chemistry [167].

The extracts should be divided into small volumes (about 1 ml) and kept at −20°C or lower temperature. Material not defatted is less stable. Sucrose (10–20%) could enhance stability. In this way we obtained the same electrophoretic pattern from fresh and two-year-old samples. Extracts lyophilized show more diffuse bands after electrophoresis compared to samples stored frozen, provided they are not thawed too often. A droplet of an undiluted ampholytes pH 7 to 10 as additive is recommended when a precipitate is difficult to dissolve. In general any lyophilization should be avoided since it leads to – sometimes irreversible – oxidation and crosslinks; only for samples to be treated with SDS it can be tolerated.

TOTAL PROTEIN

The determination of 'total protein' is still a difficult task. The usual Kjeldahl-procedure is applied despite questionable calculation factors due to the different amino acid composition in a mixture of unknown proteins. The method of Lowry *et al.* [95] is recommended for samples without any phenols. Most of the seed extracts contain such interfering components and hence this method is of limited value. Dye-binding procedures are of some practical value. The method of Bradford [23] gained reputation in recent years since it is fast and gives comparable results with different proteins from plants [144] and animals. We applied the method successfully to different seed proteins; however, we were not able to get a shift of the Coomassie Blue G color in trying to determine gelatin [43], and the determination of prolamins may also be hampered. Dye-binding with Acid Green 25 was used to track heat denaturation of wheat proteins [66]. For more or less colorless protein solutions it seems that the method of choice is a biuret procedure which can tolerate SDS and ME. All proteins including the heat-denatured and membrane-bound ones will dissolve if the samples are heated with SDS under reducing conditions. The biuret reagent is not too dependent on the kind of amino acids involved and has given highly reproducible figures for the protein content based on estimates after gel electrophoresis. The only disadvantage is its low sensitivity; less than 20 μg protein per 100 μl cannot be detected [4]. Proteins can be determined by extraction with SDS in the presence of a very large excess of starch [99], even after boiling and when swollen. A simple and fast estimation is done by spotting a protein solution onto a chromatography paper, precipitation and staining with 0.25% Coomassie blue in 20% aqueous sulfosalicylic acid for 2 min, destaining three times for 2 min in water/methanol/acetic acid (14 : 6 : 1) and comparison with a reference [147]. The adsorbed dye can be extracted, e.g. [51], for photometric determination.

PROTEIN SEPARATION

Introduction

As mentioned before extracts from plant tissue are rich in phenols and carbohydrates which interfere with the protein isolation. Especially glycoproteins interact with polysaccharides, either in a random fashion or – as with lectins –

specifically [27]. Adsorption to a solid surface is often involved which is proportional to the surface area to which the protein is exposed. Ubiquitous phenols are easily autoxidized to polyphenols which tan the proteins and render them insoluble, particularly in alkaline environment. Therefore time-consuming protein separations involving exposure to air and large surfaces tend to give unsatisfactory results. Speed, inert carriers and protective or competing additives during the separation partly overcome these problems. Electrophoretic methods using polyacrylamide gels result in negligible adsorption and high speed. To minimize the interactions between macromolecules, compounds with a very low MW competing for binding sites can be added. For this purpose compounds should be chosen which can later be removed without problem. Urea, tetramethylurea, dimethylformamide, formamide, glycerol and sugars, particular sucrose, are the most common substances added. In case of formamide, a possible contamination with formaldehyde is rarely considered which may lead to conversion of amino groups to uncharged N-hydroxymethyl groups. Formamide distillation in vacuo is required. For analytical separations in an electric field the addition of tensides like polyhydroxyethylsorbitan esters (e.g. TWEEN), nonylphenolethers (NP 40), alkylbetaines (EMPIGEN) etc. is common practice. This will often give more distinct bands in electrophoresis but it is well known that the native protein structure is not necessarily restored after removing additives, especially those with hydrophobic or longer side chains and when disulfide bonds had been reduced. A protein solution can be concentrated by dry polyethyleneglycol (e.g. Carbowax) surrounding the dialysis tubing. However, enough material penetrates the membrane and interferes in SDS-PAGE [157]. Concentration in tested collodion bags is recommended [54].

To suppress the formation of polyphenols, reducing agents such as ME, monothioglycerol, thioglycolate and various sulfites are also added during the electrophoretic run. Since they may reduce disulfide bridges it is a matter of concentration, pH and the particular reductant to suppress this reaction when native proteins are to be characterized. Thioglycolate, e.g. as Tris-salt is recommended of permanent protection of SH-groups, if put into the anodic reservoir. See also the chapter 'Preparation of Extracts' and Figure 5. For amide determination it is used in a similar way [146].

Electrokinetic Systems

Possibilities and limitations of electrophoretic methods in the chemistry of plants [58a, b, 82] and cereals [5] is reviewed and a summary for SDS-PAGE given [153]. Starch gel, the first gel electrophoretic carrier introduced by O. Smithies is still used (sometimes for historical reasons?) as well as cellulose acetate, but crosslinked polyacrylamide is replacing them due to its versatility and economy. Starch gels are easily sliced into layers; for PAA slicing is possible with a special tool (Labor-Müller, D-3540 Hann.Münden). PAGE should be performed in slabs rather than in tubes for better comparison of samples and also for having the chance to use two-dimensional separations. Continuous systems (one buffer, one gel system) are easier to handle and more reliable than discontinuous systems since, in continuous systems [153], one protomer cannot pick up different loads of SDS (Figure 2). If small sample volumes and a suitable buffer are used, PAGE with a continuous system will give excellent separations. PoroPAGE will tolerate larger volumes of the sample. We have gained distinctive patterns of all kinds of seeds and tubers in the last decade. This applies also to thin-layer gels [56].

Among many buffers tested, the continuous Tris/borate buffer pH 8.9, introduced by us for tuber proteins [93, 160] and the borate buffers for SDS-PAGE at pH 8.9 [80], at pH 7.9 [160] and at pH 7.1 [163], proved to be most suitable for a wide range of different plant extracts (Figure 2, Figure 4). The buffer pH 8.9 for SDS-PAGE seems now to be favored by many groups [29, 71, 79, 117, 140]. However, a high concentration of fructose – not glucose – in the sample may interfere [90]. An lactate buffer pH 3.6 with 6 M urea is convenient and efficient [88 b]. The lactate buffer pH 3.1 is frequently used in PAGE as Al-salt or, more recently, as Na-salt [6, 35]. We had problems with the solubility of some seed proteins. A glycine-acetate buffer of the same pH is applied for wheat [100] and 5 M urea in 35% acetic acid is used to separate barley proteins [60, 128]. We lowered the content of acetic acid to 18% in order to avoid attack on Plexiglas.

Before discussing individual techniques we will evaluate the pro and con of some electrokinetic separations, since they are occasionally not considered. The sharpest separation is achieved by PAGIF and PoroPAGE. However, the enthusiasm about sharp bands in PAGIF, especially in thin layers, should not lead to the conclusion that this pattern reflects as many characteristics of a protein mixture as does PAGE. In PAGIF, only a quarter of all possible changes of amino acids in a protein [104, 155] and no change of shape or MW can be seen. In PoroPAGE the MW of the native molecule governs the separation, but not as precisely as it is true for the charge in PAGIF pattern. In both the PAGIF and the PoroPAGE the proteins will theoretically come to an end point in contrast to PAGE with its dynamic separation. In PAGE any difference in overall charge and in conformation, including intramolecular interaction, change the rate of electrophoretic migration. Furthermore, the sorting by shape and net charge of the proteins can and must be varied at will by changing the porosity of the gel or/and the pH of the medium, respectively. Since in evolution the mutation of amino acids on the surface of a protein is more likely than inside the molecule, the chance to see a different pattern in PAGE is at least four times as high compared to PAGIF and far higher than in PoroPAGE (Figure 3). These obvious thoughts were substantiated by many findings and a recent paper gives a sound experimental proof [126]. Electrophoresis in starch or agarose is more or less a charge-dependent separation. PAGE in a SDS-containing medium has the advantage that all the proteins are solubilized, highly negatively charged and random coiled. The denatured protein (the protomer) is separated by size. In spite of

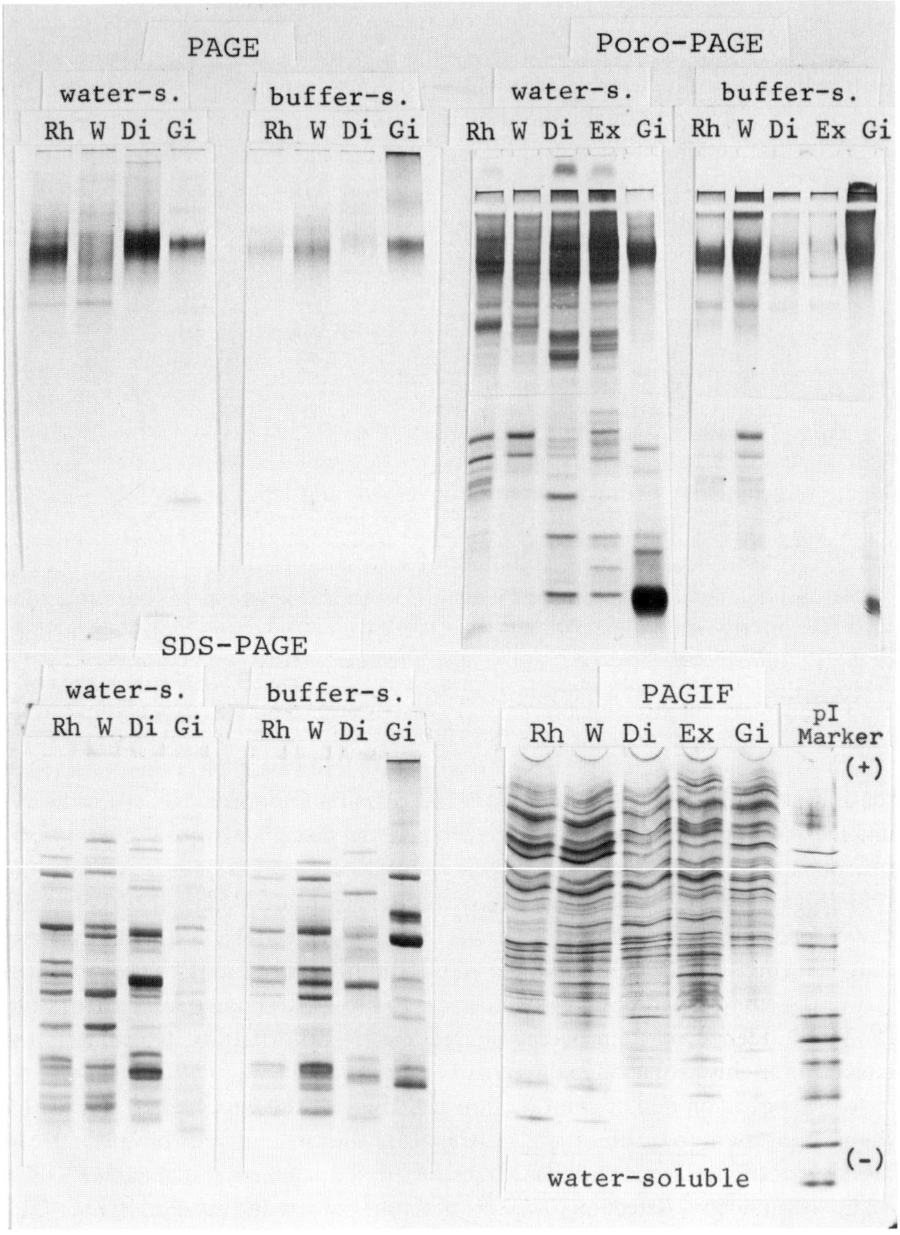

Figure 3. Comparison of 4 methods and their suitability *for distinguishing cultivars*, shown for seed proteins of leguminosae, extracted with water or buffer pH 8.9. *Standard PAGE* in Tris-borate pH 8.9 (or in other buffers): Diffuse bands. *PoroPAGE* in Tris-borate pH 8.9 and 6 M urea: Characteristic patterns in the lower MW-range for water extracts. Upper part underexposed. *SDS-PAGE* in Tris-borate pH 7.1.: Sharp bands. Characteristic patterns when water extract is taken. *PAGIF* (0.1 mm PAA on a carrier, Servalyt pH 4–9): Sharp bands, when water extracts were freed from low molecular weight compounds. Patterns not as characteristic as in Poro- or SDS-PAGE, but more typical compared with PAGIF-patterns in urea (not shown). Pisum sativum: cv Kleine Rheinländerin (Rh), cv Wunder von Kelvendon (W). Vicia faba: cv Diana (Di), cv Express (Ex). Trigonella foenum graecum: cv Giza 2 (Gi).

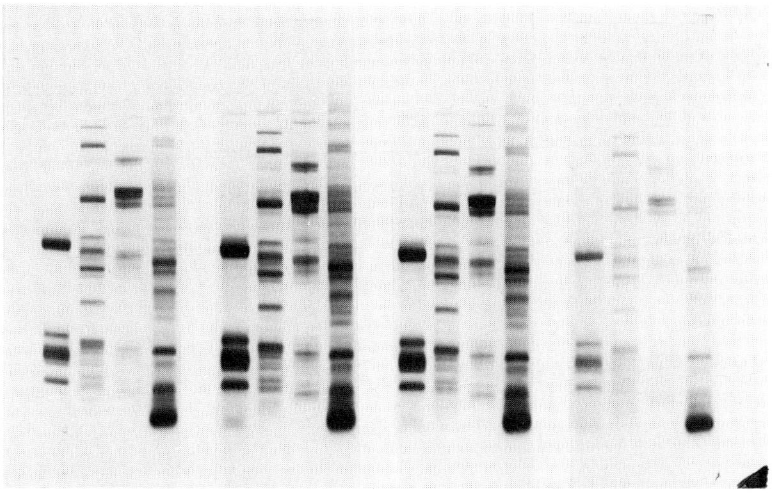

Figure 4. Influence of a second protein *staining* on enhancement and color differentiation in SDS-PAGE valid for all buffers. Shown is the separation in SDS-Tris-borate buffer pH 7.1. Extracts from potato tubers (slot 1, 5, 9, 13) seeds of peas (2, 6, 10, 14) broad beans (3, 7, 11, 15) and maize (4, 8, 12, 16) were separated in the same slab and first stained with Coomassie blue R (Supranolcyanin 6 B Bayer) in water-methanol-acetic acid containing 3% trichloroacetic acid. After destaining, sections were either not re-stained (1–4), stained with Coomassie blue R (5–8) or G (9–12) or Supranolechtrot BB Bayer (13–16). With Supranolechtrot praeprolamins (or prolamins, not shown) are preferentially stained and appear as characteristic bluish-red bands, barely reproducible in the photograph, among light-red bands.

the diminished individual characteristics, the SDS-PAGE gives quite representative patterns of extracts from seed proteins. That is possibly due to the many glycoproteins involved. The carbohydrate moiety can be characteristic in itself and since this part of the molecule does not bind SDS in the same fashion as the protein part it leads to a characteristic apparent MW in SDS-PAGE. One must conclude that only combined techniques will resolve the many components in an extract. – It is no problem to separate small molecules down to oligopeptides when using 15 to 20% PAA. Identifying them in the gel is much more laborious, but there seems to be a solution to this problem [169].

A few words on staining, evaluation and documentation. Supranolcyanin 6B Bayer (equivalent to Coomassieblue, ICI) is the favorite dye, a solution of 0.025% in water/methanol/acetic acid (80:20:7) containing 3% trichloroacetic acid resulted in the most sensitive detection (0.1 μg/band) after destaining with the same solvents (67:29:4). Re-staining is useful, particularly in SDS-PAGE. A double staining, first with Supranolechtrot BB (Bayer, Color Index Acid Red 154), followed by the above mentioned stain, enhances and stabilizes the intensity of bands, e.g. of smaller proteins with acidic isoelectric points [151 b]. Re-staining in reversed order and using only a tenth of Supranol-echtrot BB results in color differentiation of proteins [158] (Figure 4). A double staining with Coomassieblue G and R was tested [65] as well as with the usual protein dyes [174]. Electrophoretic destaining is not recommended based on our experience with AA-techniques since 1960. Even with a more careful destaining some faint bands may be lost during the washing. For some bands their gradual fading is reduced when 6% instead of 3% TCA is used for fixation. The silver stain has some merits for gels below 1 mm thickness. It seems to be 10 to 20 fold more sensitive than our Supranolcyanin technique. It is delicate and time-consuming to handle. Most methods have not yet matured. The best procedure up to now in our hands was that of Morrissey [108].

Comparison of electrophoretic patterns is best done by running an internal standard of a well-known cultivar in the same gel and by comparison of bands with respect to intensity and distance from the starting point. Densitometry of closely neighboring bands in one-dimensional separations is of little value since we have not yet found an equipment which does not confuse the picture. Scanning can be of interest when highly reproducible gels are evaluated for diagnostic work and not for individual proteins. In addition, two-dimensional patterns show how many proteins can be found behind one band. Far better is the quantitative analysis after mapping [19, 81] and a computerized version thereof [2, 76, 91]. These and some other groups pioneer in 2-D-separations, but the cost for evaluation is a serious drawback. Keeping the dried original gels or the photograph is preferred at the moment.

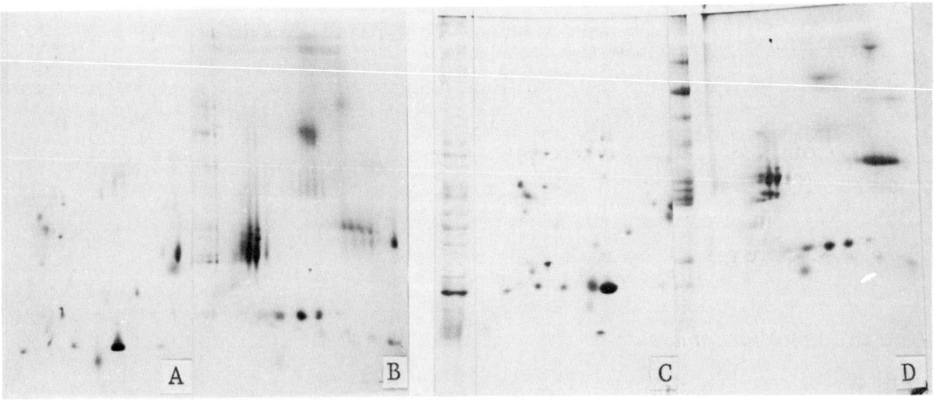

Figure 5. Influence of reducing environment on identical samples: maize extracted with water (A, C:) and buffer (B, D). Without (A, B) and with (C, D) a 'curtain' of thioglycolate from the cathodic reservoir, which realizes the continuous reduction. Samples were first focused in 2% Servalyt pH 3–7 and 6 M urea; for the samples in C and D the cathodic barrier solution contained 1.2 mg thioglycolate per tube of 5 mm diameter during PAGIF. All samples were incubated after PAGIF at pH 8.9 with 0.1% SDS in 8 M urea for 15 min at room temperature. Second dimension in 10–28% PoroPAGE and SDS-Tris-borate buffer pH 8.9. For PoroPAGE in C and D 0.5% thioglycolate was added to the cathodic reservoir.

Separation by Charge

Reviews are given [129a, b]. Focusing should be the first step in examining a mixture in order to gain some idea about the distribution of isoelectric points (IP). This is best done in PAA, possibly also in presence of urea, bearing in mind that urea will change the IP defined for aqueous solutions. All of the commercial ampholytes work well and give the same IP on a different slope (pH vs. migration distance). For cereal proteins we prefer Servalyts® (Serva, D-6900 Heidelberg), since they destain faster and technical grades are available which are often sufficient for the separation and are more economical. PAGIF is done in tubes, particularly with raw extracts and higher loads, or in flat gels when the salt concentration is low and exact comparison is needed [121] (Figure 3). A high sucrose content (25%) can improve the separation. Focusing can be done under well defined reducing conditions by adding 1.2 mg thioglycolic acid or 1 mg dithionite, $Na_2S_2O_4$, per tube (5 mm ∅) to 500 ml of the cathodic barrier solution (Figure 5). We found that this causes less loss of some proteins compared to transfering SH- to HSO_3-groups or shielding them by protective groups. Preparative focusing is performed in ampholytes by using granulated gel beds of Sephadex or Biogel [157]. In beds 15 × 13 × 1 cm the load can reach 2 g proteins. Separation in a sucrose gradient with additives to keep focused proteins in solution was not as successful [98]. For references and experimental details see [157]. Focusing without a carrier in a coil [98] and chromatofocusing with the help of an ion-exchanger [142] should be considered.

Separation by Size

PoroPAGE has still not received the reputation it deserves for seed proteins. Since the native molecules are separated, enzyme detection is possible. Reducing conditions can be applied as with PAGIF or PAGE. More often used is PoroPAGE in SDS-containing buffers, where any native protein structure is lost and only the number of protomers found or synthesized in the respective tissue is disclosed. PoroPAGE in urea is often a better alternative. Examples for PAGIF and PoroPAGE are given (Figures 3, 7, 8). For experimental hints see [157].

Separation of Protomers

The use of hydro-lipo-philic compounds with anionic, cationic or amphoteric groups, and especially of SDS, is common in protein analysis. The possibility of dissolving any protein except elastin under reducing conditions and of the MW-estimation of protomers has been shown [41, 42, 153]. The basic requirement for the reproducible performance in MW-estimation of protomers is the complete conversion of the protein structure to the so-called subunits or protomers or – misleading – polypeptides. In this process they are brought to a random coil structure and are

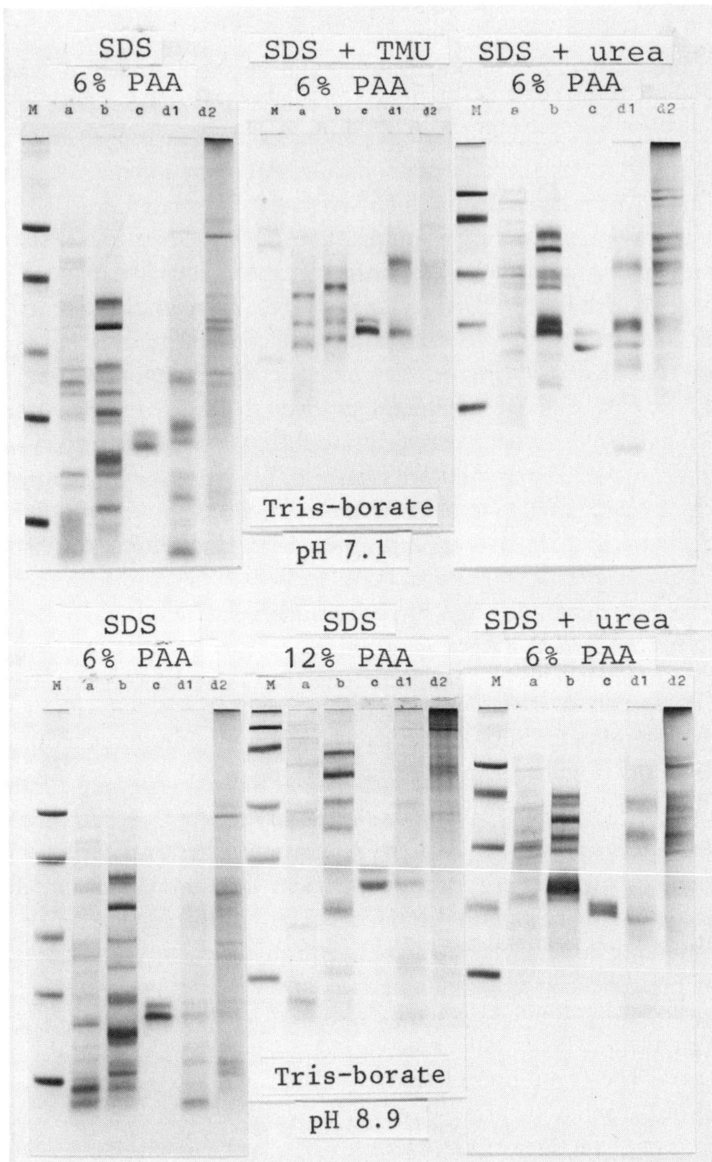

Figure 6. The *irregular dependence* of individual *migration* rates upon electrophoretic conditions. SDS(0.1%)-PAGE of maize proteins, converted to protomers. In each gel the first slot shows the MW-marker protomers. Water soluble proteins after consecutive, exhaustive extraction (a), the corresponding residues further extracted with Tris-borate buffer (b) at pH 8.9, followed by isopropanol (c) and the final residue extracted twice with SDS-ME-Servalyt (d1, d2). – MW-markers: as in 'Comments on the Figures'. TMU = Tetramethyl urea.

loaded with the respective compound. For the most popular sodium (or lithium) salt of the dodecanol-monoester of sulfuric acid, short name SDS, a recent review is given [153]. Since different proteins uncoil under different conditions, either boiling or ultrasonication at room temperature is done [61] in the presence of SDS and for a few minutes (e.g. 3 and 9 min, resp.), with or without reductants. Also SDS-incubation in 6 to 10 M urea at room temperature can be applied. When proteins still in a gel matrix have to be converted to the protomers, the gel matrix will slow down the process, particularly in gels above 6% PAA. Prolonged incubation in SDS-containing solution is therefore required when proteins focused in cylinders are prepared for SDS-PAGE in the second dimension. However, there is the risk that during the incubation with SDS partially denatured proteins within the gel are digested by still native proteinases focused in the neighborhood. This has rarely been recognized and may – according to our experience with SDS-mapping since 1971 – explain erratic results reported in the literature.

In SDS-PAGE one can observe different apparent MW for a given protomer depending on the buffer [151b, 153], regardless whether complexing (phosphate/borate) or not-complexing (chloride) anions or chelating agents (e.g. EDTA) are used (Figure 6). This was emphasized again by [87] and also shown for marker proteins by [28]. In any case SDS-PAGE has to be performed in at least three different PAA-concentrations to give some safety for MW-determination [153]. SDS-PAGE is easily overloaded; a minimum of 0.1 µg per band can be detected. Reduction of samples for SDS-PAGE either by ME or dithiothreitol does not give identical patterns (Figure 1) which possibly is due to different redoxpotentials or accessibilities of S-S-bridges within the protein.

Separation by PAGE

As mentioned in the general outline, PAGE is the most sensitive method to detect slight differences among proteins. It is hampered by the fact that the physical parameters involved cannot be defined, barely in continuous gels and buffer systems [127] and even less in the popular discontinuous ones of Ornstein and Davis [113] and followers. The separation of seed proteins can often be improved by polymerization of the gel in the presence of urea (5–10%) not concentrated enough to affect quaternary structures of 25% sucrose and of 10% propan-2-ol. [156]. During the run a continuous flow of such sulfhydryl compounds as 0.5% thioglycolate or 0.4% Na-dithionite from the cathodic reservoir may be as beneficial as in PAGIF. A preparative version is described for separation of the proteins in a gel block 16 mm thick followed by cutting out the bands and isolating them from the gel by mashing and extraction [148]. Purer proteins not contaminated with oligomers of AA are obtained by discontinuous elution of the fractions separated in a rectangular cuvette. The protein concentration in the eluate is high enough for re-electrophoresis without concentration [150, 156]. This set-up is still used in this fashion today.

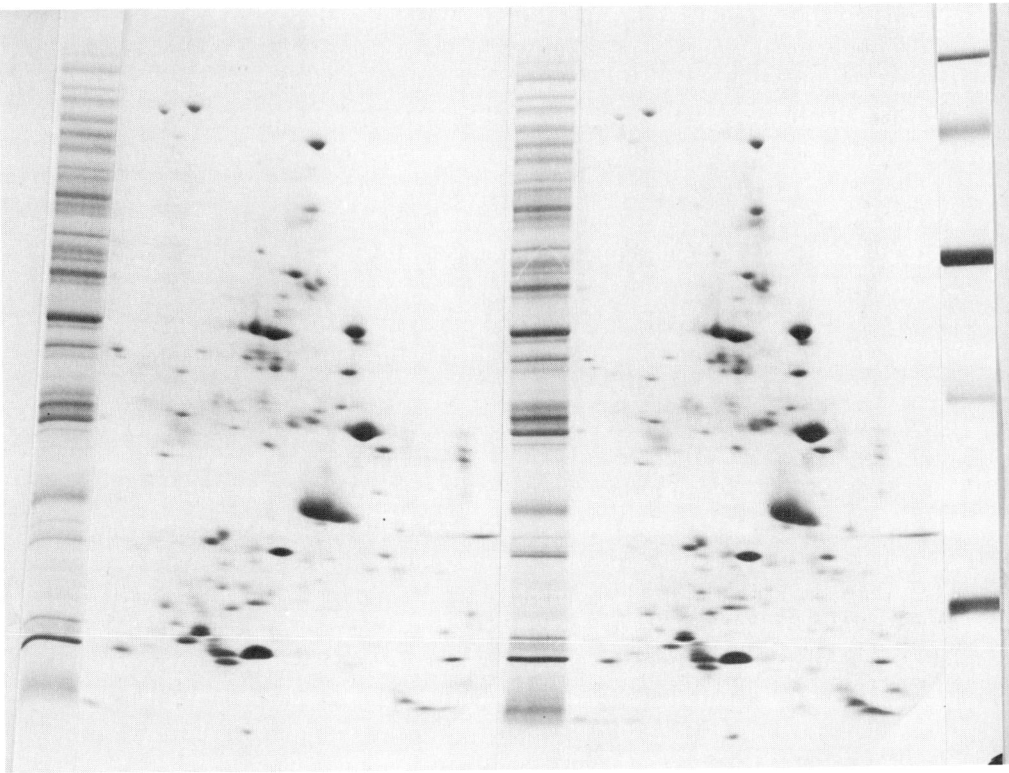

Figure 7. One-dimensional separation of protomers (water-soluble, SDS-loaded maize proteins) in the vertical lanes *compared with mapping* (first PAGIF in rods with 6 M urea and 4% Servalyt pH 5–9 placed horizontally for second dimension in PoroPAGE with 10 to 28% PAA and SDS-Tris-borate buffer pH 8.9). Even sharp bands in one-dimensional SDS-PoroPAGE (vertical) are further resolved by mapping. Moreover, nearly identical patterns are seen on the left (seeds harvested in 1978) and on the right (same cultivar, but harvested in 1977). MW-markers: far right. The 2 mappings were done in one gel.

Mapping

Two-dimensional electrophoresis of proteins, especially the combination of charge separation by PAGIF with PAGE, PoroPAGE or SDS-PAGE for which we introduced the term 'mapping' (Figure 7) [96a, b, 160] to distinguish from 'fingerprinting' for polypeptides, will become the most powerful tool in biochemical genetics. The separation benefits from at least two physical characteristics, and it seems that quantitative evaluation of mappings with more than 1000 spots is possible [2, 76]. Recent papers of mapping methods for proteins of wheat [65, 107], of peas [104] or French beans [25] or maize [175] have appeared. A short region on a regular PAGE pattern with apparently a few diffuse bands is frequently resolved [162] by mapping into more than 100 discrete spots when PAGIF was done within

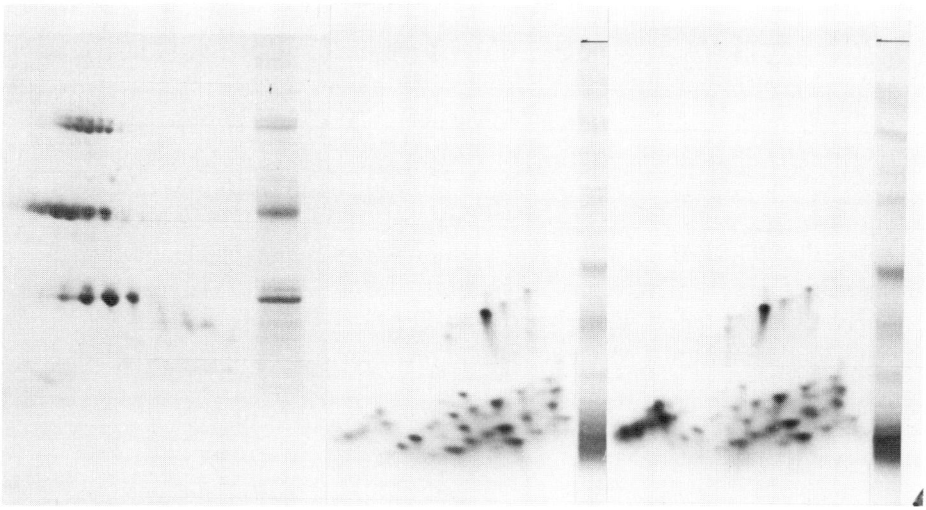

Figure 8. Further resolution of bands or regions, which are diffuse in one-dimensional PAGE, *by two-dimensional techniques*. Left: Seed proteins of coffee beans separated in rods by PAGIF (4.5% PAA, pH 4–8 in 8 M urea) followed by SDS-PoroPAGE (10–30%) in Tris-borate buffer pH 8.9; the one-dimensional SDS-PoroPAGE for comparison. Center and right: Potato proteins from slightly different genotypes separated in the first dimension by PAGIF (6% PAA) in 2% Servalyt pH 4–5 followed by vertical PAGE (7.5% PAA) in Tris-borate buffer pH 8.9; the one-dimensional PAGE is also shown for comparison.

one pH unit, e.g. between pH 4 and 5 (Figure 8). We think any nonequilibrium separation in a pH-gradient (NEPHGE) for the first dimension is of limited value since neither the charge of the protein is defined nor is this a highly reproducible method. Instead, for resolving a large number of proteins with similar IP, ampholytes of a narrower pH-range should be chosen. If not available, NEPHGE may replace it as shown for glutenin [65].

For easier comparison of spots in mapping, 3 to 4 cylinders containing PAGIF-separated proteins of different samples are placed simultaneously on the same slab gel for the second dimension. The cylinders are split longitudinally, one half is stained for reference, and the other used for the second dimension. Maximum precision for identical running conditions is also achieved if different samples are run in one gel during focusing and cut apart after focusing. Split-gel PAGIF in tubes will not work but slots somewhat apart in a rectangular gel bar will. After the PAGIF, the bar is sliced longitudinally, and the halves (2 × 2 × 100 mm) are placed parallel into long slots on top of a slab (6 × 130 × 180 mm) for the second dimension. After PAGE, SDS-PAGE or PoroPAGE the slab is split in two layers (3 × 130 × 180 mm) by a slicer provided with the PANTA-PHOR. In this way a perfect 'three-dimensional' comparison is possible [154]. This more elaborate procedure is not necessary when enough reference spots make the evaluation of two-dimensional gels unequivocal. The 'three-dimensional' comparison of known with unknown proteins, or just with a reference, is even better when the protein is first specifically split by an enzyme or a chemical (e.g. cyanogen bromide) to yield characteristic polypeptides. The splitting rate is followed by electrophoresis. More convenient, however, but less informative, is examining the products of protein splitting by one-dimensional techniques, e.g. SDS-PAGE [34a, 153, 177] or better by PAGE or PoroPAGE, which give additional information on charge properties since SDS is omitted [154]. To avoid hard-to-remove SDS, urea (10 M) can be used to uncoil most proteins for enzymatic attack. Enzymes have been selected, e.g. papain and thermolysin, which act even at this high urea concentration [154].

IMMUNO-CHEMICAL SEPARATION

Identification of proteins by immunochemical means will depend on the characteristics of the surface. Production of monospecific antiserum calls for the isolation of the protein used as the antigen. This is best achieved by preparative focusing or electrophoresis of the native extract [156]. Antibodies are raised in e.g. rabbits or goats. More recently, successful immunization is done by injecting hens, followed by the isolation of antibodies from the egg yolk [122], which is a productive and convenient source. Since all the techniques for performing immunochemical assays were recently summarized [33, 38, 114, 168] it remains merely to mention that the antigenicity of storage proteins in seeds is not very high. Injection of considerable quantities of the antigen and several booster injections are needed. Also in this field the analysis in two dimensions (e.g. PAGIF followed by immuno-PAGE or -agarose electrophoresis) is recommended, as described in [157].

SEQUENCE OF AMINO ACIDS

Sequence analysis is not covered here (see 'Introduction'). This approach is – together with X-ray analysis – the most rigorous one and the isolation of very pure proteins is necessary beforehand. For a general idea the papers of [16, 18, 138] for barley/wheat, of [17] for maize, [20] for legumes may be consulted. A new approach without isolation of single proteins is done by sequencing peptides of incomplete hydrolysates [12] to deduct characteristic properties from sequences in typical fractions separated by molecular sieving or ion exchange chromatography. The possibility of sequencing nucleic acids with simple electrophoretic methods will soon play a complementary role in spite of 'dormant' regions not coding for proteins and confusing the picture [145]. Phaseolin of French beans is one example [165].

ENZYMES

Enzymes in seeds are too numerous to deal with in this chapter. However, reference is given to a review [58a, b] and the important work of Daussant and his group [38], where one can also find the different techniques of immunochemistry applied to seed extracts for quantification of a special protein. The multiple forms of esterases, peroxidases, amylases, alcohol dehydrogenases, glutamic oxaloacetic transaminases, aminopeptidases, etc. from dry or germinated seeds are the usual enzymes investigated for taxonomic or genetic work. Out of many papers, references [4, 57, 63, 74] may serve as examples. Proteases, nucleases, polymerases, polynucleotide phosphorylases and phosphorylases can be detected in PAGE by the substrate or primer inclusion techniques [147], e.g. proteases in cereals [110] or their reaction can be followed [133]. A special interaction of glycogen phosphorylase with the primer in normal and opaque maize has been detected [151a, 163].

SOME APPLICATIONS

Electrokinetic and other methods applied to proteins of seeds have targets in the genetic and taxonomic field [72] as well as for food and processing qualities, and lately in plant pathology [141]. In taxonomy, electrophoretic patterns have added safety to morphological characterization and have often replaced the traditional evaluation due to their higher speed and independence of environmental factors. The first 'Index' of this kind, based on protein and esterase patterns, was published for all European cultivars of potatoes [161]. The first cereal index was released for wheat in France [3, 6]. Others followed in Germany [112, 121], Sweden [1a], Great Britain [47, 49], Italy [35], Austria [172], Australia [176b], New Zealand [34b], South Africa [67] and worldwide [119]; for barley in Germany [60], Canada [102], Australia [96], Great Britain [141] and even for malted barley [139]; and for oats [75]. It has been claimed that prolamin patterns are more reliable for variety identification as compared to albumins and globulins [22, 48]. We believe that this is not necessarily so, at least not for maize. One may also consider that most of the separations of wheat proteins were done by SGE. Since it is known that SGE is predominantly a separation by charge and not overly influenced by the size of the proteins as in PAGE, it is predictable that SGE does not discriminate as well as PAGE. Therefore the mobility in SGE must be a direct function of the content of the ionized amino acids. For identification of protein bands it seems convenient and more adaptable to improved methods when bands of new varieties are compared with bands of one reference variety which is readily available and has been known for a long time.

During the preparation of this chapter, we were faced with a tremendous number of publications. Less than a quarter of worthwhile papers could be cited, many of which we experimentally endorsed. We regret those papers we overlooked and hope they will be sent to us for revision of this chapter at a later date. We are indebted to the Deutsche Forschungsgemeinschaft (D-5300 Bad Godesberg) for funding a program on biochemical genetics for which this article was prepared as a start. We sincerely thank our colleagues for giving us hints and winks.

COMMENT ON THE FIGURES

All separations in slabs were done in the PANTA-PHOR (gel size 13 × 18 cm), MONO-PHOR (22 × 18 cm) or the POOMA-PHOR (two gels 15 × 20 cm). Gel thickness was 2 or 3 mm for PAGE, PoroPAGE, Mappings. Cathode is always on the top. For PAGIF in the PANTA- or MONO-PHOR the gels were 0.1 mm thick on plastic support. MW-markers for SDS-PAGE are Phosphorylase b, 97.4 kD; Serum albumin, 67 kD; Alcohol dehydrogenase, 37 kD; Chymotrypsin A, 25.7 kD; Lysozyme, 14.3 kD. Apparatus are distributed by Labor-Mueller, D-3510 Hann. Münden, West-Germany. Detailed lab-instructions for performing the separations are available on request (Messeweg 11, D-3300 Braunschweig) in German, English of Spanish, respectively. Please indicate.

REFERENCES

1a. Almgard, G, and D Clapham (1977): Swedish wheat cultivars distinguished by content of gliadins and isozymes. Swedish J. Agric. Res. 7: 137–142.

1b. Anderson, NG, and NL Anderson (1979): Molecular anatomy. Behring Instit. Mitt. 63: 169–210.

2. Anderson, NL, J Taylor, AE Scandora, BP Coulter, and NG Anderson (1981): The TYCHO-system for computer analysis of 2-D-gel electrophoresis patterns. Clin. Chem. 27: 1807–1820.

3. Anonymous (1981): Clé de détermination des variétés de blé par électrophorèse des gliadines. Caractérisation des nouvelles variétés inscrites en 1981. Bureau Interprof. Etudes Analyt. (B.I.P.E.A. eds.) F-75040 Paris: 1–17.

4. Auriau, P, J-C Autran, L Charbonnier, G Doussinault, P Feillet, B Godon, P Grignac, P Joudrier, K Kobrehel, J Koller, M Rousset, and S Rivallant (1976): Variabilité génétique de la composition des gliadines, gluténines, β-amylases, α-esterases, peroxidases et phosphatases acides du blé (*T. aestivum*). Ann. Amélior. Plantes 26: 51–66.

5. Autran, J-C, R Berrier, M-F Jeanjean, P Joudrier, and K Kobrehel (1981): Emplois de l'électrophorèse dans la filière 'cereales' possibilités et limites actuelles. Industries Céréales 8: 3–19.

6. Autran, J-C, and A Bourdet (1975): L' identification des variétés de blés: établissement d' un tableau général de détermination fondé sur le diagramme électrophorétique des gliadines du grain. Ann. Amélior. Plantes 25: 277–301.

7. Autran, J-C, W Bushuk, CW Wrigley, and RR Zillman (1979): Wheat cultivar identification by gliadin electrophoregrams. IV. Comparison of International Methods. Cereal Foods World 24: 471–475.

8. Bade, H, and H Stegemann (1982): Characterization of proteins from green coffee beans by 2-dimensional electrophoresis. Z. Acker- Pflanzenbau (J. Agron. Crop Sci.) 151: 89–98.

9. Barker, RD, E Derbyshire, A Yarwood, and D Boulter (1976): Purification and characterization of the major storage proteins of *Phaseolus vulgaris* seeds and their intracellular and cotyledonary distribution. Phytochem. 15: 751–757.

10. Barratt, DH (1980): Cultivar identification of *Vicia faba* by SDS-PAGE of seed globulins. J. Sci. Food Agric. 31: 813–819.
11. Beachy, RN (1982): Molecular aspects of legume seed storage protein synthesis. CRC Crit. Rev. Food Sci. Nutr. 16: 187–198.
12. Belitz, H-D, H Wieser, and W Seilmeier (1981): Vergleichende Untersuchungen über Kleberproteinfraktionen verschiedener Getreidearten. Getreide, Mehl, Brot 35: 118–120.
13. Bhatty, RS, AE Slinkard, and FW Sosulski (1976): Chemical composition and protein characteristics of lentils. Can. J. Plant Sci. 56: 787–794.
14. Bietz, JA (1979): Recent advances in the isolation and characterization of cereal proteins. Cereal Foods World 24: 199–207.
15. Bietz, JA, and FR Huebner (1980): Structure of glutenin: Achievements at the northern Regional Research Center. Ann. Technol. Agric. 29: 249–277.
16. Bietz, JA, FR Huebner, JE Sanderson, and JS Wall (1977): Wheat gliadin homology revealed through N-terminal amino acid sequence analysis. Cereal Chem. 54: 1070–1083.
17. Bietz, JA, JW Paulis, and JS Wall (1979): Zein subunit homology revealed through amino-terminal sequence analysis. Cereal Chem. 56: 327–332.
18a. Bietz, JA, KW Shepherd, and JS Wall (1975): Single-kernel analysis of glutenin: Use in wheat genetics and breeding. Cereal Chem. 52: 513–532.
18b. Bietz, JA, and JS Wall (1975): The effect of various extractants on the subunit composition and associations of wheat glutenin. Cereal Chem. 52: 145–155.
18c. Bietz, JA, and JS Wall (1980): Identity of high molecular weight gliadin and ethanol-soluble glutenin subunits of wheat: Relation to gluten structure. Cereal Chem. 57: 415–421.
19. Bossinger, J, MJ Miller, Kiem-Phong Vo, EP Geiduschek, and Nguyen-Huu Xuong (1979): Quantitative analysis of two-dimensional electrophoretograms. J. Biol. Chem. 254: 7986–7998.
20. Boulter, D (1979): Structure and biosynthesis of legume storage proteins. Seed Protein Improvement in Cereals and Grain Legumes I: 125–136; IAEA Vienna.
21. Boulter, D (1981): Biochemistry of storage protein synthesis and deposition in the developing legume seed. Adv. Bot. Res. 9: 1–31.
22. Boyd, WJR, and JW Lee (1967): The control of wheat gluten synthesis at the genome and chromosome levels. Experientia 23: 332–333.
23. Bradford, MM (1976): A rapid and sensitive method for the quantitation of microgram quantities of protein utilizing the principle of protein-dye binding. Anal. Biochem. 72: 248–254.
24a. Brown, JWS, and RB Flavell (1981): Fractionation of wheat gliadin and glutenin subunits by two-dimensional electrophoresis and the role of group 6 and group 2 chromosomes in gliadin synthesis. Theor. Appl. Genet. 59: 349–359.
24b. Brown, JWS, CN Law, AJ Worland, and RB Flavell (1981): Genetic variation in wheat endosperm proteins: An analysis by two-dimensional electrophoresis using intervarietal chromosomal substitution lines. Theor. Appl. Genet. 59: 361–371.
25a. Brown, JWS, TC Osborn, FA Bliss, and TC Hall (1981): Genetic variation in the subunits of globulin-2 and albumin seed proteins of French bean. Theor. Appl. Genet. 60: 245–250. see also pp. 83–88.
25b. Brown, JWS, Y Ma, FA Bliss, and TC Hall (1981): Genetic variation in the subunits of globulin-1 storage protein of French bean. Theor. Appl. Genet. 59: 83–88.
26. Burgermeister, W, and AA Shah unpublished.
27. Burridge, K (1978): Direct identification of specific glycoproteins and antigens in sodium dodecyl sulfate gels. In: Ginsburg, V (ed.): Methods in Enzymology, 50: 54–64; Academic Press New York.
28. Bury, AF (1981): Analysis of protein and peptide mixtures. Evaluation of three SDS-PAGE buffer systems. J. Chromatography 213: 491–500.
29. Bushuk, W, K Khan, and G McMaster (1980): Functional glutenin: A complex of covalently and non-covalently linked components. Ann. Technol. Agric. 29: 279–294.

30. Caldwell, KA (1979): The fractionation and purification of gliadins by hydrophobic interaction chromatography. J. Sci. Food Agric. 30: 185–196.
31. Chen, CH, and W Bushuk (1970): Nature of proteins in *Triticale* and its parental species. I–III. Can. J. Plant Sci. 50: 9–30.
32. Chung, KH, and Y Pomeranz (1978): Acid-soluble proteins of wheat flours. I. Effect of delipidation on protein extraction. Cereal Chem. 55: 230–243.
33. Clark, MF (1981): Immunosorbent assays in plant pathology. Ann. Rev. Phytopathol. 19: 83–106.
34a. Cleveland, DW, SG Fischer, MW Kirschner, and UK Laemmli (1977): Peptide mapping by limited proteolysis in SDS and analysis by gel electrophoresis. J. Biol. Chem. 252: 1102–1106.
34b. Coles, GD, and CW Wrigley (1976): Laboratory methods for identifying New Zealand wheat cultivars. N.Z. J. Agric. Res. 19: 499–503.
35. Dal Belin Peruffo, A, C Pallavicini, Z Varanini, and NE Pogna (1981): Analysis of wheat varieties by gliadin electropherograms; I. Catalogue of electropherogram formulas of 29 common wheat cultivars grown in Italy. Genetica Agraria 35: 195–208.
36. Dale, G, and AL Latner (1969): Isoelectric focusing of serum proteins in acrylamide gels followed by electrophoresis. Clin. Chim. Acta 24: 61–68.
37. Danno, G, and M Natake (1980): Isolation of foxtail millet proteins and their subunit structure. Agric. Biol. Chem. 44: 913–918.
38. Daussant, J, and A Skakoun (1981): Immunochemical approaches to studies of isozyme regulation in higher plants. In: Rattazzi, MC, JG Scandalios, and GS Whitt (eds.): Isozymes: Current Topics in Biological and Medical Research 5: 175–218; Liss Inc., New York.
39. Derbyshire, E, DJ Wright, and D Boulter (1976): Legumin and vicilin. Storage proteins of legume seeds. Phytochem. 15: 3–24.
40. de Wreede, I, and H Stegemann (1981): Bestimmung von Proteinen nach Solubilisierung. Jahresber. Biol. Bundesanstalt, H 112.
41. de Wreede, I, and H Stegemann (1981): Determination of proteins in the presence of large quantities of SDS and ME by a modified biuret-method. Z. Anal. Chem. 308: 431–433.
42. de Wreede, I, and H Stegemann (1982): Trennung von Muskelprotein, Kollagen und Elastin durch fraktionierte Extraktion mit SDS-Lösungen. Z. Lebensm. Unters. Forsch. 174: 200–207.
43. de Wreede, I, H Stegemann, and HH Heinert (1982): Proteine in Brühwurst. Löslichkeit und elektrophoretische Bewertung. Z. Lebensm. Unters. Forsch. 174: 366–373.
44. Dierks-Ventling, C. (1981): Storage proteins in *Zea mays* (L.): Interrelationship of albumins, globulins and zeins in the *opaque-2* mutation. Eur. J. Biochem. 120: 177–182.
45. Doll, H (1977): Storage proteins in cereals. In: Muhammed, A, R Aksel, and RC von Borstel (eds.): Genetic Diversity in Plants: 337–347; Plenum Publ. Corp.
46. Doll, H, and B Andersen (1981): Preparation of barley storage protein, hordein, for analytical SDS-PAGE. Anal. Biochem. 115: 61–66.
47. Draper, SR, and EA Craig (1981): A phenotypic classification of wheat gliadin electropherograms. J. Nat. Inst. Agric. Bot. 15: 390–398.
48. Dronzek, BL, PJ Kaltsikes, and W Bushuk (1970): Effect of the D genome on the protein of three cultivars of hard red spring wheat. Can. J. Plant Sci. 50: 389–400.
49. Ellis, JRS, and CH Beminster (1977): The identification of UK wheat varieties by starch gel electrophoresis of gliadin proteins. J. Nat. Inst. Agric. Bot. 14: 221–231.
50. Engelhardt, H, and D Mathes (1981): HPLC of proteins using chemically modified silica supports. Chromatographia 14: 325–332.
51. Esen, A (1980): A simple colorimetric method for zein determination in corn and its potential in screening for protein quality. Cereal Chem. 57: 129–132.
52. Feillet, P (1980): Wheat Proteins. Evaluation and measurements of wheat quality. In: Inglett, GE, and L Munck (eds.): Cereal for Food and Beverages: 183–200; Academic Press, New York.

53. Foster, JF, JT Yang, and NH Yui (1950): Extraction and electrophoretic analysis of the proteins of corn. Cereal Chem. 27: 477–487.
54. Francksen, H, and R Garadi (1974): Complete recovery of proteins in concentrators and some drawbacks, revealed by polyacrylamide gel electrophoresis. Z. Anal. Chem. 271: 340–344.
55. Galyean, RD, JA Laney, M Harden, and D Krieg (1980): Water-soluble proteins of selected pearl millet genotypes. J. Food Sci. 45: 1165–1167.
56. Goodman, MM, and CW Stuber (1980): Genetic identification of lines and crosses using isoenzyme electrophoresis. Proc. Ann. Corn Sorghum Res. Conf. 35: 10–31.
57. Görg, A, W Postel, R Westermeier, E Gianazza, and PG Righetti (1980): Gel gradient electrophoresis, isoelectric focusing and 2-D-techniques in horizontal, ultrathin PAA-layers. J. Biochem. Biophys. Meth. 3: 273–284.
58a. Gottlieb, LD (1981): Electrophoretic evidence and plant populations. In: Reinhold, L, JB Harborne, and T Swain (eds.): Progress Phytochemistry 7: 1–46.
58b. Gottlieb, LD (1982): Conservation and duplication of isozymes in plants. Science 216: 373–389.
59. Graveland, A (1980): Extraction of wheat proteins with SDS. Ann. Technol. Agric 29: 113–123.
60. Günzel, G, and G Fischbeck (1979): Die Sortendiagnose am Gerstenkorn; Technik und Zuverlässigkeit eines Elektrophoreseverfahrens für die praktische Anwendung. Brauwissenschaft 32: 226–232.
61. Hamza, M, H Stegemann, and EM El-Tabey Shehata, unpublished.
62. Harris, RH (1955): Flour particle size and its relation to wheat variety, location of growth, and some wheat quality values. Cereal Chem. 32: 38–47.
63. Hart, GE, AKMR Islam, and KW Shepherd, (1980): Use of isozymes as chromosome markers in the isolation and characterization of wheat-barley chromosome addition lines. Genet. Res. 36: 311–325.
64. Heitefuss, R, DJ Buchanan-Davidson, and MA Stahmann (1959): Stabilization of extracts of cabbage-leaf proteins by polyhydroxy-compounds for electrophoretic and immunological studies. Arch. Biochem. Biophys. 85: 200–208.
65. Holt, LM, R Astin, and PJ Payne (1981): Structural and genetical studies on the high-molecular-weight subunits of wheat glutenin. Part 2: Relative isoelectric points determined by two-dimensional fractionation in polyacrylamide gels. Theor. Appl. Genet. 60: 237–243.
66. Hook, SCW (1980): Dye-binding capacity as a sensitive index for the thermal denaturation of wheat protein. A test for heat-damaged wheat. J. Sci. Food Agric. 31: 67–81.
67. Howard, NL, and A Lehman (1980): PAGE-separation of wheat gliadin fractions for cultivar identification and registration. Gewasproduksie/Crop Prod. 9: 61–63.
68. Huebner, FR, and JS Wall (1980): Wheat glutenin: Effect of dissociating agents on molecular weight and composition as determined by gel filtration chromatography. Agric. Food Chem. 28: 433–438.
69. Hussein, KRF, N Eustatiu, and H Stegemann (1977): Comparison between buffers, ethanol, chloroethanol and isopropanol as extractants of wheat proteins used for separation procedures in polyacrylamide gel-electrophoresis. Rev. Roumaine Biochimie 14: 247–251.
70. Hussein, KRF, and H Stegemann (1978): Comparison of proteins from wheat kernels by various electrophoretic methods in polyacrylamide. Z. Acker- Pflanzenbau (J. Agron. Crop Sci.) 146: 68–78.
71a. Jeanjean, MF, and P Feillet (1978): Physicochemical properties of wheat gel proteins: Effects of isolation conditions. Cereal Chem. 55: 864–876.
71b. Jeanjean, MF, and P Feillet (1980): Properties of wheat gel proteins. Ann. Technol. Agric. 29: 295–308.
72. Jensen, U (1981): Proteins in plant evolution and systematics. In: Ellenberg, H, K Esser, K Kubitzki, E Schnepf and H Ziegler (eds.): Progress in Botany 43: 344–369; Springer-Verlag, Berlin.

73. Juliano, BO, and J Boulter (1976): Extraction and composition of rice endosperm glutelin. Phytochem. 15: 1601–1606.
74. Kahler, AL, and RW Allard (1981): Worldwide patterns of genetic variation among four esterase loci in barley. Theor. Appl. Genet. 59: 101–111.
75. Kim, SI, and J Mossé (1979): Electrophoretic patterns of oat prolamins and species relationships in *Avena*. Can. J. Genet. Cytol. 21: 309–318.
76. Klose, J (1982): Genetic variability of soluble proteins studied by 2-D-electrophoresis on different inbred mouse strains and on different mouse organs. J. Molecular Evolution 18: 315–328.
77. Klose, J, and M Feller (1981): Two-dimensional electrophoresis of membrane and cytosol proteins of mouse liver and brain. Electrophoresis 2: 12–24.
78. Kobrehel, K, and W Bushuk (1978): Studies of glutenin. XI. Note on glutenin solubilization with surfactants in water. Cereal Chem. 55: 1060–1064.
79. Kobrehel, K, and B Matignon (1980): Solubilization of proteins with soaps in relation to the bread-making properties of wheat flours. Cereal Chem. 57: 73–74.
80. Koenig, R, H Stegemann, H Francksen, and HL Paul (1970): Protein subunits in the Potato Virus X group. Determination of the molecular weights by polyacrylamide electrophoresis. Biochim. Biophys. Acta 207: 184–189.
81. Kronberg, H, HG Zimmer, and V Neuhoff (1981): Implicit modeling of spots for the evaluation of two-dimensional electrophoretograms. In: Allen, RC, and P Arnaud (eds.): Electrophoresis: 413–423; de Gruyter, Berlin.
82. Ladizinsky, G, and T Hymowitz (1979): Seed protein electrophoresis in taxonomic and evolutionary studies. Theor. Appl. Genet. 54: 145–151.
83. Landry, J. (1979): La zeine du grain de mais. Biochimie 61: 549–558.
84. Landry, J, and T Moureaux (1970): Heterogenéité des glutelines du grain de mais. Extraction sélective et composition en acides amines des trois fractions isolées. Bull. Soc. Chim. Biol. 52: 1021–1037.
85a. Landry, J, and T Moureaux (1980): Distribution and amino acid composition of protein groups located in different histological parts of maize grain. J. Agric. Food Chem. 28: 1186–1191.
85b. Landry, J, and T Moureaux (1981): Physicochemical properties of maize glutelins as influenced by their isolation conditions. J. Agric. Food Chem. 29: 1205–1212.
86. Landry, J, T Moureaux, and JC Huet (1972): Extractibilité des protéines du grain d'orge: dissolution sélective et composition en acides aminés des fractions isolées. Bios 7–8: 281–292.
87. Lanzillo, JJ, J Stevens, and BL Fanburg (1980): A comparison of commonly used discontinuous and continuous buffer systems for electrophoresis in SDS-containing PAA-gels. Electrophoresis 1: 180–186.
88a. Larkins, BA (1981): Seed Storage Proteins: Characterization and biosynthesis. In: Marcus, A (ed.): The Biochemistry of Plants; Proteins and Nucleic Acids: 449–489; Academic Press, New York.
88b. Larkins, BA, AC Mason, and WJ Hurkman (1982): Molecular mechanisms regulating the synthesis of storage proteins in maize endosperm. CRC Crit. Rev. Food Science Nutrition: 199–215.
88c. Laurière, M, and J Mossé (1982): Polyacrylamide gel-urea electrophoresis of cereal prolamines at acidic pH. Anal. Biochem. 122: 20–25.
89. Lee, JW (1968): Preparation of gliadin by urea extraction. J. Sci. Food Agric. 19: 153–156.
90. Lerch, B, and H Stegemann (1969): Gel electrophoresis of proteins in borate buffers: Influence of some compounds complexing with boric acid. Anal. Biochem. 29: 76–83.
91. Lester, EP, PF Lemkin, and LE Lipkin (1981): New dimensions in protein analysis. Anal. Chem. 53: 3; 390A–404A.
92. Lis, H, and N Sharon (1981): Lectins in higher plants. In: Marcus, A (ed.): The Biochemistry of Plants; Proteins and Nucleic Acids 6: 371–447; Academic Press, New York.

93. Loeschcke, V, und H Stegemann, (1966): Polyacrylamid-Elektrophorese zur Beurteilung von Proteinen der Kartoffel (*Solanum tuberosum* L.). Z. Naturforsch. 21 B: 879–888.
94. Lottspeich, F, A Henschen, and KP Hupe (1981): High Performance Liquid Chromatography in Protein and Peptide Chemistry. Proc. Intern. Symp. Martinsried; de Gruyter, Berlin; 388 pp.
95. Lowry, OH, NJ Rosebrough, AL Farr, and RJ Randall (1951): Protein measurement with the Folin phenol reagent. J. Biol. Chem. 193: 265–275.
96a. Macko, V, and H Stegemann (1968): Genetische Veränderungen des Proteinmusters der Kartoffelknolle nach Trennung im isoelektrischen Gradienten. Jahresbericht Biol. Bundesanstalt, A 56–58.
96b. Macko, V and H Stegemann (1969): Mapping of potato proteins by combined electrofocusing and electrophoresis; identification of varieties. Hoppe-Seylers Z. Physiol. Chem. 350: 917–919.
97. Macko, V, and H Stegemann (1970): Free electrofocusing in a coil of polyethylene tubing, Anal. Biochem. 37: 186–190.
98. Maier, G, und K Wagner (1980): Routinemethode zur Identifizierung von Weizensorten mit Hilfe der Polyacrylamidgelelektrophorese. Z. Lebensm. Unters. Forsch. 170: 343–345.
99. Mäkinen, A, J de Wreede, H Stegemann, und HH Heinert (1979): Löslichkeit und gelelektrophoretische Muster hitzedenaturierter Proteine, auch von stärkehaltigen Proben, nach Einwirkung von Na-dodecylsulfat (SDS). Z. Lebensm. Unters. Forsch. 168: 282–285.
100. Marchylo, BA, and DE LaBerge (1980): Barley cultivar identification by electrophoretic analysis of hordein proteins. I. Extraction and separation of hordein proteins and environmental effects on the hordein electropherogram. Can. J. Plant. Sci. 60: 1343–1350.
101. Marchylo, BA, and DE LaBerge (1981): Barley cultivar identification by electrophoretic analysis of hordein proteins. II. Catalogue of electropherogram formulae for Canadian-grown barley cultivars. Can. J. Plant. Sci. 61: 859–870.
102. Margolis, J, and KG Kenrick (1968): Polyacrylamide gel electrophoresis in a continuous molecular sieve gradient. Anal. Biochem. 25: 347–365.
103. Marshall, DR, and AHD Brown (1975): The Charge-State Model of protein polymorphism in natural populations. J. Mol. Evolution 6: 149–163.
104. Matta, NK, and JA Gatehouse (1981): Modification of the basic subunits of pea legumin on storage. Phytochem. 20: 2621–23.
105. McCausland, J, and CW Wrigley (1977): Identification of Australian barley cultivars by laboratory methods: Gel electrophoresis and gel isoelectric focusing of the endosperm proteins. Austr. J. Exp. Agric. Anim. Husb. 17: 1020–1027.
106. Mecham, DK, EW Cole, and H Ng (1972): Solubilizing effect of mercuric chloride on the 'gel' protein of wheat flour. Cereal Chem. 49: 62–67.
107. Mecham, DK, DD Kasarda, and CO Qualset (1978): Genetic aspects of wheat gliadin proteins. Biochem. Genet. 16: 831–853.
108. Morrissey, JH (1981): Silver stain for proteins in polyacrylamide gels: A modified procedure with enhanced uniform sensitivity. Anal. Biochem. 117: 307–310.
109. Moss, HJ, CW Wrigley, F MacRitchie, and PJ Randall (1981): Sulfur and nitrogen fertilizer effects on wheat. II. Influence on grain quality. Aust. J. Agric. Res. 32: 213–226.
110. Moureaux, T (1979): Protein breakdown and protease properties of germinating maize endosperm. Phytochem. 18: 1113–1117.
111. Nelson, OE (1980): Genetic control of polysaccharide and storage protein synthesis in the endosperms of barley, maize and sorghum. In: Pomeranz, Y (ed.): Advances in Cereal Science and Technology 3: 41–71; Am. Assoc. Cereal Chemists, St. Paul, Minn.
112. Nierle, W (1976): Versuche zur elektrophoretischen Erkennung von Weizensorten. Getreide, Mehl, Brot. 30: 208–265.
113. Ornstein, L, and BJ Davis (1964): Disc electrophoresis. I: Background and theory, II: Method and application to human serum proteins. Ann. N.Y. Acad. Sci. 121: 321–349, 404–427.

114. O'Sullivan, MJ (1981): Enzyme immunoassay. Anal. Proc. (London) 18: 104–108.

115. Padhye, VW, and DK Salunkhe (1979): Extraction and characterization of rice proteins. Cereal Chem. 56: 389–393.

116. Park, WM, and H Stegemann (1979): Rice protein patterns. Comparison by various PAGE-techniques in slabs. Z. Acker-Pflanzenbau (J. Agron. Crop Sci.) 148: 446–454.

117. Paulis, JW, and JS Wall (1979): Distribution of electrophoretic properties of alcohol-soluble proteins in normal and high-lysine sorghums. Cereal Chem. 56: 20–23.

118a. Payne, PI, and KG Corfield (1979): Subunit composition of wheat glutenin proteins, isolated by gel filtration in a dissociating medium. Planta 45: 83–88.

118b. Payne, PI, KG Corfield, and JA Blackman (1979): Identification of a high-molecular-weight subunit of glutenin whose presence correlates with bread-making quality in wheats of related pedigree. Theor. Appl. Genet. 55: 153–159.

119. Payne, PI, LM Holt, and CN Law (1981): Structural and genetical studies on the high-molecular-weight subunits of wheat glutenin. Part 1: Allelic variation in subunits amongst varieties of wheat (*Triticum aestivum*). Theor. Appl. Genet. 60: 229–236.

120. Pelikan, M, F Dudas, and M Stankova (1976): Einfluß der Änderungen von Eiweißkomplex des Weizenkornes unter dem Einfluß der Umweltfaktoren. Acta Universitatis Agriculturae, Brnĕ 24: 231–238.

121a. Pietsch, G (1980): Ultradünnschicht-isoelektrische Fokussierung, eine Möglichkeit zur Identifizierung von Winterweizensorten. Göttinger Pflanzenzüchter-Seminar 4: 87–90.

121b. Pietsch, G (1980): Isoelektrische Fokussierung als Hilfsmittel für die Identifizierung von Pflanzensorten. Elektrophorese Forum '80. Diskussionstagung TU München: 89–90.

122 Polson, A, B von Wechmar, and MHV van Regenmortel (1980): Isolation of viral IgY antibodies from yolks of immunized hens. Immunological Comm. 9: 475–493.

123. Porath, J (1981): Development of modern bioaffinity chromatography (a review). J. Chromatogr. 218: 241–259.

124. Preston, KR, W Woodbury, RA Orth, and W Bushuk (1975): Comparison of gliadin and glutenin subunits in the Triticinae by SDS-PAGE. Can. J. Plant Sci. 55: 667–672.

125. Qualset, CO, and CW Wrigley (1979): Electrophoresis and electrofocusing identify wheat varieties. California Agric. 33: 10–12.

126. Ramshaw, JAM, JA Coyne, and RC Lewontin (1979): The sensitivity of gel electrophoresis as a detector of genetic variation. Genetics 93: 1019–1037.

127. Raymond, S, and L Weintraub (1959): Acrylamide gel as a supporting medium for zone electrophoresis. Science 130: 711.

128. Razin, S, and S Rottem (1967): Identification of Mycoplasma and other microorganisms by PAGE of cell proteins. J. Bacteriology 94: 1807–1810.

129a. Righetti, PG, and AB Bosisio (1981): Applications of isoelectric focusing to the analysis of plant and food proteins. Electrophoresis 2: 65–75.

129b. Righetti, PG, E Gianazza, and K Ek (1980): New developments in isoelectric focusing. J. Chromatogr. 184: 415–456.

130. Ryan, CA, and M Walker-Simmons (1981): Plant proteinases, resp. proteinase inhibitors. In: Marcus, A (ed.): The Biochemistry of Plants; Proteins and Nucleic Acids 6: 351–350 and 351–370; Academic Press, New York.

131. Salcedo, G, J Prada, and C Aragoncillo (1979): Low MW gliadin-like proteins from wheat endosperm. Phytochem. 18: 725–727.

132. Salcedo, G, R Sanchez-Monge, A Argamenteria, and C Aragoncillo (1980): The A-hordeins as a group of salt soluble hydrophobic proteins. Plant Sci. Lett. 19: 109–119.

133. Saleemuddin, M, H Ahmad, and A Husain (1980): A simple, rapid, and sensitive procedure for the assay of endoproteases using Coomassie Brilliant Blue G-250. Anal. Biochem. 105: 202–206.

134. Scholz, G, J Richter, and R Manteuffel (1974): Studies on seed globulins from legumes. I.

Separation and purification of legumin and vicilin from *Vicia faba* L. by zone precipitation. Biochem. Physiol. Pflanzen 166: 163–172.

135. Schwerdtfeger, E (1958): Über die Extraktion pflanzlicher Proteine. Wiss. Abhandl. Deutsch. Akad. Landwirtsch.-Wiss. Berlin 37: 153–163.

136. Shah, AA, and H.Stegemann (1982): Proteins of jojoba beans. Extraction and characterization by electrophoresis. Z. Acker- u. Pflanzenbau.

137. Shapiro, AL, E Viñuela, and JV Maizel (1967): Molecular weight estimation of polypeptide chains by electrophoresis in SDS-polyacrylamide gels. Biochem. Biophys. Res. Commun. 28: 815–820.

138. Shewry, PR, JC Autran, CC Nimmo, EJL Lew, and DD Kasarda (1980): N-terminal amino acid sequence homology of storage protein components from barley and a diploid wheat. Nature 286: 520–522.

139. Shewry, PR, AJ Faulks, S Parmar, and BJ Miflin (1980): Hordein polypeptide pattern in relation to malting quality and the varietal identification of malted barley grain. J. Inst. Brewing 86: 138–141.

140a. Shewry, PR, JM Field, MA Kirkman, AJ Faulks, and BJ Miflin (1980): The extraction, solubility, and characterization of two groups of barley storage polypeptides. J. Exp. Botany 31: 393–407.

140b. Shewry, PR, JM Field, EJL Lew, and DD Kasarda (1982): The purification and characterization of two groups of storage proteins (secalins) from rye (*Secale cereale* L.). J. Exp. Botany 133: 261–268.

141. Shewry, PR, HM Pratt, AJ Faulks, S Parmar, and BJ Miflin (1979): The storage protein (hordein) polypeptide pattern of barley (*Hordeum vulgare* L.) in relation to varietal identification and disease resistance. J. Nat. Inst. Agric. Bot. 15: 34–50.

142. Sluyterman, LA (1982): Chromatofocusing: a preparative protein separation method. Trends Biochem. Sciences 7: 168–170.

143. Smithies, O (1955): Zone electrophoresis in starch gels: Group variations in the serum proteins of normal human adults. Biochem. J. 61: 629–641.

144. Snyder, JC, and SL Desborough (1978): Rapid estimation of potato tuber total protein content with Coomassie Brilliant Blue G-250. Theor. Appl. Genet. 52: 135–139.

145. Southern, EM (1982): New methods for analyzing DNA make genetics simpler. Biochem. Soc. Transactions 10: 1–4.

146. Stegemann, H (1959): Cystinreduktion als Grundlage einer vereinfachten Amidgruppen-Bestimmung für alle Proteine. Hoppe-Seylers Z. Physiol. Chemie 315: 137–140.

147. Stegemann, H (1968): Die Primereinschluß-Technik zum Enzymnachweis bis 10^{-12} g nach Polyacrylamid-Elektrophorese, dargestellt an Phosphorylasen. Z. Anal. Chemie 243: 573–578.

148. Stegemann, H (1970): Protein-Mapping, Schnell-Dialyse und MW-Bestimmung im Mikrogramm-Bereich. Z. Anal. Chem. 252: 165–169.

149. Stegemann, H (1971): Gras-Samen; Erkennung von Sorten durch Elektrophorese. Jahresber. Biol. Bundesanstalt P 63.

150. Stegemann, H (1972): Apparatur zur thermokonstanten Elektrophorese oder Fokussierung und ihre Zusatzteile. Z. Anal. Chem. 261: 388–391.

151a. Stegemann, H (1977): Identifizierung von Maissorten mit gel-elektrophoretischen Methoden. Z. Acker- Pflanzenbau (J. Agron. Crop Sci.) 144: 157–161.

151b. Stegemann, H (1977/79): Indicator proteins in potato and maize for use in taxonomy and physiology. Gel-electrophoretic patterns. Proc. Symp. Seed Proteins of Dicotyledonous Plants, Gatersleben 1977; Abhandl. Akad. Wissensch. DDR, Abt. Math./Technik N 4, (1978) Akademie-Verlag Berlin, 4: 215–224.

152. Stegemann, H (1978): Properties and use of proteins from potatoes and other sources. In: Adler-Nissen, J, BO Eggum, L Munck, and HS Olsen (eds.): Biochemical Aspects of New Protein Food 44: 11–20; Pergamon Press, Oxford.

153. Stegemann, H (1979): SDS-gel-electrophoresis in polyacrylamide, merits and limits. In: Righetti,

PG, CJ van Oss, and JW Vanderhoff (eds.): Electrokinetic Separation Methods: 313–336; Elsevier North-Holland, Amsterdam.
154. Stegemann, H (1980): Eine neue 3-D-Methode zur Charakterisierung von ähnlichen Proteinen. Jahresber. Biol. Bundesanst. H 119–120.
155. Stegemann, H (1980): Leistungsfähigkeit elektrophoretischer Methoden für Genetik und Sortendiagnose. Göttinger Pflanzenzüchter-Seminar 4: 119–130.
156. Stegemann, H (1980): Preparative electrophoresis in gel-blocks with discontinuous elution. In: Radola, BJ (ed.): Electrophoresis 1979: 571–582; de Gruyter, Berlin, New York.
157. Stegemann, H (1981): Gel-Elektrophorese und Fokussieren in Platten mit der Apparatur PANTA-PHOR. Laboranweisungen für das Institut für Biochemie, BBA, Messeweg 11, D-3300 Braunschweig. 32 Seiten. Instructions also available in English or Spanish to be sent on request.
158. Stegemann, H unpublished.
159 Stegemann, H, AM El-Tabey Shehata, and M Hamza (1980): Broad bean proteins (*Vicia faba* L.). Electrophoretic studies on seeds of some German and Egyptian cultivars. Z. Acker- Pflanzenbau (J. Agron. Crop Sci.) 149: 447–453.
160. Stegemann, H, H Francksen, and V Macko (1973): Potato proteins: Genetic and physiological changes, evaluated by one- and two-dimensional PAA-gel-techniques. Z. Naturforsch. 28c: 722–732.
161a. Stegemann, H and V Loeschcke (1973): Arbeiten für Atlas europ. Kartoffelsorten, Jahresber. Biol. Bundesanstalt P 69.
161b. Stegemann, H, and V Loeschcke (1976): Index Europäischer Kartoffelsorten/Index of European Potato Varieties, Electrophoretic Spectra: National Registers, Appraisal of Characteristics, Genetic Data. Mitt. Biol. Bundesanstalt Berlin 168: 1–362.
162a. Stegemann, H and L Schilde (1980): Proteinmuster in der Kartoffelknolle bei 3 Mutanten der Schalenfarbe. Jahresber. Biol. Bundesanstalt H 118.
162b. Stegemann, H and P Schmiediche (1981): Proteinmuster von *Oxalis tuberosa*. Jahresber. Biol. Bundesanstalt.
163. Stegemann, H und G Voss (1975, 1976): Makromolekulare Komponenten in Mais, Beziehung zu phytopathologischen und genetischen Eigenschaften. Jahresber. Biol. Bundesanstalt H 64–65 und H 66–67.
164. Stevens, DJ (1973): Reaction of wheat proteins with sulphite. III. Measurement of labile and reactive disulphide bonds in gliadin and in the protein of aleurone cells. J. Sci. Food Agric. 24: 279–283.
165. Sun, SM, JL Slightom, and TC Hall (1981): Intervening sequences in a plant gene. Comparison of the partial sequence of cDNA and genomic DNA of french bean phaseolin. Nature 289: 37–41.
166. Svensson/Rilbe, H (1962): Isoelectric fractionation, analysis, and characterization of ampholytes in natural pH gradients. III. Description of apparatus for electrolysis in columns stabilized by density gradients and direct determination of isoelectric points. Arch. Biochem. Biophys. Suppl. 1: 132–138.
167. Torchinsky, YM (1981): Sulfur in Proteins. Pergamon Press, Oxford, 294 pp.
168. Torrance, L, and RAC Jones (1981): Recent developments in serological methods suited for use in routine testing for plant viruses. Plant Pathology 30: 1–24.
169. Tsugita, A, S Sasada, R v.d. Broek, and JJ Scheffler (1982): A new PAGE-system. Separation of small peptides and proteins in a volatile buffer system after modification with a strongly acidic fluorescent NH_2-reagent. Eur. J. Biochem. 124: 171–176.
170. van Vunakis, H (1980): Radioimmuno assays: An overview. Methods Enzymol. 70: 201–209.
171. Vesterberg, O (1967): Isoelectric fractionation, analysis, and characterization of ampholytes in natural pH gradients. V. Separation of myoglobins and studies on their electrochemical differences. Acta Chem. Scand. 21: 206–216.
172. Wagner, K und G Maier (1981): Die österreichischen Weizensorten – ihre Identifizierung durch PAGE. Getreide, Mehl und Brot 35: 205–208.

173. Wall, JS, and JW Paulis (1978): Corn and Sorghum Grain Proteins. In: Pomeranz, Y (ed.): Advances in Cereal Science and Technology 2: 135–219; Am. Assoc. Cereal Chemists, St. Paul, Minn.

174. Wilson, CM (1979): Studies and critique of amido black 10 B, Coomassie Blue R, and Fast Green FCF as stains for proteins after PAGE. Anal. Biochem. 96: 263–278.

175. Wilson, CM, PR Shewry, and BJ Miflin (1981): Maize endosperm proteins compared by SDS-PAGE and PAGIF. Cereal Chem. 58: 275–281.

176. Wrigley, CW, PJ Robinson, and WT Williams (1981): Association between electrophoretic patterns of gliadin proteins and quality characteristics of wheat cultivars. J. Sci. Food Agric. 32: 433–442.

177. YuMa, FA Bliss, and TC Hall (1980): Peptide mapping reveals considerable sequence homology among the three polypeptide subunits of G1 storage protein from french bean seeds. Plant Physiol. 66: 897–902.

178. Zaman, Z, and RL Verwilghen (1979): Quantitation of proteins solubilized in sodium dodecyl sulfatemercaptoethanol-tris electrophoresis buffer. Anal. Biochem. 100: 64–69.

4. Genetics of Seed Proteins in Wheat

E. PORCEDDU, D. LAFIANDRA and G.T. SCARASCIA-MUGNOZZA

INTRODUCTION

The history of cultivated wheat has been closely interlaced with the history of civilization since the early days of agriculture. With domestication, wheat has lost its inherent ability to disseminate to such an extent that its survival today depends essentially on man. However, at the same time, it has received so much care and attention worldwide that it can rightly be regarded as the most important of all food crops. Furthermore, the range of its utilization is practically as wide as the number of regions in the world. In addition to bread, which is a staple food everywhere, wheat is used – either alone or in combination with other farm products – to prepare macaroni, spaghetti, cous-cous, fric, chappati, ingera, tortillas, rolls, crackers, cookies, biscuits, cakes, doughnuts, muffins, pancakes, waffles, noodles, piecrust, icecream cones, puddings, pizza, bulgur, rolled flakes, etc. The fact that, to this day, it provides the basic nutrition for over one billion people or about 35 to 40 percent of mankind in such a wide variety, wheat is expected to remain foremost among the staple crops for man. It is therefore appropriate, if not imperative, to improve its nutritive properties which are now limited essentially by the rather low protein content particularly compared to its high caloric level and to the low nutritive value of its proteins.

As far as the protein content is concerned, it is known that the ratio of protein energy to total energy is normally low in wheat – as, more generally, in cereals – and that the biological value of proteins, representing the amount of absorbed nitrogen retained by the body, is also low especially in comparison with animal proteins (Table 1).

The poor nutritive value, represented by the limited ability to supply amino acid patterns in concentrations similar to that required by the human body, is due primarily to low levels of lysine and more generally, of essential amino acids such as triptophan, threonine, isoleucine and methionine (Table 1) and to the high levels of

Table 1. Nutritional quality of some cereal crops (26,235 adapted).

Cereal species	Limiting amino acids: Gross deficiency	Limiting amino acids: Marginal deficiency	Chemical score (A/T)	Total protein /energy ratio A	Efficiency of utilization (NPU)(1) % B	Utilizable protein content % AxB
Wheat	Lys	Ileu, Try	44	0.134	44	5.9
Barley	Lys	Ileu, Thr	54	--	--	--
Maize	Lys	Try	41	0.110	43	4.7
Oats	Lys	Meth,Thr,Ileu	57	0.120	66	7.9
Rice	Lys	Meth;Thr,Ileu	57	0.090	54	4.9
Rye	Lys	Phe,Ileu,Try	46	--	--	--
Sorghum	Lys	Meth,Try,Phe	31	0.117	42	4.9
Egg					87-100	

(1) NPU = Biological value (N retained/N ingested)xtrue digestibility (N absorbed/N in the food)

some of them, such as leucine and arginine. It is known, in fact, that when amino acids are unbalanced:

i) limiting amino acid requirements may become higher as is the case with lysine when wheat gluten is added to the diet of mice in order to raise the amount of basic amino acids [223];
ii) if one amino acid is in excess, this may depress the utilization of another, structurally similar, e.g. excess leucine in maize depresses the utilization of isoleucine [93], while excess arginine in millet depresses the utilization of lysine [78];
iii) an excess in one amino acid over others may contribute to limited digestibility of all the proteins, e.g. an excess of proline versus lysine and arginine limits the digestibility of cereal proteins [137].

The term 'protein quality' however has meanings other than essential amino acid balance. As reported by Payne [236] wheat protein quality relates to visco-elasticity for preparing leavened bread and extensibility for biscuit-making but also for *durum* wheat, protein quality is related to the ability to resist disintegration and retain a firm structure during and after cooking.

The economic and nutritive importance of this crop and the need to improve its production and quality have stimulated the development of research. A great number of scientists are engaged in this work ranging right from those concerned with the more theoretical aspects to those working on practical problems. A considerable impetus to studies on nutritive quality has come from the identification of mutants controlling the synthesis of specific proteins in maize [207], sorghum [289] and barley [86, 101].

Unfortunately, notwithstanding the large-scale research, the broad variation existing in nature and the attempts to induce mutations of practical use, mutants having the desired protein content or amino acid composition have not been identified. The failure is possibly due, on the one hand, to the polyploid nature of most cultivated wheats which can disguise manifestation of specific genes and on the other hand, to the fact that studies on related diploid species, particularly wild species, are still limited in scope and have started rather recently. It is quite possible, in fact, that only a very limited segment of the variability existing in nature has passed on into the cultivated types which seem to have expanded their variability because of their polyploid nature rather than as a result of the variability derived from the diploid species. However, the importance of finding a solution to the nutritive problem has prompted scientists to carry on further research. The specific fields of research are assessment of variability in nature, the implications of polyploidy and of the nucleus-cytoplasm relationship, environmental effects upon the technological properties of proteins, the nature and mechanisms of genetic control, protein synthesis, the genetic basis determining the amount of protein in seeds and their nutritive value, the chemical and physical nature of those protein components responsible for rheological properties etc. In more recent times, the use of wild material, aneuploids etc. and advanced refined techniques such as electrophoresis, electrofocusing etc. have been introduced and are now expanding. Excellent reviews of existing literature on these subjects were recently published [140, 165].

The purpose of the study presented here is to sum up information on the inheritance of protein content and total amino acid composition of specific proteins and protein components on the mechanisms controlling protein synthesis – also in view of the action of specific genes and their location on the chromosomes as a basis for a further refinement of research aimed at increasing protein content and improving the nutritive and technological properties of wheat seeds.

ORIGIN AND EVOLUTION OF WHEAT

The various species of *Triticum* form a polyploid series based on $x = 7$ and consisting of 3 different levels of ploidy: diploids ($2n = 2x = 14$), tetraploids ($2n = 4x = 28$) and hexaploids ($2n = 6x = 42$) [265, 148, 149, 271].

The wild *diploid* species (Table 2) are possibly of monophyletic origin, but have considerably diversified morphologically as well as ecologically and geographically. Each of them contains a different genome whose chromosomes do not pair regularly at meiosis of interspecific hybrids, thus causing total isolation of individual species [151].

The *polyploid* species are a classic example of evolution *via* amphiploidy. Despite the close affinity among genomes [218], chromosome pairing is the same as in diploids and inheritance is disomic [227, 228] due to the presence in chromosome 5BL of the gene Ph [259] which suppresses the pairing of homoeologous chromosomes and ensures regular segregations, high fertility and genetic stability.

In wheat, there are three groups of polyploids [337], each characterized by the presence of species having one genome in common. The polyploids in group A have in common the genome of diploid wheat, *T. monococcum*, those in group D the genome of *Ae. squarrosa* and those in group C the genome of *Ae. umbellulata*.

The identification of the different genomes was and still is the subject of many accurate investigations. While the establishment that the A genome could be donated by *T. monococcum* was rather simple and occurred in the early stages of investigation [148, 149, 271], the recognition of *Ae. squarrosa* as the donor of D genome was to follow only after a number of years of research [150, 201]. Only later on, however, Riley and Chapman [260] obtained an experimental hybrid between *T. aestivum* and *Ae. squarossa* and were able to observe a good chromosome pairing.

The AB genome of hexaploid wheat could be derived either from wild or cultivated *T. turgidum*. Two convincing arguments seem to suggest that the AB genome was derived from the cultivated rather than from a wild form of *T. turgidum:*

i) a limited overlapping of the distribution areas of *T. turgidum* var. *dicoccoides* and *Ae. squarrosa* compared to the much broader overlapping of cultivated *T. turgidum* particularly var. *dicoccum* and of *Ae. squarrosa* and
ii) the difference in the mode of dissemination of tetraploid wheat and of *Ae. squarrosa* which may have prevented the dissemination of the hybrid.

If these assumptions are correct, then the origin of the AB genome from cultivated *T. turgidum* could explain the absence of wild hexaploid wheats.

The donor of the B genome has yet to be identified conclusively. Considerable morphological,

Table 2. Classification of cultivated wheats and closely related wild species.

Species	Genomes	Wild	Cultivated	
		hulled	hulled	free-threshing
Diploid (2n=14)				
T.speltoides = Ae.speltoides	S=G?	all	--	--
T.bicorne = Ae.bicornis	S^b	all	--	--
T.longissimum = Ae.longissima	S^1	all	--	--
T.urartu				
T.searsii = Ae.searsii				
T.tauschii	D	all	--	--
(= Ae.squarrosa)				
T.monococcum	A	var.boeoticum	var.monococcum	
		(wild einkorn)	(cultiv.einkorn)	
Tetraploid (2n=28)				
T.timopheevi	AG	var.araraticum	var.timopheevi	
T.turgidum	AB	var.dicoccoides	var.dicoccum	var.durum
		(wild emmer)	(cult.emmer)	var.turgidum
				var.polonicum
				var.carthlicum
Hexaploid (2n=42)				
T.aestivum	ABD		var.spelta	var.aestivum
			var.macha	var.compactum
			var.vavilovii	var.sphaerocuccum

geographic and cytological evidence seems to suggest that it can be identified with *T. speltoides*. Hence, for example, Jenkins [107] reported that chromosomes of the experimental hybrid *T. turgidum* × *Ae. speltoides* formed seven bivalents at meiosis, although *Ae. speltoides* carried characters that distinguish tetraploid from hexaploid forms; Pathak [234] found that two pairs of chromosomes of *Ae. speltoides* have large satellites as *T. turgidum*; Sarkar and Stebbins [270] described morphological similarities and Riley *et al.* [261] reported caryotypic, geographical, synaptic and morphological evidences supporting the idea that *Ae. speltoides* had donated the B genome, although Sears [275] considered some of the evidence inconclusive. Kimber and Athwal [156] and Dvorak [60] obtained cytological data that rendered these conclusions questionable. Since then, Kimber [154, 155] has indicated that the B genome is now extinct and *T. speltoides* could instead be the donor G genome to *T. timopheevi*. Johnson [115] has suggested that the donor of the B genome could be *T. urartu* whose area of diffusion overlaps with that of

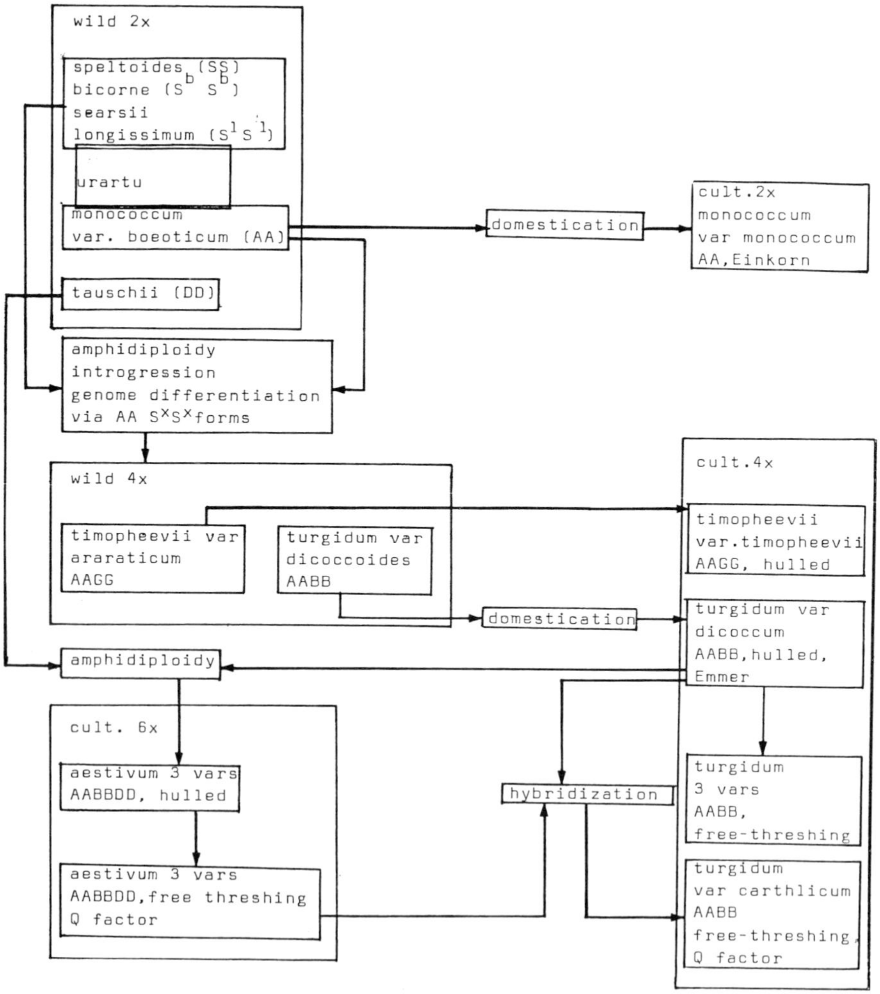

Figure 1. Evolutionary relationships of the wheats, Triticum.

T. boeoticum, while Konarev *et al.* [163] believed that *T. urartu* is the donor of the A genome to the turgid group, and Feldman [74] stated that the donor of the B genome is a *T. searsi* (S^SS^S) also from the *sitopsis* section (SS) (Figure 1).

The identification of the donor of the B genome is not a matter of pure scientific curiosity, but could have important implications in breeding programmes since it seems that the cytoplasm of cultivated polyploid wheat is similar to that of the B genome progenitor [152, 186, 182, 183, 184, 185, 298, 299, 276].

In contrast to the diploids which are genetically isolated from one another and have therefore undergone a diverging evolution, the polyploid wheats are characterized by a converging evolution. In fact, each one of them contains genetic material from one or more genomes and can exchange genes to produce a great number of genomic recombinations.

THE CARYOPSIS

Structure. The caryopsis is a small, indehiscent, dry, one-seeded fruit formed by the seed and the encasing pericarp. In turn, the seed consists of an embryo, an endosperm and a double-layer tegument – seed coat and nucellar epidermis – which contains it (Figure 2). The embryo consists of the embryo-axis, the scutellum, and the epiblast, whereas the endosperm is formed by the aleurone layer and by the starchy endosperm.

Origin. The seed derives from the fertilized ovule, whereas the pericarp originates from the ovary wall [333] and therefore consists solely of maternal tissue [258]. The latter also originates from the seed-coat produced by the ovule integuments, and the perisperm derived from the nucellus. Thus, only the embryo and the endosperm contain genes from both parents: actually, the embryo is the product of oosphere fertilization by one of the pollen reproductive nuclei and the endosperm results from the fusion of the two female polar nuclei with the other reproductive nucleus.

Components. Although the size, shape and chemical composition of the caryopsis depend on the type of wheat and pedoclimatic conditions of the cultivation environment, the relative ratios between the different parts of the caryopsis are rather stable. It is generally believed that the parts derived from the maternal tissue amount to about 8%, the embryo 7% and the endosperm 85% [278, 189]. Thus, it is the endosperm which contains the reserves from which the seedling draws its nourishment until the time it becomes autotrophic and forms the material utilized in human nutrition. Consequently, this study has been focussed essentially upon the proteins of this latter portion.

Development. The wheat kernel develops from a double fertilization process. The fusion of the reproductive nucleus with the oosphere will produce the embryo, formed by the embryo-axis, the epiblast and scutellum. At maturity, the latter will contain a considerable amount of proteins and oils and these will supply the very first nourishment to the embryo-axis as it starts germinating [246, 189].

The endosperm derives from the fusion of one of the pollen reproductive nuclei with the two polar

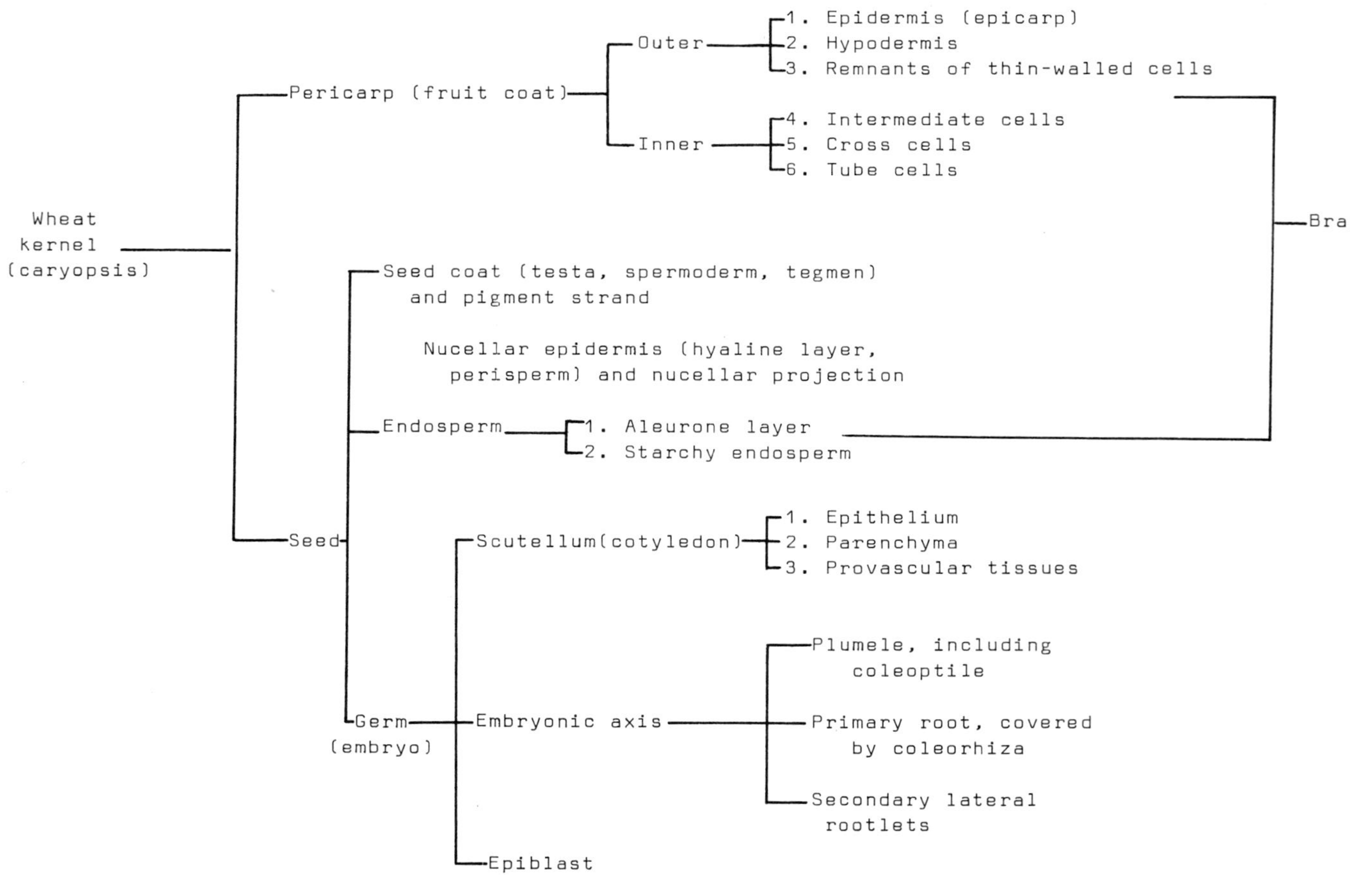

Figure 2. Parts of the wheat kernel and their relation to each other (189).

nuclei [333, 194]. Its development starts immediately after fusion and is very rapid; numerous early mitoses not accompanied by cell division occur and a large vacuole is formed; one or two days later a single large multinucleate cell is formed with a peripheral zone of cytoplasm and a large central vacuole while the nucellus disintegrates to become a thin unicellular layer [108]. An active cell division then occurs with sub-division of the vacuole into compartments, the migration of the cytoplasm and nuclei around the individual vacuoles and lastly, with complete cytokinesis to form the cell walls. The central portion of the thick cytoplasmic mass is transformed into a layer of thick-walled cells with a dense cytoplasm which will form the cells of the aleurone layer of the crease region [194], whereas the dorsal and then gradually the central segment give rise to highly vacuolated, thin walled cells governing further endosperm growth. These will lead to the formation of peripheral, prismatic cells some of which – the outer ones – may develop thickened walls and eventually form a new aleuronic or sub-aleuronic layer [287, 194]. The whole process is completed about two weeks after anthesis [267, 109, 66, 287].

Subsequent endosperm growth is only by expansion of the cell walls, these being forced by starch granules and protein bodies which begin to form and accumulate at about 7 days after anthesis [37] and continue to do so until the seed has reached maturity [108, 216].

Cell size and the length of the time during which the cells are capable of displaying a meristematic activity appear to be the main factors responsible for different protein contents in the cells of the aleurone layer and in the remaining cells. The cells in the aleurone layer which have the highest protein concentrations – 45% versus 11% in the inner portion of the endosperm – [143, 144, 96, 97, 98] are in fact smaller, have thicker walls and stay active much longer than the embryo cells which soon cease being active and form large deposits of starch grains with few reserve protein bodies [47, 66, 143, 267, 268, 296].

SEED PROTEINS

Knowledge about the nature of various proteins contained in the seed is based essentially on the classification by solubility proposed by Osborne [232] according to which proteins can be distinguished into albumins, globulins, gliadins and glutenins depending on their solubility in water, saline solutions, 70% aqueous ethanol and diluted acid of alkali solutions.

This classification is still widely used today although extractions made by chromatographic and electrophoretic techniques, gel filtration, ultracentrifugation show that the four classes are not clearly defined and are mutually overlapping. A brief account of the main physical and chemical properties of seed proteins is presented in the following points as a basis for further discussion on genetical aspects. Readers who are interested may refer to the papers by Kasarda *et al.* [140], Konzak [165], for complete and detailed presentation.

Albumins and Globulins

Albumins and *globulins*, also called soluble or cytoplasmic proteins, are essentially represented by enzymes involved in metabolic activity, are mainly located in the germ and aleurone layers and amount to about 20% of total proteins in the caryopsis. The higher the protein content, the lower the percentage [70]. Significant differences are found between *durum* and common wheat both in terms of quantity –

in common wheat soluble proteins are higher than in *durum* [7] – and in terms of quality due to the presence of specific protein components. These differences allowed to devise electrophoretic [257, 284, 73] and immunoelectrophoretic [248] methods in order to detect the presence of bread wheat flour in *durum* macaroni. Obviously, this is a problem of considerable importance in some countries where pasta should be prepared only from *durum* wheat.

Albumins are generally present in greater amounts than globulins. Molecular weight is between 12,000 and 16,000 for albumins and between 20,000 and 200,000 for globulins.

Specific soluble proteins. Two-thirds of albumins consist of protein components with an inhibiting activity on animal α amylase. They do not seem to be active on wheat enzymes [53].

When the extract of the albumin fraction is subjected to filtration on Sephadex G100 column, three absorption peaks are obtained whose molecular weight is approximately 60,000, 24,000 and 12,000 [53], respectively corresponding to peaks II, III and IV. Peak I, instead, consists of protein fractions contained in the outer layers of the caryopsis and in the embryo [247]. Peaks II and III are effective against human and *Tenebrio molitor* α amylases whereas the fraction in Peak IV is only active against the α amylases of insects [53].

By submitting the 24,000 MW albumin component (peak III) to disc electrophoresis on poliacrylamide gel, with alkaline buffer, four peaks – IIIa, IIIb, IIIc and IIId are obtained, all with the same α amylases inhibiting activity. This albumin fraction is also designated 0.19 on the basis of the electrophoretic mobility of its main component (IIId) relative to bromophenol blue. It would even appear that the other components of peak III are derived from this 0.19 fraction by simple de-amination [53]. By treating 0.19 albumin with dissociating solvents, like guanidine hydrochloride or S.D.S., two sub-units are obtained with an MW close to 13,000 which [53] are believed to form a family of strictly correlated inhibitors.

Silano *et al.* [285] have shown that upon gel electrophoresis at pH 8.5, the 12,000 MW fraction breaks down into five components which have by now been isolated and characterized. Again in this case, it is believed that these proteins form a correlated family. As much as 80% of the 60,000 MW component seems to be formed by 12,000 MW sub-units.

The α-amylase inhibitory activity seems to be due to the disulphide bonds [247] and can be irreversibly interrupted by reducing agents.

Chloroform-methanol-soluble proteins. Proteins with electrophoretic mobility similar to that of albumins and an MW less than 25,000 [266] can be extracted with mixtures of chloroform and methanol [204, 205, 206, 80, 256].

The electrophoretic pattern at pH3 of extracts, obtained with $CHCl_3$/MeOH (2:1 v/v) from endosperms of *Triticum aestivum* shows three main bands designated

CM1, CM2 and CM3, while extracts of *Triticum turgidum* only have the CM2 and CM3 bands [263]. Aragoncillo [4] purified a faster variant of CM3, called CM3[1], which is only present in certain varieties of *Triticum durum*.

Rodriguez Loperena *et al.* [263], Redman and Ewart [256] Salcedo *et al.* [266] suggest that these proteins could also be classified as proteolipids because they are usually extracted from flours as a lipid complex; after dissociation of the lipid complex, by exposure to acid buffers, they become water-soluble and are still soluble in the chloroform/methanol mixture; in addition, they are soluble in 70% ethanol [263]. Although their molecular weight is approximately 13,000 and although they are classified as globulins and albumins, there is increasing evidence that they cannot be included in any of the solubility classes defined by Osborne [266].

Purothionins. This is another group of proteins of considerable interest although their function in the endosperm is still unknown. Purothionins were first obtained and crystallized by Balls *et al.* [10] and are so designated because of their high sulphur content.

Purothionins consist of two fractions, called α and ß according to their electrophoretic mobility in polyacrylamide gel at pH 3.1. Their molecular weight is approximately 4900 [191, 192], and they can be extracted from flours with petroleum ether, as a lipoproteic complex, and with sodium chloride solutions and diluted acids [140, 75, 255]. Purothionins have high cystine, lysine and arginine, low aspartic and glutammic acid and proline and no hystidine, methionine and triptophan [255, 229].

The amino acid sequences of α and ß purothionins were determined by Jones and Mak [133, 192]. These scientists reported that α purothionins are formed by two proteins, called α_1 and α_2, having different amino acid composition and sequence but the same electrophoretic mobility. Although their role in the endosperm is unknown, they have a toxic effect on small animals when injected intravenously *via* the intraperinoneal route. In addition, they inhibit bacterial and yeast growth [140].

Gliadins and Glutenins

Gliadins and glutenins, also called storage proteins, are the main components of gluten which is obtained by washing flour in water. These classes of proteins are each composed of many and different molecular species and it is from their structure and interaction that the viscoelastic properties of the gluten are derived.

Gliadins are made of single polypeptide chain stabilized by intrachain disulfide binding [226, 16] with particular amino acid composition, that includes large amounts of glutamine (38–56% of all amino acid residue), proline (15–30%) and a small amount of lysine [140]. They have been classified into α, ß, γ and ω [325] according to their relative mobility upon gel electrophoresis with aluminium lactate buffer with pH ranging from 3.1 to 3.3; this nomenclature was related to the gliadin fractions obtained on moving-boundary electrophoresis [136].

While α, ß and γ gliadins have about the same amino acid composition, molecular weight ranging from 30,000 to 45,000 and similar N-terminal sequences, ω gliadins show differences for molecular weight, amino acid composition and N-terminal amino acid sequences. In fact, ω gliadins have molecular weight ranging from 65,000 to 80,000 as determined by ultra-centrifugation, SDS-PAGE or gel filtration and some components contain no cystine and methionine [23, 16, 43, 44].

In turn, each group of gliadins comprises many molecular species when subjected to one dimensional PAGE; each band may contain more than one component. Wrigley [326] was able to detect 46 different components in one single variety, by combining two dimensional techniques, isoelectrofocusing in the first dimension with electrophoresis in starch gel at pH 3.2 in the second dimension.

Another two-dimensional technique has been described recently [203] for the separation of the gliadins; this approach uses separation at two largely different pH (3.2 for the first dimension and 9.2 for the second) in the same polyacrylamide slab gel.

There are possibly hundreds of gliadin components that could be differentiated by means of two dimensional techniques and the complexity could be even greater since evidence has been reported for apparently single components being composed of different components differing only in the substitution of a neutral amino acid in their primary structure and this would not affect the separation of these components with electrophoretic techniques [140, 19, 138].

Glutenins may be defined as a complex of proteins having high molecular weight, made up of numerous polypeptidic sub-units, joined together by both covalent and non-covalent links [15]. They represent about 30–40% of total flour proteins and are considered the most important contributors to the strength and viscoelastic properties of the wheat dough [18].

Native glutenins have a broad spectrum of molecular weights, ranging from about 40,000 to over several millions with mean values from 150.000 to 3.000.000 [134, 225, 302]. There is considerable evidence that these proteins are extremely dispersed and that the molecular weight is an important variety characteristic, involved in the bread-making quality [147]; rheological properties may in fact derive from the distribution of molecular weights.

The use of SDS-PAGE which separates the reduced glutenin on the basis of their molecular weight, enabled to obtain additional information on the nature of the sub-units. Bietz and Wall [16] obtained 15 different sub-units with molecular weight ranging from 11.600 to 133.000. These values were confirmed by Orth and Bushuk [231] who also reported that 2 high molecular weight sub-units are not present in *durum* wheat. Later on, Bietz and Heubner [15] showed that slow-moving components have a light molecular weight and indicated the possibility of grouping the sub-units into two distinct fractions on the basis of their solubility in 70% ethanol: the insoluble fraction containing high molecular weight glutenins and the soluble fraction containing sub-units with molecular weight ranging from 36.000 to 44.000 and so very similar to those of high molecular weight reduced gliadins. These low molecular weight glutenins are present in alcoholic extracts and represent 5–10% of the total proteins extracted by gel filtration on Sephadex G 100.

Under PAGE in pH 3.2 aluminium lactate buffer, ethanol-soluble reduced glutenins and these polypeptides differ from low molecular weight gliadins (mainly α and ß) but present sub-units having mobility intermediate between γ and ω.

N-terminal amino acid sequences of high molecular weight gliadins are similar to those obtained from ethanol-soluble glutenins but are different from sequences observed in low molecular weight gliadins [19]. These findings enabled Bietz and Wall [17] to affirm that many of these polypeptides are homologous and may be derived from a few ancestral genes.

Attempts to determine N-terminal sequences of ethanol-insoluble or purified high molecular weight glutenin sub-units have so far been unsuccessful.

PROTEIN SYNTHESIS AND ACCUMULATION IN SEEDS

The processes governing the synthesis and accumulation of proteins cover several aspects, the crucial points of which have been summed up [210] as follows:

the plant obtains nitrogen from the soil absorbing it as nitrates and in small amount as ammonium;

nitrate can be reduced to ammonium in the roots or translocated as such through the xylem to the green parts where it is reduced; also, it can be provisionally stored in the vacuoles in any part of the plant;

ammonium is usually assimilated in the roots, is rarely found in the xylem and is not usually stored in the vacuoles;

amino acids and particularly glutamin and asparagin can act as transport metabolites for nitrogen from the roots to the leaves and vice versa;

most of the nitrogen reaches the seed in the form of amino acids.

The first step in the process, once the nitrogen has re-entered the plant, is related to the nitrate-reductase activity (Figure 3) which limits the scope of the whole process [13]. The enzyme is synthesized in the 80S ribosomes, located in the cell cytoplasm, is induced by the presence of nitrate in the cell, [95] and its activity is controlled by the amount of produce via a feedback system. Further reduction of nitrate to ammonium is performed by nitrate-reductase localized in chloroplasts. Nitrite-reductase activity is dependent on reduced nicotinammide adenine dinucleotide (NADH) [262] which is supplied by the transformation of 3 phosphoglyaceraldehyde, an intermediate produce of CO_2 assimilation, to 1, 3 disphosphoglyaceraldehyde.

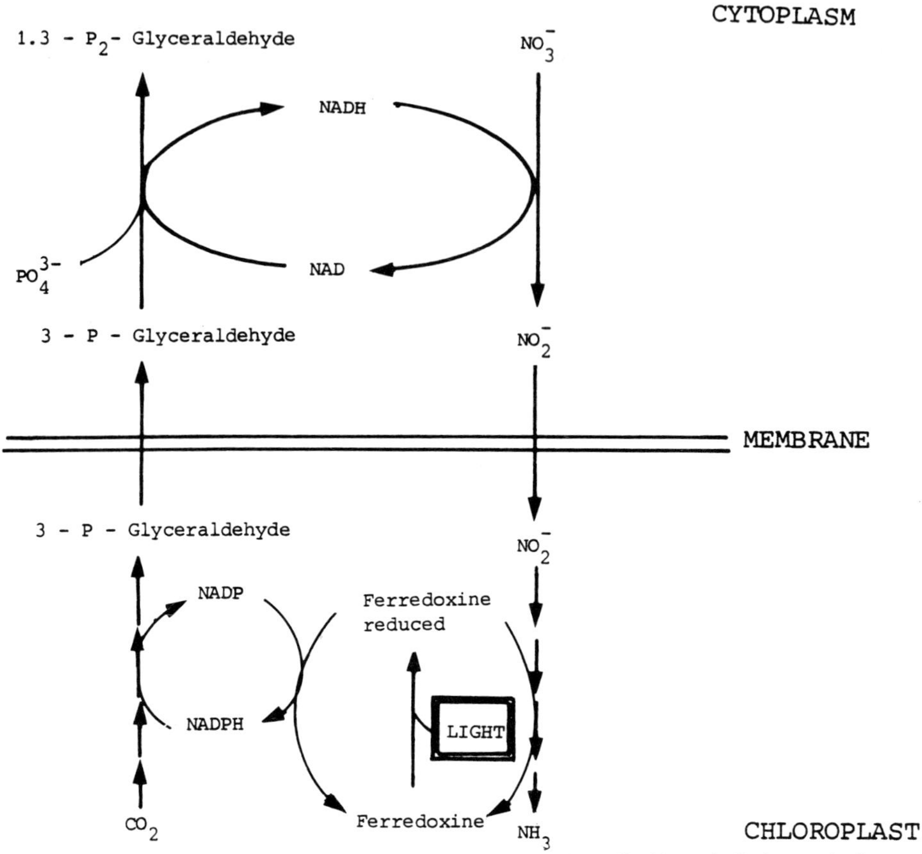

Figure 3. Relationship between photosynthetic process, carbon dioxide assimilation and nitrate reduction.

Originating in the chloroplasts, the 3 phosphoglyaceraldehyde passes into the cytoplasm where it is transformed by NAD which is simultaneously reduced to NADH. The nitrite formed by this reduction enters the chloroplast for further reduction to NH_3 by nitrite-reductase which receives electrons from the reduced ferredoxine, a product of photosynthesis. The reduced ferredoxine supplies electrons also to the NADP enzyme which is responsible for the reduction of CO_2 [145, 146, 309, 181, 233] so that the protein and sugar synthesizing processes are in competition and are self-regulating via the NAD. Nitrite reduction also occurs in the root tissues although the process and the controlling enzymes are not yet known [210].

Ammonium nitrogen is first transformed into glutamate via the glutamate dehydrogenase, although Miflin and Lea [211, 212] have shown that in most bacteria and higher plants an alternative route might operate via glutamine synthetase and glutamate synthetase and then to other amino acids via the transaminases. Synthesis of protein occurs at different parts of the plant.

Once fertilization occurs, both the newly synthesized amino acids and those derived from the hydrolysis of proteins from different parts of the plant migrate towards the developing seeds in whose cells also the genes specifying the synthesis of reserve proteins are transcribed and translated in addition to those governing the synthesis of functional proteins.

The pericarp, which in the early days after fertilization forms a conspicuous portion of the developing seed [108] also contains chloroplasts that may have a significant role in the production of amino acids, sugar and other compounds: these are carried into the endosperm and cooperate in protein and starch synthesis [267]. As a matter of fact, the maternal tissues also contain enzymes that are important in protein synthesis, such as glutamic synthetase which is at peak value about 12 days after fertilization and is not present in the endosperm [258].

A remarkable rise in amino nitrogen content is observed during the 7 days following anthesis after which it stays constant for about 35 days whereupon it undergoes a drastic drop. Protein synthesis starts after formation of the uninucleate cells and goes on – though not steadily – until maturity [37, 108, 216].

Formation of water-soluble proteins begins at the same time as zygote development and a constant value is attained about 40 days after anthesis; their synthesis occurs primarily in the cells of the aleurone and sub-aleurone layer which eventually will contain few protein bodies and small starch granules [267, 268, 296]. Conversely, formation of insoluble proteins begins only during the third week after anthesis and their biosynthesis rises gradually until the 45th day and is mostly located in the innermost portion of the endosperm.

At least two classes of ribosomes should be involved in protein synthesis [85]. One would be very active during the early developmental stages and would be responsible for synthesis in the proteoplasts, namely, plastids consisting of a lipoproteic membrane encasing both the ribosomes, which synthetize the proteins and the protein bodies in which they are stored [221, 222, 11]. According to Buttrose [37], the Golgi apparatus is involved in the accumulation process while Campbell *et al.* [39] believe that extrusion into the vacuoles is through a system similar to that of pinocytosis. The protein bodies formed during this early stage would seem to be comparatively small [84] and to have an amino acid composition similar to that of the albumin and globulin fractions [109].

The other mechanism would be activated later, about 15–20 days after anthesis. In it, protein synthesis appears to be carried out by the ribosomes associated to the rough endoplasmic reticulum within the lumen of which the proteins are stored [11, 39, 40]. These proteins form larger protein bodies with an amino acid composition comparable to that of the prolamins and glutelins even though they also contain some 15–20% functional proteins [109].

Campbell *et al.* [40] studied the development of wheat grain from intact plants and from detached ears grown in a medium containing 0.25 and 4% sucrose. They, therefore, claim that the protoplast theory by Morton *et al.* [221] is lacking in any evidence whatsoever. During the development stage protein accumulation inside the cistern would seem to determine the formation of protein bodies similar to those of maize [31] and of barley [213] and also – though in smaller amounts – in the structure of vacuoles which may be closely associated to the rough endoplasmic reticulum. Therefore, he denies the existence of two parallel storage systems and contends that the protein synthesis is carried out by the polyribosomes attached to the rough endoplasmic reticulum and that proteins either accumulate inside the lumen of the rough endoplasmic reticulum or are transferred into the vacuoles. At the beginning of endospermic development, proteins accumulate preferably in the vacuoles whereas later on the distension of the endoplasmic reticulum is more common. This interpretation seems to be supported by the findings of Barlow *et al.* [11] according to which the membrane of the endoplasmic reticulum forms a kind of barrier for the protein bodies originating during the later stage of development. In close association with the protein bodies there also seem to exist osmiophilic inclusions [37, 40] whose origin has not yet been entirely clarified, although it is known that deposits of similar electronic density are also present in the cytoplasm. These inclusions could be equivalent to the arginine-rich proteins localized in the aleurone and sub-aleurone layers of wheat and other graminaceous plants [76].

The mechanism by which wheat proteins are accumulated does not seem to differ much from that observed in other seeds although there are differences concerning the level of organization at maturity.

GENETIC CONTROL OF PROTEIN CONTENT AND AMINO ACID COMPOSITION

Considerable amount of information is reported in literature concerning the variation and genetic control of protein content and amino acid composition in wheat with greater emphasis being laid on bread wheat because of its importance and diffusion.

The available information can be summarized as follows.

Variation in Protein Content

Among Species. A systematic analysis of protein content in a large group of varieties and wild species of wheat was first carried out by Lawrence *et al.* [173]. The values they found on over 230 wheat samples including 12 different species in addition to the commercial samples, indicate (Table 3) that the highest protein contents were detected in *T. dicoccoides* (29.6%) and the lowest in *T. turgidum* (9.8%). Values of interest, above 17%, were also shown by samples of *T. macha*, *T. spelta* and *T. vavilovi*. The values of *T. vavilovi* were instead the lowest in the findings

Table 3. Protein content (%) in different species of *Triticum* grown at different locations.

	Pullman (1)	Aberdeen (173)		Fargo (173)		Bari (310)		
		Max	Min	Max	Min	$\bar{x}$	Max	Min
T. boeoticum	--	19.40 -	24.08	17.81 -	22.21	23.5	26.4	18.0
T. monococcum	16.1	15.46 -	18.99	16.09		21.0	27.5	18.2
T. dicoccoides	29.6	12.75 -	27.00	14.37 -	21.70	18.8	23.6	12.7
T. dicoccum	16.0	15.64 -	17.06	18.91		17.6	24.3	12.5
T. turgidum	9.8	12.83 -	14.82	13.33		12.1	16.2	10.0
T. polonicum	17.8	15.33 -	18.28	18.74		15.9	18.1	13.5
T. carthlicum	16.9	12.65 -	18.30	--		17.0	17.9	15.8
T. timopheevi	15.0	12.90 -	20.65	12.04 -	16.66	20.3	22.3	17.8
T. orientale	14.0	12.32 -	17.24	16.42 -	17.15	12.5	--	--
T. ispahanicum	--	--	--	--		14.4	--	--
T. spelta	19.1	11.55 -	15.71	--		15.3	20.3	12.7
T. macha	18.2	10.64 -	16.46	--		13.3	15.2	10.9
T. zuckovskyi	--	--	--	22.26 -	28.16	24.7	--	--
T. durum	12.9	--	--	--		14.4	16.7	10.6
T. vavilovii	17.5	9.96		--		--	--	--
T. spheorococcum	16.0	14.86 -	19.17	--		--	--	--
T. piramidale	15.4	11.80 -	13.36	17.87 -	20.01	--	--	--
T. paleocolchicum	15.9	--	--	16.63		--	--	--

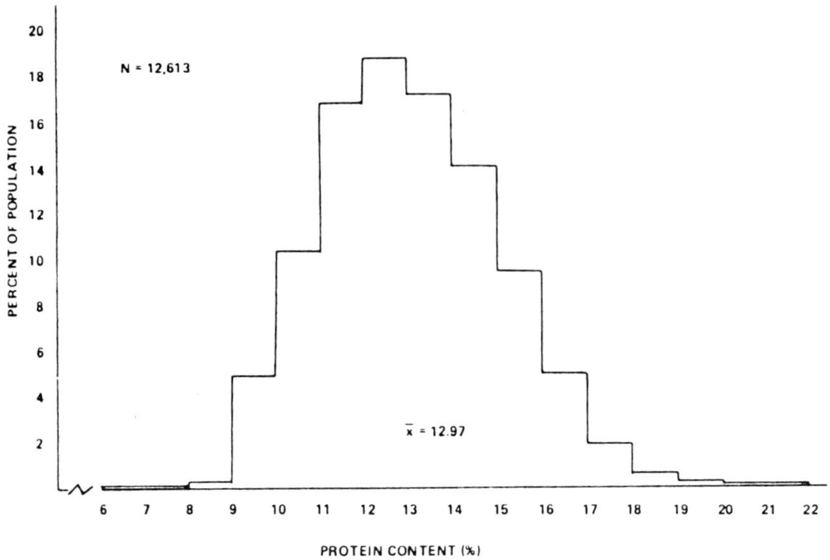

Figure 4. Frequency distribution for grain content among 12,613 wheats in the USDA World Collection (311).

of Villegas *et al.* [310] who had analyzed variation for protein content in different wheat species cultivated at two locations, Aberdeen and Fargo. Of the 64 samples cultivated at Aberdeen, those of *T. dicoccoides* were prominent for high protein (27%) and those of *T. vavilovi* for low protein (9.96%). Of the samples cultivated at Fargo, the highest values were observed in *T. zukovskii* (28.16%) and the lowest ones in *T. timopheevi* (12.04%). Also the analysis made by Porceddu [250] on 228 samples from 14 different species confirmed that the highest mean levels are those for *T. zukovski* (24.7%) and the lowest for *T. turgidum* (12.1%). However, the absolutely highest levels were those of the two diploids, *T. monococcum* (27.5%) and *T. boeticum* (26.4%) while the lowest ones were, here again, those of *T. turgidum* (10%). It should be noted, however, that *T. dicoccum* and *T. dicoccoides* were the two species showing the highest variability among samples.

The high values observed in diploid wheats were confirmed by Muntz *et al.* [224] who also confirmed the lowest values for *T. durum* and *T. aestivum*. The general tendency of the more primitive or wild types for higher protein content, already claimed, among others, by Harlan [91] corresponds to the data reported by Avivi [9] who found values of about 24% in seeds of *T. dicoccoides* collected in a natural environment and values around 33% in seeds from plants grown in green house as against 14.1% and 18.6% in seeds of *T. durum* grown under the same conditions. Avivi [9] also confirms the high variability of samples of *T. dicoccoides* which has values between 17% and 28% in the open air and between 24% and 43% in the

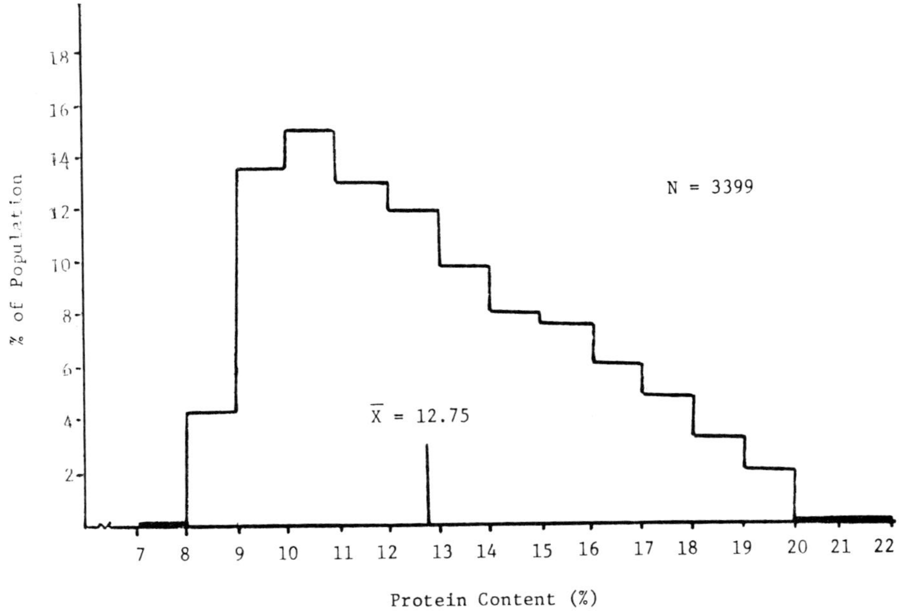

Figure 5. Frequency distribution for whole grain protein content among 3399 durum wheats in the USDA World Collection (323).

green house. Sharma *et al.* [277] found averages of 28.5% and 29.3% for *T. dicoccoides* grown in irrigated and dry environments respectively.

Within Species. Results of analysis performed on over 12,000 samples from the ARS-USDA common wheat collection indicate that protein content within this species may range between 6.9% and 22% (Figure 4) with a mean value around 13% and a standard deviation of about 2% [311].

This wide range in protein content suggests that there are significant differences among the wheats in the world collections and has made it possible to identify over 500 samples with protein values above 17%; since the seeds were derived from a single cultivation location, it may be concluded that they contain a potential source of germplasm.

Variation in *durum* wheat is rather similar; in fact, the analysis of a collection of almost 3,400 samples indicates that protein content varies between 7.3% and 21.3% (Figure 5) the mean being 12.75% and the standard deviation 2.86 [323]. These wide ranges of values suggest that there must exist, also for this species, tremendous amounts of variability despite the fact that the number of available samples is only about 1/4 that of common wheat. About 4% of the samples in the *durum* wheat collection has revealed a protein content higher than 18.5%, namely, two standard deviations above the mean thus pointing to a true wealth of genes for high protein content. Analyses made on 200 *durum* wheat samples selected for protein content and cultivated at different locations in Ethiopia and in the U.S.A., have made it possible to bring out the strong environmental effect upon protein content and the existence of strong genotype × environment interaction for character expression [323].

However, results obtained by Johnson *et al.* [127], Johnson and Mattern [122] in common wheat, and Porceddu *et al.* [253] in *durum* wheat showed that marked differences in protein content stability may exist in different materials. Johnson *et al.* [131] and Vogel *et al.* [312] found that over 5% of the variation in the world collection can be ascribed to genetic causes and issued a list of germplasm sources for high protein content (Table 4). Part of this list was confirmed by Porceddu [252] when the *durum* wheat world collection was grown in Italy.

Genetic control of Protein Content

Findings on the nature of genetic control are rather controversial although all scientists agree that this is undoubtedly a complex subject and one that is difficult to study due to the strong influence of the environment upon its expression.

Some scientists [21, 46, 1, 324, 169, 297] contend that protein content is a

Table 4. Protein and lysine content in selected samples of *durum* wheat grown at Mesa (Arizona, USA) (311) and Bari (Italy) (252).

P.I.	Protein %		Lysine (Mesa)	
	Bari	Mesa(Az)	% Dry weight	% Prot.
58786	18.0	--	--	--
134935	17.6	19.7	0.54	2.74
165117	18.2	19.6	0.54	2.80
166816	17.6	19.5	0.54	2.81
185196	18.1	19.7	0.55	2.82
185721	21.1	21.1	0.62	2.95
185731	19.3	19.7	0.54	2.76
185765	17.6	19.7	0.56	2.86
185764	18.0	19.5	0.56	2.89
191089	18.8	19.7	0.60	3.09
191608	20.3	19.6	0.52	2.69

character governed by a complex polygenic system, with genes distributed in all the chromosomes. Conversely, there are others [90, 122, 165] who claim that the character is governed by a few major genes without, however, excluding the action of many other genes controlling the intensity.

Gene effects seem to be mainly additive [214, 14, 303], while the non-additive ones would seem to be of minor importance though statistically significant. Among the latter, a prevalence of the effects due to partial dominance was reported [51, 177, 301, 42, 41, 131, 90] although effects of superdominance [100] and complete absence of dominance [297, 94, 142] have also been found.

Heretability estimates also offer material for contrasting opinions: h^2 estimates range from 0.15–0.26 [300] to about 0.90 [142]; intermediate values are, however, very common [51, 94, 297, 1, 180, 83, 82, 55]. Sometimes contradictory values have been reported, such as 0.37–0.79 [177] and 0.66–0.79 and 0.89 [142] in the same paper.

Actually the way in which protein content is expressed, the experimental design being used, the relationship between the analyzed materials, the parental materials used in crosses from which the analyzed lines are derived, and so on, all these have a strong influence upon estimate and fully justify the discrepancies mentioned above.

Concerning the first point, it should be noted that protein content is usually expressed as the amount of protein in 100 g of seed. Thus, this reflects not just the effects of the environment upon the synthesis and storage of protein but also upon other substances among which starch is of primary importance. Indeed, the amount of starch strongly conditions the percentage value to be attributed to one and the same amount of protein. In their study of the effect of grain size and grain filling upon the protein content of wheat in the world collection, Johnson *et al.* [127] find that the size of the caryopsis does not affect protein content whereas grain filling does so remarkably. For these reasons, Favret *et al.* [68] and Jain *et al.* [105] suggest that protein content should be expressed as the amount in one seed, not as a percentage of seed weight. In fact, they find [105] h^2 levels of 0.69 when these values are expressed as the amount of protein in one seed and only 0.22 when they are expressed as the amount of protein in 100 g of seed.

Concerning the experimental design, Bhullar *et al.* [14] note that the more complex it is and the more generations are considered in the analysis, the lower are the estimates of additive effects and the higher those of dominance, very likely because there is a possibility to detect the effects of linkage. Moreover, in analyzing 21 generations from a cross between Kaliansona and PI 170926 cross, they find h^2 values of 0.79 in estimates from F_2's and of 0.97 when they also include backcrosses, while Kaul and Susulski [142] find estimates of 0.66 in F_3/F_2 regressions and 0.89 in F_4/F_3 and F_5/F_4 regressions. It is likely that these increases are due to the lines responding more homogeneously to the environment as their inbreeding level increases. Variability for g $\times$ e interactions for protein content was in fact found to be less for materials with similar genetic background [283].

Lastly, concerning the choice of parental material used to derive lines for the analysis, much of the research has been done by using limited sets of crosses, often between related parents, mostly in order to study specific situations. Only few studies were carried out with the aim to obtain an overall view of the inheritance problem.

From this point of view, the most widely studied material is no doubt Atlas 66 which, together with Atlas 50, was identified for its high protein content in a series of segregating lines of the Redhart/No.11.2/Frondoso cross which were selected for their rust resistance [208]. In addition, to clarify the character's inheritance in Atlas 66, these studies also provide an example of peculiar aspects that can be encountered when investigating the inheritance of a given character. Therefore, it may be appropriate to report the salient points in some detail.

High protein in Atlas 66 is due to two major genes which express dominance for low protein content [42] and were inherited from Brazilian Frondoso. One of them is located on 5D chromosome [217] and is closely linked to one gene for rust resistance [130]. Being extensively used in crosses, Atlas 66 has shown to be able to supply productive lines like the other commercial varieties but with about 2% more proteins [126, 128, 209].

Some of these lines, e.g. those selected from the cross Atlas 66/Comanche/Lancer have already been under normal cultivation for a long time in several states of the USA under the name 'Lancota' [131]. Lancota does not have a definite protein level expressed as a percentage of weight but has been tested at various locations for several years with different amounts of nitrogen fertilizer, and has proved to maintain its superiority over the other types in all environments [131].

The action of the two genes seems to occur *via* the combination of a higher and more long-lasting NO_3 reductase activity with more active nitrogen uptake from the soil and a better ability to translocate the assimilates to the grain [131].

It is interesting to note that the higher protein content of Atlas 66 and its derivatives, usually measured in whole seed, is transferred to its white flour which indicates that the 'high protein content' effect lies in the endospermic portion of the caryopsis [128].

Crosses between Atlas 66 and Nap Hal – another important gene source for high protein content [317] have made it possible to discover that the genes from Atlas 66 are different from those of Nap Hal and act additively upon one another, which is a further contribution to their protein superiority in terms of protein content [131]. It should also be noted that the high protein content in Nap Hal can be observed both in the endosperm and in fractions other than endospermic [132].

Crosses between these two lines and different varieties also indicate that in addition to these major genes, there is a whole series of genes which could be accumulated through current breeding procedures. Thus, for example, the cross Favorit/5/Cirpiz/Jang Kwang/4/Atlas 66/Comanche/3/Velvet has been used for selection of lines having the same yield as Lancota and a protein level higher than

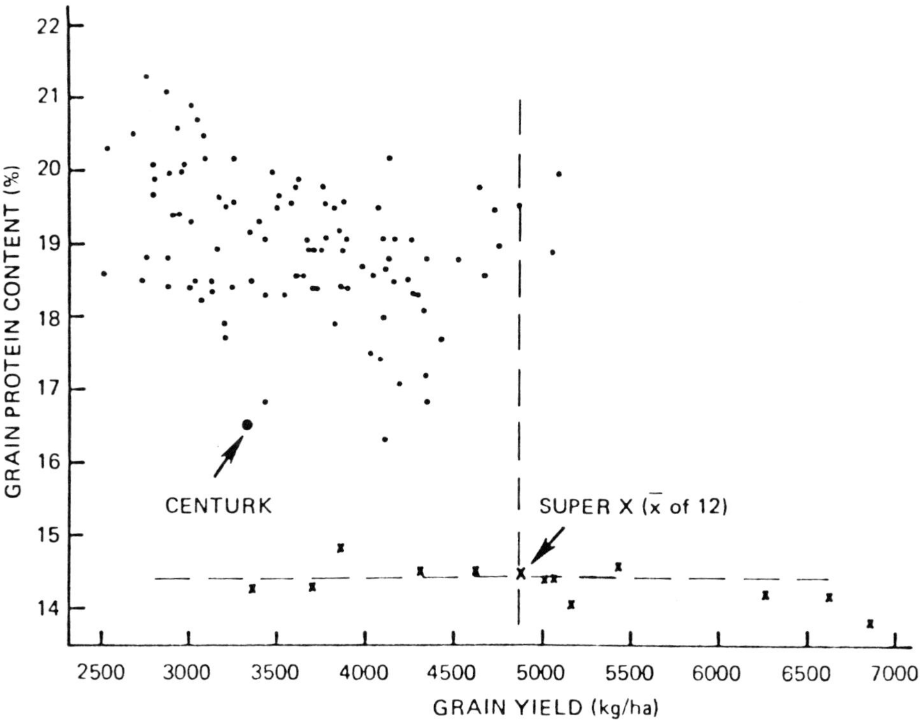

Figure 6. Relationship of grain yield and grain protein content among lines selected by L. Klepper for high protein in 19 CIMMYT spring wheat populations. Data are from a replicated yield nursery grown at Yuma, Arizona, in 1978 (123).

19% while under the same conditions the protein content in normal varieties was 15.9% and in Lancota 17.5% [123].

Other genes for high protein content, different from those of Atlas 66 and Nap Hal, and showing additive effects with such varieties, were identified in April Bearded [124] which has the same yield as Justin but 5% more protein [121]. Further evidence of the presence of minor genes in many wheats comes from the experience of Klepper [157] at CIMMYT who was able to obtain, by means of selection cycles, high protein lines from parents with normal protein contents (Figure 6).

A useful source of genes for high protein content is also provided by wild species related to wheat. This is shown by the results obtained at South Dakota State University where three high protein lines (SD 69103, Hand and Flex) were released in 1976, all derived from the cross Hume°2//Nehed°4/Agatha and which derive their property from genes of *Agropyron elongatum* via the parent Agatha [319]. An equally interesting result is obtained by the Seed Research Associates, Scott City, Kansas, who released line 'Plainsman v' which owes its high protein content to genes from *Aegilops* [12].

Another important source, which may be liable to interesting implications, is provided by *T. timopheevi* whose cytoplasm is known to yield a higher protein content than the ordinary varieties [272]. Even the Cmst A lines have a consistently higher protein content than the maintenance lines which obviously have a different cytoplasm.

Role of Cytoplasm. The above-mentioned examples, obviously, recall the role of nuclear and cytoplasmic systems in protein synthesis. The comprehension of this mechanism may, in fact, represent an important point in designing breeding strategies for higher protein content and/or better protein composition. The existence of nuclear cytoplasmic interactions in wheat has been indicated by many scientists but only few of them were able to document the results in an indisputable manner and even among them few were able to give good explanations. Kihara and Tsunewaki [153], for example, have reported that nuclear genes of the bread wheat 'Salmon' perform better in the *Ae. caudata* cytoplasm than in its own. Similar results were obtained by Bush and Maan [32] in alloplasmic lines of *T. aestivum*, having *T. macha*, *T. dicoccoides* and *Ae. squarrosa* cytoplasms but the reasons of this behaviour were not definitely identified.

Evidence from different species gives rise to different hypotheses. The first hypothesis concerns, undoubtedly, the interaction between products of ribosomes controlled by chloroplast DNA and products of ribosomes controlled by nuclear DNA in synthetizing RuDP-Case [321]. It is known, in fact, that RuDP-Case proteins coded for by chloroplast DNA are similar in *Ae. speltoides*, *T. dicoccum*, *T. turgidum*, *T. timopheevi* and *T. aestivum* and that they have an electrophoretic mobility different from those of *T. boeoticum*, *T. monococcum*, *T. urartu* and *Ae. squarrosa* [45].

Another hypothesis concerns the fact that nuclear DNA also codes for two of the numerous ribosome proteins composing the 50S subunit of chloroplast ribosome [24] and for proteins to which chlorophyll pigments are attached.

Yet another hypothesis concerns physical interaction between chloroplast and cytoplasm via *mitocondria*, the control via chloroplast DNA of nitrite reductase and/or ferrodoxine that, as mentioned, is involved in nitrite reductase and RuDP-Case activity.

Nitrate Reductase Activity (NRA). Johnson *et al.* [131] mention that the wheat variety Lancota owes its high protein content to the combination of higher and longer-lasting NO_3 reductase activity with more active nitrogen uptake from the soil and a better ability to translocate the assimilates to the grain. Experimental results suggest that the three aspects are independent [49] and that varieties may differ in all of them [125, 131, 128, 122, 62, 63, 95].

Very few attempts have so far been made to relate the three factors to the genetic ability to accumulate nitrogen under field conditions [52]. Preliminary evidence suggests that NR is possibly the trait which limits the nitrogen assimilation and could be chosen as a physiological criterion for selection [157].

Since electrophoretic analysis has indicated the existence in wheat of only one isozyme for NR with no difference for band migrations in NR extracts from roots or shoots [307]. Possibilities for selection depend on genetic mechanisms involved in enzyme synthesis and degradation, endogenous inhibitors or activators and the Km of the enzyme for the substrate.

Genetic studies have shown that in wheat NRA is controlled by a polygenic system with high heritability values [62] and that shoots appear to be the place where maximum NRA occurs [215] and the onset of tillering is the best period for sampling for NRA [64, 95, 49]. NRA appears to be controlled by genes located at several chromosomes. Studies performed using Chinese Spring aneuploids showed that inhibitors or regulators of NRA are located on chromosomes 2A, 7A and 7B [62, 63]. Further evidence on the involvement of group 2 chromosomes in the control of regulation has been provided by Jagannath and Bhatia [104] and by Warner and Konzak [315].

Studies performed using genotypes with similar translocation efficiency and the same NR isozymatic pattern indicated that better N responsive lines for grain yield and total protein differed from the others in NRA [254] and that differences were controlled by major dominant genes identified as Nra/nra [77]. However, this finding does not exclude that several genes may control the amount of NRA.

Variation for Lysine Content.

Among Species. Research concerning amino acid in seed proteins is no doubt more limited than studies on protein content. Except for a very small number which is concerned with the whole amino acid spectrum, the majority of these studies is restricted to the determination of lysine which is that amino acid limiting the nutritional value of materials that are rich in protein.

As is the case with proteins, also the first systematic review of lysine content in different wheat species is the work of Lawrence *et al.* [173]. Their values (Table 5) indicate that the samples having the highest lysine content belonged to the species *T. pyramidale*, *T. sphaerococcum*, *T. persicum* and *T. timopheevi* and some of the *T. durums*, while those lowest in lysine belong to the species *T. dicoccoides*. The high values for *T. pyramidale* and *T. persicum* are confirmed by the subsequent experiments made by Villegas *et al.* [310] who found intermediate levels in *T. sphaerococcum* and low values in *T. dicoccum*.

When values are expressed as the amount of lysine in wheat, rather than as percentage of the protein, the highest values are those of *T. boeoticum* followed by *T. timopheevi* whereas the lowest values are obtained in *T. vavilovi*. By cultivating a large number of these samples at another location, it has been possible to confirm

Table 5. Lysine content (% of protein) in different species of *Triticum* grown at different locations.

	Pullman(173)	Aberdeen (173)	Fargo(310)
T. boeoticum	--	3.05 - 3.50	2.39 - 2.65
T. monococcum	3.00	2.59 - 2.96	3.05
T. dicoccoides	2.21	2.62 - 3.25	2.30 - 2.81
T. dicoccum	2.91	2.09 - 2.85	2.63
T. turgidum	3.27	2.89 - 3.35	2.75
T. polonicum	2.94	2.80 - 3.00	2.48
T. carthlicum	3.24-3.31	3.21 - 3.99	
T. timopheevi	3.24	2.86 - 3.04	2.63 - 2.82
T. orientale	2.67	2.67 - 3.08	2.57 - 2.68
T. ispahanicum	--	--	--
T. spelta	2.83	2.87 - 3.52	
T. macha	2.76	2.77 - 3.63	
T. zukowski	--	--	2.48 - 2.87
T. durum	2.70-3.30	--	--
T. vavilovii	--	3.30	--
T. sphaerococcum	3.04-3.48	2.66 - 2.79	--
T. piramidale	3.09-3.50	3.28 - 3.59	2.34 - 2.54
T. paleocolchicum	--	--	2.52
T. aestivum	2.46-3.84	--	--

(1) Lawrence *et al.* 1958.

(2) Villegas *et al.* 1968.

another finding of Lawrence *et al.* [173], namely, the limited effect of the environment upon lysine levels.

A confirmation for the low levels of *T. dicoccoides* comes from the experience of Avivi [9] with material collected in natural environments and cultivated in the green house, and of Lafiandra *et al.* [171] who report values for the entire amino acid spectrum in a series of wheat samples belonging to di- tetra- and hexaploid species and to samples from three species of *Aegilops* (Table 6).

Table 6. Amino acid composition in different species of *Triticum* (171)

Amino acid	*T.monococcum*	*T.durum*	*T.aestivum*	*T.speltoides*	*T.Tauschii*	*T.longissimum*	*T.dicoccoides*
Lysine	2.81 ab	3.01 a	2.84 ab	2.54 b	2.78 ab	2.81 ab	2.54 b
Histidine	2.19 c	2.38 abc	2.33 bc	2.88 abc	2.63 a	2.45 ab	2.40 abc
Arginine	4.52 a	4.96 a	4.88 a	4.38 b	4.70 a	4.74 a	4.46 a
Aspartic acid	5.35	5.37	5.25	4.94	5.05	5.44	5.09
Threonine	2.82 b	3.06 ab	2.98 ab	3.12 ab	3.16 a	3.16 ab	2.93 ab
Serine	4.33	4.78 e	4.80 de	5.32 abc	5.37 a	5.36 ab	5.16 abcd
Glutamic acid	28.50 de	28.64 de	28.94 d	37.80 a	36.20 ab	34.66 bc	34.41 bc
Proline	9.31 d	9.71 cd	9.77 cd	11.65 ab	12.53 a	11.29 b	10.90 bc
Glycine	3.49 c	3.94 bc	4.08 ab	3.73 bc	4.37 a	3.98 b	3.75 bc
Alanine	3.37 b	3.72 a	3.63 ab	3.36	3.63 ab	3.56 ab	3.34 b
Valine	4.27 ab	4.16 ab	4.08 b	4.26 ab	4.75 a	4.15 b	3.95 b
Isoleucine	3.65	3.40	3.27	3.39	3.57	3.34	3.22
Leucine	6.43	7.06 abcd	6.97 bd	7.48 a	7.46 ab	7.29 abcd	7.38 ab
Tyrosine	2.84	2.84	2.96	3.23	3.08	2.96	3.19
Phenylalanine	4.63 bc	4.51 c	4.46 c	5.39 a	5.15 ab	5.24 ab	5.03
Protein content	14.52	13.29	13.92	21.35	17.00	21.30	19.55

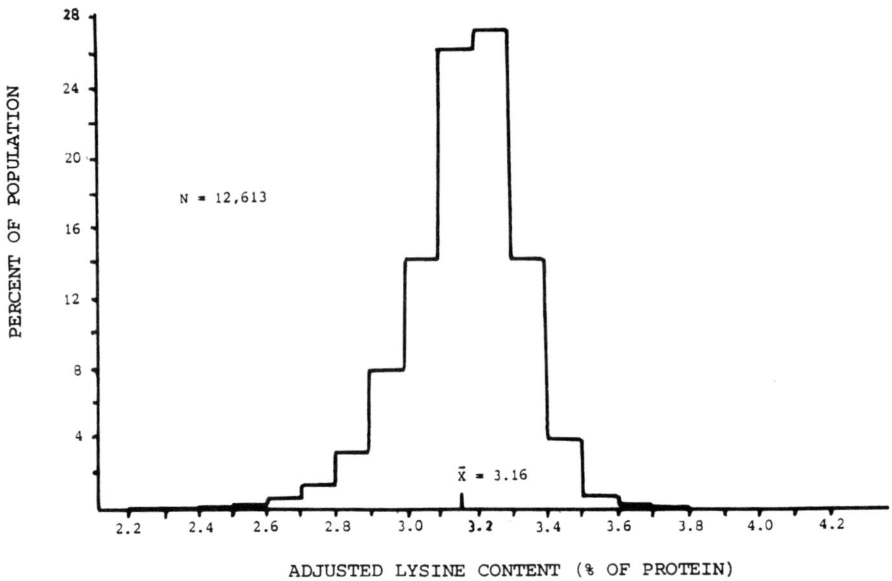

Figure 7. Frequency distribution of lysine adjusted to 12.97 percent protein among 12,613 wheats in the USDA World Collection (311).

Within Species. Lysine variation within the *durum* and the bread wheat species is similar to that among species. Analysis of over 12,000 bread wheat samples indicates that lysine content ranges between 2.25 and 4.26 (Figure 7), the mean being 3.16 and the standard deviation 0.49 when protein content is expressed as protein percentage, but varies between 0.25 and 0.66 with a mean equal to 0.40 and standard deviation 0.049 when the content is expressed as a percentage of the sample [311].

The analysis of about 3400 *durum* wheat samples [323] also indicates that lysine levels range between 2.43 and 4.29 (Figure 8) with a mean 3.15 and standard deviation 0.27 when expressed as a percentage of the proteins, and between 0.26 and 0.60% with mean 0.39 and standard deviation 0.06 when expressed as a percentage of the dry matter content.

Relationship between lysine and proteins. Results obtained both within and among species indicate that protein content and lysine per unit protein in the various wheats are inversely correlated. Similarly, in common wheats with less than 13.5% protein, Lawrence *et al.* [173] had observed a negative correlation between lysine (expressed as percentage of protein content) and protein percentage while no correlation was observed in wheats with protein levels above the said value. This correlation was further confirmed by an analysis (Figure 9) on more than 12,000 bread wheat samples by Vogel *et al.* [311] who found that the limiting value for such a correlation was to be fixed at 15%. Similar relationships were found by Simmonds

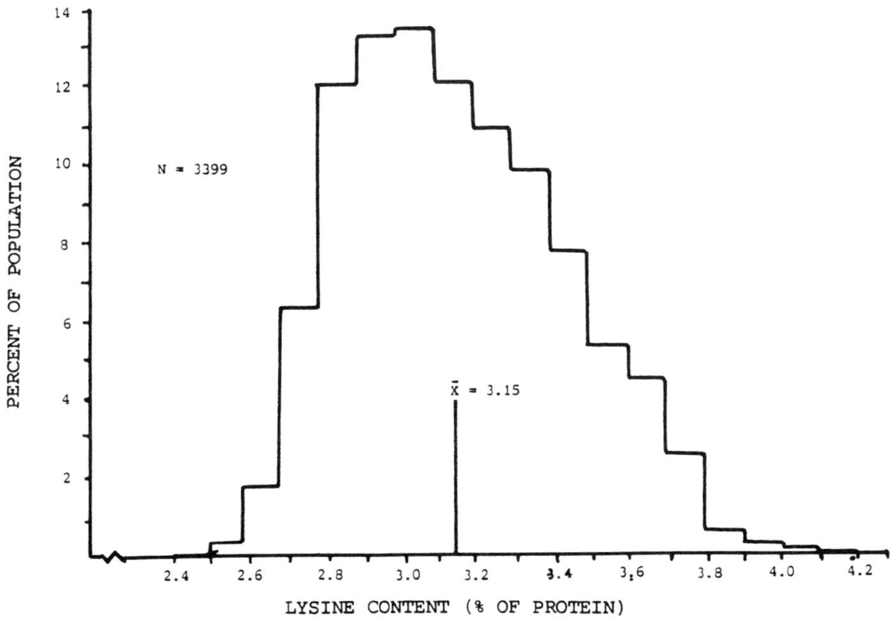

Figure 8. Frequency distribution for lysine expressed as percent of whole grain protein among 3399 durum wheats in the USDA World Collection (323).

[286] and by McDermoth and Pace [200] in commercial bread wheat varieties and by Villegas *et al.* [310] and by Worede [323] in *durums.*

However, it should be noted that when lysine is expressed as a percentage of seed weight, the correlation becomes positive. The results of analysis on the bread and *durum* wheat collection give correlation values of 0.90 and 0.92, respectively [311, 323]. These values are especially important as they show that high protein wheats can yield more lysine per seed unit weight that those that are high in lysine (as protein percentage) but low in proteins.

Experiments conducted on high protein wheats and on high lysine wheats selected from the world collection indicated that the former contained only 2.75% lysine per protein but yielded more lysine, methionine and threonine per seed unit weight in wheats having high lysine content. Conversely, the low protein content of wheats high in lysine did not produce any increase in the three amino acids per seed weight [129]. Subsequent evaluations made at different locations of high lysine wheat show that contrary to the findings of Lawrence *et al.* [173], most of the variation was induced by the environment and that the genetic component in the variation only accounted for 0.5% which is an indication that the lysine content can be raised by about 17%. This low variability for lysine is ascribed [129] to the presence of more than one genome in cultivated wheats. In other words, it might be

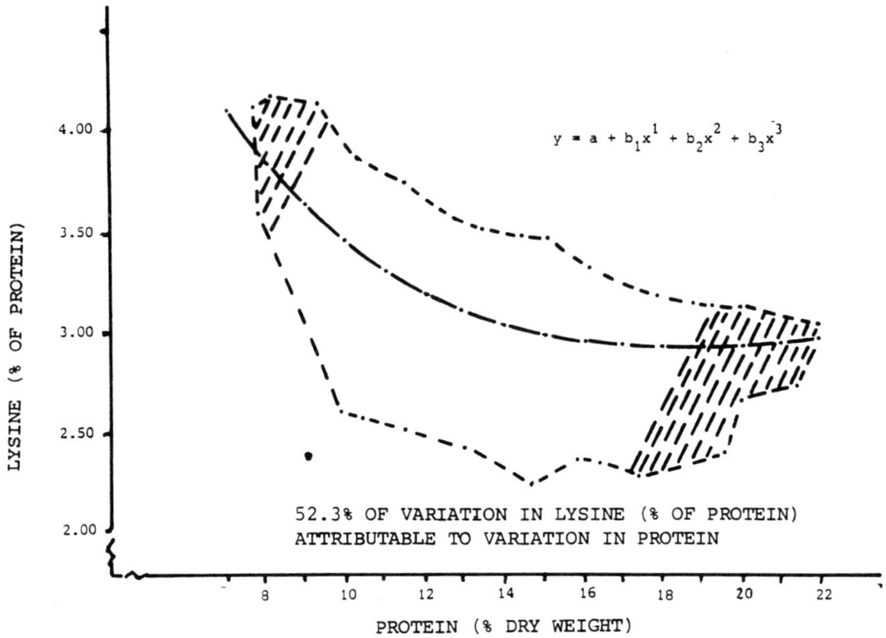

Figure 9. Curvilinear regression (third degree polynomial model) of lysine percent of protein on protein and the range of dispersion of lysine values about the regression line computed from the analysis of 12,613 common wheats from the USDA World Collection (311).

possible that one gene (with ample effect upon lysine content), present in one genome, be disguised by genes in another genome even though – as pointed out again by Johnson *et al.* [132, 123, 196] later on – the range of variation for lysine content which exists in the di- and tetraploid wheats of the world collection is not more than that of the hexaploid species.

As described for protein content, also for lysine, the environmental effects seem to act primarily through kernel morphology. Analyses made in different environments on samples from the world collection differing widely in terms of kernel size, protein content and lysine indicate that kernel size does not affect lysine content which is instead highly correlated with shrivelling ($r = 0.75$). Clearly, this relationship can be ascribed to variability for lysine content which exists in the different parts of the caryopsis [189].

Generally, the endosperm has a lower protein and lysine content than the embryo and the epidermis. Studies for investigating the nature of the proteins which are responsible for variations in amino acid composition and on the protein/lysine relationship have thrown more light on these aspects. Ratios among the four solubility fractions making up seed proteins are not fixed but vary from wheat to wheat. Pence *et al.* [243] analyzed flours from 32 wheats and found that, for example, the albumin + globulin fractions had values between 13 and 22% of total protein and that there was a direct correlation between the amount of these proteins and total protein in the kernel. However, the correlation was negative when the two fractions were expressed as a percentage of protein content. Similar results were obtained by Ulmer and Mattern [306].

Furthermore, Simmonds [286] and Mattern *et al.* [197, 198] found that albumins and globulins have a higher lysine content than glutenins and gliadins, which contain as much as 50% of the proteins in the kernel and are relatively low in lysine. Lysine variation and the lysine (%)/protein relationships could then be attributed to the proportional variation of the two protein fractions [200, 286].

Increase in lysine as the protein concentration diminishes could, in other words, be regarded as a reflexion of the increased albumin + globulin/gluten ratio. Wheats with low protein content contain higher amount of soluble protein and therefore their lysine content per unit protein is higher. This seems to explain why lysine increases with shrivelling; moreover, since with protein levels higher than 15% – as pointed out earlier – there is no relationship between protein and lysine content (protein percentage) it would seem that at such levels the albumin + globulins/gluten ratio also remains stable [311]. Consequently, in selecting for lysine content, the emphasis should be on increasing the two soluble fractions which should be allowed to vary considerably with the others remaining at constant levels.

Selection Possibilities. The genetic variation, however small, for lysine content observed in the world collections seems adequate enough to overcome the drop in lysine which occurs at increasing protein levels. Rather, if genetic variation for lysine content (0.5%) is associated with genetic variation for protein levels (5%) it would seem possible to develop high-yielding varieties with high protein and with lysine levels equal to those of the low protein varieties. This possibility is supported by the results obtained by the group working at ARS-Nebraska. Analyses made on the world collection enabled the group to identify CI 13449, a soft wheat cultivated for pastry flour and high in lysine released at Pullman, Washington and also to find out that Nap Hal was higher in lysine (0.3–0.4) than the other varieties. Crosses between these wheats have produced lines which, despite the moderate transgressive segregation shown in F_2 [131], had been able to store, after some generations, a considerable number of genes for lysine content without lowering either protein content or productivity. In F_6 (Figure 10) lines were present with 15% protein and 3.4% lysine, as an average for over 16 cultivation sites and with yields equal to those of the best-yielding parent [132].

The presence of genes with additive effects was confirmed by research by Diehl *et al.* [56] in crosses of April Bearded with Atlas 66, two varieties which have normal lysine contents. These scientists also found the presence of genes with epistatic effects. These facts suggest that it would be appropriate – as already implied in the results obtained in different segregating generations by Johnson *et al.* [132] – to start selecting for high lysine only after F_3, that is to say, when the frequency of

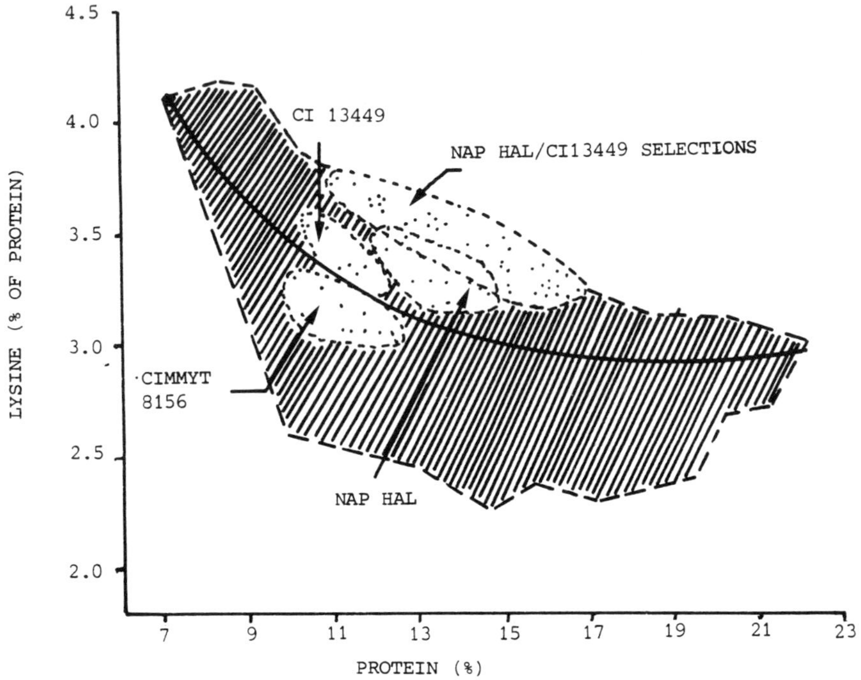

Figure 10. Breeding advances from Nap Hal/CI 13449 cross in relation to World Collection lysine-protein regression (123).

favourable recessive gene combinations increases and epistatic combinations become possible.

GENETIC CONTROL OF ENDOSPERMIC PROTEINS.

Research on protein and lysine content in wheat indicates that these two characters are controlled by a great number of genes which can be accumulated in a single line by breeding procedures. At the same time, research started by Simmonds [286] and carried on by several other workers [67, 34] on amino acid content of the different protein fractions has made it possible to explain the lysine-protein relationship and to suggest new possible methods for selection.

One more step forward has been made by the use of electrophoretic techniques [136, 325] which have shown, as already assumed by Pence *et al.* [244], that the solubility fractions were heterogeneous and were made up of a considerable number of protein components. Each component resulted as the primary product of a structural gene [292, 141, 203, 162] and so served as a marker for one gene, one chromosome and/or one entire genome. This evidence has promoted the use of protein components in studies aimed at sorting out the complexities of the solubility fractions, better comprehension of the relationship between protein fractions and components and their pasta and bread-making quality, understanding the biochemical development and genetic control of proteins, solving genetic, taxonomic and breeding problems, etc. Thus, it is obvious that the comprehension of mechanisms of regulation, inheritance and genetic control is extremely important from practical as well as from theoretical points of view.

Albumins and Globulins The inheritance of albumins, globulins and other endospermic proteins, with the exception of gliadins and glutenins is yet obscure [81]. Differences in extraction and fractionation procedures and the lack of characterization of single components create marked discrepancies in experimental results and give unprecise conclusions. Table 7 shows a tentative list of results of experiments carried out to study the chromosomal location of genes which control different protein fractions in this group.

Gliadins Gliadins, undoubtedly, are proteins which have so far received the greatest attention from scientists. This is due to their technological significance – since, together with the glutenins, they are responsible for the viscoelastic properties of the gluten – and their impact on the nutritional value. They are in fact poor in essential amino acids, and so responsible for the low nutritional value of total wheat proteins. There are obviously many other reasons which justify the fact that research on these proteins has become increasingly attractive and unfolds a wealth of information. A peculiar aspect of gliadins is their high level of heterogeneity which is strictly determined by the genotype – the genotype produces an electrophoretic pattern of the gliadins which differs from that of any other genotype with regard to the number, mobility and intensity of the components – and wholly independent of cultivation environment [178, 326, 336]. Thanks to this fact, they are widely used for variety identification [65, 330, 6], for detecting species and genera in food preparations [327], for checking off-types in pure seed production [3], for building up pedigree [331] and more generally in a wide range of breeding studies.

Different methods have been proposed for expressing the gliadin electrophoretic profile by means of formulae [162, 6, 35, 135, 269, 328]. Current activities of the study group No. 6 of the International Association for Cereal Chemistry are aimed at establishing a uniform procedure for electrophoretic identification. Achievement of this goal would open the way for a world-wide catalogue of electrophoretograms to be compiled for wheat varieties.

An interesting research subject is the chromosomal location of genes coding for gliadin components. As far as we know, Boyd and Lee [25] were the first scientists to deal with this subject. By using ditelocentric lines of Chinese Spring they were able to prove that chromosome 1D was certainly involved in the synthesis of some slow-

Table 7. Chromosomal location of genes that control globulins, albumins, and low molecular weight hydrophobic proteins (81).

Protein class	Chromosomes		
Purothionins (apoprotein	1AL	1BL[a]	1DL[a]
(digalactosyl diglyceride)	5AL	5B	5DS
Globulins (fastest 'doublet')	1AL	1BL[a]	1DL[a]
Albumins	5AL	5BL	5DL
Albumins (PCS)	-	-	3D
(Mb 0.19)	-	-	4D
Albumins	-	1B [a]	-
(soluble in Tris buffer pH 8.7)	3AS	3BS	3D
	4A [a]	4B [a]	-
	6A [a]	-	-
Albumins	3A	3B	3D
(extract. 0.8M salt; 3M salt)			
CM-proteins (CM1, CM2)	-	7B	7D
(CM3)	4A [a]	-	-
70% EtOH extract	-	3B	3D [a]
(non-gliadins)	4A [a]	4B [a]	4D [a]
	-	-	7D
70% EtOH extract (albumins)	-	3BS	3D
	4A [a]	-	4D
	-	-	5D
	-	6B	-
(CM proteins)	4A	-	?D
	-	7B	7D

a: absence of chromosome only decreases band or spot extra doses enhance.

moving components even though they could not exclude effects of other chromosomes. Indeed, the electrophoretic pattern of DT1D showed that 2 components had altogether disappeared whereas others only retained differences in intensity. Boyd and Lee [25] excluded any action by the entire D genome – no further change in components was observed when this genome was removed from a 'reconstructed tetraploid' – but could not exclude the action of genes at A and B genomes.

Subsequently, Shepherd [279] associated some components of the electrophoretic pattern to the chromosomes of the homologous groups 1 and 6 in a stock of nulli-tetra. Together with the complete disappearance of protein components associated with substitution of chromosomes of the homologous groups 1 and 6, he observed that a number of components could be controlled by genes at more than one chromosome. Results obtained by the above-mentioned authors are today explained by the fact that certain bands which seem to be homogeneous under one-dimensional electrophoretic technique are actually formed by different proteins having similar mobility which can only be separated by two-dimensional electrophoresis. The different components that resulted were obviously coded by different genes. Using substitution lines of Tatcher in Chinese Spring, Solari and Favret [292] noted changes associated to chromosomes 1A, 7D (slow-moving), 1B, 2B (intermediate migration) and 2D and 6D (faster migration). Some points remained, however, to be clarified and some of the components were not associated to any chromosome.

In the meantime, ultra-sensitive methods were introduced, such as the two-dimensional technique, by which gel electrofocusing in the first dimension is followed by electrophoretic separation as a second dimension. Thus, Wrigley [326] was able to give even more prominence to the heterogeneity present in wheat prolamins. By using the two-dimensional technique on a nulli-tetra stock of Chinese Spring, Wrigley and Shepherd [329] were able to assign almost all gliadin components to specific chromosomes and confirmed that the genes coding for these proteins are located on the short arms of the homoeologous chromosomes 1 and 6. More specifically, chromosomes 1A, 1B and 1D essentially control ω and γ components whereas α and ß components are controlled by chromosomes 6A, 6B and 6D. The control of chromosomes from groups 1 and 6 on gliadin components was confirmed for varieties other than Chinese Spring, thus supporting the idea that synthesis and control of these proteins are governed by a single common model in different materials.

By using substitution lines of Cheyenne in Chinese Spring, Kasarda *et al.* [141] confirmed these findings and assigned to specific chromosomes in the groups 1 and 6 as many as 13 out of the 25 components that the proteins of Cheyenne exhibit on electrophoresis in aluminium lactate 8.5 mM pH 3.2. In particular, Kasarda *et al.* [141] showed that the synthesis of A gliadins (a fraction of the α gliadins with unusual aggregating properties which are present only in a few wheat varieties [139]) is controlled by genes on the short arm of chromosome 6A. More recently,

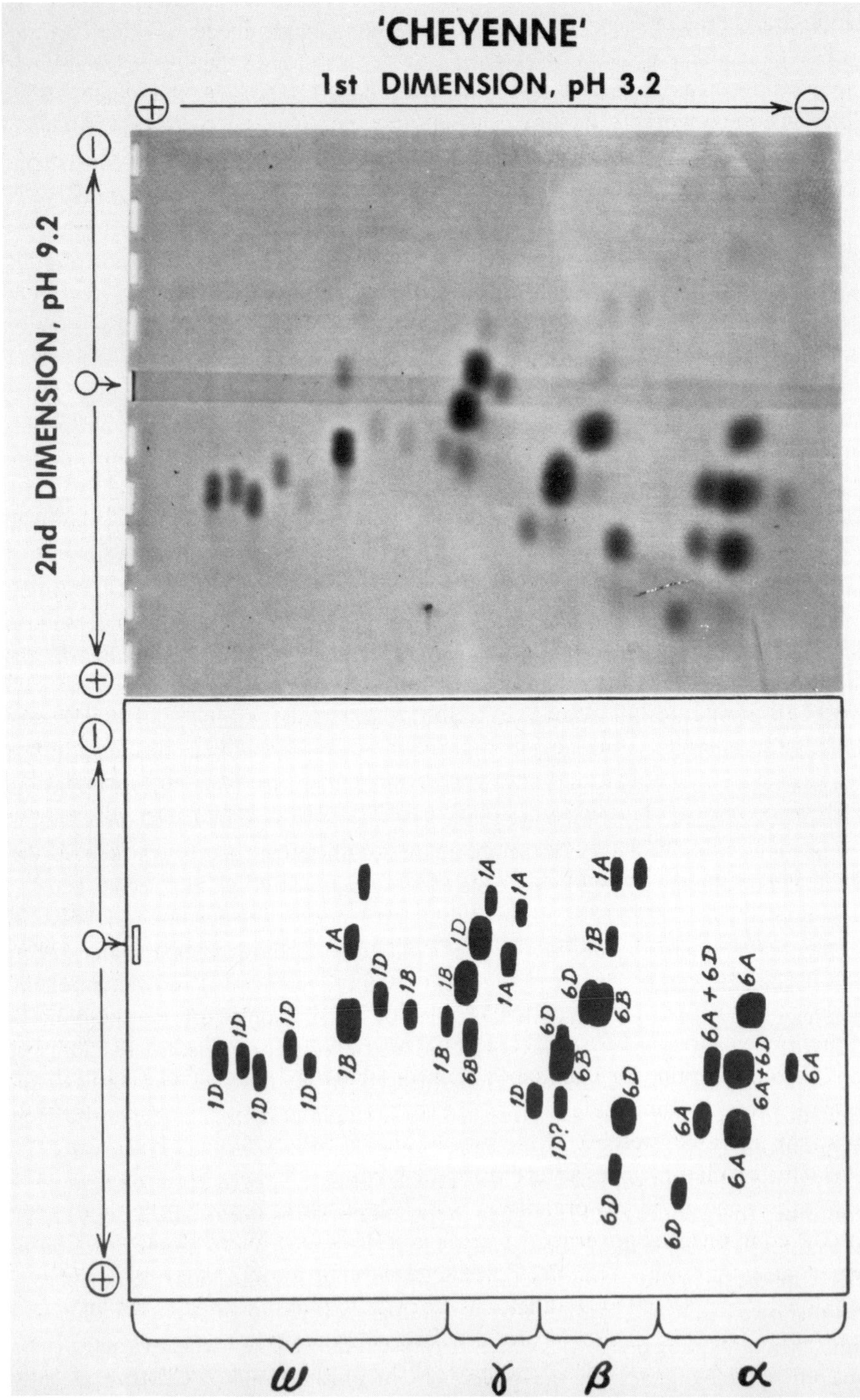

Figure 11. Chromosomal location of genes which control gliadin synthesis in Cheyenne.

Lafiandra (unpublished results), by using the two-dimensional technique developed by Mecham *et al.* [203], has completed the allocation of the gliadin components to the chromosomes of Cheyenne (Figure 11). Similar results were achieved by Sozinov and co-workers [294] by using entirely different materials and methods of genetic analysis. They in fact preferred the monosomic analysis to the toilsome and time-consuming development of intervarietal chromosome substitution lines. The analysis of gliadins of the F1's and F2's obtained by crossing the monosomic series of Chinese Spring to USSR varieties, enabled them to confirm that gliadin synthesis is controlled by genes on chromosomes 1 and 6 and to shape a concept of 'block of components' meaning a group of components inherited in Mendelian, linked form, considered as a single allele [294]. By using 'allelism tests', they were thus able to determine the genetic control of synthesis in several varieties without repeating the monosomic analysis. In order to utilize the various blocks as genetic markers, they started mapping these blocks and so were able to map the Gld1B block [264].

Thus, it seems to be a definite finding that the chromosomes of groups 1 and 6 and more particularly their short arms contain the genes responsible for gliadin control. Still, this does not exclude the fact that genes from other chromosomes may also be involved. In fact, first Shepherd [279] and then Solari and Favret [292] found that genes of the chromosomes in group 2 were also involved. Indeed, Shepherd [279] had observed that the N2DT2A stock would produce a phenotype which was similar to that of the 6D chromosome deficient stock, namely, N6DT6A and N6DT6B. Similar results were also obtained with tetra 2A. Waines [314] and Bietz *et al.* [20] also obtained similar evidences. Shepherd [279] assumed that chromosome 2A was liable to contain a gene whose presence in four doses inhibited gene activity in chromosome 6D. This hypothesis has recently received support from Brown and Flavell [28] who used the two-dimensional techniques and observed the dose-dependent regulatory effect of the chromosomes in group 2.

This particular aspect calls attention to the manner in which the genes coding for gliadin components are expressed and controlled. The comprehension of these mechanisms represents in fact an indispensable prerequisite for a correct genetic manipulation of specific components aimed at improving the quality and quantity of proteins. Solari and Favret [291] and Favret *et al.* [69] were the first to approach this subject. Adopting the Haldane [87] concept that the condition for a substance to be regarded as the immediate product of a gene is that the presence of one allele must entail the potential presence of the substance, and assuming the presence of silent alleles and regulatory genes according to the Jacob and Monod [103] idea to explain gene expression in micro-organisms, they suggested that the synthesis of each specific component is governed by genes at at least one locus. Results in F1 and segregant generations were thus explained assuming that the synthesis of one component is regulated by the allele dosage in the triploid endosperm so that single, double and triple doses should produce different phenotypes.

A few years later, Doekes [57] grouped the electrophoretic patterns of the gliadins into 6–7 sections each of which was inherited *en bloc*, and suggested that 6 genetic factors were respectively involved in the control of α, β_1, β_2, γ, ω_1, and ω_2 fractions. However, only subsequent studies [203, 295, 293] enabled to ascertain that gliadin components are coded for by single dominant genes with the exclusion of any possible implication of recessive genes. By means of electrophoretic analysis of gliadin profiles, performed on single seeds from three varieties, their F1's (including reciprocals) and F2's, Mecham *et al.* [203] showed that all the components of either parent were present in the patterns of the F1 seed, that the specific components of one parent were clearly visible in the offspring even when the genes were in single dose and that the intensity of single bands reflected the gene dosage, depending on whether the genes had been derived from either the mother or the father in agreement with the triploid nature of the endosperm. They used the two-dimensional analysis to find that bands showing segregation ratios different from 3:1 were composed of two or more components, each segregating independently. In addition, linkage analysis revealed the presence of co-dominant alleles and of closely linked genes coding for components in both coupling and repulsion situations; presence of bands in single segregation lines was such that the components seem to be coded for clusters of structural genes [203], the 'blocks' described by Sozinov and Poperelya [293]. The same results were obtained by Sozinov *et al.* [293, 294, 295] again on bread wheat and by Lafiandra (unpublished results) on *durum* wheat.

The information that is being obtained from this group of proteins seems to be truly inexhaustible. Analyses made on *durum* wheat populations have brought out the presence of a considerable amount of variability within the populations in terms of gliadin profiles. This variability has been shown to differ in different populations and in different regions (Damania, personal communication). In agreement with reports by Kasarda *et al.* [140], there would be hundreds of different gliadin components which can be identified by means of electrophoretic techniques. This seems due to the fact that the structure and function of these proteins are not crucially related, so that gene mutations for certain components might not be as dangerous as those in enzymes.

This wealth of forms in the same population could be taken as an indication of the variability existing in the population also for other characters that are of greater agronomic or qualitative value. The existence of clusters or blocks of genes or anyhow non-recombinant parts of chromosomes may play an important role in preserving genes and shaping characters. Thus, for instance, according to their nomenclature, the G1d1B3 block is to be regarded as having a potential value for the identification of stem rust resistance in wheat [293]. Similar possibilities have been revealed in barley for mildew resistance [110].

Such a wealth of components may partly be ascribed to the polyploid nature of cultivated wheats as confirmed by the fact that the components are controlled by genes at all three genomes [141, 329]. Polyploidy, however, may have played a

Table 8. N-terminal sequences of α, β, γ and ω -gliadins (44)

Wheats	Gliadins	Residues 1	2	3	4	5	6	7	8	9	10
Golden seal flour	α_2										
Scout 66	α_T										
Ponca	α_8 to α_{12}	Val–	Arg–	Val–	Pro–	Val–	Pro–	Glu–	Leu–	Glu–	Pro
Ponca	β_5										
Capelle	β_{22}										
Ponca	γ_1										
Ponca	γ_2	Asn–	Ile–	Gly–	Val–	Asp–	Pro–	Trp–	Gly–	Gln–	Val
		Pro		Gln		Val	Gln		Leu		
Ponca	γ_3	Asn–	Met–	Gly–	Val–	Asp–	Pro–	Trp–	Gly–	Gln–	Val
		Pro		Gln	Gln	Val	Gln	Gln			
Kolibri	γ	Val–	Ile–	Val–	Gln–	Val–	Arg–	Gln–	Leu–	Gln–	Val
Justin	ω_{1D}	Lys–	Glu–	Leu–	Gln–	Ser–	Pro–	Gln–	Gln–	Ser–	Phe
Justin	ω_{1B}	Ser–	Arg–	Leu–	Leu–	Ser–	Pro–	Arg–	Gly–	Lys–	Glu

secondary and rather recent role as suggested by the fact that diploid wheats also have rather complex electrophoretic patterns. Mecham *et al.* [203] believe that what is more important from this point of view is the process of gene duplication to form many identical copies of a gene, some of which may have experienced asymmetrical crossing-overs to form genes that code for different components. Subsequently, the modified genes may have undergone further events, such as reiteration or DNA amplification leading to the formation of new types, hence of new species with differences – occasionally remarkable – in the gliadin components [48]. Valuable information in this respect is provided by N-terminal amino acid sequences. On the basis of sequences reported so far, gliadins have been divided into three main sub-families, characterized, respectively, by the presence of α, β or ω type of N-terminal sequences [8, 138]. The α type is referred to the N-terminal sequence found in α, β and γ_1 gliadins; the γ type to the N-terminal sequence of γ_2, γ_3 and γ gliadins and the ω type to the sequences present in the ω gliadins (Table 8).

As can be seen, a strong homology exists between α and β gliadins; γ_2 and γ_3 gliadins show similarity to one another, while the homology between the two ω components is restricted to the residue 3(Leu), 5(Ser), and 6(Pro). Peptide maps of peptic digest of reduced and alkylated α, β, and γ_1 gliadin are similar, while maps of γ_3 show a different pattern. Also, triptic digest of two ω gliadins show similarities between themselves but are different from other gliadin components.

γ gliadins are coded by chromosomes of homologous group 1, while α gliadins are coded by group 6. The homology observed in the α and γ family suggested to Bietz *et al.* [19] that a group 6 gene coded a protein like α, β or γ_1, while a group 1 gene coded for protein similar to γ_2, γ_3; homology of γ_2 and γ_3 to the other gliadin proteins further suggests that genes at group 1 and 6 chromosomes arose through duplication and mutation of a single gene in a still remote ancestor.

Since N-terminal analysis of ω gliadins from variety Justin produced an amino acid sequence:

PRO-GLN-GLN-PRO-TYR-PRO-GLN-GLN-PRO-TYR

with a repeated five member sequence and a predominance of glutamine and proline, Kasarda [138] speculates that amplification or reiteration of a short fragment of DNA, carrying codons for few amino acids, produced the basis for an ancestral gene coding for α, β and γ gliadins. This hypothesis is also supported by the Thompson and Murray [304] concept that DNA sequence amplifications, rearrangements and deletions have been frequent events in plant evolution. In addition, it can be quoted that the mechanism is similar to that proposed for explaining the origin of genes coding for proteins having repetitive sequences [334, 22, 158]. So, reiteration of different sequences may have given rise to the ω gliadins; and the process may have occurred in relatively recent times since the limited number of amino acids in the sequence suggests that there was no time for further evolution.

Glutenins. As reported earlier in this paper, it has now become possible to separate with the SDS-PAGE the glutenin sub-units and to investigate their genetic control. By means of this technique it is also possible to observe the presence of a considerable amount of variation in high molecular weight sub-units from different varieties and to show that these sub-units are genetically controlled and independent of the cultivation environment [231].

Actually the first observations were rather contradictory. Contrary to Orth and Bushuk [231] evidences, Bietz *et al.* [20] found that, in fact, glutenin sub-unit patterns were rather similar in 80 varieties of hexaploid bread wheat whereas a great amount of variation was detected in tetraploid wheats. In more recent times, detailed analyses have been carried out, simultaneously but independently, by two research groups utilizing aneuploid and substitution lines, in addition to common varieties [174, 175, 240, 241, 242, 99]. Also, the results obtained by them so far are similar.

An examination of glutenin sub-unit banding patterns from 98 wheat cultivars by Lawrence and Shepherd [174] revealed that a total of 20 sub-units were present in the examined material, that with few exceptions each variety contained between 3 and 5 different sub-units, and that the banding patterns of different varieties helped in the identification of wheat cultivars. They also observed that some bands or band combinations never occur together in the same cultivar (i.e. they behave as alter-

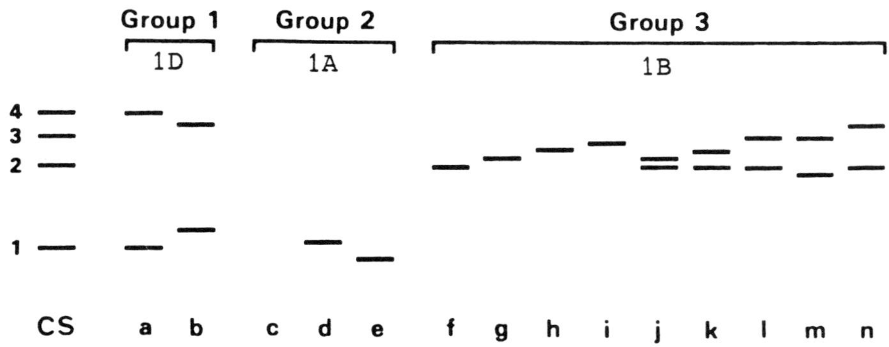

Figure 12. Diagram showing the three groups of glutenin subunit bands, or band combinations, where the bands, or band combinations, within each group are alternatives to each other. Chinese Spring bands 1–4 are shown for comparison. Amongst a world collection of 98 wheat cultivars the number of cultivars possessing each band, or band combination, was as follows – Group 1: a = 43, b = 54; Group 2: c = 24, d = 37, e = 37; Group 3: f–n = 23, g = 5, h = 13, i = 12, j = 74, k = 2, l = 29, m = 7 (174).

natives to each other) and that the bands could be clustered into three groups, so that each cultivar possesses one of the alternative forms in each group (Figure 12): two pairs of bands occurring as alternatives in group 1, three patterns, one of which with a null form, in group 2 and nine banding patterns in group 3.

These findings and results from analyses carried out using Chinese Spring aneuploids and substitution lines indicated that alternative bands within each group are controlled by genes of the same chromosome and more precisely that bands in group 1 are controlled by genes on the long arm of chromosome 1D, those in the second group by gene(s) on the long arm of chromosome 1A and those in the third group by genes on the long arm of chromosome 1B [174, 175]. These scientists also obtained evidences that bands in each of the groups 1 and 3 are controlled by genes at two loci, that these are closely linked so that any recombination could be detected and that bands behaving as alternatives are controlled by alternative forms of the same gene (allele).

Findings of Payne *et al.* [240, 241, 242] are similar except for minor differences probably due to the different materials analyzed. For instance, the number of subunit patterns controlled by chromosome 1B was equal, although a few were different and some were faintly stained; Payne *et al.* [241] found four banding patterns for sub-units controlled by chromosome 1D instead of two, etc. Both groups confirmed the Bietz *et al.* [20] finding that Nap-Hal does not possess any pattern in group 1 (1D), although they were unable to explain whether this behaviour is due to a deletion or to an unexpressed gene.

Deletion was also one of the mechanisms claimed by Lawrence and Shepherd [174] and by Payne *et al.* [241] to explain the presence of a null phenotype in the group 2(1A). Other possible mechanisms were that genes were repressed or that unequal crossing-over could have caused the loss of part of the gene, with an alteration of an initiator sequence or insertion of a terminator sequence causing the production of a low MW protein which is not detected in the PAGE system. Payne *et al.* [241], however, were able to exclude one of the Lawrence and Shepherd [174] hypotheses that the two bands are controlled by genes at separate but closely linked loci, each of which possesses a null allele.

Both research groups found that most of the variation was due to sub-units coded by genes at chromosome 1B, while that due to sub-units coded by genes at 1A was very low. The large variation controlled by 1B gives support to the hypothesis of polyphiletic origin of the B genome, as it will be presented later on, while analysis of various diploid species having AA genome revealed that most of them, with the exception of *T. urartu*, actually contained two high MW sub-units: one with mobility similar to either one of those controlled by 1A chromosome in hexaploids and the other, weaker, having mobility similar to the 1By sub-units [241]. These findings suggested that sub-units at the three different groups could be controlled by homoeoallelic genes at the three genomes, i.e. that chromosomes 1A, 1B and 1D derive from a common ancestral chromosome, and glutenin sub-units, controlled by genes at each of them, are derived from the same ancestral protein species. Interesting also is the hypothesis about the origin of different sub-units controlled by alleles at the same locus. On the basis of Dunker and Rueckert [59] and Weber and Osborn [318] indication that a linear relationship exists between lg of polypeptide MW and electrophoretic mobility of most protein species, Lawrence and Shepherd [174] speculated that alleles at the same locus differ only in so far as they code for polypeptides of different lengths, as a consequence of deletion or addition of an amino acid sequence during splicing or post-translational cleavage of a precursor, as found in maize [172]. They did not exclude, however, that allelic genes of unequal size may have derived from unequal crossing-over of repeated DNA sequences or that different mobility is due to the presence or absence of an attached moiety, such as a carbohydrate. Still at speculative stage is also the mechanism hypothesized to justify the fact that sub-unit 6, controlled by 1B, stains much less than any of its allelic counterparts. On the basis of results obtained in studies with aneuploids and segregant wheat lines from reciprocal F_1 crosses Payne *et al.* [241] proposed that some sub-units are controlled by a number of reiterated genes and the weak staining of sub-unit 6 is related to the number of structural genes present.

Studies on chromosomal control of endospermic proteins in species other than wheat showed that rye, barley, *Ae. elongatum* and *Ae. umbellulata* all have one chromosome that seems to be similar to the group 1 chromosomes in wheat [280] and carry genes that control the synthesis of prolamines on the short arm and genes that control high MW proteins on the long arm [176]. Since barley carries the genes that control the synthesis of these proteins on chromosome 5, they confirm that this chromosome may be regarded as being homoeologous with the chromosomes of group 1 in wheat.

Analysis of alien substitution lines of *Ae. umbellulata* in Chinese Spring indicated that chromosome $1C^u$ of *Ae. umbellulata* can be considered homoeologous to the chromosomes in group 1 in wheat [29].

PROTEINS IN WHEAT EVOLUTIONARY STUDIES

Analysis of protein composition is proving to be useful not only to devise new approaches for nutritional and technological improvements but also to study the relationship among species and to trace the evolutionary history of genera. The main observations of *Triticum* and related genera are summarized here with the aim of reviewing the possibilities offered by protein analysis and to present a picture of relationship among species of *Triticum* and *Aegilops* as a prerequisite for enlarging the genetic base of breeding materials.

According to Simpson [288], the basic criterion for ascertaining philogenetic relationship is the gene homology which, in general, cannot be measured directly because of reproductive barriers existing between species. Until recent times, indirect estimates of gene homologies were only obtained through comparisons of morphological features or chromosome pairing in F_1. Morphological features, in fact, reflect genetic differences and provide an ample base for inferences on genome relationship, whereas chromosome pairing is based on the premise that homologous genes are involved. Both methods, however, have some limitations. The fact that morphological characters are the result of the interaction of a very large number of genes causes difficulties in equating them to the number of genes and in distinguishing gene homology from gene analogy [2]. Chromosome pairing is usually expressed as bivalents or chiasma frequencies; but the possibilities of inaccuracy due to non-homologous pairing or translocations involving chromosomes belonging to different genomes, prevent definite conclusions, especially when long-established polyploids are analyzed [274]. In more recent times, actually in the last 20 years, the development of biochemical methods permitted in-depth studies on protein structure and to have a more direct approach to estimating gene homology both at the species and the genus levels. Variations in amino acid sequence of protein molecule are in fact a direct result of variations in nucleotide sequence of DNA and so reflect the genotype. As mentioned in the previous pages, gene differences existing among species can be detected, at least in part, through their effects on the mobility of specific components of the protein spectrum, thus facilitating the analysis.

Concerning more specifically the wheat, which has species at different ploidy levels, protein spectrum can provide indications on relationship between genomes of different species and evidence on the identity of genome contributors to the amphiploids. Pioneer studies on this subject were conducted by Hall, Johnson and co-workers [88, 89, 111, 112, 118, 119, 120]. On the basis of results obtained on plasma proteins of amphibians and reptile species and subspecies [54], on white-egg proteins of avian species [282], on green-crabs from different areas [322], they started a series of studies whose results indicated that electrophoretic patterns of seed proteins, extracted with 70% ethanol, show little intra-specific variation, are genome-specific and so are able to discriminate various degrees of affinity between genomes [113, 114, 313, 170] and that the protein spectrum of polyploids essentially represents the sum of parental spectra, so that amphiploids genome origin may be deduced from comparison of their protein spectra to those of their putative donors [88, 118].

By adopting these concepts, Johnson [113] was able to demonstrate that all subspecies of *T. aestivum* have a very uniform profile and that it is simulated by the pattern produced by mixing proteins from *T. dicoccum* and *Ae. squarrosa*, thus confirming that *Ae. squarrosa* contributed to the D genome and *T. dicoccum*, and not *T. dicoccoides*, donated the AB genome. Electrophoretic profiles also confirmed that the D genome of *Ae. squarrosa* is present, together with that of *Ae. caudata*, in *Ae. cylindrica* [116], a wild tetraploid species having a very uniform profile [111], considered as the donor of the D genome until Kihara [150], McFadden and Sears [201] showed that it was an allopolyploid.

The natural conclusion of these findings was that hexaploid wheats have had a monophiletic origin [113] so favouring the concept that cultivated hexaploids derive from the so-called primitive spelta complex by mutation of a single gene (q $\rightarrow$ Q) which governs the free threshing character [202] and not by independent crosses between *Ae. squarrosa* and two different tetraploids, one bearing the q gene and the other the Q gene [168, 187]. The only tetraploid species carrying the Q gene, *T. carthlicum*, which has an albumine profile similar to that of *T. dicoccum*, in fact did not simulate the profile of hexaploids when its proteins were mixed with those from various types of *Ae. squarrosa*.

The only exception to this general behaviour was represented by *T. macha*, whose profile was not simulated by mixtures of proteins from *T. dicoccum* and *Ae. squarrosa*; its albumine profiles were quite similar to those of the spelta types, but its gliadin pattern was much simpler, suggesting that the contributors of its AB and D genomes were probably different from those involved in the origin of most of the primitive aestivums.

Analysis of albumine electrophoretic profiles evidenced the existence of two groups of tetraploids, the emmer (AABB) and the *timopheevi* (AAGG) one, showing such dissimilarities between them, to suggest that they differ with respect to both genomes [120], while a number of accessions had aberrant profiles or uncertain affinity to either group; all tetraploid cultivars, however, had a common protein profile, and wild diploid showed to harbour the genes required to account for some of the more complex tetraploid protein profiles. These findings indicated to Johnson [114] that tetraploids endemic to S.W. Asia had a polyphyletic origin derived from various combinations of diploid *Triticums* which evidently contributed to the A as well as to the B genome, while cultivated types essentially derived from one of the several wild types. Later analysis of protein electrophoretic profiles of seeds from more than 500 accessions of wild wheats from S.W. Asia allowed Johnson [115] to distinguish between two different diploid wheats, *T. boeoticum* and *T. urartu*, whose proteins, when mixed, accounted for all the albumin bands in each of the tetraploid patterns. This finding and the evidence that heterogeneous albumin profiles favoured the Lilienfeld and Kihara [179] idea that *T. dicoccoides* and *T. araraticum* had independent origins, that is, that tetraploids represent different combinations of AA and BB parental biotypes from different areas from Lebanon to Transcaucasia and Western Iran, convinced Johnson [115] to hypothesize that *T. urartu* was involved in the origin of the B genome. The hypothesis was also supported by morphological features, hybrid sterility reactions, overlapping in areas of endemism, etc.

Results of crosses between *T. urartu* and ditelosomic lines of *T. aestivum* cv. Chinese Spring, however, showed that *T. urartu* chromosomes paired with those of A genome, indicating that this diploid could have contributed a second A genome but not the B genome to the amphiploids [61]. This same conclusion was reached by Konarev *et al.* [163] through protein analysis. On the basis of results from immunochemical analysis of alcoholic fractions of seed protein they proposed *T. urartu* as the donor of the A genome of emmer and hexaploid wheats while *T. boeoticum* or *T. monococcum* could have contributed the A genome to the *timopheevi* wheats and Iraqi *T. dicoccoides* [164]. Immunochemical reactions indicate in fact that this latter is composed of two groups: one of Iraqi origin, similar to *T. araraticum*, and the other of Syrian and Palestinian origin, identical to *T. dicoccum*; some forms of this second group also appear to be similar to *T. durum*.

As far as the origin of the second genome, B and/or G, is concerned, immunochemical studies of genome-specific proteins confirmed hypotheses of alloploid origin of tetraploid wheats and indicated the partial homology of B and G genomes thus supporting the Kostov [167] hypothesis that the genome of *T. monococcum* is not homologous to that of emmer and *aestivum* wheats. They also indicate the possibility that members of *Sitopsis* section of *Aegilops* participated in the formation of polyploid wheats; more specifically *Ae. speltoides* could have provided the G genome to *timopheevi* wheats, while *Ae. longissima*, or a form close to it, might have donated the B genome to the other polyploids.

As regards the D genome, the antigens of alcohol soluble proteins are in line with evidences indicating that it was contributed by *Ae. squarrosa* ssp. *strangulata* [161]. Mentioned analysis provided the basis for representing the phylogeny of polyploid wheats, as in figure 13 [163].

Important practical implications also derive from results obtained in diploid species. *Ae. squarrosa* has an antigen spectrum much more distinct than other diploids, bearing A and B genomes, so reflecting the later origin of hexaploid wheats compared with the tetraploid ones. The antigen composition of the genome-specific proteins of alcohol fraction also indicates that *T. boeoticum* and *T. monococcum* have common genome, so that the greater number of components of electrophoretic

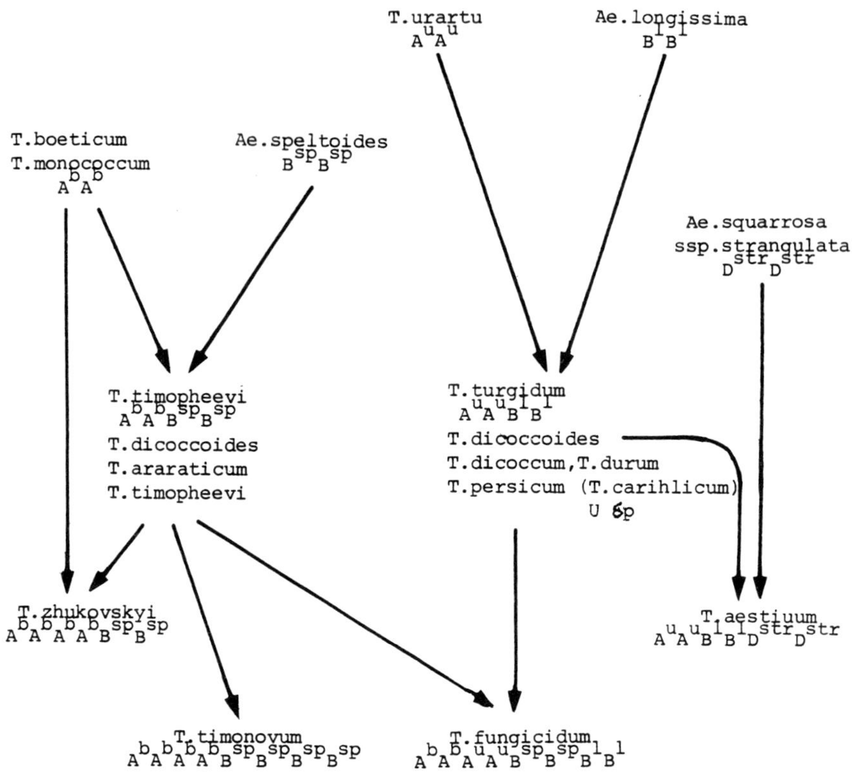

Figure 13. Phylogenesis of polyploid wheats (163).

spectrum of gliadins in accessions of *T. monococcum* might be ascribed to the additional experiences it had under cultivation, which have promoted an increase in replications of a number of genes and genetic systems [193], but there is no matter for considering them as two independent species [188]. The same kind of analysis indicated that *T. urartu* is qualitatively distinct from the preceding species [160], as is also evident from differences in the spectrum of acid phosphatase isoenzymes [102], supporting the hypothesis that it is an independent species [305, 335, 79].

Concerning the situation of *Aegilops* in section *Sitopsis*, Peneva and Migushova [245] have found that they possess three antigens; the first is characteristic of *Ae. bicornis*, the second of *Ae. longissima* and the third of *Ae. speltoides* and *Ae. aucheri*. The first two antigens are also present in *Ae. sharonensis* which might be the result of an introgressive hybridization and so more recent than the bearers of the single antigens.

Genomes of *Ae. speltoides* and *Ae. aucheri*, which have a similar antigen composition and gliadin spectra, revealed an unhomogeneous genome, with samples having simplified and others having an enriched or amplified genome. In addition, their antigen composition shows some similarities with those of *T. monococcum* and *T. boeoticum* [163].

A third approach has been explored by Cole *et al.* [48]. On the basis of the fact that migration of proteins in SDS-PAGE mainly depends on the molecular weight of their polypeptidic chain and may provide quantitative as well as qualitative differences among species, they analyzed total proteins, extracted from seeds of various species of *Triticum* and *Aegilops*, according to their molecular weight by SDS-PAGE. Protein patterns indicate that *T. boeoticum* and *T. monococcum* show similar qualitative and quantitative aspects and differ in qualitative aspects from *T. urartu*, so supporting the concept that, although closely related, this latter is in process of becoming reproductively isolated [117]. Also, *Ae. squarrosa* appears to be quantitatively similar to the three mentioned diploids. Protein patterns also showed that the four species were both qualitatively and quantitatively dissimilar to *Ae. longissima*, *Ae. speltoides* and *Ae. searsii* which show quantitative similarities among them. Another important result of this analysis was that the genome of tetraploids did not appear as the simple summation of intact diploid genomes, but rather as the ultimate result of more than two diploids [114], which during the polyploidization process should have experienced some exchange of genetic material [308, 92]. The formation of tetraploids could have occurred through the combination of reduced and unreduced gamets to form a triploid, which by a similar process gave origin to tetraploid in the following generation [92]. Concerning more specifically the different species, analyses indicate that *T. boeoticum*, *T. urartu* and *Ae. squarrosa* could have been involved in the origin of *T. dicoccoides* and *T. polonicum*, which show similar patterns, supporting the Johnson [115] theory that tetraploid wheats derived from *T. boeoticum* × *T. urartu*, but not excluding the Maan [185] concept that the cytoplasm of tetraploid wheats has been contributed by *Ae. squarrosa*, together with some genetic material. This latter hypothesis is also supported by Caldwell and Kasarda [38] who found that albumins and globulins of tetraploid *T. dicoccoides* are similar to those of *T. aestivum* and so that probably the albumins and globulins characteristic of *Ae. squarrosa* were already present in the tetraploid. Protein patterns of tetraploids suggested that more than two diploids have contributed to their genome and that the amount of contribution from any diploid varies among species and even samples; in general, however, *T. dicoccum* shows some similarities to *Ae. searsii* and *T. dicoccoides* to *Ae. longissima*.

Analysis also confirmed the monophiletic origin of bread wheats, which should result from one or few crosses between some cultivated tetraploids and *Ae. squarrosa*, although some modifications might have occurred over the years through introgression, as suggested by the lack of identity among cultivars [48].

Finally, the finding that the MW of prolamin components is species-specific induced Cole *et al.* [48] to speculate that differences among species arose during speciation of *Triticinae* and may be connected to the process itself. In fact, amplifications, transpositions and deletions of portions of DNA have been well-known evolutionary events [304].

In the past few years, the study of evolutionary relationship in cereals has also been approached by the analysis of N-terminal amino acid sequencing of prolamins. For example, Kasarda (personal communication) has found that N-terminal sequences of the ω-gliadins contributed by the B and D genomes differ substantially, making it likely that useful information about the B genome donor might be obtained by sequencing similar proteins from diploid species that may have contributed the B genome. Conversely, the protein from *Ae. squarrosa* shows good homology with the protein controlled by chromosome ID, as expected. In addition, rearrangement of the sequences to bring different parts of the sequences into register may reveal interesting homologies indicating the role for transposition and amplifications of DNA fragments in the evolution of genes coding for proteins and may provide support for the multiple gene hypothesis.

The analysis has also provided evidence for close homology between wheat and barley [281], while results obtained by Autran *et al.* [8] in *Triticum*, *Aegilops* and *Secale* indicate that the first two genera are so closely related to support the proposal of Morris and Sears [218] to incorporate *Aegilops* into the genus *Triticum* and that *Secale* is evolutionarily closer than the other two to the common ancestor that gave rise to the sub-tribe *Triticinae*, as proposed by Smith and Flavell [290].

RELATIONSHIP BETWEEN GLIADIN AND GLUTENIN COMPONENTS AND THE BREAD- AND PASTA-MAKING QUALITY

It has been known since long that the amount and quality of proteins have considerable influence on the technological properties of wheat. Several reports, in fact, indicate that bread-making properties of flour and pasta-making properties of semolina are related to the amount of gluten protein and more directly to the gliadin glutenin component [36, 33, 249, 230, 190, 195, 316].

Experiments conducted on common wheat utilizing substitution lines enabled Schmidt *et al.* [273] to demonstrate that bread quality, as detected from milling experiments, mixographic and bread preparation tests, was clearly controlled by a limited number of genes, showing partial dominance, localized in chromosomes 4B, 7B and 5D. Results also indicated that genes at chromosomes 4A, 3B and 4D could be involved.

Presence of genes controlling the dough-mixing strength in chromosome 1D was repeatedly reported [320, 219, 220, 198, 199]. The relations obtained, however, were not so high to exclude the involvement of genes at other chromosomes.

During these past years, attempts have been made to correlate the presence or absence of specific gliadin or glutenin components to the pasta- and bread-making quality of *durum* and bread wheats. Relationships between the electrophoretic profiles of gliadins and the pasta-making quality of several *durum* wheat varieties have been investigated by Damidaux *et al.* [50]. The results provided a criterion for dividing *durum* wheat varieties into two groups: one group characterized by the presence of one component in the γ-gliadin region, identified as 45, with good pasta-making qualities and the other group including varieties that are characterized by a component called 42 with poor pasta-making qualities. The results were confirmed by Kosmolak *et al.* [166] when analyzing Canadian cultivars. Wasik and Bushuk [316] pointed out that the glutenins extracted from good pasta-making varieties had a predominant component of molecular weight 53,000 compared to another component of molecular weight 60,000, while more recently, Autran [5] observed a correlation between the pasta-making quality and the sub-units present in the range of molecular weights between 62,000 and 70,000.

In bread wheat, Payne *et al.* [238] found a positive correlation between the presence of a high molecular weight glutenin sub-unit controlled by chromosome 1A, and the bread-making quality. Later, Payne *et al.* [239] correlated the bread-making quality with the presence of two sub-units (5 and 10), both controlled by chromosome 1D. They have also found that these latter sub-units have a stronger effect upon the bread-making quality than those controlled by chromosome 1A and finally, that the two groups of sub-units (1A, 1D) exhibited additive effects. Bournof and Bouriquet [30] found that bread-making quality of a stock of 47 genetically related varieties was correlated with two sub-units of molecular weight 122,000 and 108,000. Bread-making quality has also shown to be correlated to gliadins.

Relationships between blocks of gliadin components and quality were shown by Sozinov and Poperelya [294], while Wrigley *et al.* [332] pointed out the relationship between certain gliadin bands and the characteristics of hardness and dough-mixing strength. In order to explain these results two different hypotheses were produced [238, 5]. The first hypothesis, which, however, does not meet with much approval, assumes that the gliadin- and glutenin-coding genes are closely linked to the genes that control the bread- and pasta-making quality, while according to the second hypothesis which receives more credit, specific gliadin or glutenin components should be directly involved in the bread-making quality. Studies by Jeanjean *et al.* [106] indicate that during heating certain gluten fractions aggregate more than others showing that proteins' ability to associate into insoluble complexes and to form an insoluble network is a characteristic of wheat varieties with high bread-making quality. Kobrehel [159] showed that aggregates are stronger in the varieties with good bread-making qualities. Similarly, pasta quality would be related to the ability of proteins to aggregate and form a network that prevents loss of gelatinized starch during cooking [71, 72]. These characteristics could be ascribed to the larger hydrophobic surface area and to the higher sulphur content associated to the component 45 [106, 5].

Additional information, obtained by Kobrehel [159] indicates that α, β and γ gliadins would be related to the bread-making quality whereas ω gliadins do not play any important role. Similar results confirmed that α and β gliadins would be the most important fractions acting upon the bread-making quality [27].

In conclusion, the results so far obtained indicate that both gliadins and glutenins are involved in determining the bread- and pasta-making quality of common and *durum* wheats. However, further research is needed before the properties and structure of different components are fully understood.

DIRECTIONS FOR FURTHER INVESTIGATIONS

The previous points underline the considerable progress made during the past decades in the knowledge of seed proteins in wheat. Improvement in the quality and quantity of food prepared from wheat has been obtained as a result of the application of these findings to the wheat breeding and processing technologies. Additional improvements are expected in the future. However, it would be extremely naive to think that all problems have been solved. On the contrary, we are only now beginning to understand the physiological basis of production and the genetics of different physiological phenomena. Consequently, it seems worthwhile to outline the major deficiencies in our present day knowledge and hypothesize how they can be corrected.

The first factor limiting the protein content concerns the availability and utilization of the soil nitrogen. The increasing cost of fertilizers limits the supply of nitrogen in the soil and urges the search for nitrogen-fixing bacteria, which could be a long-term research. Concerning the utilization of nitrogen by the plant, little is known about the biochemical processes involved, in spite of the considerable amount of experimental results available. Nitrate reductase activity has been claimed as an important factor but it is not the only biochemical process involved and in practice cannot be used as a unique criterion for selection. A clearer picture of the biochemical processes involved in nitrogen assimilation should be obtained and their genetic basis cleared up before proposing a criterion for an efficient selection.

The presence of high amounts of reduced nitrogen in plant organs does not mean high amounts of storage protein in the seeds. In addition, nitrogen to be stored in the seed enters the grain in a limited range of compounds which, once on the grain, are metabolized to give all the 20 amino acids in the seed. Knowledge of the regulation of biosynthetic pathways is a prerequisite for devising a selection method aimed to identify the protein composition. Protein storage, in fact, is the final step of nitrogen assimilation and depends, in addition to the mentioned points, also on the rate of arrival of nitrogen, the identity of substrate, etc.

Some of the points are known but many still need to be clarified. It is known that high protein genotypes can incorporate amino acids more rapidly than low protein ones [58] but the role of different steps is unknown. The interactions of numerous physiological and biochemical processes involved are such that probably a clear picture of the situation will emerge only through the analysis of a system containing processes at different parts of the plant like roots, leaves, spike and seed. Aneuploidy and other forms of genetic engineering can greatly help in this research.

Analysis of world collections of different species of *Triticum* and mutant genotypes for constraints and limiting factors would allow the detection of material carrying genes to be recombined in superior genotypes. Very promising, in this respect, appears the joint utilization of genetic engineering which would allow the insertion of specifically isolated and/or modified genes with conventional plant breeding.

Another interesting line of research is the assessment of the relative role of nucleous versus cytoplasm compounds in the synthesis of seed proteins. This would include not only the base mechanisms and rate of the protein synthesis but also its composition, solubility groups, amino acid distribution and the number and characteristics of ribosomal proteins. These findings would permit greater exploitation of the variability existing among species, the possible use of a cytoplasm or cytoplasmic parts different from those of *durum* and common wheat and the tailoring of better nucleous cytoplasm combinations.

In recent times, considerable progress has been made on the structure of proteins, on the genetic control of different components and on the role they play in the pasta- and bread-making process but many points still remain unsolved. One of the major deficiencies concerns the genetics of LMW glutenin sub-units which represent about a third of the total seed protein in wheat. Genes controlling these sub-units are still to be located in the chromosomes and also their variability among varieties is unknown.

Some information is available on genes controlling gliadins and HMW glutenins, although they have not yet been precisely located on the arms, either with respect to the centromere or to other genes. Also the linkage relationships among genes on the same arm are unknown.

Comparisons between the situation in wheat and that in other species, especially barley, raised some questions on the evolution of different genomes. For example, it seems important to determine whether genes coding for gliadins, located on chromosomes 6, are duplications or translocations of those in chromosome 1 [237]. Questions such as these need to be solved before exploring possibilities offered by advanced techniques such as genetic engineering. New or modified genes should in fact be inserted in a specific site. For example, genes which have to be only expressed in the seed should be placed under the regulation of existing sequences in the genome or the same control sequences inserted with the genes. Further studies on both gene control sequences and amino acid sequences analysis of specific protein species could provide this information.

REFERENCES

1. Aamodt, OS and JH Torrie (1935): Studies on the inheritance and the relation between kernel texture and protein content in several spring wheat crosses. Can. J. Res. 13: 202–219.
2. Anfinsen, CB (1959): The Molecular Basis of Evolution. John Wiley and Sons, New York.
3. Appleyard, DB, J McCausland, and CW Wrigley (1979): Checking the identity and origin of off-types in the propagation of pedigreed wheat seed. Seed Sci. Technol. 7: 459–466.
4. Aragoncillo, C (1973): Proteinas CM en *Triticum* ssp. purification, caracterizacion y regulacion genetica. Thesis Polytechnical Univ. Madrid.
5. Autran, JC (1981): Recent data on the biochemical basis of *durum* wheat quality. (In press).
6. Autran, JC, and A Bourdet (1975): L'identification des variétés de blé: Etablissement d'un tableau général de détermination fondé sur le diagramme électrophorétique des gliadines du grain. Ann. Amélior. Plantes 25: 277–301.
7. Autran, JC, B Fleury, P Joudrier, and A Bourdet (1974): Blés hexaploides, blés tétraploides et espèces sauvages. Etude comparée de leur composition protéique et de l'hétérogénéité de certaines fractions. Symp. franco-soviétique sur l'amélioration génétique de la qualité des blés durs, Montpellier.
8. Autran, JC, EJL Lew, CC Nimmo, and DD Kasarda (1979): N-terminal amino acid sequencing of prolamins from wheat and related species. Nature 282: 527–529.
9. Avivi, L (1978): High grain protein content in wild tetraploid wheat *T.dicoccoides* corn. Proc. 5th Int. Wheat Genet. Symp., New Delhi: 372–380.
10. Balls, AK, WS Hale, and TH Harris (1942): A crystalline protein obtained from a lipoprotein of wheat flour. Cereal Chem. 29: 279–288.
11. Barlow, KK, JK Lee, and M Vask (1974): Morphological development of storage protein bodies in wheat. In: Bialeski, RL, AR Ferguson, and MM Creswell (eds.): Mechanisms of Regulation of Plant Growth, 763–797; Bull. 12 Rag. Soc. N.Z. Wellington.
12. Bates, LS, EG Heyne, and TC Roberts (1977): Wheat: Current situation and future outlook. Baker's Dig. 51(5): 87.
13. Beevers, L, and RH Hagemann (1969): Nitrate reduction in higher plants. Ann. Rev. Pl. Physiol. 20: 495–522.
14. Bhullar, BS, KS Gill, and GS Mahl (1978): Genetic analysis of protein in wheat. Proc. 5th Int. Wheat Genet. Symp., New Delhi: 413–625.
15. Bietz, JA, and FR Huebner (1980): Structure of glutenin: Achievements at the northern regional research center. Ann. Technol. Agric. 29(2): 249–277.
16. Bietz, JA, FR Huebner, JE Sanderson, and JS Wall (1977): Wheat gliadin homology revealed through N-terminal amino-acid sequence analysis. Cereal Chem. 54: 1070–1083.
17. Bietz, JA, FR Huebner, and JS Wall (1973): Glutenin the strength protein of wheat flour. Baker's Dig. 47: 26–34.
18. Bietz, JA, KW Sheperd, and JS Wall (1975): Single-kernel analysis of glutenin: use in wheat genetics and breeding. Cereal Chem. 52: 513–532.
19. Bietz, JA, and JS Wall (1972): Wheat gluten subunits: Molecular weights determined by sodium dodecyl sulfate-poly-acrylamide gel electrophoresis. Cereal Chem. 49: 416–430.
20. Bietz, JA, and JS Wall (1980): Identity of high-molecular gliadin and ethanol-soluble glutenin subunits of wheat: relation to gluten structure. Cereal Chem. 57: 415–421.
21. Biffen, RH (1909): On the inheritance of strength in wheat. J. Agric. Sci. 3: 86–91.
22. Black, JA, and GH Dixon (1967): Evolution of protamine: a further example of partial gene duplication. Nature 216: 152–154.
23. Booth, MR, and JAD Ewart (1969): Studies on four components of wheat gliadins. Biochim. Biophys. Acta 181: 226–233.

24. Bourque, DP, and SG Wildman (1972): Evidence that nuclear genes code for several chloroplast ribosomal proteins. Biochem. Biophys. Res. Commun. 50: 532–537.
25. Boyd, WJR, and JW Lee (1967): The control of wheat gluten synthesis at the genome and chromosome levels. Experientia 23: 332–333.
26. Bozzini, A, and V Silano (1978): Control through breeding methods of factors affecting nutritional quality of cereals and grain legumes. In: Friedman, M (ed.): Nutritional Improvement of Food and Feed Protein: 249–274; Plenum Press, New York.
27. Branlard, G, and M Rousset (1980): Les caractéristiques électrophorétiques des gliadines et la valeur en panification du blé tendre. Ann. Amélior. Plantes 30(2): 133–149.
28. Brown, JWS, and RB Flavell (1981): Fractionation of wheat gliadin and glutenin subunits by two-dimensional electrophoresis and the role of group 6 and group 2 chromosomes in gliadin synthesis. Theor. Appl. Genet. 59: 349–359.
29. Brown, JWS, RJ Kemble, CN Law, RB Flavell (1979): Control of endosperm proteins in *Triticum aestivum* (var. Chinese Spring) and *Aegilops umbellulata* by homoeologous group 1 chromosomes. Genetics 93: 189–200.
30. Burnouf, T, and R Bouriquet (1980): Glutenin subunits of genetically related European hexaploid wheat cultivars: their relation to breadmaking quality. Theor. Appl. Genet. 58: 107–111.
31. Burr, B, and FA Burr (1976): Zein synthesis in maize endosperm by polyribosomes attached to protein bodies. Proc. Nat. Acad. Sci. USA 73: 515–519.
32. Bush, RH, and SS Maan (1974): Possible use of cytoplasmic variability in wheat improvement. Wheat Newsletter 20: 163–166.
33. Bushuk, W, KG Briggs, and LH Shebeski (1969): Protein quantity and quality as factors in the evaluation of bread wheats. Can. J. Plant Sci. 49: 113–122.
34. Bushuk, W, and CW Wrigley (1971): Glutenin in developing wheat grain. Cereal Chem. 48: 448–455.
35. Bushuk, W, and CW Wrigley (1974): Proteins: Composition, structure and function. In: Inglett, GE (ed.): Wheat: Production and Utilization: 119–145; Avi, Westport, Conn.
36. Bushuk, W, and RR Zillman (1978): Wheat cultivar identification by gliadin electrophoretograms. I. Apparatus, method and nomenclature, Can. J. Plant. Sci. 58: 505–515.
37. Buttrose, MS (1963): Ultrastructure of the developing wheat endosperm. Austr. J. Biol. Sci. 16: 305–317.
38. Caldwell, KA, and DD Kasarda (1978): Assessment of genomic and species relationships in *Triticum* and *Aegilops* by PAGE and by differential staining of seed albumins and globulins. Theor. Appl. Genet. 52: 273–280.
39. Campbell, WP, JW Lee, TP O'Brien, and MG Smart (1981): Endosperm morphology and protein body formation in developing wheat grain. Aust. J. Plant Physiol. 8; 5–19.
40 Campbell, WP, JW Lee, and DH Simmonds (1974): Protein synthesis in the developing wheat grain. In: Papers and Minutes of the 24th Ann. Conf. (Melbourne). Ray. Austr. Chem. Inst. Cereal Chem. Div. 6.
41. Ceccarelli, S, E Piano, GMB Gianoni, and S Arcioni (1973): Gene action and selection progress in four crosses of soft wheat. Genet. Agr. 27: 378–395.
42. Chapman, SR, and FH MacNeal (1970): Gene effects for grain protein in five spring wheat crosses. Crop Sci. 10: 45–46.
43. Charbonnier, L (1974): Isolation and characterization of gliadin fractions. Biochim. Biophys. Acta 359: 142–151.
44. Charbonnier, L, T Terce-Laforgue, and J Mossé (1980): Some physicochemical properties of *Triticum vulgare* ß, γ and ω gliadins. Ann. Technol. Agric. 29(2): 175–190.
45. Chen, K, JC Gray, and SC Wildman (1975): Fraction I protein and the origin of polyploid wheats. Science 190: 1304–1306.
46. Clark, JA (1926): Breeding wheat for high protein content. J. Am. Soc. Agron. 18: 648–661.

47. Cobb, NA (1905): Universal nomenclature of wheat. Misc. Publ. 539; Dept. Agr. New South Wales, 75 pp.
48. Cole, EW, JG Fullington, and DD Kasarda (1981): Grain protein variability among species of *Tricitum* and *Aegilops*: Quantitative SDS-PAGE studies. Theor. Appl. Genet. 60: 17–30.
49. Dalling, MJ, GM Halloran, and JH Wilson (1975): The relation between nitrate reductase activity and grain nitrogen productivity in wheat. Aust. J. Agric. Res. 26: 1–10.
50. Damidaux, R, JC Autran, P. Grignac, and P Feillet (1978): Mise en évidence de relations applicables en sélection entre l'électrophorégramme des gliadines et les propriétés viscoélastiques du gluten de *T.durum* Desf., C.R. Acad. Sci. Paris 287: 701–704.
51. Davis, WH, GK Middleton, and TT Herbert (1961): Inheritance of protein, texture and yield in wheat. Crop Sci. 1: 235–278.
52. Deckard, EL, and RH Bush (1978): Nitrate reductase assays as a prediction test for crosses and lines in spring wheat. Crop. Sci. 18: 289–293.
53. De Ponte, R, R Parlamenti, T Petrucci, V Silano, and M Tomasi (1976): Albumin α–amylase inhibitor families from wheat flour. Cereal Chem. 53: 805–820.
54. Dessauer, HC, and W Fox (1956): Characteristic electrophoretic patterns of plasma proteins of orders of *Amphibia* and *Reptilia*. Science 124: 225–226.
55. Dick, PL, and RJ Baker (1975): Variation and covariation of agronomic and quality traits in two spring wheat populations. Crop Sci. 15: 161–165.
56. Diehl, AL, VA Johnson, and PJ Mattern (1978): Inheritance of protein and lysine in three wheat crosses. Crop Sci. 18: 391–395.
57. Doekes, GJ (1973): Inheritance of gliadin composition in bread wheat *Triticum aestivum* L. Euphytica 22: 28–34.
58. Donovan, GR, JW Lee, and RD Hill (1977): Compositional changes in the developing grain of high- and low-protein wheats. II Starch and protein synthetic capacity. Cereal Chem. 54: 646–656.
59. Dunker, AK, and RR Rueckert (1974): Observations on molecular weight determinations on polyacrylamide gel electrophoresis. J. Biol. Chem. 244: 5074–5080.
60. Dvorak, J (1972): Genetic variability in *Aegilops speltoides* affecting homoeologous pairing in wheat. Can. J. Genet. Cytol. 14: 371–380.
61. Dvorak, J (1976): The relationships between the genome of *Triticum urartu* and the A and B genomes of *Triticum aestivum*. Can. J. Genet. Cytol. 18: 371–377.
62. Edwards, IB (1973): Physiologic and genetic studies of nitrate reductase activity and nitrogen distribution in spring wheat (*Triticum aestivum* L.). Thesis, North Dakota State University, Fargo.
63. Edwards, IB (1974): Heritability estimates of nitrate reductase activity in spring wheat and the chromosomal location of genes effecting nitrogen reduction. 5th South Afr. Genet. Congr. (Abstr.).
64. Eilrich, GL, and RH Hageman (1973): Nitrate reductase activity and its relationship to accumulation of vegetative and grain nitrogen in wheat. Crop Sci. 13: 59–66.
65. Ellis, RP (1971): The identification of wheat varieties by the electrophoresis of grain proteins. J. Nat. Inst. Agric. Bot. 12: 223–235.
66. Evers, AD (1970): Development of the endosperm of wheat. Ann. Bot. 34: 547–555.
67. Ewart, JAD (1968): Fractional extraction of cereal flour proteins. J. Sci. Food Agric. 19: 241–245.
68. Favret, EA, L Manghers, R Solari, A Avila, and JC Monsiglio (1970): Gene control of protein production in cereal seeds. Improving Plant Protein by Nuclear Techniques: 87–95; IAEA Vienna.
69. Favret, EA, R Solari, L Manghers, and A Avila (1969): Genetic control of the qualitative and quantitative production of endosperm proteins in wheat and barley. New Approaches to Breeding for Plant Protein Improvement: 87–107; IAEA Vienna.
70. Feillet, P (1976): Les albumines et globulines du blé. Ann. Technol. Agric. 25(2): 203–216.
71. Feillet, P (1977): La qualité des pâtes alimentaires. Ann. Nutr. Diet. 12: 299.
72. Feillet, P and V Abecassis (1976): Valeur d'utilisation des blés durs. Semaine d'étude de céréaliculture, Gembloux, pp. 551.

73. Feillet, P, and K Kobrehel (1972): Recherche et dosage des produits de blé tendre dans les pâtes alimentaires par électrophorèse des protéines solubles. Ann. Technol. Agric. 21: 17–24.
74. Feldman, M (1978): New evidence on the origin of the B genome of wheat. Proc. 5th Intern. Wheat Genet. Symp. New Delhi: 120–132.
75. Fisher, M, DG Redman, and GAH Elton (1968): Fractionation and characterization of purothionin. Cereal Chem. 45: 48–57.
76. Fulcher, RG, TP O'Brien, and DH Simmonds (1972): Localization of arginine-rich proteins in mature seeds of some members of the Gramineae. Aust. J. Biol. Sci. 25: 487–497.
77. Gallagher, LW, KM Soliman, CO Qualset, KC Huffaker, and DW Rains (1980): Major gene control of nitrate reductase activity in common wheat. Crop Sci. 20: 717–721.
78. Ganapathy, SN, and RG Chitre (1970): Factors affecting the utilization of millet protein by rats. Fed. Proc. Fed. Amer. Soc. Exp. Biol. 29: 761 Abs.
79. Gandilyan, PA (1972): Wild-growing species of *Triticum* of the Armenian SSR. Both. Zh. 57, 2.
80. Garcia-Olmedo, F, and P Carbonero (1970): Homeologous proteins synthesis controlled by homeologous chromosomes in wheat. Phytochem. 9: 1495–1497.
81. Garcia-Olmedo, F, P Carbonero, C Aragoncillo, and G Salcedo (1978): Chromosomal control of wheat endosperm proteins: A critical review. Seed Protein Improvement by Nuclear Techniques: 555–566; IAEA Vienna.
82. Gill, KS, SS Bains, G Sing, and KS Bains (1973): Partial diallel test crossing for yield and its components in *T. aestivum* L. Proc. 4th Int. Wheat Gen. Symp., Columbia: 29–33.
83. Gill, KS, and GS Brar (1973): Genetic analysis of grain protein and its relationship with some economic traits in *T. aestivum* L. Indian J. Agric. Sci. 43: 173–176.
84. Graham, JSL, RK Morton, and JK Raison (1963): Isolation and characterization of protein bodies from developing wheat endosperm. Aust. J. Biol. Sci. 16: 375–383.
85. Graham, JSD, RK Morton, and JK Raison (1964): The *in vivo* uptake and incorporation of radioisotopes into proteins of wheat endosperm. Aust. J. Biol. Sci. 17: 102–114.
86. Hagberg, A, KE Karlsson, and L Munck (1970): Use of hiproly in barley breeding. Improving Plant Protein by Nuclear Techniques: 121–132; IAEA Vienna.
87. Haldane, JBS (1942): New Paths in Genetics. Harpers, New York and London.
88. Hall, O (1959): Immuno-electrophoretic analyses of allopolyploid rye wheat and its parental species. Hereditas 45: 495–504.
89. Hall, O, and BL Johnson (1962): Electrophoretic analysis of the amphiploid of *Stipa viridula* × *Oryzopsis hymenoides* and its parental species. Hereditas 48: 530–535.
90. Halloran, GM (1975): Genetic analysis of grain protein percentage in wheat. Theor. Appl. Genet. 46: 79–86.
91. Harlan, J (1976): Genetic resources in wild relatives of crops. Crop Sci. 16: 329–333.
92. Harlan, JR, and JM de Wet (1975): On O Winge and a prayer. The origin of polyploidy. Bot. Rev. 41: 361–390.
93. Harper, AE, DA Benton, and CA Elvehjem (1955): L-leucine, an isoleucine antagonist in the rat. Arch. Biochem. Biophys. 57: 1–12.
94. Haunold, A, VA Johnson, and JW Schmidt (1962): Genetic measurements of protein in the grain of *T. aestivum* L. Agron. J. 54: 203–206.
95. Hernandez, HH, DE Walsh, and A Bauer (1974): Nitrate reductase of wheat: Its relation to nitrogen fertilization. Cereal Chem. 51: 330–336.
96. Hinton, JJC (1947): The distribution of vitamin B-1 and nitrogen in the wheat grain. Proc. Roy. Soc. (London) 13: 134–148.
97. Hinton, JJC (1955): Resistance of the testa to entry of water into the wheat kernel. Cereal Chem. 32: 296–306.
98. Hinton, JJC (1959): The distribution of ash in the wheat kernel. Cereal Chem. 36: 19–31.
99. Holt, LM, R Astin, and PI Payne (1981): Structural and genetical studies on the high-molecular-weight subunits of wheat glutenin. Theor. Appl. Genet. 60: 237–243.

100. Hsu, CS, and FW Susulski (1969): Inheritance of protein content and sedimentation value in diallel crosses of spring wheat (*T. aestivum* L.). Can. J. Genet. Cytol. 11: 967–976.
101. Ingversen, J, B Køie, and H Doll (1973): Induced seed protein mutant of barley. Experientia 29: 1151–1152.
102. Jaaska, VV (1970): Biochemical data on the origin of Transcaucasian endemic wheats. Eesti KSV Tea, Akad. Toim. Biol. 19(4): 344.
103. Jacob, F, and J Monod (1961): Genetic regulatory mechanisms in the synthesis of proteins. J. Mol. Biol. 3: 318–356.
104. Jagannath, DR, and CR Bhatia (1972): Effect of rye chromosome 2 substitution on kernel protein content of wheat. Theor. Appl. Genet. 42: 89–92.
105. Jain, HK, NC Singhal, MP Singh, and A Austin (1975): An approach to breeding for higher protein content in bread wheat. Breeding for Seed Protein Improvement using Nuclear Techniques: 39–46; IAEA Vienna.
106. Jeanjean, MF, R Damidaux, and P Feillet (1980): Effect of heat treatment on protein solubility and viscoelastic properties of wheat gluten. Cereal Chem. 57: 325–331.
107. Jenkins, JA (1929): Chromosome homologies in wheat and *Aegilops*. Amer. J. Bot. 16: 238–245.
108. Jennings, AA, and RK Morton (1963): Changes in carbohydrate, protein and non protein nitrogenous compounds of developing wheat grain. Aust. J. Biol. Sci. 16: 318–331.
109. Jennings, AC, and RK Morton (1963): Changes in the nucleic acids and other phosphorus containing compounds of developing wheat grain. Aust. J. Biol. Sci. 16: 332–341.
110. Jensen, J, JH Jorgensen, HP Jensen, H Giese, and H Doll (1980): Linkage of the Hordein loci Hor1 and Hor2 with the powdery mildew resistance loci M1-k and M1-a on barley chromosome 5. Theor. Appl. Genet. 58: 27–31.
111. Johnson, BL (1967a): Confirmation of the genome donors of *Aegilops cylindrica*. Nature (London) 216: 859–862.
112. Johnson, BL (1967b): Tetraploid wheats: seed protein electrophoresis pattern of the emmer and timopheevi groups. Science 158: 131–132.
113. Johnson, BL (1972): Seed protein profiles and the origin of the hexaploid wheats. Amer. J. Bot. 59: 952–960.
114. Johnson, BL (1972): Protein electrophoretic profiles and the origin of the B genome of wheat. Proc. Nat. Acad. Sci. USA 69: 1398–1402.
115. Johnson, BL (1975): Identification of the apparent B-genome donor of wheat. Can. J. Genet. Cytol. 17: 21–39.
116. Johnson, BL (1975): Seed protein patterns and the gene resources of wheat. In: Scarascia Mugnozza, GT (ed.): Genetics and Breeding of *Durum* Wheat: 153–164.
117. Johnson, BL, D Barnhart, and O Hall (1967): Analysis of genome and species relationships in the polyploid wheats by protein electrophoresis. Amer. J. Bot. 54: 1089–1098.
118. Johnson, BL, and HS Dhaliwal (1976): Reproductive isolation of *T. boeoticum* and *Triticum urartu* and the origin of the tetraploid wheats. Amer. J. Bot. 63: 1088–1094.
119. Johnson, BL, and O. Hall (1965): Analysis of phylogenetic affinities in the Triticinae by protein electrophoresis. Amer. J. Bot. 52: 506–513.
120. Johnson, BL, and O Hall (1966): Electrophoretic studies of species relationships in *Triticum*. Acta Agric. Scand. Suppl. 16: 222–224.
121. Johnson, VA, and PJ Mattern (1972): Improvement of the nutritional quality of wheat through increased protein content and improved amino acid balance. Summary report of research findings. July 1, 1966–Dec. 31, 1972. Contract AID/Csd-1208. Agency for Internat. Developm. Departm. of State. Washington, D.C.
122. Johnson, VA, and PJ Mattern (1975): Improvement of the nutritional quality of wheat protein through increased protein content and improved amino acid balance. Report of research findings. Jan. 1, 1973–March 31, 1975. Contracts AID/Csd-1208 and AID/ta-C-1093. Agency for Internat. Developm. Departm. of State. Washington, D.C.

123. Johnson, VA, PJ Mattern, and SL Kuhr (1978): Genetic improvement of wheat protein. Seed Protein Improvement in Cereals and Grain Legumes II: 165–181; IAEA Vienna.
124. Johnson, VA, PJ Mattern, and JW Schmidt (1967): Nitrogen relations during spring growth in varieties of *T. aestivum* L. differing in grain protein content. Crop Sci. 7: 664–667.
125. Johnson, VA, PJ Mattern, and JW Schmidt (1970): The breeding of wheat and maize with improved nutritional value. Proc. Nut. Soc. 29: 20–31.
126. Johnson, VA, PJ Mattern, and JW Schmidt (1972): Genetic studies of wheat protein. In: Inglett, GE (ed.): Symposium: Seed Proteins: 126–136; AVI Publ. Co. Westport, Conn.
127. Johnson, VA, PJ Mattern, and KP Vogel (1975): Cultural, genetic and other factors affecting quality of wheat. In: Spencer, A (ed.): Bread: Social, Nutritional and Agricultural Aspects of Wheat Bread: 127–140; Appl. Sci. Publ. London.
128. Johnson, VA, PJ Mattern, DA Whited, and J Schmidt (1969): Breeding for high protein control and quality in wheat. New Approaches to Breeding for Improved Plant Protein: 29–40; IAEA Vienna.
129. Johnson, VA, JW Schmidt, and PJ Mattern (1971): Protein improvement in wheat. In: Proc. 3rd FAO/Rockfeller Foundation Wheat Seminar, Ankara: 166–172.
130. Johnson, VA, JW Schmidt, PJ Mattern, and A Haunold (1963): Agronomic and quality characteristics of high protein F2 derived families from a soft red winter-hard red winter cross. Crop Sci. 3: 3–10.
131. Johnson, VA, JW Schmidt, and JE Stroike (1973): Genetic advances in wheat protein quantity and composition. Proc. 4th Intern. Wheat Genet. Symp., Columbia: 547.
132. Johnson, VA, KD Wilhelmi, SL Kuhr, PJ Mattern, and JW Schmidt (1978): Breeding progress for protein and lysine in wheat. Proc. 5th Int. Wheat Genetic Symp., New Delhi: 825–835.
133. Jones, BC, and AS Mak (1977): Amino acid sequences of the two α-purothionins of hexaploid wheat. Cereal Chem. 54: 511–523.
134. Jones, RW, GE Babcock, NW Taylor, and FR Senti (1961): Molecular weights of wheat gluten fractions. Arch. Biochem. Biophys. 94: 483–493.
135. Jones, RW, GL Lookhart, SB Hall, and KF Finney (1980): Polyacrylamide gel electrophoretic pattern of gliadin proteins from the 80 most commonly grown U.S. wheat varieties. 65th Ann. Meet. Amer. Assoc. Cereal Chem.
136. Jones, RW, NW Taylor, and FR Senti (1959): Electrophoresis and fractionation of wheat gluten. Arch. Biochem. Biophys. 84: 363–376.
137. Kakade, ML (1974): Biochemical basis for the differences in plant protein utilization. J. Agric. Food Chem. 22: 550–555.
138. Kasarda, DD (1980): Structural and properties of α-gliadines. Ann. Technol. Agric. 29(2): 151–173.
139. Kasarda, DD, JE Bernardin, and CC Nimmo (1976): Wheat proteins. Adv. Cereal Sci. Technol. 1: 158–236.
140. Kasarda, DD, JE Bernardin, and CO Qualset (1974): Relationship of gliadin protein components to chromosomes through the use of substitution lines. Cereal Sci. Today 19: 403–?.
141. Kasarda, DD, JE Bernardin, and CO Qualset (1976): Relationship of gliadin protein components to chromosomes in hexaploid wheats (*T. aestivum* L.). Proc. Nat. Acad. Sci. (Wash.) 73: 3646–3650.
142. Kaul, AK, and FW Susulski (1965): Inheritance of flour protein content in a Selkirk × Gabo cross. Can. J. Gen. Cytol. 7: 12–17.
143. Kent, NI (1966): Subaleurone cells of high protein content. Cereal Chem. 43: 585–601.
144. Kent, NI, and AD Evers (1969): Variation in protein composition within the endosperm of hard wheat. Cereal Chem. 46: 293–300.
145. Kessler, E (1955): Role of photochemical processes in the reduction of nitrate by green algae. Nature 176: 1069–1070.

146. Kessler, E (1957): In: Research in Photosynthesis. Interscience Publishers. New York, pp. 250.
147. Khan, K, and W Bushuk (1977): Glutenin: structure and functionality in bread making. Proc. 10th Nat. Conf. Wheat Utilization Research, Tucson: 101–115.
148. Kihara, H (1919): Über cytologische Studien bei einigen Getreidearten. I. Spezies-Bastarde des Weizens und Weizenroggen-Bastarde. Bot. Mag. (Tokyo) 33: 17–38.
149. Kihara, H (1924): Cytologische und genetische Studien bei wichtigen Getreidearten mit besonderer Rücksicht auf das Verhalten der Chromosomen und die Sterilität in den Bastarden. Mem. Coll. Sci. Kyoto Imp. Univ. Ser. B Vol. 1.
150. Kihara, H (1944): Die Entdeckung der DD-Analysatoren beim Weizen. Agr. Hort. (Tokyo) 19: 889–890.
151. Kihara, H (1954): Considerations on the evolution and distribution of *Aegilops* species based on the analyser method. Cytologia 19: 336–357.
152. Kihara, H (1966): Factors affecting the evolution of common wheat. Indian J. Gen. Pl. Breed. 26a: 14–28.
153. Kihara, H, and K Tsunewaki (1968): Some fundamental problems underlying the program for hybrid wheat breeding. Seiken Zihô 16: 1–14.
154. Kimber, G (1973): The relationships of the S genome diploids to polyploid wheats. Proc. 4th Int. Wheat Genet. Symp., Columbia: 81–85.
155. Kimber, G (1974): A reassessment of the origin of the polyploid wheat. Genetics 78: 487–492.
156. Kimber, G, and Athwal (1972): A reassessment of the course of evolution in wheat. Proc. Natl. Acad. Sci. USA 69: 912–915.
157. Klepper, LA (1976): Nitrate assimilation enzymes and seed protein in wheat. Proc. 2nd Int. Winter Wheat Conf. Zagreb: 334–340.
158. Kobayashi, K, and JL Fox (1978): The evolution of protein sequences by repetitions gene duplication: clostridial flavodoxin. J. Mol. Evol. 11: 233–243.
159. Kobrehel, K (1980): Extraction of wheat proteins with salts of fatty acids and their electrophoretic characterization. Ann. Technol. Agric. 29(2): 125–132.
160. Konarev, AV (1975): Differentiation of the first genomes of polyploid wheats, based on data from immunochemical analysis of the alcohol fraction of the grain protein. Bull. VIR 47.
161. Konarev, AV, IP Gavrilyuk, and EP Migushova (1974): Differentiation of diploid wheats as indicated by immunochemical analysis. Dokl. Vses Acad. Skh. Nauk. (USSR) 6: 12.
162. Konarev, AV, EP Migushova, IP Gavrilyuk, and VG Konarev (1971): On the nature of the genome of wheats of the *T. timopheevi* group as indicated by eletrophoretic and immunochemical analysis. Dokl. Vses. Akad. Skh. Nauk. 4: 13.
163. Konarev, VG, IP Gavrilyuk, NK Gubareva, and TI Peneva (1979): Seed proteins in genome analysis, cultivar identification and documentation of cereal genetic resources: a review. Cereal Chem. 56: 272–278.
164. Konarev, VG, IP Gavrilyuk, TI Peneva, AV Konarev, AG Khakimova, and EP Migushova (1976): The nature and origin of polyploid wheat genomes based on data from grain protein biochemistry and immunochemistry. Agric. Biol. 11(5): 656–665 (Russian).
165. Konzak, CF (1977): Genetic control of the content, amino acid composition and processing properties of proteins in wheat. Adv. Gen. 19: 407–582.
166. Kosmolak, FG, JE Dexter, RR Matsuo, D Leisle, and BA Marchylo (1980): A relationship between *durum* wheat quality and gliadin electrophoregrams. Can. J. Pl. Sci. 60: 427–432.
167. Kostov, D (1940): Origin and selection of wheats from the cytogenetic point of view. Izv. Akad. Nauk. USSR Ser. Biol. 1.
168. Kuckuck, H (1964): Experimentelle Untersuchungen zur Entstehung der Kulturweizen. Z. Pflanzenzüchtg. 51: 97–140.
169. Kuspira, J, and J Unrau (1957): Genetic analysis of certain characters in common wheat using whole chromosome substitution lines. Can. J. Pl. Sci. 37: 300–326.

170. Ladizinsky, G, and BL Johnson (1972): Seed protein homologies and the evolution of poliploidy in *Avena*. Can. J. Genet. Cytol. 14: 875–888.
171. Lafiandra, D, E Porceddu, and G Colaprico (1979): Aminoacid composition and species relationships in genus *Triticum*. Wheat Int. Serv. 50: 51–55.
172. Larkins, BA, and WJ Hurkman (1978): Synthesis and deposition of zein in protein bodies of maize endosperm. Pl. Physiol. 62: 256–263.
173. Lawrence, GJ, and KW Shepherd (1980): Variation in glutenin protein subunits of wheat. Aust. J. Biol. Sci. 33: 221–233.
174. Lawrence, GJ, and KW Shepherd (1981): Inheritance of glutenin protein subunits of wheat. Theor. Appl. Genet. 60: 333–337.
175. Lawrence, GJ, and KW Shepherd (1981b): Chromosomal location of genes controlling seed proteins in species related to wheat. Theor. Appl. Genet. 59: 25–31.
176. Lawrence, JM, KM Day, E Huey, and B Lee (1958): Lysine content of wheat varieties, species and related genera. Cereal Chem. 35: 169–178.
177. Lebsock, KL, CC Fifield, GM Gurney, and W Greenaway (1964): Variation and evaluation of mixing tolerance, protein content and sedimentation value in early generations of spring wheat, *T. aestivum* L. Crop Sci. 4: 171–174.
178. Lee, JW, and JA Ronalds (1967): Effect of environment on wheat gliadin. Nature 213: 844.
179. Lilienfeld, F, and H Kihara (1934): Genomanalyse bei *Triticum* und *Aegilops* V. *Triticum timopheevi* Zhuk. Cytologia 6: 87–122.
180. Lofgren, JR, KF Finney, EG Heyne, LC Bolte, RC Hoseney, and MD Shogren (1968): Heritability estimates of protein content and certain quality and agronomic properties in bread wheats (*T. aestivum* L.). Crop Sci. 8: 563–567.
181. Losada, M, JM Paneque, and FF Rodriguez del Campo (1963): Mechanism of nitrite reduction in chloroplasts. Biochem. Biophys. Res. Commun. 10: 298–310.
182. Maan, SS (1973): Cytoplasmic and cytogenetic relationships among tetraploid *Triticum* species. Euphytica 22: 287–300.
183. Maan, SS (1973): Cytoplasmic variability in Triticinae. Proc. 4th Int. Wheat Genet. Symp. Columbia: 367–373.
184. Maan, SS (1975): Cytoplasmic variability of speciation in Triticinae. In: Wali, MKJ (ed.): Prairie: A multiple View: 255–281; Ground Forks, N.D., Univ. N. Dakota Press.
185. Maan, SS (1975): Cytoplasmic male-sterility and male-fertility systems in wheat. In: Scarascia Mugnozza, GT (ed.): Genetics and Breeding of *Durum* Wheat: 117–137.
186. Maan, SS, and KA Lucken (1972): Interacting male sterility-fertility restoration systems for hybrid wheat research. Crop Sci. 12: 360–364.
187. MacKey, J (1966): Species relationship in *Triticum*. Hereditas Suppl. 2: 237–276.
188. MacKey, J (1968): Relationships in the Triticinae. 3rd Int. Wheat Genet. Symp. Canberra.
189. MacMasters, MM, JJC Hinton, and D Bradbury (1971): Microscopic structure and composition of the wheat kernel. In: Pomeranz, Y (ed.): Wheat Chemistry and Technology: 51–113; AACC Cereal Chem. Inc. St. Paul, Minn.
190. Mac Ritchie, R (1973): Conversion of a weak flour to a strong one by increasing the proportion of its high molecular weight gluten protein. J. Sci. Food Agric. 24: 1325–1329.
191. Mak, AS, and BL Jones (1976): Separation and characterization of chymotyptic peptides from α and ß-purothionins of wheat. J. Sci. Food Agric. 27: 205–213.
192. Mak, AS, and BL Jones (1976): The amino acid sequence of wheat ß-purothionin. Can. J. Biochem. 54: 835.
193. Makhlayeva, PF, and SL Tyuterev (1973): Study of wheat genomes and their wild relatives, using the method of molecular hybridization of DNA-DNA. Tr. VIR 52(1).
194. Mares, DJ, K Horstog, and BA Stone (1975): Early stages in the development of wheat endosperm. 1) The change from free nuclear to cellular endosperm. Aust. J. Biol. 23: 311–326.

195. Matsuo, RR, and GM Irvine (1975): Rheology of *durum* wheat products. Cereal Chem. 52: 131–135.
196. Mattern, PJ, R Morris, JW Schmidt, and VA Johnson (1973): Locations of genes for kernel properties in the wheat variety Cheyenne using chromosome substitution lines. Proc. 4th Int. Wheat Genet. Symp., Columbia: 703–707.
197. Mattern, PJ, A Salem, VA Johnson, and JW Schmidt (1968): Amino acid composition of selected high protein wheats. Cereal Chem. 45: 437–444.
198. Mattern, PJ, A Salem, and GH Volkmer (1968): Modification of the Maes continuous-extraction process for fractionation of hard red winter wheat flour proteins. Cereal Chem. 45: 319–328.
199. Mattern, PJ, JW Schmidt, R Morris, and VA Johnson (1968): A feasibility study of the use of a modified Maes protein extraction process and chromosome substitution lines for bread wheat quality identification. Proc. 3rd Int. Wheat Genet. Symp., Canberra: 449–456.
200. McDermoth, EE, and J Pace (1960): Comparison of the amino acid composition of the protein in flour and endosperm from different types of wheat with particular reference to variation in lysine content. J. Sci. Fd. Agr. 11: 109–115.
201. McFadden, ES, and ER Sears (1944): The artificial synthesis of *Triticum Spelta*. Records Genet. Soc. Am. 13: 26–27 (Abstr.).
202. McFadden, ES, and ER Sears (1946): The origin of *Triticum spelta* and its free threshing hexaploid relatives. J. Hered. 37: 81–89: 107–116.
203. Mecham, DK, DD Kasarda, and CO Qualset (1978): Genetic aspects of wheat gliadin proteins. Biochem. Genet. 16: 831–853.
204. Meredith, P (1965): On the solubility of gliadinlike proteins. I. Solubility in nonaqueous media. Cereal Chem. 42: 54–63.
205. Meredith, P (1965): On the solubility of gliadinlike proteins. III. Fractionation by solubility. Cereal Chem. 42: 149–160.
206. Meredith, P, HG Sammons, and AC Frazer (1960): Examination of wheat gluten by partial solubility methods. I. Partition by organic solven. J. Sci. Food Agric. 11: 320–328.
207. Mertz, ET, LS Bates, and OE Nelson (1964): Mutant gene that changes protein composition and increases lysine content in maize endosperm. Science 145: 279–280.
208. Middleton, CE, E Bode, and BB Bayles (1954): A comparison of the quality of protein in certain varieties of soft wheat. Agron. J. 46: 500–502.
209. Miezen, K, EG Heyne, and KF Finney (1977): Genetic and environmental effects on the grain protein content in wheat. Crop Sci. 17: 591–593.
210. Miflin, BJ (1978): Energy considerations in nitrogen metabolism. In: Miflin, BJ, and M. Zoschke (eds.): Carbohydrate and Protein Synthesis. A Seminar held in Giessen (Germany), September 7-9/1977: 13–32.
211. Miflin, BJ, and PJ Lea (1976): The pathway of nitrogen assimilation in plants. Phytochemistry 15: 873–885.
212. Miflin, BJ, and PJ Lea (1977): Amino acid metabolism. Ann. Rev. Pl. Physiol. 28: 299–329.
213. Miflin, BJ, and PR Shewry (1979): The biology and biochemistry of cereal seed prolamins. Seed Protein Improvement in Cereals and Grain Legumes I: 137–158; IAEA Vienna.
214. Mihaljev, I, and M Kovocev-Djolai (1978): Inheritance of grain content in a diallel wheat cross. Proc. 5th Intern. Wheat Genet. Symp. New Delhi: 755–761.
215. Minotti, LA, and WA Jackson (1970): Nitrate reduction in the roots and shoots of wheat seedlings. Planta 86: 267–271.
216. Mitra, R, and CR Bhatia (1973): Studies on protein biosynthesis in developing wheat kernels. Nuclear Techniques for Seed Protein Improvement: 379–389; IAEA Vienna.
217. Morris, R, JW Schmidt, PJ Mattern, and VA Johnson (1966): Chromosomal locations of genes for flour quality in the wheat variety Cheyenne using substitution lines. Crop Sci. 6: 119–122.
218. Morris, R, JW Schmidt, PJ Mattern, and VA Johnson (1968): Quality tests for six substitution lines involving Cheyenne wheat chromosomes. Crop Sci. 8: 121–122.

219. Morris, MR, JW Schmidt, PJ Mattern, and VA Johnson (1973): Chromosomal locations of genes for high protein in the wheat cultivar Atlas 66. In: Proc. 4th Int. Wheat Genet. Symp. Columbia: 715–718.
220. Morris, R, and ER Sears (1967): The cytogenetics of wheat and its relatives. In: Quinsbery, and Reitz (eds.): 19–87; Madison.
221. Morton, RK, BA Palk, and JK Raison (1964): Intracellular components associated with protein synthesis in developing wheat endosperm. Biochem. J. 91: 252–258.
222. Morton, RK, and JK Raison (1963): A complete intracellular unit for incorporation of amino acid into storage protein utilizing adenosine triphospate generated from phytate. Nature 200: 429–433.
223. Munaver, SM, and AE Harper (1959): Amino acid balance and imbalance. II. Dietary level of protein and lysine requirement. J. Nutr. 69: 58–64.
224. Müntz, K, K Hammer, C Lehmann, A Meister, A Rudolph, and F Scholz (1979): Variability of protein and lysine content in barley and wheat specimens from the world collection of cultivated plants at Gatersleben. Seed Protein Improvement in Cereals and Grain Legumes. II: 183–200; IAEA Vienna.
225. Nielsen, HC, GE Babcock, and FR Senti (1962): Molecular weight studies on glutenin before and after disulfide-bond splitting. Arch. Biochem. Biophys. 96: 252–258.
226. Nielsen, HC, AC Beckwith, and JS Wall (1968): Effect of disulfide-bond cleavage on wheat gliadin fractions obtained by gel filtration. Cereal Chem. 45: 37–47.
227. Nilsson-Ehle, H (1909): Kreuzungsuntersuchungen an Hafer und Weizen. Lund Univ. Arskr. Afd. 2, 5: 122.
228. Nilsson-Ehle, H (1911): Kreuzungsuntersuchungen an Hafer und Weizen. Lund Univ. Arskr. Afd. 2, 7: 1–82.
229. Nimmo, CC, DD Kasarda, and EJL Lew (1974): Physical characterization of the wheat protein purothionin. J. Sci. Food Agric. 25: 607–617.
230. Orth, RA, and W Bushuk (1972): A comparative study of the proteins of wheats of diverse baking qualities. Cereal Chem. 49: 268–275.
231. Orth, RA, and W Bushuk (1973): Studies of glutenin. Relation of variety, location of growth and baking quality to molecular weight distribution of subunits. Cereal Chem. 50: 191–197.
232. Osborne, TB (1907): The proteins of the wheat kernel. Carnegie Inst. Washington Publ. 84: 1–119.
233. Paneque, A, FF Del Campo, and M Losada (1963): Nitrite reduction by isolated chloroplasts in light. Nature 198: 90–91.
234. Pathak, GN (1940): Studies in the cytology of cereals. J. Genet. 39: 437–467.
235. Payne, PR (1978): Human protein requirements. In: Norton, G (ed.): Plant Proteins: 247–263; Butterworths, London.
236. Payne, PI (1981): Breeding for protein quantity and protein quality in seed crop. In: Int. Symp. on Seed Proteins, Versailles, (in press).
237. Payne, PI, KG Corfield, and JA Blackman (1979): Identification of a high-molecular-weight subunit of glutenin whose presence correlates with bread-making quality in wheats of related pedigree. Theor. Appl. Genet. 55: 153–159.
238. Payne, PI, KG Corfield, LM Holt, and JA Blackman (1981b): Correlations between the inheritance of certain high molecular weight subunits of glutenin and bread-making quality in progenies of six crosses of bread wheat. J. Sci. Food Agric. 32: 51–60.
239. Payne, PI, PA Harris, CN Law, LM Holt, and JA Blackman (1980b): The high-molecular weight subunits of glutenin: structure, genetics and relationship to bread-making quality. Ann. Technol. Agric. 29: 309–320.
240. Payne, PI, LM Holt, and CN Law (1981a): Structural and genetical studies on the high-molecular subunits of wheat glutenin. Part I. Allelic variation in subunits amongst varieties of wheat. (*T. aestivum*). Theor. Appl. Genet. 60: 229–236.
241. Payne, PI, LM Holt, GJ Lawrence, and CN Law (1982): The genetics of gliadin and glutenin, the major storage proteins of the wheat endosperm. Qual. Plant. Mat. Veg. (in press).

242. Payne, PI, CN Law, and EE Mudd (1980a): Control by homoeologous group 1 chromosomes of the high-molecular-weight subunits of glutenin, a major protein of wheat endosperm. Theor. Appl. Genet. 58: 113–120.
243. Pence, JW, NE Weinstein, and DK Mecham (1954): The albumin and globulin contents of wheat flour and their relationships to protein quality. Cereal Chem. 31: 303–311.
244. Pence, JW, NE Weinstein, and DK Mecham (1954): A method for the quantitative determination of albumins and globulins in wheat flour. Cereal Chem. 31: 29–37.
245. Peneva, TI, and EF Migushova (1973): The structure of genome (S) (B) in *Aegilops* of the group *Sitopsis* according to the data of electrophoretic and immunochemical analysis of gliadins. Tr. Prikl. Bot. Genet. Sel. (USSR) 52: 178–192.
246. Percival, J (1921): The Wheat Plant. Duckworth and Co. London.
247. Petrucci, T, M Tomasi, P Cantagalli, and V Silano (1974): Comparison of wheat albumin inhibitors of alpha-amylase and trypsin. Phytochem. 13: 2487–2495.
248. Piazzi, SE, and F Cantagalli (1969): Immunochemical analysis on soluble proteins of wheat. Cereal Chem. 46: 642–646.
249. Pomeranz, Y (1971): Composition and functionality of wheat flour components. In: Pomeranz, Y (ed.): 585–674; Am. Ass. Cereal Chem., St. Paul, Minn.
250. Porceddu, E (1973): Moltiplicazione e valutazione della collezione di frumenti selvatici. In: Laboratorio del Germoplasma, C.N.R., Relazione annuale per il 1972: 31–34.
251. Porceddu, E (1974): Moltiplicazione e valutazione della collezione dei frumenti. Germoplasma, C.N.R. Relazione annuale per il 1973.
252. Porceddu, E (1976): Analisi della collezione di frumento duro per il contenuto in proteine. In: Laboratorio del Germoplasma, C.N.R., Relazione annuale per il 1975.
253. Porceddu, E, G Pacucci, P Perrino, C Della Gatta, and J Maellaro (1975): Protein content and seed characteristics in populations of *T. durum* grown at three different locations. In: Scarascia Mugnozza, GT (ed.): Genetics and Breeding of *Durum*-Wheat: 217–224.
254. Rao, KP, DW Rains, CO Qualset, and RC Huffaker (1977): Nitrogen nutrition and grain protein in two spring wheat genotypes differing in nitrate reductase activity. Crop Sci. 17: 283–286.
255. Redman, DG, and M Fisher (1968): Fractionation and comparison of purothionin and globulin components of wheat. J. Sci. Food Agric. 19: 651–655.
256. Redman, DG, and JAD Ewart (1973): Characterization of three wheat proteins found in chloroform-methanol extracts of flours. J. Sci. Food Agric. 24: 629–636.
257. Resmini, P (1968): Un nuovo metodo per identificare a dosare gli sfarinati di grano tenero presenti in quelli di grano duro e nelle paste alimentari. Tec. Molitoria 19: 145–168.
258. Rijven, AHGC, and CA Banbury (1960): Role of the grain coat in wheat grain development. Nature (London) 188: 546–547.
259. Riley, R (1965): Cytogenetics and evolution of wheat. In: Hutchinson, JB (ed.): Essay on Crop Plant Evolution: 103–122; Cambridge.
260. Riley, R, and V Chapman (1960): The D genome of hexaploid wheat. Wheat Inf. Serv. 11: 18–19.
261. Riley, R, J Unrau, and V Chapman (1958): Evidence on the origin of the B genome of wheat. J. Hered. 49: 90–98.
262. Ritenour, GL, KW Joy, JJ Bunning, and RH Hageman (1967): Intracellular localization of nitrate reductase, nitrite reductase and glutamic acid dehydrogenase in green leaf tissue. Plant Physiol. 42: 233–237.
263. Rodriguez-Loperena, MA, C Aragoncillo, P Carbonero, and F Garcia-Olmedo (1975): Heterogeneity of wheat endosperm proteolipids (C.M. proteins). Phytochem. 14: 1219–1223.
264. Rybalka, AI, and AA Sozinov (1979): Mapping the locus of Gld 1B, which controls the biosynthesis of reserve proteins in soft wheat. Cytol. Genet. 13: 276–282.
265. Sakamura, T (1918): Kurze Mitteilung über die Cromosomenzahlen und die Verwandtschaftsverhältnisse der *Triticum* Arten. Bot. Mag. (Tokyo) 32: 151–154.

266. Salcedo, G, MA Rodriguez-Loperena, and C Aragoncillo (1978): Relationships among low MW hydrophobic proteins from wheat endosperm. Phytochem. 17: 1491–1494.
267. Sandstedt, RM (1946): Photomicrographic studies of wheat starch. I. Development of the starch granules. Cereal Chem. 23: 337–359.
268. Sandstedt, RM, and OC Beckord (1946): Photomicrographic studies of wheat starch. II. Amylolytic enzymes and the amylase inhibitor of the developing wheat kernel. Cereal Chem. 23: 548–559.
269. Sapirstein, HD (1981): Wheat cultivar identification by computer analysis of gliadin electrophoretograms. Thesis, Univ. Manitoba, Winnipeg, Canada.
270. Sarkar, P, and GL Stebbins (1956): Morphological evidence concerning the origin of the B genome in wheat. Amer. J. Bot. 43: 297–304.
271. Sax, K (1922): Sterility in wheat hybrids. II. Chromosome behaviour in partially sterile hybrids. Genetics 7: 49–68.
272. Schmidt, JM (1971): Cytoplasmic male sterility and fertility restoration. Seiken Zihô 22: 113.
273. Schmidt, JW, PJ Mattern, VA Johnson, and R Morris (1974): Investigations on the genetics of bread wheat baking quality. Genetic Lectures 3: 83–101, Oregon State University Press.
274. Sears, ER (1948): The cytology and genetics of wheats and their relatives. Adv. Genet. 2: 239–270.
275. Sears, ER (1969): Wheat cytogenetics. Ann. Rev. Genet. 3: 451–468.
276. Shands, HL, and G Kimber (1973): Reallocation of the genomes of *T.timopheevi* Zhuk. Proc. 4th Int. Wheat Genet. Symp. Columbia: 101–108.
277. Sharma, HC, JG Waines, and KW Foster (1981): Variability in primitive and wild wheats for useful genetic characters. Crop. Sci. 21: 555–559.
278. Shellenberger, JA, and AB Ward (1967): Experimental milling. In: Quinsenberry, KS, and LP Reitz (eds.): Wheat and Wheat Improvement: 445–469; Am. Soc. Agron., Mad., Wisconsin.
279. Shepherd, KW (1968): Chromosomal control of endosperm proteins in wheat and rye. Proc. 3rd Int. Wheat Genet. Symp. Canberra: 86–89.
280. Shepherd, KW (1973): Homoeology of wheat and alien chromosomes controlling endosperm protein phenotypes. In: Proc. 4th Int. Wheat Genet. Symp. Columbia: 745–760.
281. Shewry, PR, JC Autran, CC Nimmo, EJL Lew and DD Kasarda (1980): N-terminal amino acid sequence homology of storage protein components from barley and diploid wheat. Nature 286: 520–522.
282. Sibley, CG (1960): The electrophoretic patterns of avian egg-white proteins and taxonomic characters. Ibis 102: 215–284.
283. Siddiqui, KA (1972): Protein content and quality of wheat chromosome substitution lines. Hereditas 71: 157–160.
284. Silano, V, U De Cillis, F Pocchiari (1969): Varietal differences in albumin and globulin fractions of *T.aestivum* and of *T.durum*. J. Sci. Food Agric. 20: 260–261.
285. Silano, V, F Pocchiari, and DD Kasarda (1973): Physical characterization of α–amylase inhibitors from wheat. Biochim. Biophys. Acta 317: 139.
286. Simmonds, DH (1962): Variation in the amino acid composition of Australian wheats and flour. Cereal Chem. 39: 445–455.
287. Simmonds, DH (1974): The structure of the developing and mature triticale kernel. In: Tsen, CC (ed.): Triticale: First man-made Cereal: 105–121; AACC, St. Paul, Minn.
288. Simpson, GG (1945): Principles of classification and classification of mammals. Bull. Amer. Mus. Nat. Hist. 83: 1–350.
289. Singh, R, and U Axtell (1973): High lysine mutant gene (*hl*) that improves protein quality and biological value of grain sorgum. Crop Sci. 13: 535–539.
290. Smith, DB, and RB Flavell (1974): The relatedness and evolution of repeated nucleotide sequences in the genome of some gramineae species. Biochem. Genet. 12: 243–255.
291. Solari, RM, and EA Favret (1968): Genetic control of protein constitution in wheat endosperm and its implication on induced mutagenesis. Mutations in Plant Breeding II: 219–231; IAEA Vienna.

292. Solari, RM, and EA Favret (1970): Chromosome location of genes for protein synthesis in wheat endosperm. Biol. Genet. Inst. Fitotec. Castelar. 7: 23–26.
293. Sozinov, AA, and FA Poperelya (1979): Polymorphism of prolamines and breeding. J. Agric. Sci. 10: 21–34.
294. Sozinov, AA, and FA Poperelya (1980): Genetic classification of prolamines and its use for plant breeding. Ann. Technol. Agric. 29: 229–245.
295. Sozinov, AA, AF Stelmakh, and AJ Rybalka (1978): Genetic analysis of gliadins in common wheat varieties. Genetika USSR 14: 1955–1967.
296. Stevens, DJ (1973): Reaction of wheat proteins with sulphite. III. The accessibility of disulphide and thiol groups in flour. J. Sci. Food Agric. 17: 202–204.
297. Stuber, CE, VA Johnson, and JW Schmidt (1962): Grain protein content and its relationship to other plant and seed characteristics in the parents and progeny of a cross of *T. aestivum* L. Crop Sci. 2: 506–508.
298. Suemoto, H (1968): The origin of the cytoplasm of tetraploid wheat. Proc. 3rd Int. Wheat Genet. Symp. Canberra: 141–152.
299. Suemoto, H (1973): The origin of the cytoplasm of tetraploid wheats. Proc. 4th Int. Wheat Genet. Symp. Columbia: 141–152.
300. Sunderman, DW, M Wise, and EM Sneed (1965): Interrelationships of wheat proteins content, flour sedimentation value, farinograph peak time and dough mixing and baking characteristics in the F2 and F3 generations of winter wheat, *T.aestivum* L., Crop Sci. 5: 537–540.
301. Tandon, JP (1967): Inheritance of protein content in an intervarietal cross of wheat. J. Res. PAU. Ludhiana 4: 348–352.
302. Taylor, NW, and JE Cluskey (1962): Wheat gluten and its glutenin component: viscosity, diffusion and sedimentation studies. Arch. Biochem. Biophys. 97: 399–405.
303. Thakur, SK, and GS Sethi (1977): Genetic analysis of protein content in wheat. In: Gupta, AK (ed.): Genetics and Wheat Improvement: Oxford and IBN Publ. Co., New Delhi.
304. Thompson, WF, and MG Murray (1980): Sequence organization in pea and mung bean DNA and a model for genome evolution. Proc. 4th John Innes Symp. Norwich: 31–45.
305. Tumanyan, NT (1938): A new species of wild wheat. *Tr. Arm.* Fil. Akad. Nauk. USSR 2.
306. Ulmer, RL, and PJ Mattern (1972): The composition of flour protein from wheat cultivars differing in genetic ability to produce protein. Agron. Abstr. Amer. Soc. Agron: 71.
307. Upcroft, JA, and J Done (1974): Starch gel electrophoresis of plant NADH-nitrate reductase and nitrite reductase. J. Exper. Bot. 25: 503–508.
308. Vanecko, S, and JE Varner (1955): Studies on nitrite metabolism in higher plants. Plant Physiol. 30: 388–391.
309. Vardi, A (1973): Introgression between different ploidy levels in the wheat group. In: Proc. 4th Int. Wheat Genet. Symp. Columbia: 131–141.
310. Villegas, E, CE McDonald, and KA Giles (1970): Variability in the lysine content of wheat, rye and triticale protein. Cereal Chem. 47: 746–757.
311. Vogel, KP, VA Johnson, and PJ Mattern (1973): Results of systematic analysis for protein and lysine composition of common wheats (*T.aestivum* L.) in the USDA world collection. Agr. Exp. Stn. Univ. Nebraska, College of Agriculture, Res. Bull: 258–271.
312. Vogel, KP, VA Johnson, and PJ Mattern (1975): Re-evaluation of common wheats from the USDA world wheat collection for protein and lysine content. Nebraska Res. Bull. 272: 36.
313. Waines, JG (1976): Electrophoretic-systematic studies in *Aegilops*. Thesis Univ. California, Riverside.
314. Waines, JG (1973): Chromosomal location of genes controlling endosperm protein production in *Triticum aestivum* cv. Chinese spring. Proc. 4th Int. Wheat Genet. Symp., Columbia: 873–877.
315. Warner, RL, and CF Konzak (1975): Nitrate reductase activity in wheat substitution lines. Wheat Newsl. 21: 160–161.

316. Wasik, RJ, and W Bushuk (1975): Relation between molecular weight distribution of endosperm proteins and spaghetti-making quality of wheats. Cereal Chem. 52: 322–328.
317. Watson, CA, and JR Welsh (1966): Monosomics. Cereal Sci. Today 11: 286–290.
318. Weber, K, and M Osborn (1969): The reliability of molecular weight determinations by dodecyl sulfate-polyacrylamide gel electrophoresis. J. Biol. Chem. 244: 4406–4412.
319. Wells, DG, and CR Cowley (1976): Registration of SD69103, Hand and Flex winter wheat germplasm. Crop Sci. 16: 888.
320. Welsh, JR, and ER Hehn (1964): The effect of chromosome 1D on hexaploid wheat flour quality. Crop Sci. 4: 320–323.
321. Wildman, SG, K Chen, JC Gray, SD Kung, P Kwanyeun, and K Sakano (1975): Evolution of ferredoxin and fraction I protein in the genus *Nicotiana*. In: Birky, CW jr, PS Perlman, and JT Byers (eds.): Genetics and Biogenesis of Mitochondria and Chloroplasts: 310–329; Ohio State Univ. Press, Columbus.
322. Woods, KR, EC Paulsen, RL Engle, and JH Pert (1958): Starch gel electrophoresis of some invertebrate sera. Science 127: 519–520.
323. Worede, M (1974): Genetic improvement of quality and agronomic characteristics of *durum* wheat for Ethiopia. Thesis, Univ. Nebraska, Lincoln.
324. Worzella, WW (1942): Inheritance and inter-relationship of the components of quality, cold resistance and morphological characters in wheat hybrids. J. Agric. Res. 65: 501–522.
325. Woychik, JH, JA Boundy, and RJ Dimler (1961): Starch gel electrophoresis of wheat gluten proteins with concentrated urea. Arch. Biochem. Biophys. 94: 477–482.
326. Wrigley, CW (1970): Protein mapping by combined gel electrofocusing and electrophoresis. Application to the study of genotypic variations in wheat gliadins. Biochem. Gen. 4: 509–516.
327. Wrigley, CW (1977): Characterization and analysis of cereal products in foods by protein electrophoresis. Food Technol. Aust. 29: 17–20.
328. Wrigley, CW, JL Autran, and KW Bushuk (1982): Identification of cereal varieties by gel electrophoresis of the grain proteins. (In press).
329. Wrigley, CW, PJ Robinson, and WT Williams (1981): Association between electrophoretic patterns of gliadin proteins and quality characteristics of wheat cultivars. J. Sci. Food Agric. 32: 433–442.
330. Wrigley, CW, and KW Shepherd (1973): Electrofocusing of grain proteins from wheat genotypes. Ann. N. Y. Acad. Sci. 209: 154–162.
331. Wrigley, CW, and KW Shepherd (1974): Identification of Australian wheat cultivars by laboratory procedures: Examination of pure samples of grain. Aust. J. Exp. Agr. Anim. Husb. 14: 796–804.
332. Wrigley, CW, and KW Shepherd (1977): Pedigree investigation using biochemical markers: the wheat cultivar Gabo. Aust. J. Exp. Agric. Anim. Husb. 17: 1028–1031.
333. Yampolsky, C (1957): Wheat. Wallerstein Lab. Commun. 20: 343–358.
334. Ycas, M (1972): *De Novo* origin of periodic proteins. J. Mol. Evol. 2: 17–21.
335. Zhukovsky, PM (1964): Cultured Plants and their Relatives (in Russian). Leningrad.
336. Zillman, RR, and W Bushuk (1979): Wheat cultivar identification by gliadin electrophoretograms. III. Catalogue of electrophoregram formulas of Canadian wheat cultivars. Canad. J. Plant Sci. 59: 287–298.
337. Zohari, D, and M Feldman (1962): Hybridisation between amphidiploids and the evolution of polyploids in the wheat (*Aegilops-Triticum*) group. Evolution 16: 44–61.

5. Characterization and Synthesis of Barley Seed Proteins

P.R. SHEWRY and B.J. MIFLIN

INTRODUCTION

Many ancient religions taught that barley was the first crop; recent archaeological discoveries support this in that they show that barley was used as a food source 17,000 to 18,000 years ago [185]. The major use of the crop in developed countries today is for feeding domestic animals but considerable quantities are still used for direct human consumption in the Third World. As a food for non-ruminant animals it is limited chiefly by its content of essential amino acids although for young animals its total protein content is also slightly too low. However, when barley is supplemented by added lysine and threonine then the biological value of the protein is high and the growth rate of pigs is increased [48, 49]. The normal supplement in feeding rations is not synthetic amino acids but rather soybean or other high protein meal. Because all of these supplements are not readily available in Europe but have to be imported (for example, in 1979 Europe imported 13 million tonnes of soyabean meal worth 3.3×10^9 U.S. dollars) there has been a considerable economic stimulus to the study of barley seed proteins. This upsurge in research has paralleled similar studies on maize seed proteins (which are also deficient in lysine). In contrast there has been a long and continuous study of wheat proteins probably because of their importance in breadmaking. In this article we intend to bring together many of the results of recent work on barley grain proteins and consider these in the broader context of past studies and, briefly, in relation to other cereal seed proteins. Only passing reference will be made to effects of mineral nutrition and high lysine mutations on seed proteins as these are covered in another chapter. The main emphasis will be on the chemical characteristics of the seed proteins and their contributions to the total N content and nutritional quality of the seed rather than on any catalytic roles that they may have. This will mean that we shall be particularly concerned with the storage proteins; since these are the predominant proteins in the seed they have the greatest influence on amino acid composition and are thus of most importance in determining the nutritional quality.

Early studies on the protein components of barley seeds were made by Einhof [38] who showed that alcohol-soluble proteins were present. However the first systematic extraction and classification of barley seed proteins was probably made by Osborne [120] and Osborne and Harris [123]. On the basis of similar studies of other plant and seed proteins Osborne [121] put forward his system of protein classification. In this they are divided into groups according to their solubility in water (albumins), dilute salt solutions (globulins), aqueous alcohols (prolamins) and dilute acid or alkali (glutelins). This classification has remained in use and we shall broadly follow it although the albumins and globulins will be treated together as a salt-soluble group. Osborne and other contemporary workers originally gave names to all of the different fractions of each species; the only one remaining in common use for barley is the term 'hordein', originally used by Proust [131], for the prolamin fraction. The term prolamin was suggested as a generic term because all of the alcohol soluble proteins of cereal seeds have a large proportion of proline and amide-N, chiefly present in glutamine.

Over the period between Osborne's studies and the present day many advances have been made in methods for protein separation. Most of these have been applied to barley proteins. In earlier studies workers used ultracentrifugation techniques or Tiselius moving boundary electrophoresis. More recently electrophoresis of proteins in polyacrylamide gels (PAGE) has been the most widely used technique. This type of electrophoresis in the presence of the detergent sodium dodecyl sulphate (SDS) has been particularly valuable for separating the generally insoluble hordeins and glutelins. However, any single method is insufficient to resolve all the protein components of any one fraction and multiple separation systems (e.g. two-dimensional gel electrophoresis) have also been used.

In seeking to understand the nature of barley seed proteins, and perhaps to alter their balance within the seed, it is not sufficient to isolate and chemically characterize them. It is also important to understand their function within the seed, to know how, when and where they are synthesized; to recognize their fate during seed development, maturation and germination; to determine their genetics and to define other factors that regulate their abundance within the seed. Where relevant information exists we will also discuss the major groups of proteins in these terms.

SALT-SOLUBLE PROTEINS

Extraction and Amino Acid Composition

Osborne [120] extracted a salt-soluble fraction from milled grain using a solution of 10% sodium chloride. This solution was then saturated with ammonium sulphate and the precipitated proteins redissolved in salt-solution and dialysed against water. The precipitated globulin was removed by filtration to give a clear solution of albumin which was then coagulated by heating at 65–70°C. Osborne chose this preparation procedure, rather than sequential extraction with water followed by sodium chloride solution, because he appreciated that the presence of salts in all plant materials made it impossible to extract a true albumin fraction with water. He determined the amounts of albumin and globulin present in the grain as 0.3% and 1.95% by weight respectively.

Table 1. Amino acid composition of albumins, globulins and total salt-soluble proteins.

Amino Acid	Albumin	Globulin	Protoplasmic protein	Albumin	Globulin	Albumin	Globulin	Total salt-soluble proteins	
	← [1]Folkes & Yemm (1956) →			[2]Waldschmidt-Leitz & Brutscheck(1955)		[3]Brandt (1976)		[4]Landry et al (1972)	[5]Shewry et al (1979)
Asp	10.7	9.2	8.8	12.2	8.6	10.4	8.1	9.9	8.2
Thr	4.5	3.9	5.3	3.8	4.4	4.6	4.4	4.9	4.1
Ser	5.5	6.4	4.6	5.9	6.7	5.4	5.4	6.5	5.0
Glu	9.6	11.4	10.6	20.7	23.1	14.5	15.5	13.4	13.9
Pro	5.6	4.4	5.2	3.5	5.6	7.3	9.3	6.9	8.4
Gly	9.0	17.5	9.2	8.4	7.5	8.9	9.6	10.9	9.6
Ala	9.6	1.1	11.0	included in Glu		10.6	7.9	9.3	7.9
Cys	2.0	4.3	1.2	1.7	4.6	0.5	0	2.2	5.3
Val	7.8	6.7	6.8	8.9	7.9	7.2	7.6	6.2	6.5
Met	1.9	1.5	1.6	1.6	1.2	1.9	1.8	1.6	1.7
Ile	5.5	3.6	6.0	6.7	4.0	4.1	3.8	3.4	3.3
Leu	7.6	7.4	8.0	8.0	8.1	7.9	7.9	7.2	7.1
Tyr	3.3	2.3	3.0	4.8	2.1	2.4	2.8	2.5	2.7
Phe	4.0	3.4	4.1	4.2	4.2	3.7	3.6	3.5	2.9
His	1.9	1.7	1.9	1.5	1.7	1.9	1.7	1.9	2.3
Lys	5.3	5.2	5.2	4.2	2.7	5.1	5.3	4.9	4.0
Arg	4.3	9.0	5.0	4.1	7.7	3.8	5.5	4.8	7.1
Trp	1.7	1.1	2.5	n.r.	n.r.	n.r.	n.r.	n.r.	n.r.

Results are expressed as mole %.   n.r. = not recorded.
Values for glutamate and aspartate include the amides glutamine and asparagine respectively.

References
1 Recalculated from Folkes & Yemm (1956). Data originally expressed as N in % total N. The protoplasmic proteins are prepared from barley shoots.
2 Recalculated from Waldschmidt-Leitz & Brutscheck (1955). Data originally expressed as % total N.
3 Brandt (1976). Fractions are from 28 day old (physiologically mature) endosperms. Values for Cys and Met are from acid hydrolyses.
4 Landry et al. (1972). Values for Cys and Met are from acid hydrolyses.
5 Shewry et al. (1979b). Cysteine was determined as pyridylethylcysteine.

Fractions in 3 & 4 were prepared from endosperms only, other fractions were from milled whole seed.

Further major advances were made by Bishop [13] who established conditions for the reproducible extraction of the protein fractions, notably the use of multiple extractions with efficient stirring and the use of potassium sulphate instead of sodium chloride. Waldschmidt-Leitz and Brutscheck [181] and Folkes and Yemm [43] reported amino acid analyses of albumin and globulin fractions prepared by dialysis of a total salt-soluble fraction prepared from whole seeds, removal of the precipitated globulins by centrifugation and finally heat-coagulation of the clear albumin solution. These analyses are given in Table 1. There are some differences between the two sets of data, which are to be expected considering that they were made in the early days of amino acid analysis. With the exception of the glycine content of the globulin fraction, these differences are generally minor and both sets of data show characteristic differences between the compositions of the albumin and globulin fractions, notably more cysteine and arginine in the globulins. Folkes and Yemm [43] further showed that a preparation of protoplasmic (soluble) proteins from barley shoots had a very similar composition to the seed albumin fraction (Table 1).

More recent analyses of sequentially extracted albumin and globulin fractions showed little difference between the two ([76]; Table 1) both compositions being similar to those reported for the combined fractions from whole seeds [82] or endosperms [156]. It is probable that the sequential extraction procedure results in considerable cross-contamination of the two fractions and Bishop [13] commented on the difficulty of reproducibly separating albumin and globulin fractions. This has become increasingly apparent since, as is shown by the wide variation in the relative amounts of these two groups extracted by different workers. Whereas Osborne [120] reported that albumins and globulins represented 2.8 and 18.1% respectively of the total seed N, Rhodes and Gill [138], using similar preparation procedures, have recently shown that albumins represented between 8.5 and 12.6% and globulins between 2.3 and 5.7% of the total grain N. The values of Bishop [13] fall between these two extremes. He showed that about half of the total salt-soluble proteins were albumins as determined by heat coagulation at pH 4.6. When the heat coagulation procedure was used to fractionation an aqueous extract, however, the amount of albumins found were only approximately 60% of those determined in the total salt-soluble fraction. Bishop concluded that some globulins were present in the albumin fraction prepared from the salt-extract. Other contrasting results come from two recent studies of the protein fractions of the same line (NP113) and using grain of comparable protein content (12.5 and 13.8%). Bansal *et al.* [7] showed that albumins and globulins represented 20.8 and 11.2% of the total grain proteins while Singh and Sastry [166] determined these two fractions as 13.9 and 11.0% respectively.

These difficulties in fractionating albumins and globulins, and evidence from electrophoresis of the presence of common polypeptides in the two fractions (see below), has resulted in recent emphasis on the extraction and characterization of total salt-soluble protein fractions. The proportion of the total grain N extracted in the salt-soluble fraction, as also the proportions in the other protein fractions, varies depending on the total N content of the grain (see the chapter by Doll in this volume). Bishop [13] reported the amount to vary between 24 and 36% of the total grain N while our recent study (Kirkman *et al.*, [72]) has shown lower amounts, between 17 and 26%. More importantly, the relative amounts of N extracted in the salt-soluble and glutelin fractions are affected by the chemical conditions used for extraction, such as the salt concentration and pH, the presence or absence of a reducing agent, and also the fineness of grinding of the meal. There is an inverse relationship between the N recovered in these fractions indicating that un-extracted salt-soluble proteins are recovered in the glutelins. It is also probable that denaturation during grain storage or during extraction may result in the insolubility of salt-soluble proteins and their presence in the glutelins. These aspects have been comprehensively reviewed by Djurtoft [33] and by Préaux and Lontie [130] and will not be considered further here.

All of the N in the salt-soluble fraction is not present in proteins. Bishop [13] first noted the presence of non-protein components and showed that 58% of his salt-soluble N was precipitated by 5% trichloroacetic acid (TCA) and 50% by 3% sulphosalicylic acid. Djurtoft [33] showed that 50%–60% of his salt-soluble fraction was non-dialysable. We have made a detailed study of the salt-soluble fraction of cv. Bomi [156]. This fraction represented 23% of the total grain N, of which 70% was retained during dialysis and 61% precipitated by 5% TCA. Addition of ethanol to 70% v/v precipitated less protein (39%), indicating that some of the protein components were alcohol-soluble. It is the failure to consider these non-protein components which has probably resulted in a number of reports of unusually high amounts of salt-soluble proteins. The nature of some of these non-protein components is not completely known. However, they certainly include peptides and free amino acids (see below).

Characterization

a) *General Properties*

Osborne [120] separated total salt-soluble proteins into albumins and globulins by dialysis against water and other workers tried to improve this by using differential precipitation by low pH, mineral salts or heating [13, 14, 144, 178]. Major advances in the separation and characterization of these proteins only came, however, with the application of ultracentrifugation. Much of the early work using this procedure has been comprehensively reviewed and evaluated by Djurtoft [33] and only a brief outline will be given here. Quensel [133] extracted salt-soluble proteins using 1M NaCl in 0.1M phosphate buffer at pH 7.0. The globulin components of this fraction (initially precipitated by dialysis against water) were separated

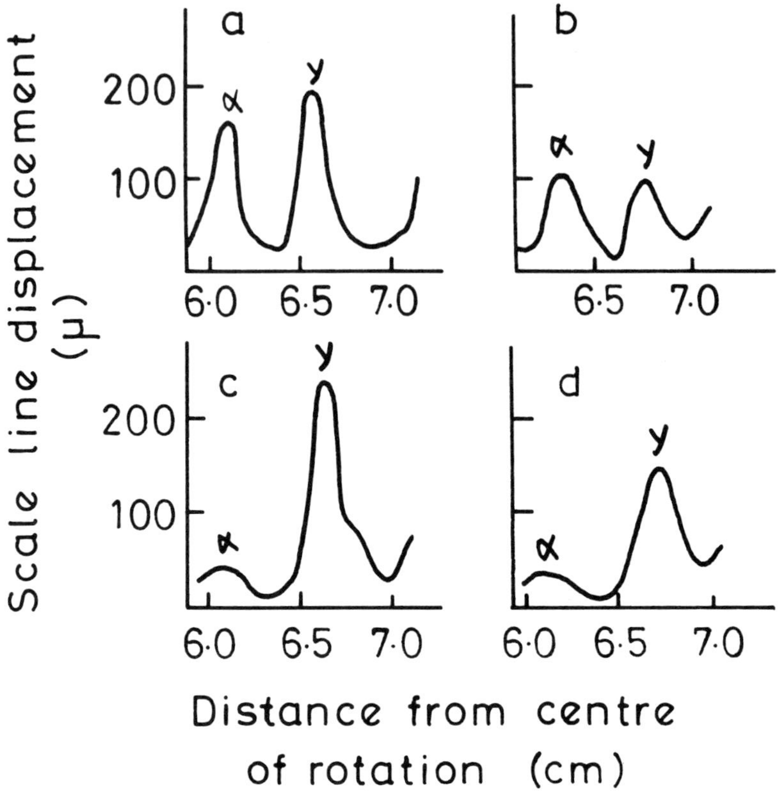

Figure 1. Sedimentation diagrams of globulin solutions prepared from seed of barley (a), rye (b), oats (c), and maize (d). Redrawn from Danielsson (1949).

by ultracentrifugation into four components termed α, β, γ and δ globulins, with sedimentation constants of 2.49, 6.21, 8.30 and 12.0 respectively. Of these the β and δ components were often present in small amounts and were only reliably detected after preliminary fractionation of the preparations by salt-precipitation. The α and γ components, however, were both present in considerable quantities (see Figure 1). Although Quensel [133] did not obtain pure preparations of any of the fractions he was able to estimate molecular weights of 26,000, 100,000, 166,000 and 300,000 for the α, β, γ, and δ globulins respectively. Danielsson [30] used Quensel's methods to compare the globulin components of barley and other cereals. This showed the presence of α and γ globulin components in the seed of rye (*Secale cereale*), wheat (*Triticum vulgare*), maize (*Zea mays*), and oats (*Avena sativa*), although there was relatively less α globulin in wheat, maize and oats (Figures 1, 2). He was unable, however, to detect β and δ globulins in any species except barley. In contrast to Quensel [133] he reported that β globulin was present in high concentration in barley seed, although he did not label this component in his sedimentation diagrams

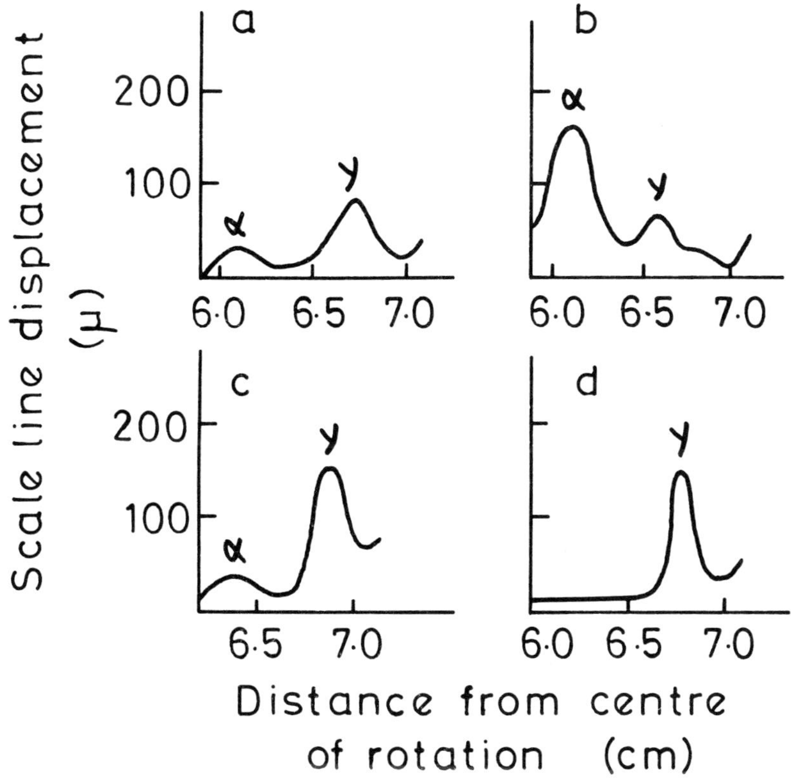

Figure 2. Sedimentation diagrams of globulin solutions prepared from wheat grain a, whole grain; b, flour (endosperm); c, bran (aleurone); d, embryo.

(Figure 1). He also showed that the γ component alone occurred in the embryos of wheat and barley. The α component appeared to be concentrated in the endosperms of both cereals, although the bran fraction of wheat (which contains the aleurone) contained a higher concentration of γ globulin than α globulin (Figure 2). The β globulin of barley also appeared to be located in the endosperm. Danielsson [30] determined the molecular weights of purified α globulin from barley endosperms and γ globulin from wheat embryos as 29,000 and 210,000 respectively. Neither Quensel [133] nor Danielsson [30] reported any investigations of the albumin proteins.

More recently Djurtoft [33] has reported a detailed re-examination of methods for the extraction of salt-soluble proteins and their analysis by ammonium sulphate fractionation and ultracentrifugation. His initial salt-soluble extract contained 24.8% of the total grain N, of which 48.8% was lost on dialysis. Sub-fractions of the total salt-soluble proteins were prepared by precipitation with ammonium sulphate and dialysis of the precipitates against water. Ultracentrifugation of these fractions

Table 2. Comparison of amino acid compositions of total globulins with α globulin, β globulin and < α-component.

	Total Globulin		α Globulin	β Globulin	<α-Component
	1	2	3	4	5
Asp	9.2	8.6	9.8	7.2	11.7
Thr	3.9	4.4	4.6	5.3	5.8
Ser	6.4	6.7	6.8	4.4	6.7
Glu	11.4	23.1	9.4	13.1	9.8
Pro	4.4	5.6	5.4	12.5	9.1
Gly	17.5	7.5	10.8	6.0	8.7
Ala	1.1	included in Glu	9.7	7.9	9.6
Cys	4.3	4.6	5.3	8.4	5.6
Val	6.7	7.9	8.3	8.8	6.1
Met	1.5	1.2	1.8	2.3	1.7
Ile	3.6	4.0	2.3	2.6	4.0
Leu	7.4	8.1	6.4	9.0	6.7
Tyr	2.3	2.1	} 2.3	5.1	2.4
Phe	3.4	4.2		3.1	2.9
His	1.7	1.7	2.0	1.1	1.1
Lys	5.2	2.7	5.6	1.5	3.6
Arg	9.0	7.7	9.6	1.7	4.5
Trp	1.1	n.r.	n.r.	n.r.	n.r.

Results are expressed as mole %.   nr. = not reported
Values for glutamate and aspartate include the amides glutamine and asparagine respectively.

References

1. Recalculated from Folkes & Yemm (1956) (see Table 1).
2. Recalculated from Waldschmidt-Leitz & Brutscheck (1955). (see Table 1)
3. Recalculated from Djurtoft (1961). Data originally expressed as g amino acid residue/100 g protein
4. Recalculated from Jensen (1952). Data originally expressed as N in % of total chromatographic N.
5. Recalculated from Djurtoft (1961) (fraction E20DC9/10BDC). Data originally expressed as N in % of total chromatographic N.

gave five purified groups of proteins. Three of these had similar sedimentation coefficients to the α, β, and γ globulins of Quensel [133], although Djurtoft called them α, β and γ components because he considered it doubtful that they contained only globulins. He also purified a group with sedimentation coefficients below those of α globulin (called < α-components) and a group of albumins. These five groups accounted for 46.5% of the salt-soluble protein (23.8% of the total salt-soluble N or 5.8% of the total seed N). He calculated the yields of the groups from 1 kg (dry wt.)

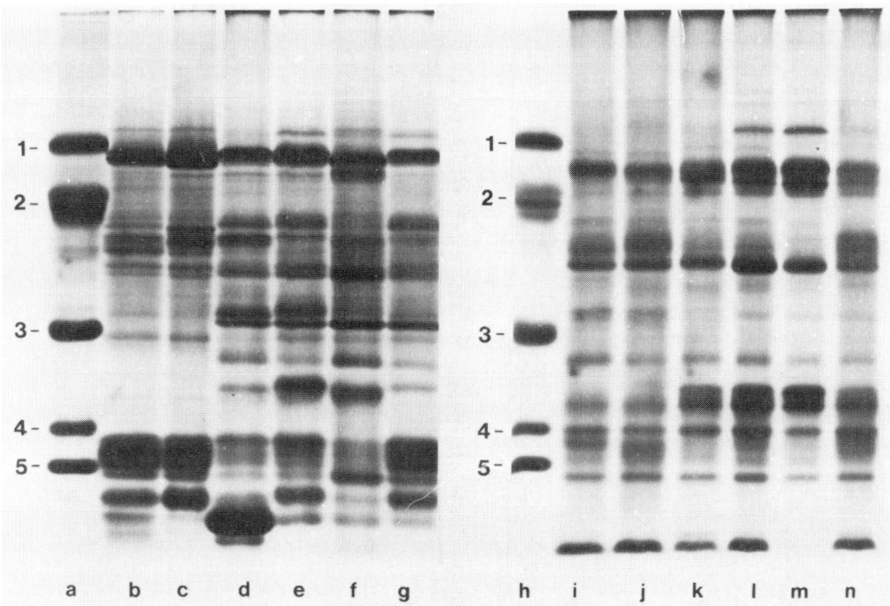

Figure 3. SDS-PAGE of albumin (b–g) and globulin (i–n) fractions from various cultivars of barley. b, i Vada; c, j Ark Royal; d, k Hiproly; e, l Risø 7; f, m Risø 1508; g, n Bomi. a, h are molecular weight standards: 1, 66,000; 2, 45,000; 3, 24,000; 4, 18,400; 5, 14,300. From Rhodes and Gill (1980).

barley to be 3.38 g α-components, 0.84 g β-components, 0.47 g γ-components, and about 0.1 g and 1.3 g of albumin and <α-components respectively. Djurtoft [33] also reported amino acid analyses of three of his fractions, one of which contained only components with sedimentation coefficients corresponding to the <α-components. One of the other fractions was a mixture of components with sedimentation coefficients equivalent to those of the α and β globulins of Quensel [133]. Comparison of the amino acid composition of this fraction with a composition previously reported for β globulin by Jensen [61] enabled him to calculate the amino acid composition of α globulin. The amino acid compositions of α globulin, β globulin and the <α-components are given in Table 2. The fractions differ considerably in their amino acid compositions, notably cysteine, leucine, glutamate + glutamine, proline and aromatics are higher in β globulins, basic amino acids are higher in α globulin, and the <α-component has a composition which is in many respects intermediate between the two. Variation in the ratios of these components could account for many of the differences in composition of the globulin preparations of Folkes and Yemm [43] and Waldschmidt-Leitz and Brutscheck [181] (Table 2).

Although other preparative procedures such as Tiselius electrophoresis and ion exchange chromatography were in use by some workers at the same time as the

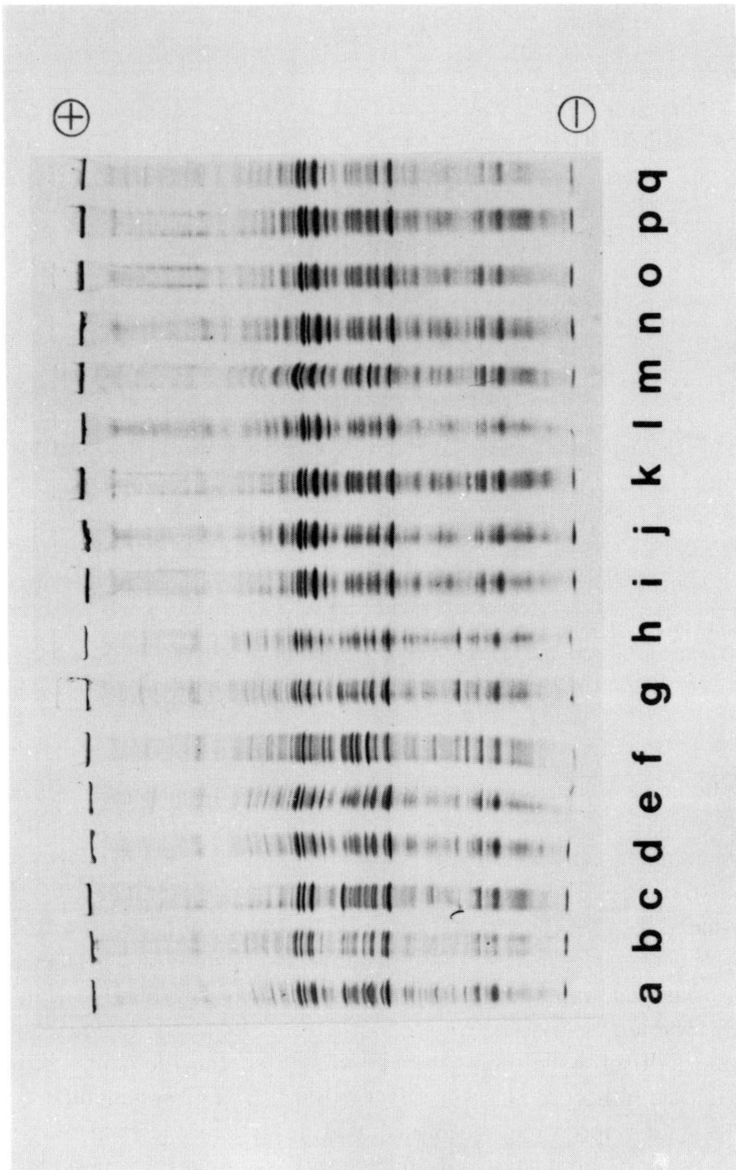

Figure 4. IEF (pH 3.5–10) of reduced and pyridylethylated salt-soluble protein fractions from exotic and commercial barley lines. a–h Exotic lines from the World Barley Collection. a, C1 1916 (Canada); b, C1 2010 (Hungary); c, C1 2045 (China); d, C1 2087 (Ethiopia); e, C1 2103 (USSR); f, C1 2239 (India); g, C1 2245 (China); h, C1 2260 (Japan). i–q Commercial varieties (Western Europe) i, Varunda; j, Golden Promise; k, Malta; 1, Harkra; m, Hoppel; n, Jupiter; o, Maris Mink; p, Birgitta; q, Julia. The World Barley Collection lines were selected to give a wide range of geographic origins. The commercial varieties were selected on the basis of differences in their hordein patterns (see p. 000), indicating the possible presence of other genotypic differences. (Authors unpublished results.)

ultracentrifugation studies, none achieved general acceptance or resulted in the purification of fractions comparable to the groups of globulins separated by ultracentrifugation.

The next major advance was the use of gel electrophoretic procedures; although the introduction of these coincided with a decline in the use of ultracentrifugation and as a result the α, β, γ and δ globulins (and the <α-component and albumins of Djurtoft [33]) have not been subjected to electrophoretic analysis. PAGE and isoelectric focusing (IEF) have, however, been extensively used to analyze total preparations of albumins (Figure 3), globulins (Figure 3) and total salt-soluble proteins (Figure 4), especially in investigations of high-lysine mutants and the cultivars from which they were derived [20, 103, 104, 111, 112, 138, 152, 162, 164, 166, 175]. Studies show that albumin and globulin fractions contain a large number of polypeptides (Figures 3, 4). This is especially apparent when the fractions are separated by IEF [104, 152, 164]. There is also sometimes evidence for similar polypeptides in the albumin and globulin fractions [20] indicating the presence of common components in the crude fractions analysed in this type of study.

PAGE has also been valuable in the purification of certain individual components of the salt-soluble fraction, by enabling preparative procedures to be monitored and the homogeneity of the final isolated protein to be assessed. The purification and characterization of certain specific salt-soluble components will now be described.

b) *Protease inhibitors*

The high-lysine line Hiproly was discovered by Munck *et al.* [113] who screened the world barley collection using the dye-binding capacity technique of Mossberg [110]. The high-lysine character appears to be controlled by a single recessive gene, calles *lys*, which has been mapped on chromosome 7 [66]. Early studies using PAGE showed that Hiproly differed from its normal-lysine sister line (C1 4362) in the polypeptide compositions of its albumin and globulin fractions, notably the presence of increased amounts of certain albumin components [111, 112, 175]. This has been confirmed in a recent study by Rhodes and Gill [138]. Ingversen and Køie [55] fractionated salt-soluble proteins from a number of lines, including Hiproly, and concluded that although Hiproly and Carlsberg II (a normal line) contained identical major components (as demonstrated by gel filtration and PAGE), certain lysine-rich proteins were present in increased amounts in Hiproly. Hejgaard and Boisen (1980) have recently shown that Hiproly has a 4 to 7-fold increased content of four major salt-soluble proteins identified by crossed immunoelectrophoresis as 'free' ß amylase, 'free' protein Z and chymotryptic inhibitors 1 and 2, the four proteins having lysine contents of 5.0, 7.1, 9.5 and 11.5 g per 100 g protein respectively. They showed that the four proteins together accounted for about 7% of the grain lysine in normal varieties, 17% in Hiproly lines and more than 50% of the increase in grain lysine due to the *lys* gene. In an independent study Jonassen and co-workers at the Carlsberg Institute purified and characterized two lysine-rich albumins, called SPIIA AND SPIIB from Hiproly [62, 63, 172]. The complete amino acid sequence of the smaller component (SPIIB) was determined showing the presence of 6 lysine residues out of a total of 72 [172] (Table 3). The sequence also showed the presence of only one methionine and no cysteine residues. The calculated mol. wt. was 8072. The larger component, SPIIA, was blocked to Edman-degradation but analysis of cyanogen bromide and tryptic peptides indicated that it was identical to SPIIB, except for the presence of an N-terminal extension which probably consisted of between 8 and 11 residues including 2 lysines (Table 3). The authors suggested that SPIIB was a fragment produced by limited proteolysis during purification. The total amounts of SPIIA and SPIIB in Hiproly, its normal lysine sister-line (CI 4362) and the normal commercial line Bomi were determined by immunoassay [63]. This showed that Hiproly contained 24 times as much SPII albumin as Bomi and 3.6 times as much as CI 4362. The amounts present in the three lines were 0.284, 0.077 and 0.012 g per 100 g dry seed of Hiproly, CI 4362 and Bomi respectively. The author calculated that the elevated content of SPII albumin accounted for 37% of the difference in lysine content between Bomi and Hiproly and for 19% of that between CI 4362 and Hiproly.

More recently Svendsen *et al.* [171] showed that the SPII proteins were immunologically identical to the chymotrypsin inhibitor 2 of Hejgaard and Boisen [51] and that they were strongly inhibitory to subtilisin, weakly inhibitory to chymotrypsin but not inhibitory to trypsin. Comparison of the amino acid sequences of SPIIB and a protease inhibitor from potatoes (potato inhibitor 1) showed the presence of considerable sequence homology, 45% of the amino acids being in identical positions [171]. This is a striking similarity in view of the wide evolutionary divergence of the two species. The SPII proteins also resemble potato inhibitor 1 in their inhibitory activity towards subtilisin and chymotrypsin, but differ in not inhibiting trypsin [95]. The role of the SPII proteins in Hiproly grain is not known.

Mikola and Suolinna [108] purified a trypsin inhibitor from the water-soluble fraction of barley grain. The purified protein gave a single band on PAGE at pH 3.5 and had a mol. wt. of 14,400 as determined by equilibrium ultracentrifugation. It had 127 residues per mole as calculated from amino acid analysis (Table 3), and a formula weight of 14,055. The composition was notable for the high content of cysteine, and this all appeared to be in the disulphide linked form in the unhydrolysed protein. The high disulphide content is in contrast to the SPII proteins of Hiproly (see above and Table 3), but is characteristic of other proteinase inhibitors from plant sources [139, 140]. The purified protein was inhibitory to trypsin but not to a number of other proteolytic enzymes including chymotrypsin and subtilopeptidase A. The authors calculated that the inhibitor accounted for 4.5% of the water soluble proteins, or 0.045% of the total grain. Mikola and Enari [106] have shown that barley grain also contains two other types of protease inhibitors which are active against malt endopeptidases, microbial proteases and chymotrypsin. Inhibi-

Table 3. The amino acid compositions (expressed as residues per mole) of high-lysine albumins (SPII proteins), Trypsin inhibitor and phytohaemagglutinin from barley grain.

	← Hiproly high-lysine proteins →			Trypsin inhibitor[c]	Phyto-haemagglutinin[d]
	SPIIB[a]	SPIIB[b]	SPIIA[a]		
Asp	7.1	7	7.8	10	22.6
Thr	3.2	4	3.8	7	16.8
Ser	1.1	1	3.2	8	14.2
Glu	10.3	9	11.4	14	22.5
Pro	4.2	4	4.8	11	15.3
Gly	3.8	5	5.9	10	25.6
Ala	3.4	4	4.0	10	29.8
Cys	0	0	0	10	0
Val	10.9	10	11.2	6	18.5
Met	1.0	1	1.0	2	4.1
Ile	6.0	6	5.6	5	8.4
Leu	6.4	6	5.8	9	22.8
Tyr	1.1	1	1.0	5	4.8
Phe	1.0	1	1.0	3	9.6
His	0.9	1	0.8	3	6.6
Lys	6.2	6	7.5	2	19.9
Arg	4.9	5	4.9	9	10.7
Trp	0.8	1	0.8	3	n.r.

n.r. = not reported

Values reported for aspartate and glutamate include the amides asparagine and glutamine respectively.

[a] Composition from amino acid analysis (Jonassen 1980a)

[b] Composition from amino acid sequence (Svendsen *et al.* 1980b)

[c] Nearest integer values calculated from 24 and 72 hr hydrolyses (Mikola & Suolinna, 1969).

[d] Calculated from 24, 48 and 72 hr hydrolyses (Partridge *et al.* 1976).

tors of all three types occur in the embryo, aleurone and starchy endosperm, but they differ in their relative activities in the three tissues [73]. Also the trypsin inhibitor purified by Mikola and Suolinna [108] appears to be only present in the aleurone and starchy endosperm; different trypsin inhibitors are present in the embryo. The embryonal and endospermal inhibitors were separated by gel-filtration on Sephadex G75 [107].

The function of those inhibitors in the grain is not known, although it has been suggested that they might provide protection by inhibiting the proteases of predators, or prevent premature proteolysis of the storage proteins by the seed's own proteases. Where present in large amounts they may also have an additional function as storage proteins for germination, when the protease inhibitor activity in the endosperm disappears within 4 to 5 days [73].

c) *Phytohaemagglutinins*

Foriers *et al.* [45, 46] purified a protein from barley embryos (barley germ) which agglutinated trypsinized rabbit erythrocytes. The yield was between 120 and 240 μg per g of embryos. Isoelectric focusing showed the presence of two major components, pI ~ 5.8 and 6.1, which were present in approximately equal amounts. Isoelectric focusing in the presence of 3M urea again gave two bands (pI ~ 5.4 and 6.4) while SDS-PAGE showed the presence of two non-covalently linked polypeptide chains with apparent molecular weights of 17,000 and 23,000. Dansylation showed two N-terminal amino acids, alanine and glycine. The authors suggested that the protein was a dimer consisting of one α (mol. wt. 17,000) and one ß (mol. wt. 23,000) subunit. Partridge *et al.* [125] also purified a phytohaemagglutinin from whole milled barley grain. This gave single bands on electrophoresis at pH 8.3, SDS-PAGE and IEF (pI 4.95) and equilibrium ultracentrifugation gave a molecular weight of 31,000. Amino acid analysis (Table 3) showed the absence of cysteine and the presence of four residues of methionine. There were also about 20 residues of lysine, which represented almost 8% of the total. The protein did not appear to be a glycoprotein and, in contrast to the preparation of Foriers *et al.* [45, 46], did not agglutinate trypsinized rabbit erythrocytes. The location of the protein in the embryo or endosperm was not determined.

Thus it appears that barley seeds contain at least two proteins which have phytohaemagglutinin activity. The agglutinin of Foriers *et al.* [45, 46] may be homologous to the wheat germ agglutinin. The latter is a dimer (mol. wt. 36,000) of two identical monomers (mol. wt. 18,000) which can be dissociated by extremes of pH, by urea or by high salt concentrations. Although the molecule is rich in cysteine (about 37 residues per monomer) it does not appear to be disulphide bonded [50].

d) *Purothionins*

The purothionins are a group of proteins which are extracted as lipoprotein complexes in petroleum ether extracts of grain of wheat, barley, rye and related species of *Triticum* and *Aegilops* (see Carbonero *et al.*, [29] for a detailed bibliography). Their presence in barley grain was first demonstrated by Redman and Fisher [137] who reported that petroleum ether extracts of milled grain contained two proteins (which they called α and ß hordothionins) which, after acid treatment to remove the lipid components, had similar electrophoretic characteristics to the purothionins previously characterized from wheat. They purified the major component (α hordothionin) and showed that it was related to wheat purothionins in its amino acid composition (Table 4), tryptic peptide map, immunological reaction and C-terminal amino acid (lysine). The presence of purothionin-like proteins in barley was confirmed by Hoseney *et al.* [53]. Although Redman and Fisher [137] calculated a molecular weight of 13,000 for α hordothionin, it is more probably close to 5,000 as recently reported for wheat purothionins [89, 90]. The amino acid compositions of purothionins are characterized by large amounts of lysine, arginine and cysteine and no methionine, or histidine (Table 4).

Table 4. Amino acid compositions (expressed as mole %) of α hordothionin from barley, α purothionin from wheat and a structurally related globulin fraction (globulin A) from barley.

	α hordothionin	α purothionin	Globulin A
Asp	5.2	4.6	4.5
Thr	5.9	5.6	7.4
Ser	9.9	10.9	9.0
Glu	2.6	2.6	2.4
Pro	4.8	4.5	4.2
Gly	9.1	9.8	10.6
Ala	5.3	4.7	4.4
Cys	16.9	17.0	18.4
Val	3.4	1.9	3.8
Met	0	0	0
Ile	0	1.1	0
Leu	10.0	10.1	8.9
Tyr	1.7	1.8	1.8
Phe	2.3	2.5	2.5
His	0	0	trace
Lys	11.5	11.3	11.2
Arg	11.3	11.5	10.9

Recalculated from Redman & Fisher, 1969.

Results are expressed as mole %.

Values reported for glutamate and aspartae include the amides glutamine and asparagine respectively.

The functions of purothionins in the grain are not known, but Carbonero *et al.* [29] have recently shown that hordothionins co-sediment with protein bodies during sucrose gradient centrifugation, where they constitute 3–4% of the fraction. They showed that the hordothionine could be extracted with dilute sulphuric acid without apparently affecting the structure of the protein body, and concluded that it was externally associated with the protein bodies and was possibly part of the protein-lipid matrix in which they were embedded.

It has also been shown that proteins present in the globulin fractions of barley and wheat are structurally identical or almost identical to purothionins prepared from the lipoprotein complexes extracted in organic solvents [136, 137] and it is possible that the complexes are in fact artefacts as suggested by Mak and Jones [90]. The amounts of hordothionins present in the seed are not precisely known. However Redman and Fisher [137] reported yields of 550 mg α-hordothionin and 35 mg ß hordothionin from 10 kg flour and larger yields (over 3 g/10 kg) of impure structurally-related globulins.

Purothionins have been shown to be toxic to animals, bacteria and yeasts, and Mak and Jones [90] have shown a similarity in the primary structures of ß purothionin and viscotoxin A3, a low molecular weight toxic protein from mistletoe.

Functions

It is probable that most components of this fraction represent the basic metabolic machinery in the cell, especially enzymes. This is consistent with the similarity of the amino acid compositions of the seed albumins and the soluble proteins of the barley shoot [43]. However, at least some of the globulin components are probably storage proteins from protein bodies present in the embryo and aleurone. The major storage protein component of oats is a globulin which has been purified and characterized by Peterson [129], who reported that the molecular weight was 322,000 and the sedimentation constant 12.1. This is in good agreement with the δ globulin component of barley (MW = 300,000; S = 12.0), although Danielsson [30] reported this component to be absent from oats and the main storage component to be γ globulin (MW = 210,000; S = 8.1). The reason for this disparity is not known but variation in sedimentation velocity for the legume globulins, which have been studied in more detail, is well documented [170].

Genetic control

Analysis of fractions from a range of genotypes shows the presence of considerable variation in the electrophoretic patterns of the total salt-soluble proteins and the individual albumin and globulin fractions. Przybylska and Kapala [132] made an extensive study of total salt-soluble proteins from 50 lines, including high-lysine mutants and exotics. These fractions had between 8 and 10 bands on the basic (pH 8.3) gel systems of Davis [32], and the patterns showed some correlation with certain morphological characters (hulled or naked, colour of hull or, in naked barleys, colour of pericarp). Some lines also had characteristic additional bands. The fractions were also separated on acidic gels (pH 4.5) when between 9 and 12 bands were observed. The patterns on this system did not appear to be related to the morphological features, but most of the bands observed showed great variation in relative intensity. Some minor bands were also absent from some of the lines. Although most of this study was carried out using milled grain samples, the authors did show that there was little or no variation between individual grain of a line. Kapala *et al.* [65] also made a serological examination of globulins from 21 lines by double diffusion and immunoelectrophoresis against antisera raised against globulins from two varieties. Although most of the varieties examined appeared to be serologically similar, some definite differences were observed. These differences did not appear to be related to those demonstrated in the previous electrophoretic study [132]. El Negoumy *et al.* [39] have also shown differences in the SDS-PAGE patterns of total salt-soluble proteins from 23 lines. They recognized the presence of 12 different bands with 6–8 present in any one line, and classified the lines into 6 groups on the various combinations of these. Genetic variation has also been reported in the electrophoretic patterns of separate albumin and globulin fractions [76, 138, 142] (Figure 3). We have analyzed total salt-soluble protein fractions by isoelectric focusing, which resolves many more components than PAGE. Figure 4 shows fractions from 8 exotic lines and 9 commercial varieties analyzed by this procedure. It shows the presence of differences in both relative intensity and the absence or presence of bands between the exotic lines. Only minor differences are present between the commercial lines, and these generally involve band intensity rather than absence or presence.

Most studies of isoenzymes in barley have been made using leaf tissue. However Nilson and Hermelin [118] studied liquid endosperms and reported differences in the patterns of esterases, leucine aminopeptidases, peroxidases, cytochrome oxidases and catalases in 12 barley varieties.

Although the presence of genetic variation in the salt-soluble fractions is well-established, relatively little is known about the genetic control of the individual components. Kreft *et al.* [79] have shown that the progeny from crosses between varieties with different patterns of water-soluble proteins exhibit differences in band intensity rather than absence or presence of bands, and interpret this as indicating differences in regulatory rather than structural genes. As would be expected the results also indicated that several genes were involved.

Free Amino Acids

The salt-soluble nitrogen fraction also contains small amounts of free amino acids. Analysis of these by ion-exchange chromatography shows the presence of

Table 5. The free and total amino acids of grains of normal barley (cv. Bomi) and mutant R2501.

	cv. Bomi					Mutant R2501			
	Free amino acids			Total amino acids		Free amino acids		Total amino acids	
	Brandt (1976)	Bright *et al* (1981)		Bright *et al* (1981)		Bright *et al* (1981)		Bright *et al* (1981)	
	mole %	mole %	nmole/ mg N	mole %	nmole/ mg N	mole %	nmole/ mg N	mole %	nmole/ mg N
Asp	5	27.0	185	5.4	1891	17.6	131	5.5	1848
Thr	9	1.6	11	4.1	1437	19.7	147	4.6	1519
Ser	12	2.9	20	5.2	1830	3.1	23	5.3	1759
Glu	27	19.0	130	24.1	8412	15.0	112	22.8	7617
Pro	5	23.4	160	14.6	5101	19.9	148	13.9	4632
Gly	0	5.6	38	6.3	2202	5.9	44	6.5	2160
Ala	24	6.9	47	6.0	2107	3.9	29	6.3	2103
Cys	0	n.d.	n.d.	n.d.	n.d.	n.d.	n.d.	n.d.	n.d.
Val	5	3.6	25	5.7	1981	2.5	19	5.8	1953
Met	0	trace	trace	1.1	378	trace	trace	1.2	400
Ile	3	n.d.	n.d.	4.0	1397	n.d.	n.d.	4.2	1393
Leu	3	n.d.	n.d.	7.7	2681	n.d.	n.d.	7.9	2642
Tyr	2	1.0	7	2.3	806	0.8	6	2.3	778
Phe	2	1.8	12	4.7	1657	1.2	9	4.6	1533
His	trace	1.8	12	2.0	701	3.1	23	2.1	708
Lys	2	1.8	12	3.0	1059	2.5	19	3.2	1081
Arg	1	3.8	26	3.7	1306	4.7	35	3.6	1217
Total			683		34945		743		33341

n.d = not determined. Values reported for glutamate and aspartate by Bright *et al* (1981) include the amides glutamine and asparagine respectively. The analyses of Brandt (1976) are for non-hydrolyzed extracts and values for glutamine and asparagine are not given. The data of Brandt (1976) is for physiologically mature endosperms 28 days after flowering. The data of Bright *et al* (1981) are for milled dry grain and are the mean of 12 (R2501) and 15 (Bomi) determinations. The data are expressed only as nmole.mg N in the original publications. The data for mole % has been calculated from these.

most of the amino acids commonly found in proteins, the non-protein amino acid γ-amino butyric acid (which is probably a universal component of all plant tissues) and other unidentified non-protein amino acids. Amino acid compositions of the free amino acids of cv. Bomi have been published by Brandt [20] and Bright *et al.* [24]. These differ in a number of respects and both are presented in Table 5. The differences may be due to the methods used. Bright *et al.* [24] analyzed hydrolyzed samples and consequently their analyses probably include some small peptides. Brandt [20] analyzed non-hydrolyzed extracts and did not determine glutamine and asparagine. It can be calculated from the data of Bright *et al.* [24] that free amino acids only account for about 2% of the total grain amino acids.

Bright *et al.* [24] have also reported the characterization of a mutant barley line, called R2501, in which the amount of free threonine is increased by over tenfold (Table 5). As a result the threonine present in the soluble fraction represents 9.6% of the total grain threonine compared to only 0.8% in the parental variety Bomi. This increase is sufficient to result in an increase in the total grain threonine from 4.1 to 4.6% of the total amino acids. This is of considerable interest as threonine is the second most important nutritionally-limiting amino acid in barley grain [48, 49].

HORDEIN

Extraction and Amino Acid Composition

Osborne [120] showed that the hordein extracted in hot 75% ethanol accounted for 37.2% of the total seed N and that although it had almost identical physical and chemical properties to gliadin from wheat and rye it differed in its elemental composition. Osborne and Harris [123] showed that hordein contained 23.3% amide N, 1.3% humin N, 70.0% non-basic N and only 4.5% basic N. Osborne and Clapp [122] and Kleinschmidt [75] both published partial amino acid analyses which showed high glutamate + glutamine (36 to 41%); low lysine (0%) and, in the analysis of Osborne and Clapp only, high proline.

Bishop [13] made a detailed study of the conditions for the extraction of the protein fractions of barley grain. He showed that it was necessary to use a standardized grinding procedure and to make multiple extractions (at least three times) with continuous shaking with each solvent. He also showed that hordein was not very soluble in cold aqueous ethanol and therefore extracted hordein with hot 70% ethanol in sealed tubes. Bishop [16] also showed that more hordein was extracted when bisulphite was added to the solvent. This observation remained unnoticed until Lontie *et al.* [87] showed that after the extraction of salt-soluble proteins and of prolamins with 60% propan-2-ol at 60°C, a further fraction could be extracted by the addition of 0.2% metabisulphite to the alcoholic solvent. Lontie and Voets [88] showed that 1% 2-mercaptoethanol, 1% thioglycollic acid or 1% sodium thiogly-

Table 6. Amino acid compositions of hordein-I, hordein-II and total hordein fractions from barley cv. Julia.

	Hordein-I	Hordein-II	Total Hordein
Asp	2.1	1.5	2.0
Thr	2.8	2.4	2.9
Ser	5.5	4.8	5.3
Glu	32.1	31.9	32.6
Pro	23.1	22.4	22.3
Gly	3.4	2.9	4.4
Ala	3.0	3.0	3.4
Cys	ND	ND	ND
Val	4.4	5.5	4.7
Met	1.0	1.2	1.1
Ile	3.5	4.2	3.4
Leu	6.5	7.1	6.6
Tyr	2.0	2.8	2.6
Phe	5.6	4.2	4.5
His	1.5	1.4	1.4
Lys	0.8	0.7	0.9
Arg	2.0	3.2	2.0

Expressed as mole %
ND = not determined. Values reported for glutamate and aspartate include the amides glutamine and asparagine respectively.
Partial amino acid compositions of these fractions were reported by Shewry et al . (1978b).

The hordein-I and II fractions were extracted by shaking at 20°C with 55% propan-2-ol and 55% propan-2-ol + 0.6% 2-mercaptoethanol respectively. (Modified from Landry and Moureaux, 1970).
The total hordein fraction was extracted by shaking at 60°C with 55% propan-2-ol + 2% 2-mercaptoethanol.

collate gave similar results for the extraction of this additional fraction and suggested that their effect was due to the reduction of disulphide bonds. They also showed that the fraction had an amino acid composition similar to that of hordein, although Waldschmidt-Leitz and Mindemann [184] had extracted and analyzed a similar fraction and called it glutelin.

More recently an analogous fraction has been extracted from maize [81, 109, 167]. Although Landry and Moureaux called this fraction glutelin-1, Sodek and Wilson called it zein-2 because it had a similar amino acid composition and some of the same polypeptides as the zein fraction extracted without a reducing agent (which they called zein-1). We have adopted a similar nomenclature for the two alcohol-soluble protein fractions from barley, although using Roman numerals to avoid confusion with the hordein structural loci which have been designated *Hor 1* and *Hor 2*. Despite the studies referred to above many workers have continued to extract hordein fractions without using reducing agents and have thus obtained low recoveries [2, 6, 7, 19, 20, 56, 77, 84, 112, 166]. Because the residual hordein polypeptides are usually extracted in the glutelin fraction, this also leads to inaccurately high recoveries of glutelins and affects the glutelin amino acid composition and polypeptide pattern. This is discussed in detail in the glutelin section.

The amounts of N recovered in hordein-I and hordein-II fractions extracted by a sequential Landry and Moureaux [81] procedure are approximately 24% and 9% of the total grain N respectively [156]. Amino acid analysis of these fractions (Table 6) shows great similarity, notably in the high glutamate and proline and the low lysine. Both compositions are similar to a total hordein fraction extracted in the presence of 2-mercaptoethanol at 60°C and accounting for 40% of the grain N.

Early Work on the Separation and Characterization of Components

Early studies by ultracentrifugation [134] showed hordein to be homogeneous thus indicating the presence of a single component. However, the application of electrophoretic procedures demonstrated that hordein was a mixture of proteins, although the number of components separated by the early procedures was small. Biserte and Scriban [12] showed 5 components by Tiselius electrophoresis, Klaushofer *et al.* [74] 3 by paper electrophoresis and Waldschmidt-Leitz and Brutscheck [182] separated 4 components by Tiselius electrophoresis which they called α, β, γ and $\delta + \varepsilon$ hordein. The same four fractions were subsequently separated by continuous carrier-free electrophoresis according to Hannig [57, 183], and were characterized by these authors as to their amino compositions. Mesrob and co-workers introduced gel filtration and characterized the fractions obtained in relation to the electrophoretic fractions [97, 98, 99]. They also used PAGE at low pH [96] to resolve hordein into at least six major and three or four minor bands. Hordein fractions prepared by gel filtration and by Hannig electrophoresis were shown to be mixtures of components on PAGE [96, 99].

During the last 10 years hordein has been subjected to intensive analysis in a number of laboratories. The following account describes the present state of knowledge with no attempt to quote all published reports. Most of the data is derived from studies in our own laboratory but similar results have been obtained by other workers.

Classification and Characterization of Groups

SDS-PAGE of hordein (Figure 5) shows a number of components which are usually classified into three groups called A, B and C hordein [77, 161]. There is also a single high molecular weight band which varies in intensity depending on extraction conditions and which was not considered in previous studies. We have recently purified and characterized this polypeptide and have designated it as 'D' hordein (Figure 5) [102]. Comparison of hordein-I and II fractions shows differences in the ratios of the four groups, hordein-I being particularly enriched in 'C' hordein and hordein-II in 'B' and 'D' hordein. Pure groups of 'A', 'B', 'C' and 'D' hordein have been prepared by different workers and their characterization will now be described.

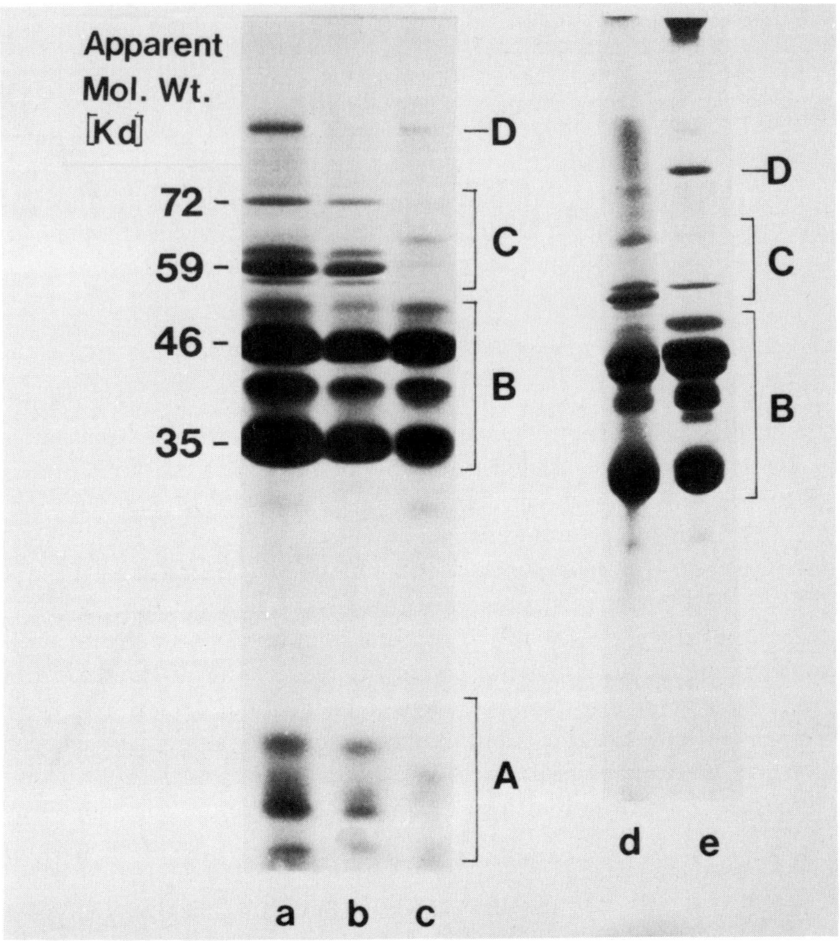

Figure 5. SDS-PAGE of reduced and pyridylethylated hordein fractions and total protein body proteins. a–c, total hordein, hordein I and hordein II fractions respectively from cv. Julia (from Shewry *et al.*, [151]); d, total hordein and e, total protein body proteins from cv. Julia (from Miflin *et al.*, [101]). Apparent mol. wts. are from Faulks *et al.*, [42].

'A' Hordein. Salcedo *et al.* [141] purified a low molecular weight (under 25,000 daltons) fraction by gel filtration of hordein-I. This gave four bands by 2-D isoelectric focusing/pH 3.2 PAGE and had an amino acid composition which differed considerably from that of total hordein, notably containing 3.1% lysine, 6.4% cysteine and only 13.6% glutamate and 10.1% proline (Table 7). They estimated the molecular weights of the components as between 10 and 16,000 daltons. More recently Aragoncillo *et al.* [3] have shown that the low molecular weight fraction prepared by gel filtration of a chloroform/methanol (2:1 v/v) extract can be separated, by a second gel filtration procedure, into three groups of

Table 7. Amino acid compositions of hordein fraction.

	'A'[a] hordein	LMW[b] hordein	'B'[c] hordein	'C'[c] hordein	'D'[d] hordein
Asp	6.5	2.3	1.4	1.0	1.5
Thr	5.5	6.6	2.1	1.0	8.0
Ser	6.5	5.6	4.7	4.6	9.4
Glu	13.6	25.5	35.4	41.2	29.6
Pro	10.1	17.0	20.6	30.6	11.4
Gly	8.7	5.5	1.5	0.3	13.6
Ala	7.5	5.3	2.2	0.7	3.4
Cys	6.4	5.3	2.5	trace	1.7
Val	6.2	4.3	5.6	1.0	4.8
Met	1.9	3.7	0.6	0.2	0.6
Ile	3.0	4.1	4.1	2.6	1.2
Leu	8.3	4.6	7.0	3.6	3.9
Tyr	3.0	3.2	2.5	2.3	3.8
Phe	3.3	4.4	4.8	8.8	1.3
Lys	3.1	0.1	0.5	0.2	0.8
His	1.4	0.4	2.1	1.1	3.1
Arg	5.0	2.1	2.4	0.8	1.8

Expressed as mole %
Values reported for glutamate and aspartate include the amides glutamine and asparagine respectively.

[a] Salcedo et al. (1980). Low mol. wt. fraction (under 25,000 daltons) prepared from hordein I from cv. Zephyr.

[b] Aragoncillo et al. (1981). Prepared by gel filtration of a chloroform/methanol (2:1) extract from cv. Zephyr.

[c] Shewry et al. (1980e). Fractions prepared by chromatography of total hordein from cv. Julia on CM cellulose.

[d] Miflin et al. (1982a). Prepared by ion-exchange chromatography, gel filtration and preparative isoelectric focusing of total hordein from cv. Sundance.

components. Two of these contained four 'A' hordein-like components. These were the quantitatively major components of the fraction and had amino acid compositions similar to that given for 'A' hordein in Table 7. The third, quantitatively minor, fraction had low mobility on starch gel electrophoresis at pH 3.2 and an amino acid composition more similar to 'B' hordein (Table 7) except for the presence of relatively large amounts of cysteine (5.3%) and methionine (3.7%). Separation of the fraction on SDS-PAGE gave three bands with apparent mol. wts. between 10,000 and 16,000 daltons. The authors suggested that the latter components should be called low molecular weight (LMW) hordein. They also suggested that since the amino acid composition of the major 'A' hordein differed considerably from that of the other hordein groups, that their designation as hordein was inappropriate. Other characters which suggest that these should not be considered as hordein are their lack of genetic variability [151, 162, 165] (Figure 6) and their apparent absence from protein bodies [101, 104, 141] (Figure 5). Although the low molecular weight bands sometimes stain intensely on SDS-PAGE gels, quantitatization based on light-scattering by the fixed protein rather than dye binding shows they account for only 1–2% of the total hordein fraction [157]. It is probable that their relatively high content of basic amino acids results in more intense staining, and that their relative amount is therefore overestimated by visual inspection of stained gels. Similar polypeptides are, however, also present in the salt-soluble protein fraction [141] and the total amount present in the whole grain has not been determined. Although they are not now thought to have a storage function, their actual function is not known.

'B' and 'C' Hordein. 'B' and 'C' hordein together form approximately 95% of the total hordein fraction [157], although the actual ratio of the two groups varies depending on the nitrogen content of the seed. Not only does increased seed N result in increased synthesis of hordein, but also in an increase in the relative amount of 'C' hordein within the fraction. Thus in seed containing 1.13% N we found that 39% of that N was present in hordein whereas in seed with 2.57% N there was 53% in hordein; the ratio of 'B' hordein: 'C' hordein in the two samples was 12.8 and 3.8. This resulted from about 2.5 and 8.6 fold increases in the total amounts of 'B' and 'C' hordein polypeptides respectively [72]. This aspect of hordein composition will be dealt with in greater detail by Doll in a separate chapter in this volume.

We have purified total 'B' and 'C' hordein fractions by ion-exchange chromatography on CM-cellulose and compared their properties [155]. Molecular weights of the fractions were determined by sedimentation equilibrium ultracentrifugation. In both cases these were appreciably lower than those determined by SDS-PAGE. Thus, whereas 'B' and 'C' hordein polypeptides had mol. wts. of between approximately 35,000 and 46,000 and 55,000 and 75,000 daltons respectively on SDS-PAGE (Figure 5), equilibrium ultracentrifugation analyses of the purified 'B' and 'C' groups showed molecular weights of approximately 32,000 and 52,000 daltons respectively. It is possible that the anomalously high mol. wts. by SDS-PAGE are related to the high proline content of the polypeptides, as similar discrepancies have been reported for proline-rich proteins from other sources [36]. Amino acid analysis (Table 7) of the two fractions showed distinct differences, notably higher concentrations in 'C' hordein of glutamate + glutamine (41% compared to 35%), proline (31% compared to 21%) and phenylalanine (9% compared to 5%). 'C' hordein also contained little or no cysteine and only small amounts (approximately 0.2 mole %) of lysine and methionine. Calculations based on the molecular weight by ultracentrifugation indicated the presence of approximately 440 residues/mole, and thus the amounts of methionine and lysine present corresponded to one or less residues per mole. Similar calculations showed approximately 270 residues/mole of 'B' hordein and this contained more methionine (0.6 mole %), lysine (0.5 mole %) and cysteine (2.5 mole %). Approximately 90% and 92% of the total aspartyl and glutamyl residues in 'B' and 'C' hordein respectively were present as the amides.

'D' Hordein. Only a single band is present in this region in SDS-PAGE sep-

Table 8. N-terminal amino acid sequences of 'C' hordein.

Sample	1	2	3	4	5	6	7	8	9	10	11	12	13	14	15	16	17	18	19	20	21	22	23	24	25	26	27	28	29	30
[1]Total Hordein cv. Hiproly	Arg	Gln Met	Leu	Asn Val	Pro	Ser	Ser	Gln Val	Glu	Leu	Gln	(Ser)	Pro	Gln	Gln	(Pro) (Asn)	Tyr	Leu	Gln	Gln	Pro	Tyr Val	Pro	(Gln) (Trp)	Asn Phe	(Pro)	Tyr	Leu	Pro(Gln)	
[2]Total Hordein cv. Bomi	Arg	Gln	Leu	Asn	Pro	Ser	Ser Gln	Gln	Glu	Leu	Gln	Ser Pro Gly	Pro	Gln	Gln Val Leu	Pro Ser	Tyr	Leu	Gln	Gln	Pro	Tyr	Pro	Gln	Asn	X	Tyr Pro Gln	Leu	Glu	X
[3]Total 'C' Hordein cv. Julia	Arg	Gln	Leu	Asn	Pro ?	(Ser)	(Ser)	Gln	Glu	Leu	Gln	(Ser) Pro	Pro	Gln	Gln	(Ser) Pro	Tyr	Leu	Gln	Pro Phe	Pro Gln	Tyr	Pro Phe	Gln	Asn Gln	(Pro)	Tyr Pro	Leu	Pro	Pro Thr
[2]C-2 Hordein cv. Bomi	Arg	Gln	Leu	Asn	Pro	Ser	Ser	Gln	Glu	Leu	X	Ser	Pro	Gln	Gln	Pro	Tyr	Leu	Gln	Gln	Pro	Tyr	Pro	Gln	Asn					
[4]'C' Component cv. Julia	Arg	Gln	Leu	Asn	Pro	Ser	Ser	Gln	Glu	Leu	Gln	Ser	Pro	Gln	Gln	Ser	Tyr	Leu	Gln	Gln	Pro	Tyr	Pro	Gln	Asn	Pro	Tyr	Leu		

Residues in brackets are subject to uncertainty. Minor residues are given below the major sequence.

[1] Bietz (1981)
[2] Schmitt and Svendsen (1980a)
[3] Shewry et al. (1980f)
[4] Shewry et al. (1980a)

arations of European barley varieties. The band accounts for approximately 2% or less of the total hordein fraction extracted with aqueous alcohol and reducing agent. However, the addition of 1% acetic acid to the solvent appears to increase the extraction of this component while having little or no effect on the other hordein polypeptides, and when extracted in this way 'D' hordein may account for over 2% of the resulting total hordein fraction [102]. Amino acid analysis (Table 7) shows high glutamate + glutamine (28%) but relatively low proline (12.7%). There is, however, a very large amount of glycine (13.6%) compared to the amounts present in 'A' (8.7%), 'B' (1.5%) and 'C' (0.3%) hordein. This fraction has an apparent molecular weight of 105,000 on SDS-PAGE but the mol. wt. determined by equilibrium ultracentrifugation is again considerably smaller, 56,000 [102]. The discrepancy between the molecular weights determined by the two procedures is greater than for 'B' and 'C' hordein which appears to be inconsistent with the lower proline content of 'D' hordein. It has, however, been suggested that the anomalous behaviour on SDS-PAGE is due to localized concentrations of proline rather than a high total amount [36].

Polymorphism of Hordein Polypeptides

SDS-PAGE of hordein fractions from a range of barley varieties shows great variation in the band patterns in the 'B' and 'C' hordein regions (Figure 6). There is little or no variation in the 'A' or 'D' hordein regions, although the latter band is sometimes difficult to see because of its relatively low concentration and its in-

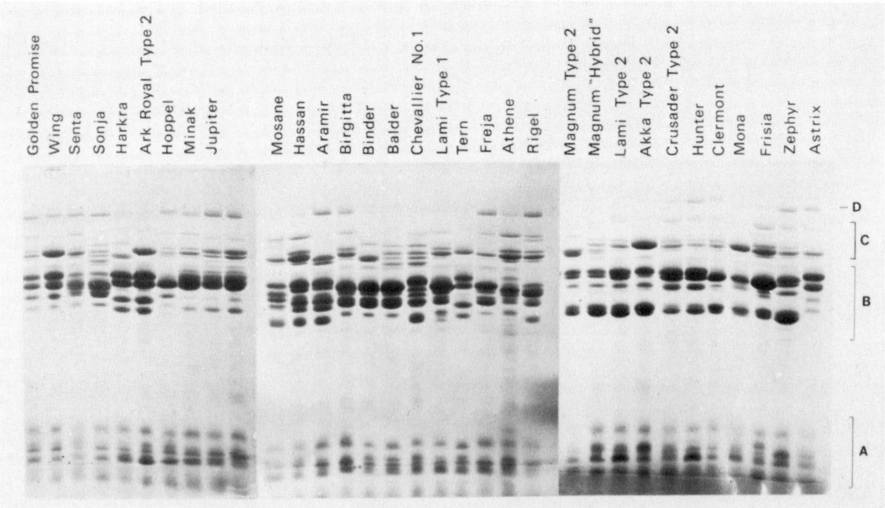

Figure 6. SDS-PAGE of reduced and pyridylethylated hordein fractions extracted from single seeds of a range of European cultivars of barley. (From Shewry *et al.*, [165]).

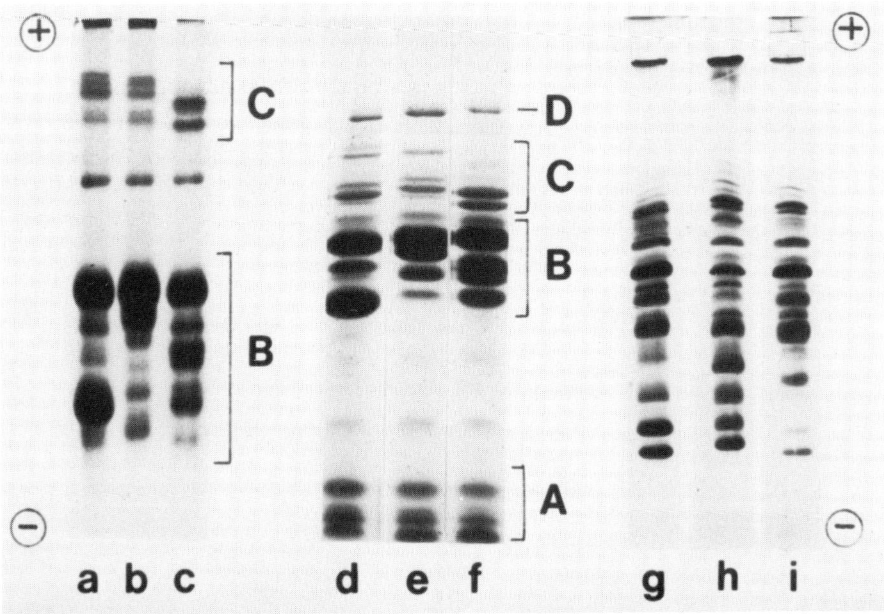

Figure 7. One-dimensional analysis of reduced and pyridylethylated hordein fractions. a–c are separated by PAGE at pH 4.6 in the presence of 6 M urea, d–f by SDS-PAGE at pH 8.9 and g–h by isoelectric focusing (IEF) in the pH range 5 to 9. a, d, g are cv. Julia; b, e, h are cv. Proctor; c, f, i are cv. Maris Mink. (From Shewry *et al.*, [151]).

complete and sometimes variable extraction in the absence of acetic acid. Although the range of variation in the 'B' and 'C' regions is great, careful consideration of the fractions from a large number of varieties shows that certain bands are commonly associated to give recognizable 'polypeptide patterns' which may be present in a number of varieties. We have studied the hordein polypeptide patterns of 139 spring and 24 winter barley varieties which have been bred and cultivated in Western Europe [151, 162, 165] and have recognized 17 different 'B' and 8 different 'C' hordein patterns. The 164 varieties could be classified into 32 groups, containing between 1 and about 40 varieties, on the basis of these patterns.

Separation of the hordein fractions by other procedures shows that the extent of this polymorphism is even greater than that shown by SDS-PAGE. A number of other analytical procedures have been used including starch gel electrophoresis (SGE), electrophoresis at pH 4.6 in the presence of 6M urea (urea-PAGE), electrophoresis in aluminium lactate buffer at pH 3.2 (lactate-PAGE), gradient gel electrophoresis in sodium lactate buffer (gradient PAGE) and isoelectric focusing (IEF). Because these separate on the basis of different chemical criteria a hordein fraction from the same variety can give widely different patterns on different gel systems, as illustrated by the comparison of SDS-PAGE, Urea-PAGE and IEF separations in

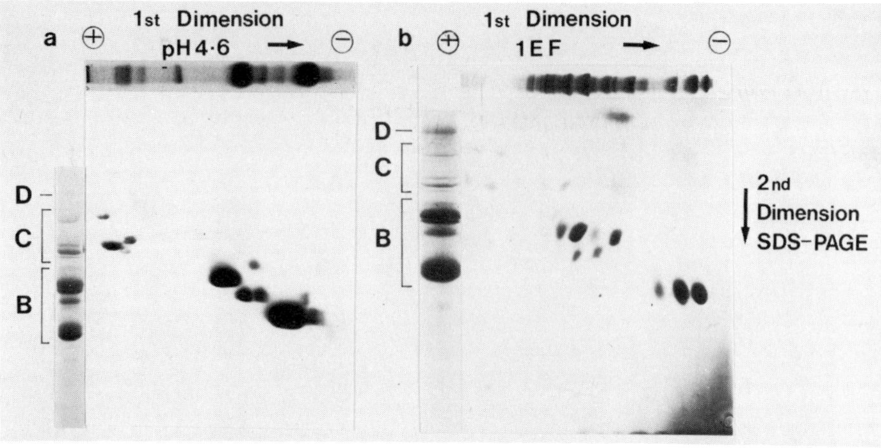

Figure 8. Two-dimensional analysis of reduced and pyridylethylated hordein fractions from barley cv. Bomi. a was separated first by PAGE at pH 4.6 in the presence of 6 M urea and then by SDS-PAGE at pH 8.9 (from Shewry *et al.*, [151]). b was separated by IEF in the pH range 5–9 followed by SDS-PAGE at pH 8.9 (from Shewry *et al.*, [152]).

Figure 7. It is often possible, using different separations, to show differences between hordein polypeptide patterns which appear to be identical on SDS-PAGE [151, 158a, 165]. Procedures for the varietal identification of barley grain based on the hordein polypeptide patterns exhibited in different electrophoretic systems have been published by a number of workers [35, 78, 91, 143, 148, 158a, 162, 165, 177, 179] and these procedures have also been successfully applied to commercial malted grain [5, 148, 153].

The range of variation in the hordein polypeptide pattern is even greater when exotic lines of *H. vulgare* or wild samples of *H. spontaneum*, the cross-fertile relative of *H. vulgare*, are analyzed ([34]; authors unpublished results). Apart from variation in the 'B' and 'C' hordein region, we have also shown that some lines from the World Barley Collection differ in the 'D' hordein pattern, although the variation is of limited extent [160].

Further resolution of the hordein fraction is given by two-dimensional (2-D) analyses combining separations based on different criteria. Most success has been achieved using a combination of IEF as the first and SDS-PAGE as the second dimension (Figure 8b) [42, 103, 104, 152], although a combination of Urea-PAGE and SDS-PAGE has also been used (Figure 8a) [151, 161]. These analyses show that many of the 'B' and 'C' hordein bands present on 1-D gels are composed of several components and the total number of polypeptides present exceeds that shown by any 1-D system. In some cases 2-D gels show differences between 'B' and 'C' hordein patterns which appear identical on 1-D gels [162], and so have some potential value for studies of varietal distinctness and identification, although only small numbers of seed can be analyzed. It is notable that the 'D' hordein band present in commercial barley varieties is partially separated into a number of spots by the IEF/SDS system (Figure 8b). This component was not apparent in the Urea-PAGE/SDS-PAGE separation, possibly due to the small amount present in the hordein sample analyzed.

Genetic Control

The existence of polymorphism in the hordein fraction makes it possible to study the genetic control by following the inheritance of bands in intervarietal crosses. The interpretation of the gel patterns is, however, made more difficult by the triploid nature of the endosperm which results in two possible types of heterozygote which differ in the dosage of alleles derived from the maternal (2 doses) and paternal (1 dose) parents. Crosses of this type were analyzed by Solari and Favret [168] who concluded that the hordein polypeptide pattern was controlled by a short segment of chromosome 5, designated *Pr-a*, which was linked to *Mla*, a complex mildew resistance locus. They suggested that several loci were present in this segment, although they did not determine the number. Oram *et al.* [119] analyzed the progeny of crosses between Bomi, its high lysine mutant Risø 1508 and Sultan. They showed that the 'B' hordein pattern was controlled by co-dominant alleles at a locus linked ($17.4 \pm 2.8\%$) to *Mla*. Because Bomi and Sultan have similar 'C' hordein patterns they could not determine the inheritance of this group.

We therefore analyzed a cross between two varieties which had different 'B' and 'C' hordein patterns and showed that the two groups of polypeptides were controlled by co-dominant alleles at separate loci, which were located approximately 10–16% recombination units apart [163]. Doll and Brown [34] confirmed this observation and designated the two loci *Hor-1* ('C' hordein) and *Hor-2* ('B' hordein) which they estimated were separated by $11 \pm 2\%$ recombination (Figure 9). These controlling loci are considered to be the structural genes for hordein because the polypeptides are present in amounts corresponding to the dosages of the alleles in heterozygous triploid endosperms. Further evidence comes from analysis of the protein phenotypes of wheat lines possessing added chromosomes of barley (see below). However, Sozinov *et al.* [169] reported the presence of not two but five linked hordein loci which they designated *Hrd A* to *Hrd E*. The recombination percentages were 6.87 ± 0.74 from *Hrd A* to *Hrd B*, 2.35 ± 0.55 from *Hrd B* to *Hrd C*, 0.36 ± 0.21 from *Hrd C* to *Hrd D* and 0.18 ± 0.11 from *Hrd D* to *Hrd E*. *Hrd A* and *Hrd B* almost certainly correspond to *Hor-1* and *Hor-2* respectively. The nature of the proteins coded by *Hrd C*, *D* and *E* is less clear, although the authors described the proteins as being present in low concentration in the 'C' hordein region of starch gels. These five loci were based on the frequency of phenotypic classes in a homozygous F_{15} population, but the linkage between *Hrd A* and *Hrd B* was confirmed by analysis of F_2 grain (the estimate being $6.34 \pm 0.90\%$). Netsvetaev [116] analyzed

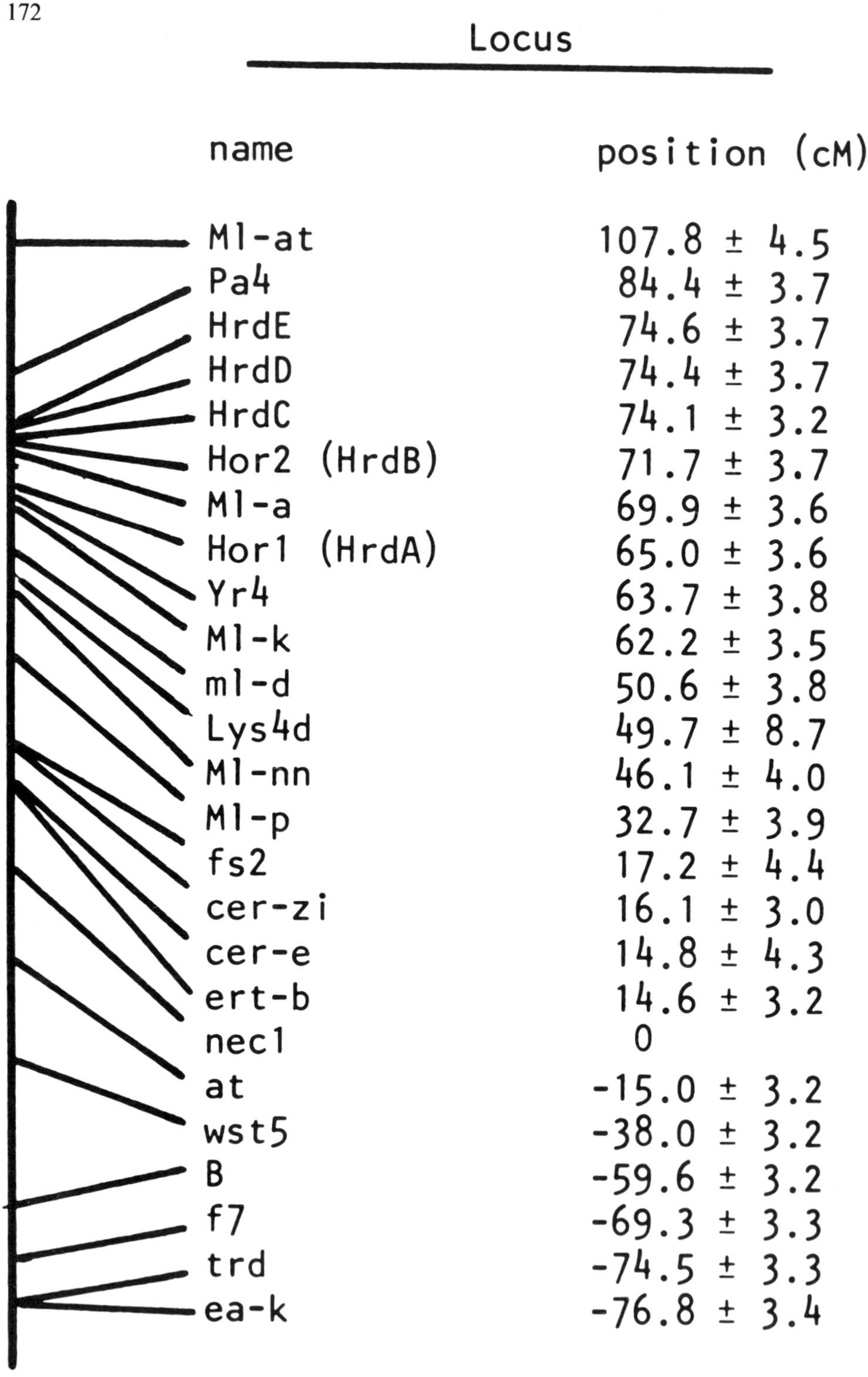

Figure 9. The barley chromosome 5 linkage map (redrawn from Jensen, [59]). The position of the centromere is probably slightly above *fs2*.

two crosses involving translocations and estimated the distance between *Hrd A* and *Hrd B* as 13.44 ± 1.11 and 13.93 ± 1.13%. He also reported that the loci were on the short arm of chromosome 5 with *Hrd A* closest to the centromere.

None of these studies revealed any recombination within either the *Hor-1* (*Hrd A*) or *Hor-2* (*Hrd B*) locus. It is possible, however, that some recombination did occur but was not noticed either because of co-migration of polypeptides in the 1-D gel systems or the difficulty of scoring heterozygotes in the triploid endosperm. Oram *et al* [119] tried to eliminate the latter problem by analyzing a cross of homozygous doubled monoploid lines produced from F_1 grain using the *Hordeum bulbosum* technique. To confirm that recombination was not occurring we made a further detailed study using 2-D and high-resolution 1-D gel analyses of doubled monoploid lines and of F_2 seed [154]. Although we analyzed the equivalent of 347 gametes (253 by 2-D analysis, 211 doubled monoploid lines) we found no evidence of recombination within either locus. However, even with this number of analyses it is unlikely (5% probability) that we would have detected recombination of less than 1%. This study did, however, show that the *Hor-1* and *Hor-2* loci were located either side of *Mla* with recombination percentages of 6.0 ± 1.8% between *Hor-1* and *Mla*, 7.6 ± 2.0% between *Hor-2* and *Mla* and 13.8 ± 1.9% between *Hor-1* and *Hor-2*. The latter is in good agreement with the previous estimates [34, 116, 151, 156, 157, 163, 165].

This was confirmed by Jensen *et al* [60], although their recombination between *Hor-1* and *Hor-2* (7.4 ± 0.9%) was closer to that of Sozinov *et al.* [169]. Shewry *et al.* [152] and Jensen *et al* [60] also used additional marker genes (for rust resistance *Yr4* and mildew resistance *Mlk* respectively) to show that *Hor-1* locus was located closest to the centromere, which is in agreement with Netsvetaev's [116] results from translocation crosses. A map of the loci on chromosome 5 resulting from these and other studies is given in Figure 9.

Lawrence and Shepherd [86] have recently studies the seed protein patterns of wheat lines possessing added chromosomes of barley. They were unable to obtain a simple addition line with chromosome 5 only as this was sterile. However, an addition line with chromosomes 5 and 6 possessed three major barley protein bands which were not present in an addition line with chromosome 6 alone. Two of these presumably corresponded to 'B' and 'C' hordein, although it is not possible to be certain. Analysis of a translocation line which possessed the short arm of chromosome 5 attached to a wheat chromosome showed that these were coded by the short arm. The third band, which appeared to be coded by the long arm of chromosome 5, almost certainly corresponds to our 'D' hordein band. We have been studying the inheritance of 'D' hordein in collaboration with Dr. R.A. Finch of the PBI, Cambridge. Our initial crosses showed that the 'D' hordein locus (which we have provisionally designated *Hor-3*) is not linked to *Hor-1*, *Yr 4* or *B*, the latter being located towards the distal end of the long arm of chromosome 5 (Figure 9). We are currently making further crosses with additional marker genes on the long arm of chromosome 5.

Because of the linkage relationships between the *Hor* loci and disease resistance genes on chromosome 5 (Figure 9) a number of hordein polypeptide patterns are associated with specifec mildew or rust resistances. We have made a detailed study of the origin and distribution of the hordein patterns present in commercially grown European barleys and have shown that, although some patterns are derived from ancestral land races, others have been introduced from exotic varieties which have been used as sources of resistance alleles [162]. Because of the close linkage, selection for the resistance allele also results in selection of the hordein pattern. In a recent study we showed that one cultivar, Hood, was a mixture of seed that differed in their hordein patterns [158a]. Two different 'B' hordein patterns were associated in some seed with each of two different 'C' hordein patterns, making a total of four different seed types. This variety was also heterogeneous with respect to mildew resistance, some plants have the *Mla4* allele (located at *Mla*) and others having the *Mla7* allele (located at *Mlk*) or both alleles. The presence of the *Mla7* allele was associated with one of the two 'C' hordein patterns but not specifically with either of the 'B' hordein patterns. This provides further evidence for the close linkage of *Hor-1* and *Mlk*.

Structural Homology

Recent studies indicate a high degree of structural homology between the polypeptides within the 'B' and 'C' hordein fractions.

'C' hordein. Shewry *et al.* [159] prepared a total 'C' hordein fraction and determined the N-terminal sequence by automated Edman degradation (Table 8). Single amino acids only were recovered from each of the first 12 positions and from 12 of the next 18, indicating a high degree of sequence homology of the component polypeptides. This procedure did not work on total 'B' hordein, presumably because the protein is N-terminally blocked. Consequently the N-terminal sequences reported for total hordein fractions by Bietz [10] and Schmitt and Svendsen [146] probably represent 'C' hordein polypeptides, although the crude fractions used may have contained some non-hordein proteins. The sequences obtained (Table 8) showed good agreement with total 'C' hordein although there was some variation in the minor residues and in the major residues at position 16, 20, 24, 29 and 30. The differences in the later positions are possibly due to analytical problems and may not represent true sequence differences. Schmitt and Svendsen [146] also analyzed a 'C' hordein fraction, termed C-2, which consisted of two SDS-PAGE bands while Shewry *et al.* [150] analyzed a 'C' hordein component which gave single bands on SDS-PAGE and lactate-PAGE, although this was probably still a mixture of components. The sequences (Table 8) were identical except for position 10 and no secondary residues were found. Shewry *et al.* [158] subsequently purified three further 'C' hordein components from a different cultivar, Maris Mink. These all gave single bands on lactate PAGE and single major bands on SDS-PAGE,

Table 9. Amino acid compositions of purified 'C' hordein components.

	1	2	3	4	5	6
Asp	1.2	0.8	1.0	1.0	1.3	0.9
Thr	1.2	1.0	0.9	1.4	1.3	1.1
Ser	2.1	2.6	3.0	3.3	3.3	2.9
Glu	41.1	41.1	38.9	38.3	38.1	37.6
Pro	27.4	31.9	32.5	30.9	30.5	33.1
Gly	1.2	0.4	0.7	0.7	1.2	1.1
Ala	1.0	0.7	0.7	0.9	1.0	0.9
Cys	0.9	trace	0	0	0	0
Val	1.6	1.1	1.3	1.0	1.2	1.1
Met	0.6	0.2	0.2	0.2	0.2	0.3
Ile	3.7	3.0	3.2	2.8	2.7	3.1
Leu	5.2	4.3	4.0	4.5	4.5	3.9
Tyr	2.5	2.3	2.0	2.2	2.2	2.2
Phe	8.0	9.0	8.6	9.3	8.5	8.6
His	0.7	0.6	0.8	1.1	1.1	1.0
Lys	0.2	0	0.1	0.2	0.3	0.4
Arg	1.7	0.8	0.8	1.0	1.0	1.3
Trp	n.r.	n.r.	1.3	1.3	1.5	0.5

Results are expressed as mole %.
n.r. - not reported
Values given for glutamate and aspartate include the amides glutamine and asparagine respectively.

1 C-2 hordein from cv. Bomi (Schmitt, 1979).
2 Component from cv. Julia (Shewry *et al.*, 1980a)
3-5 Components from cv. Maris Mink (Shewry *et al.*, 1981)
6 Component from cv. Maris Otter (recalculated from Ewart 1980a)

although IEF revealed that all were mixtures of components. The N-terminal amino acid sequences were determined for 22 residues for one component and for 10 residues for the other two. All were identical to that previously reported for the component from cv. Julia (Table 8) [150].

Schmitt and Svendsen [146] reported that the C-terminal amino acid sequence of C-2 hordein, determined by digestion with carboxypeptidase Y, was (Ser-Ile)-Ser-Met-Val-COOH. Shewry *et al.* [158] used a similar procedure to analyze the three 'C' hordein components from cv. Maris Mink. Two components gave sequences (Ser-Ile-Ser-Met-Val-COOH) similar to that of C-2 hordein, but digestion of the third component showed the presence of a mixture of isoleucine and leucine in an approximate molar ratio of 1.5 to 1.0 as the fourth amino acid from the C-terminus. This shows the presence of heterogeneity at this position in the component polypeptides. The three components from Maris Mink were also cleaved with specific enzymes (trypsin, chymotrypsin and *Staphylococcal* V8 protease) and the fragments separated by SDS-PAGE. The presence of several polypeptides in each fraction and the probable occurrence of non-specific cleavage made it difficult to interpret the peptide maps in relation to the distribution of individual amino acids along the polypeptide chains. However, the differences in the fragmentation patterns of the

```
a) NH2-SER ILE PHE LEU GLN GLU GLN PRO GLN GLN LEU VAL GLN GLY VAL SER GLN PRO GLN GLU GLN LEU(TRP)
b) NH2-VAL ILE LEU LEU GLN GLU GLN  X  GLX GLX LEU VAL GLX GLY VAL  X  GLN PRO  X   X  GLN LEU(TRP)
                                                       c) NH2-GLY VAL SER GLN PRO GLN GLN GLN LEU TRP-

   - PRO GLN GLN(VAL)GLY GLY  X  LEU

d) NH2 LEU GLN GLN SER SER  X  HIS VAL LEU GLN GLN GLN LYS GLN GLU LEU PRO GLN

         e) NH2-ALA THR  X  SER ILE ALA LEU ARG THR LEU PRO VAL.f)NH2-X SER VAL ASN VAL PRO-
i) NH2-gln leu glu ala thr thr ser ile ala leu arg thr leu pro thr met cys ser val asn val pro-

   - LEU TYR ARG ILE(LEU)
   - leu tyr arg ile val pro leu ala ile asp thr arg val gly val-COOH
                                                        g) GLY VAL-COOH
                                                     h) VAL GLY VAL-COOH
```

Figure 10. Amino acid sequences of B hordein. Directly determined amino acid sequences are given in capitals. Sequences a–f are from Schmitt and Svendsen [147] and are peptides derived from B1 hordein. They are identified as follows: a, trypsin peptide III; b, trypsin peptide IV; c, *Staphlococcus* V8 peptide II; d, CNBr fragment III; e, *Staphlococcus* V8 peptide IV; f, CNBr fragment V. Sequence g is the C-terminus of B1 hordein determined by Schmitt and Svendsen [146]. Sequence h is the C-terminus of B1 hordein determined by the authors (see text). Sequence i, given in lower case, is that predicted from the nucleotide sequence of a cDNA clone pcHvE16 that hybridizes to mRNA directing the synthesis of B hordein polypeptides (Forde *et al.*, [44]).

three fractions indicated the presence of sequence differences between the component polypeptides.

Amino acid analysis of purified 'C' hordein components also showed some differences (Table 9), notably in the content of basic amino acids. This is to be expected as most of the fractions were purified the basis of their charge at low pH, a character which is determined by the total content of basic amino acids.

'B' Hordein. Because 'B' hordein is blocked to Edman degradation at the N-terminus [146, 147, 155, 159] this approach cannot be used to compare the component polypeptides. There is, however, some evidence of sequence heterogeneity at the C-terminus. Schmitt [145] prepared a B-hordein fraction (called B-1 hordein) from cv. Bomi which contained at least two polypeptides and possibly more. Digestion with carboxypeptidase Y showed that the C-terminal amino acid was valine. The second amino acid appeared to be glycine which rendered the protein resistant to further digestion by this enzyme. We have prepared a similar fraction from a related cultivar, Sundance, and digested this with an enzyme with a different specificity (carboxypeptidase B). This showed the C-terminal sequence to be – Val-Gly-Val-COOH. However, when the same enzyme was used to digest a total 'B' hordein preparation from the same cultivar there was a release of valine followed by

Table 10. Amino acid compositions of 'B' hordein components.

	1	2	3	4	5	6
Asp	1.1	1.1	0.9	1.5	0.8	1.5
Thr	2.5	2.5	2.2	1.8	2.0	2.3
Ser	4.9	4.4	4.3	2.9	4.3	5.1
Glu	31.5	29.5	30.2	34.2	32.0	35.1
Pro	21.6	23.4	23.8	16.8	20.8	21.6
Gly	3.5	3.9	3.6	3.8	2.9	1.8
Ala	2.3	2.4	2.4	3.0	2.5	2.6
Cys	2.4	2.3	2.5	2.8	2.7	2.2
Val	5.4	5.5	5.6	6.1	6.3	5.7
Met	0.7	0.8	0.6	1.1	0.9	1.0
Ile	4.0	4.0	4.0	4.1	4.1	4.4
Leu	7.8	7.8	7.7	8.8	8.2	6.4
Tyr	2.7	2.6	2.6	2.5	2.3	2.2
Phe	5.0	4.9	5.0	4.7	5.1	4.1
His	1.2	1.3	1.2	1.6	1.6	1.4
Lys	1.0	1.0	1.0	1.0	0.8	0.2
Arg	2.1	2.3	2.3	2.8	2.5	2.3
Trp	0.2.	0.2	0.2	n.r.	n.r.	n.r.

Results are expressed as mole %.
n.r. - not reported. Values given for glutamate and aspartate include the amides glutamine and asparagine respective.

1-3 Components from cv. Maris Otter (recalculated from Ewart, 1980b).
4 B-1 hordein from cv. Bomi (Schmitt, 1979)
5 35 kd hordein band (B-1) from cv. Sundance (Miflin et al., 1982a)
6 46 kd hordein band (B3) from cv. Hoppel (Miflin et al., 1982a).

glycine and then a mixture of arginine, leucine, and threonine or glutamine (which we were unable to separate). Further sequence studies have been carried out (see Figure 10), both by amino acid sequencing of peptides derived from B1 hordein [147] and by nucleotide sequencing of cDNA clones derived from hordein mRNA [44]. The amino acid sequence predicted from the cDNA includes the C-terminal region and agrees well with the above results. It also includes sequences corresponding to two of the peptides. Some of the other peptides have overlapping sequences (Figure 10) and although there are some small differences in these overlaps, it would be premature to suggest that this is a result of sequence microheterogeneity in different B1 polypeptides. In all, these sequences represent about one third of the total expected in a B1 polypeptide.

Some minor differences are also apparent in the amino acid compositions of purified 'B' hordein components. Ewart [41] purified three 'B' hordein components from cv. Maris Otter (Table 10) and showed minor differences in amino acid composition, notably in the amount of arginine. He calculated that this represented a difference of 1 residue of arginine per mole between one of his components (1 in Table 10) and the other two, and suggested that this accounted for its different mobility on SGE. The molecular weights used for these calculations were based on the amino acid analyses and were higher than those determined by SDS-PAGE in the same study. Since even the SDS-PAGE mol. wts. can be expected to be overestimates (see p.   ) it is doubtful whether the difference in arginine represent 1 residue/mole. Also the validity of attempting to determine the exact numbers of amino acid residues per mole of components which almost certainly contain more than one polypeptide must be questioned.

We have recently made a detailed study of the polymorphism and structural homology of 'B' hordein polypeptides from eight barley varieties [42]. We initially co-migrated fractions from these varieties in pairs in all possible combinations using 2-D gel separations (IEF followed by SDS-PAGE) (Figure 11). Polypeptides which co-migrated under these conditions were assumed to be identical. Only polypeptides that accounted for about 0.5% or more of the total protein were considered, and the positions of 47 of these were mapped (Figure 12a). The polypeptides differed considerably in their distribution in the varieties and in their relative contributions to the total 'B' hordein fraction, and this is apparent from the gel separations shown in Figure 11. The structural relationships of the polypeptides were compared by excising spots from 2-D gels, eluting and cleaving the protein with cyanogen bromide (CNBr) in 70% formic acid and separating the cleavage products by SDS-PAGE using an acrylamide gradient of 5 to 30%. Cyanogen bromide cleaves at methionine residues and hence this procedure should give a comparison of the number and distribution of methionine residues in the polypeptide chain. In all cases the procedure gave a mixture of bands which probably represented a mixture of uncleaved polypeptide, partially cleaved peptides and completely cleaved peptides. Hence it was not possible to construct structural maps of the polypeptides

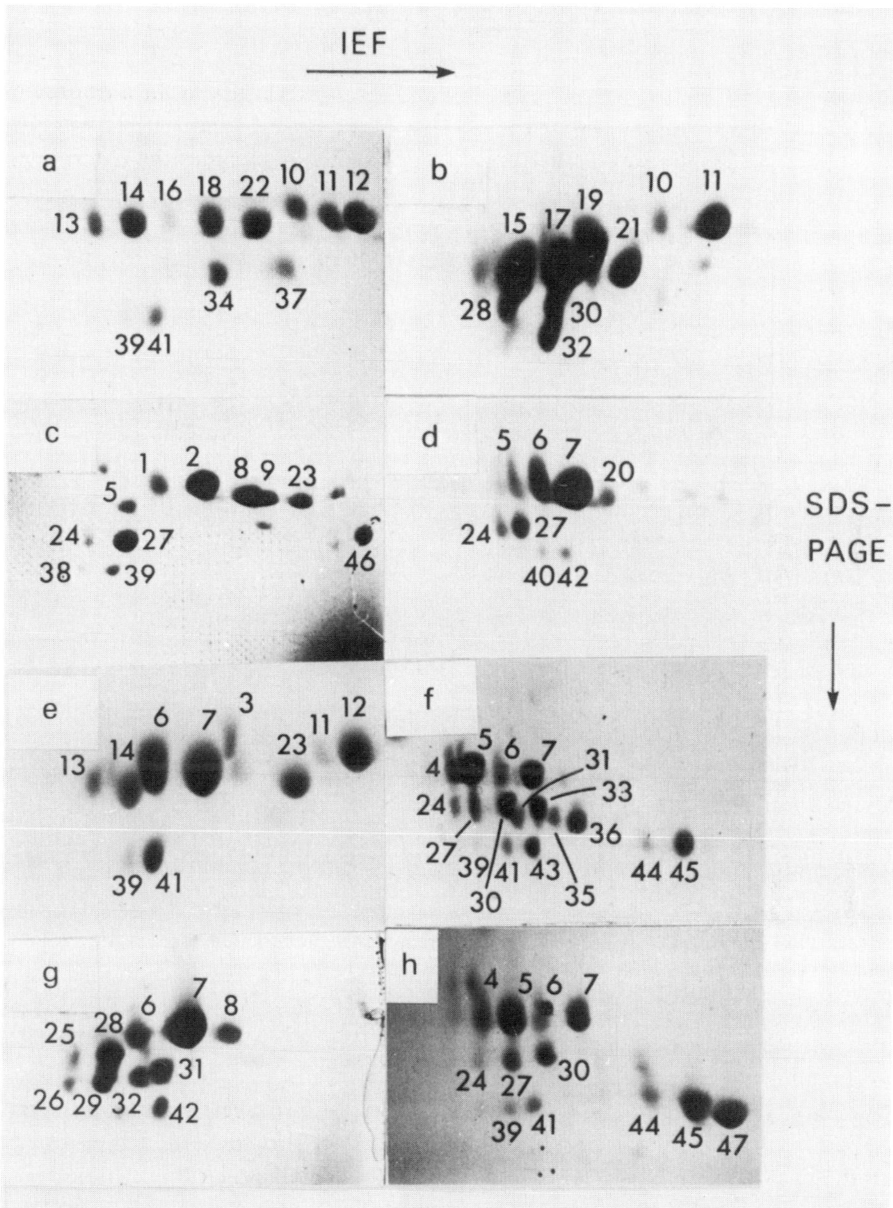

Figure 11. 2-D analysis (IEF followed by SDS-PAGE, see Figure 8) of reduced and pyridylethylated 'B' hordein from 8 barley cultivars. a, Imber; b, Malta; c, Dram; d, Hoppel; e, Jupiter; f, Maris Mink; g, Carlsberg II; h, Julia. The polypeptide numbers correspond to those in Figure 12. Polypeptides with the same number co-migrate when hordein fractions from different varieties are mixed (from Faulks *et al.*, [42]).

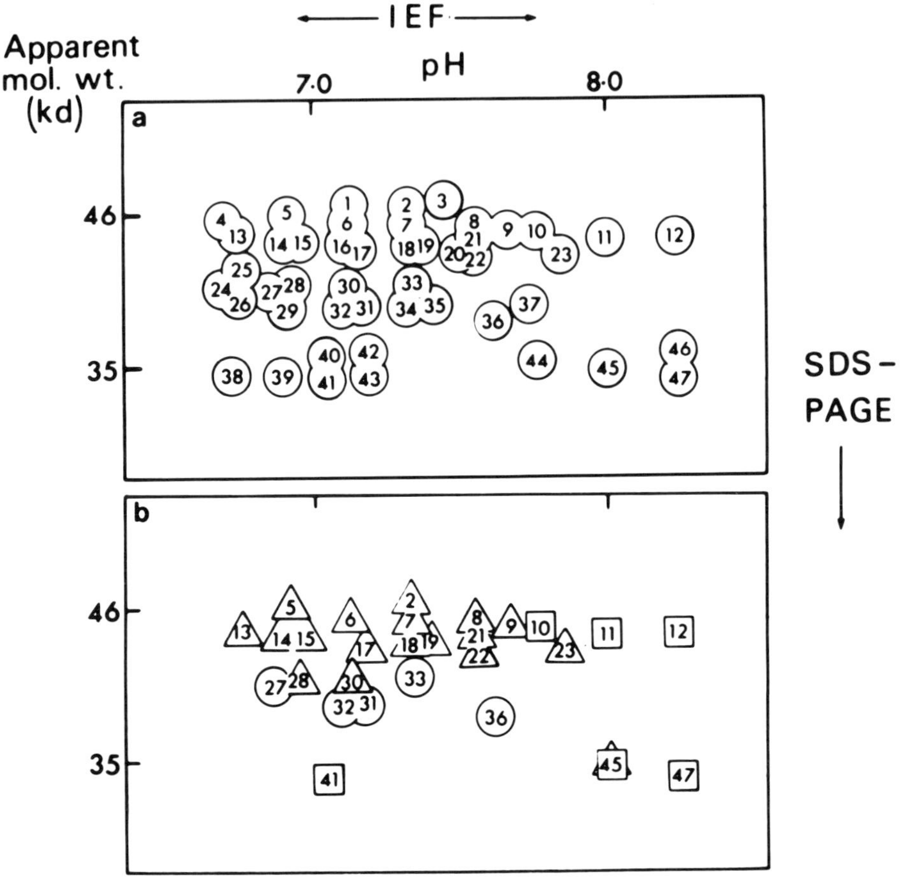

Figure 12. a, Diagrammatical representation of the positions on 2-D gels of 'B' hordein polypeptides present in the 8 cultivars shown in Figure 11. b, Diagrammatical representation of the positions on 2-D gels of 'B' hordein polypeptides with cyanogen bromide cleavage patterns I (□), II (○), and III (△). (From Faulks *et al.*, [42]).

by adding together the mol. wts. of the fragments. The patterns were, however, completely reproducible for any polypeptide and were classified into three groups on the basis of the number of low mol. wt. peptides obtained; this criterion was selected because it was assumed that these bands represented complete cleavage products and would therefore be the best indicators of structural relationships. The three groups, designated class I, II and III polypeptides, had 2, 4 and 5 low mol. wt. bands respectively. With three exceptions polypeptides migrating in the same positions in different varieties gave identical patterns, showing that in most cases identical mobility on 2-D gels is a good indicator of a close structural relationship. In two of the exceptions, polypeptides 7 and 14, the differences in cleavage pattern were present in the medium mol. wt. region and did not affect the assignment of the polypeptide to a class. The third, polypeptide 45, gave a class I pattern when extracted from Julia but a class III pattern from Maris Mink. The three classes of polypeptide differed in their distribution in the varieties. Class III polypeptides occurred in all 8 varieties where they accounted for between 40 and 70% of the total 'B' hordein. In Hoppel this was the only class of polypeptide detected, although in this variety we were only able to analyze the two major polypeptides (which accounted for about 70% of the fraction). In the other 7 cultivars polypeptides of class I (Imber, Malta, Jupiter, Julia) or class II (Dram, Maris Mink, Carlsberg II) were also present although these always accounted for a quantitatively smaller part of the fraction. We did not detect polypeptides of all three classes among the major components of any one variety.

The three classes of polypeptides also showed characteristic distributions on the 2-D maps (Figure 12b). Whereas class II and III polypeptides had similar ranges of isoelectric points (pH 6.5–8), the class II polypeptides were generally of lower mol. wt. The class I polypeptides had a wider range of mol. wt. and, with one exception (polypeptide 41), all had alkaline isoelectric points (above pH 7.5). It is not certain whether these classes of polypeptides represent structurally-related families but this is currently being investigated using other approaches.

Homology of 'B' and 'C' hordein. Holder and Ingversen [52] reported similarities in the peptide maps of chymotryptic digests of ^{35}S-methionine labelled 'B' and 'C' hordein components, prepared by *in vitro* synthesis using membrane-bound polysomes. The number of ^{35}S-labelled C-hordein peptides was, however, surprisingly large considering the presence of only one [155, 158] or two [145] methionine residues per mole and their location close to the C-terminus [146, 158]. The homology of the 'B' and 'C' hordein polypeptides has not been confirmed.

Evidence for Complex Loci. The results discussed here are all consistent with the hypothesis that the *Hor-1* and *Hor-2* loci are complex families of genes which code for polypeptides which are structurally related but in most cases probably not identical. It is probably that each family is derived from the duplication and divergence of a single ancestral gene. The data indicate that point mutations (resulting in both charge and isoelectric point isomers) and deletions and/or insertions (resulting also in mol. wt. isomers) have contributed to the range of polypeptides observed today. It can be suggested that there is little constraint on the evolution of storage proteins (probably only in relation to deposition and mobilization) and hence the vast majority of the mutations are expressed in the protein patterns. The number of genes present at each locus is not known, however a minimum number can be taken as the number of polypeptides which can be separated by electrophoresis and IEF; the actual number may be considerably greater. Over 90% of the total amino acids of 'B' and 'C' hordein are uncharged and many mutations would be expected to result in substitutions of other uncharged amino acids for these. The presence of substitutions of this type, which would not affect the migration of the polypeptides on IEF or SDS-PAGE, is indicated by the C-terminal analysis of the 'C'-hordein fractions of Maris Mink ([158]) and similar substitutions have been demonstrated by sequence analysis of gliadin proteins of wheat [67, 150]. Viotti *et al.* [180] have presented evidence from nucleic acid hybridization of up to 120 zein genes per haploid genome in maize, and a similar estimate for barley may not be unreasonable.

Structural Relationships of Hordein and Storage Proteins from Related Species

The evolutionary ancestor of barley is now generally thought to be *H. spontaneum*, a species which is cross-fertile with *H. vulgare* and is a common weed in the Middle East. Doll and Brown [34] established the genetic homology of the hordein fractions from the two species while we have shown that natural cross-fertilization between cultivated *H. vulgare* and weedy *H. spontaneum* in north India gives rise to hybrid plants [115] which have been described as *H. agriocrithon* [187], a species of uncertain status that has been suggested as the wild ancestor of *H. vulgare* [1]. We have purified 'B' and 'C' hordein fractions from *H. spontaneum* (seed originating from N. India) and shown that they are similar, if not identical, in their C-terminal and N-terminal regions to fractions from *H. vulgare*. The 'B' hordein was blocked at the N-terminus and had (Arg/Leu/Thr or Gln)-Gly-Val-COOH at the C-terminus. The 'C' hordein fraction had an identical N-terminal sequence for the first 18 residues (with the possible exception of 5 residues which could not be positively identified (Table 11)) and C-terminal sequence for the first 3 residues (Ser-Val-Met-COOH). We have also analyzed seed of other barley species including *H. murinum*, *H. marinum*, *H. hystrix*, *H. glaucum* and *H. bulbosum*. All appeared to have groups of polypeptides corresponding to B, C and D hordein.

'C' hordein polypeptides are homologous with groups of ω-gliadins present in diploid wheat and rye. Shewry *et al* [150] compared the N-terminal sequences of a 'C' hordein component from cv. Julia and an ω-gliadin from *Triticum monococcum*, a diploid wheat thought to be related to the progenitor of the A genome of tetraploid and hexaploid wheats (Table 11). The sequences were identical at 23 out of 27 positions, although the *T. monococcum* fraction appeared to be a mixture of components, one of which had an additional N-terminal alanine which was not present in 'C' hordein. Kasarda *et al.* [68] subsequently showed that an additional N-terminal amino acid was also present in two other ω-gliadins purified in small amounts from *T. monococcum*, but was not present in ω-secalins from rye (*Secale cereale*). The ω-secalins from rye differed from 'C' hordein and ω-gliadins from wheat in several positions, notably 7 and 16 where different amino acids were present in the components from the three species. The components also have similar amino acid compositions, rich in glutamate + glutamine (over 40%), proline

Table 11. N-terminal amino acid sequences of 'C' hordein from barley ω-gliadins from wheat and ω-secalins from rye.

	-1	1	2	3	4	5	6	7	8	9	10	11	12	13	14	15	16	17	18	19	20	Ref.
H. vulgare 'C' hordein	-	Arg	Gln	Leu	Asn	Pro	Ser	Ser	Gln	Glu	Leu	Gln	Ser	Pro	Gln	Gln	Ser	Tyr	Leu	Gln	Gln	1
H. spontanteum 'C' hordein	-	X	Gln	Leu	Asn	Pro	X	X	Gln	Glu	Leu	Gln	X	Pro	Gln	Gln	X	Tyr	Leu			2
Rye ω-secalin	-	Arg	Gln	Leu	Asn	Pro	Ser	Glu	Gln	Glu	Leu	Gln	Ser	Pro	Gln	Gln	Pro	Val				3
T. monococcum	Ala	Arg	Gln	Leu	Asn	Pro	Ser	Asp	Gln	Glu	Leu	Gln	Ser	Pro	Gln	Gln	Leu	Tyr	Pro	Gln	Gln	1
ω-gliadin	-	Arg	Gln	Leu	Asn	Pro	Ser	Asp	Gln	Glu	Leu	Gln	Ser	Pro	Gln	Gln	Leu	Tyr	Pro	Gln	Gln	1

Variant positions are underlined.

[1]Shewry *et al.* (1980a)

[2]Unpublished results of the authors in collaboration with Dr. W. Manson and Dr. W. D. Annan, Hannah Research Institute

[3]Kasarda *et al.* (1982)

(25–32%) and phenylalanine (7–9%) and poor in lysine (0 → 0.3%), cysteine (trace amounts or not detected) and methionine (not detected to 0.2%). They do, however, vary in their mol. wts. by SDS-PAGE: 'C' hordein between 55,000 and 70,000, ω-secalins between 48 and 53,000 and the ω-gliadins of *T. monococcum* between 44 and 63,000. The ω-gliadins and ω-secalins appear to be coded by genes on chromosomes 1R of rye and 1 of wheat [9, 69, 149, 188], and it is probable that these chromosomes are homologous to chromosome 5 of barley.

'D' hordein is clearly homologous with high mol. wt. polypeptides, commonly called glutenin, which are present in protein bodies of wheat and rye [102]. These have similar amino acid compositions to 'D' hordein, notably high glycine [70, 102], and are coded for by genes on the long arms of chromosomes 1 of wheat and 1R of rye [85, 86, 126, 127].

The chemical and genetic relationships of 'B' hordein are not known. However it can be hypothesized that they are homologous with γ-secalins which are coded by genes on chromosome 1R of rye [9, 188] and therefore presumably also with the γ gliadins of wheat.

GLUTELIN AND RESIDUAL PROTEINS

Extraction and Characterization

The glutelins are the final fraction extracted from the grain, after removal of the salt-soluble proteins, non-protein nitrogen and prolamins. It is the most poorly characterized of the grain protein fractions for several reasons.

1. Because it is the last fraction to be extracted it is frequently contaminated with residual un-extracted polypeptides from other fractions, especially 'B' hordein. This is especially so if the hordein fraction is extracted without a reducing agent, but may still occur even if a reducing agent is used (see below). Consequently most electrophoretic analyses show a strong contamination with hordein polypeptides, masking the glutelin polypeptides [20, 84, 166]. This is also apparent in the amino acid compositions which frequently show a strong similarity to hordein with high glutamate plus glutamine and proline and very low lysine [20, 84] (Table 12).
2. The use of extreme conditions to extract the previous protein fractions may denature the glutelins and render them difficult to extract and separate. This was first noted by Osborne [120] who stated 'it was not possible to extract more than a very small amount of this residual proteid with dilute potash water as the treatment for removal of the other proteids rendered it insoluble, if it were not so already'.
3. Even if not denatured during the extraction of the other protein fractions, the glutelins are not readily soluble and consequently special solvents and pro-

Table 12. Amino acid compositions of glutelin fractions prepared by different procedures.

	1	2	3
Asp	1.7	4.2	8.2
Thr	2.5	3.8	4.0
Ser	4.8	5.7	4.0
Glu	33.1	25.9	12.8
Pro	20.7	14.1	7.7
Gly	3.2	6.3	9.8
Ala	2.7	5.1	8.3
Cys	2.5	0.5	1.1
Val	5.2	6.5	7.5
Met	1.3	1.3	0.6
Ile	4.0	4.2	5.2
Leu	6.6	8.2	9.6
Tyr	2.5	2.7	trace
Phe	4.9	4.3	4.7
His	1.3	1.9	3.6
Lys	0.8	2.2	7.3
Arg	2.1	3.2	5.6

Expressed as mole %.
Values reported for glutamate and aspartate include the amides glutamine and asparagine respective.

[1] Lauriere et al (1976). A mixture of glutelins and prolamins were extracted from milled whole grain with 1% acetic acid and the glutelins precipitated by the addition of ethanol to 45% and standing at 3°C.

[2] Brandt (1976). Glutelins extracted from 28-day-old (physiologically-mature) endosperms of cv. Bomi with 0.2N NaOH after extraction of hordein with 55% propan-2-ol at 20°C.

[3] Shewry et al (1979b) Glutelins extracted from 7-week-old (physiologically mature) endosperms of Bomi with 0.05M borate buffer pH 10.0 + 1% 2-mercaptoethanol + 1% SDS after extraction of hordein with 55% propan-2-ol + 2% 2-mercaptoethanol at 60°C.

cedures must be used for their extraction and separation. In most early and some recent studies (for example, Brandt, [20]) the glutelins were extracted using dilute alkali which almost certainly results in deamidation and possibly in some peptide bond cleavage. The more recent introduction of solvents containing detergents (notably SDS or cetyltrimethyl ammonium bromide) or chaotropic agents, such as urea, has minimized the possibility of degradation, and these solvents can also be used to solubilize the proteins during electrophoretic analysis.

A number of the problems in the quantitative extraction of glutelins are illustrated by the study of Landry *et al.* [82], who compared the effects of extraction with different sequences of solvents on the total amounts and amino acid compositions of the protein fractions. Variations in the amino acid composition (3.2 to 4.6 mole % lysine; 17.9 to 24.4 mole % glutamate + glutamine and 9.4 to 14.1 mole % proline) of the glutelins (extracted with SDS) indicated a variable contamination with hordein which was dependent on the conditions used to extract the previous

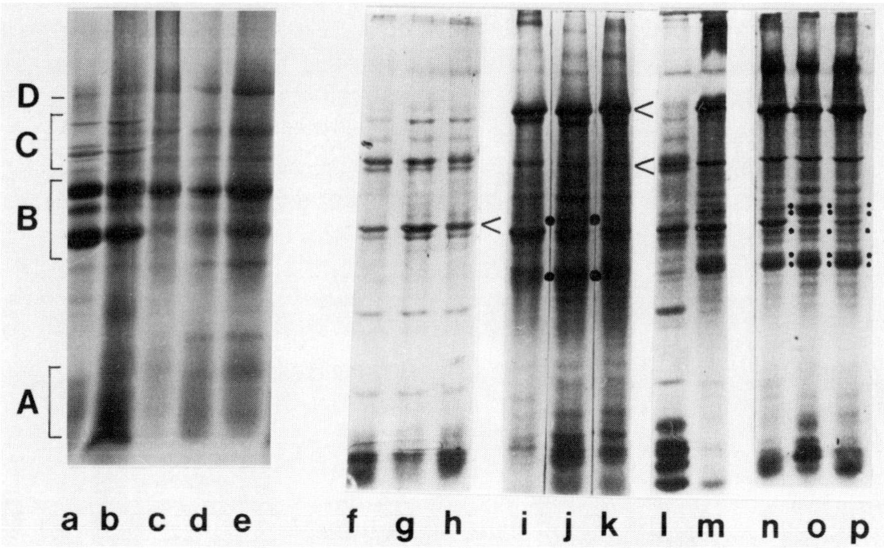

Figure 13. SDS-PAGE of glutelins, salt-soluble proteins and hordein. a, reduced and pyridylethylated total hordein; b, PE acid alcohol-soluble glutelin; c, PE acid alcohol-insoluble glutelin; d, glutelin I; e, glutelin II. a–e are all from cv. Bomi (from Shewry *et al.*, [156]). f–h are salt-soluble proteins extracted with no reducing agent, 1% 2-mercaptoethanol and 0.05% dithiothreitol respectively. i–k are glutelin fractions extracted with electrophoresis sample buffer containing 6 M urea, 1% SDS and 1% 2-mercaptoethanol after salt-soluble fractions a, b and c respectively. l is PE salt-soluble proteins extracted without reducing agent (cf. f). n, o, p are PE glutelins prepared by direct alkylation in the presence of 6 M urea after salt-soluble fractions f, g, h respectively. m is the same sample as n. The arrow in h indicates a polypeptide which is extracted in the salt-soluble fraction only in the presence of a reducing agent. The arrows in k indicate two glutelin polypeptides which are extracted more efficiently in the presence of 2-mercaptoethanol than in the presence of dithiothreitol. The dots indicate the positions of residual hordein polypeptides in the glutelin fractions (from Wilson *et al.*, [186]).

fractions. The percentage of N extracted in the glutelins (38% to 7%) varied inversely with the percentage of N previously extracted (54% to 66%). Also, increases in the percentage of N extracted in the non-glutelin fractions resulted in an increase in the percentage of residual N (from 8 to 27%) indicating that the glutelins had been made insoluble.

Several years ago we reported a detailed comparison and evaluation of procedures for the extraction of hordein and glutelin fractions [156]. The fractions extracted were monitored for purity by amino acid analysis (relative amounts of glutamate, proline and lysine) and SDS-PAGE. Data from two of these procedures is given in Table 13 and Figure 13. Procedure 1 is based on the sequential extraction procedure of Landry and Moureaux [81] with the glutelins extracted in two fractions, both in pH 10 buffer containing reducing agent but differing in the absence or presence of SDS (termed glutelin I and glutelin II respectively). Although hordein

Table 13. Amino acid compositions of glutelin and residual fractions prepared from endosperms of barley cv. Bomi by different procedures.

	Procedure 1			Procedure 2			Hordein[a]
	Glutelin I	Glutelin II	Residue	A-S glutelin	A-I glutelin	Residue	
%total grain N	2.8	24.5	9.3	4.1	9.8	11.9	
Amino Acid							
Asp	7.4	7.6	7.8	6.1	9.2	8.5	1.7
Thr	5.4	5.8	6.3	4.9	5.5	5.5	2.7
Ser	7.8	7.5	7.9	7.1	7.2	7.2	5.1
Glu	14.2	15.5	17.9	18.6	12.7	11.8	32.6
Pro	9.3	9.5	10.2	11.6	6.1	6.5	21.8
Gly	12.6	9.2	6.0	8.6	8.9	10.7	3.9
Ala	8.3	8.8	8.4	7.0	9.1	10.1	3.1
Cys	ND	ND	ND	1.3	1.2	1.2	ND
Val	6.2	2.4	6.6	6.6	6.9	7.0	5.5
Met	1.3	1.5	1.4	1.1	1.4	1.4	0.9
Ile	2.9	3.8	3.0	3.4	4.1	4.0	3.4
Leu	7.1	8.6	7.3	7.7	8.7	7.8	7.3
Tyr	3.4	3.3	3.0	2.9	2.2	2.1	2.8
Phe	2.3	3.9	3.3	3.3	4.5	4.4	4.6
His	2.9	2.8	3.0	2.3	2.5	2.3	1.7
Lys	4.8	5.4	4.0	3.4	5.3	5.3	0.9
Arg	4.0	4.5	3.7	3.9	4.5	4.2	2.1

Results are expressed as mole %.

Values given for glutamate and aspartate include the amides glutamine and asparagine respective.

ND = not determined A-I = acid insoluble A-S = acid soluble

[a] Hordein fraction extracted with 55% propan-2-ol + 2% 2-mercaptoethanol at 60°C as part of procedure 2.

Partial amino acid compositions of these fractions have been previously reported (Shewry <u>et al</u>, 1978b) and polyacrylamide gel separations are given in Fig. 13.

Procedure 1 is based on Landry and Moureaux (1970). Two hordein fractions were extracted with 55% propan-2-ol (hordein I) and the same solvent + 0.6% 2-mercaptoethanol (hordein II) followed by two glutelin fractions with 0.05M borate buffer, pH 10, + 0.6% 2-mercaptoethanol (glutelin I) and the same solvent + 1% SDS (glutelin II), all extractions being carried out at 20'C.

Procedure 2 is based on Bietz and Wall (1973). The hordein fraction was extracted with 55% propan-2-ol + 2% 2-mercaptoethanol at 60°C and the glutelins by direct alkylations of the seed meal. The alkylated glutelins were then shaken with 70% ethanol + 0.7% acetic acid at 60°C to give acid-soluble and acid-insoluble fractions.

was previously extracted in the presence of reducing agent, there is still evidence of contamination of 'B' hordein polypeptides in both fractions (Figure 13d, e). The two glutelin fractions had similar amino acid and polypeptide compositions demonstrating that they probably do not contain unique groups of proteins. The residual fraction from this procedure had more glutamine + glutamate and proline and less lysine than the glutelin fractions, indicating the presence of some un-extracted hordein. In procedure 2 the hordein was extracted by a more exhaustive procedure and the glutelins by direct alkylation of the seed meal in the presence of 8M urea. Despite the recovery of 47.4% of the seed N in the hordein fraction (compared to only 39.5% in procedure 1), extraction of the glutelins with hot acid-alcohol gave a fraction containing a mixture of hordein and glutelin polypeptides (Figure 13b). The insoluble glutelin fraction had an amino acid composition similar to the salt-soluble proteins (Table 13) with low glutamate (12.7%) and proline (6.1%) and high lysine (5.3%). The residual fraction from this procedure had a similar amino acid composition to the glutelins. The SDS-PAGE patterns of all the glutelin fractions prepared were, however, unsatisfactory with diffuse bands and an intensely-stained background.

More recently we have re-examined of the glutelin fraction, and its relationship to salt-soluble proteins, using procedures which were selected to minimize damage [186]. The endosperms used for the fractionations were physiologically mature but not dry, thus eliminating the possibility of protein damage occurring during grain dehydration. The salt-soluble proteins were extracted at 3°C in phosphate buffer, pH 7.8, containing 1.0 M NaCl, and hordein at 20°C with 50% propan-1-ol + 2% 2-mercaptoethanol. Glutelin fractions were extracted either with 0.05 M Tris buffer, pH 7.0, containing 1% SDS, 2 mM EDTA, 10% glycerol and 1% 2-mercaptoethanol (sometimes also 6 M urea) (called SDS-glutelins), or by direct alkylation of the residual meal as described previously [156] (called alkylated glutelins). For the direct alkylation the concentration of urea was reduced from 8 M to 6 M to prevent the solubilization of too much starch. Although the resulting SDS-PAGE patterns of the glutelins were contaminated with hordein, the positions of the hordein bands were determined by co-migration with authentic protein and are indicated by dots in Figure 13. The alkylated glutelins (Figure 13m–p) had similar band patterns to the SDS-glutelins (Figure 13i–k) but the bands were sharper with less background staining. It was also easier to recognize the residual hordein polypeptides in the alkylated samples. Comparison of the band patterns of the salt-soluble protein and glutelin fractions showed little similarity, with the exception of one band (arrowed in Figure 13h) which was present in both fractions but was preferentially extracted in the salt-soluble fraction in the presence of a reducing agent (cf. Figure 13f–h, i–k, n–p). The presence of a reducing agent during the salt-extraction also appeared to affect the relative intensity of other glutelin bands (note band arrowed in Figure 13k) and the amounts of residual hordein present (cf. Figure 13i–k, n–p).

It is clear from this study that good PAGE separations of glutelin fractions can be obtained if care is taken to avoid rendering the protein insoluble during grain drying and during the extraction of the other fractions. It is also apparent that the patterns obtained are affected by the conditions used to extract both the glutelin fraction itself and the salt-soluble proteins, and the presence of residual hordein bands must be expected and recognized. There were, however, relatively few polypeptides which were present in the salt-soluble protein and glutelin fractions suggesting that the glutelins contain a separate and distinguishable group of polypeptides and are hence a valid subdivision of the seed proteins. The absence of any quantitatively major bands and their wide range of mol. wts. is consistent with their function as structural and metabolic proteins rather than storage proteins, and this is also suggested by their absence from preparations of protein bodies [101].

Little is known of the genetic control or degree of genetic variability of the individual glutelin polypeptides, although we have shown that glutelins from two cultivars, Julia and Bomi, have similar SDS-PAGE patterns [156]. If the glutelin fraction does contain metabolic and structural proteins it can be assumed to be under the control of many genes and relatively little genetic variability can be expected.

SYNTHESIS OF BARLEY SEED PROTEINS

In vivo synthesis

Bishop [15] was probably the first person to study the developmental sequence in which the protein fractions of the barley grain accumulate. His results are shown in Figure 14a and differ little from the more recent results of Rahman *et al.* [135] (Figure 14b). These results indicate that the salt-soluble proteins are made earlier in development than the other two fractions and soon reach their maximum amount. The hordein fraction, as may be expected from its storage function, is present in measurable amounts only after about 18 days after anthesis by which time the endosperm has attained about 10% of its final dry weight [135]. The glutelins appear to be formed earlier than the hordeins but, unlike the salt-soluble proteins, they continue to accumulate for a longer period during development. The results for the salt-soluble fraction in Figure 14a and b are probably an oversimplification because when the albumins and globulins are followed separately [20] they have entirely different developmental patterns (Figure 14c); the globulins behave more like the hordeins in that they accumulate late in development and the albumins actually decrease in amount in the latter stages of development.

The kinetics of accumulation of the protein fractions do not indicate the degree of turnover that may be occurring within them. Although there are now several elegant techniques for measuring protein turnover (e.g. see Davies [31]) these do not appear to have been applied to the developing seed. It is probable that the only results which

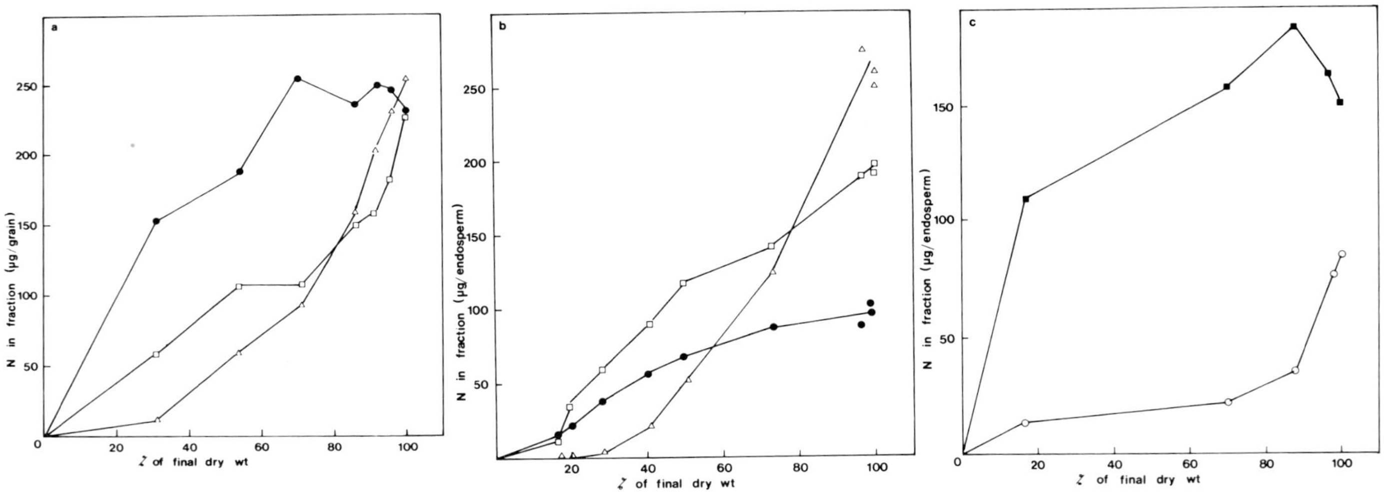

Figure 14. The accumulation of different protein fractions in the developing grain of barley. a, redrawn from Bishop [15] for grain grown at Rothamsted in 1929; b, redrawn from Rahman *et al.* [135] for grain grown at Rothamsted in 1980; c, redrawn from Brandt [20] ● Total salt-soluble N, ■ albumin, ○ globulin, △ hordein, □ glutelin.

relate to turnover are those from the labelling studies we carried out to compare the metabolism of the high-lysine mutant Risø 1508 and its normal parent, Bomi. In this study ears were exposed to $^{14}CO_2$ for a period of 1 h and then harvested and analyzed after either 24 h or at maturity (7 weeks after anthesis). These crude 'chase' experiments indicated that whereas ^{14}C, incorporated into salt-soluble proteins at the time of their maximum rate of accumulation, decreased to about half at maturity, that incorporated into the hordein fraction increased considerably during the interval between the time of isotope administration and maturity. The results are thus indicative of turnover occurring in the salt-soluble protein fraction but not in the hordeins.

Each of the protein fractions consists of many individual polypeptides which are the products of different genes and it would be unrealistic to expect them all to behave in an identical manner. Some studies have been done on the individual components of the fractions. Where the albumins have been studied it has usually been in terms of changes in the amounts of various enzyme activities rather than in the total amounts of any individual proteins. In general they follow the pattern for the albumins shown in Figure 14c, although β amylase continues to accumulate throughout development [80]. In this case the enzyme is stored for subsequent release during germination.

Some controversy exists in the literature as to whether there are changes [23] or not [20] in the rate of accumulation of the different hordein polypeptides. We have recently reinvestigated this question and have shown that there are indeed differences in the composition of the hordein fraction of cv. Sundance at different stages of development (Figure 15, [135]). 'C'-hordein polypeptides form a greater proportion of the total in the early stages of development, but during the latter stages there is an increase in the relative amount of 'B' hordein and in one specific group of polypeptides (called B1 hordein) within this fraction. This latter result is interesting as it could indicate differences in the relative rates of synthesis of polypeptides specified by the *Hor-2* locus.

Protein Deposition

Proteins are not evenly distributed within the grain, disproportionally large amounts being present in the embryo, the aleurone and the subaleurone layers. The solubility fractions are also differentially distributed with the prolamins being located solely in the starchy endosperm and the albumins and globulins largely in the embryo and aleurone. The proteins may be expected to be distributed in the soluble phase, in membranes, in organelles and in protein bodies. Because the latter are particular features of seeds and are important in the storage of protein they will be dealt with in some detail. Within a seed more than one type of protein body can be recognized by morphological characteristics. The distribution of the different types varies according to the tissue studied.

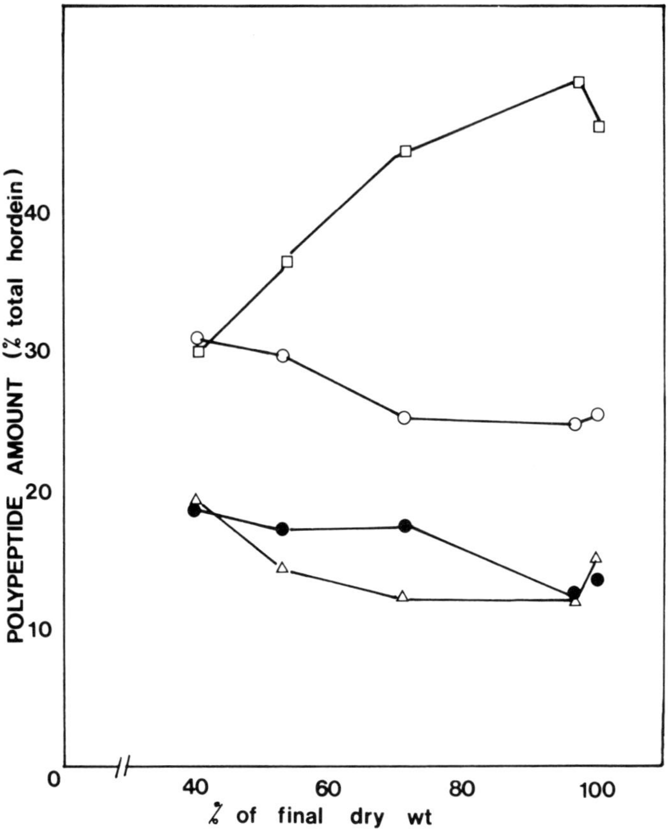

Figure 15. Changes in the relative amounts of different hordein polypeptides during barley grain development □ B1, ● B2, ○ B3, △ C1. Redrawn from Rahman *et al.* [135].

Aleurone and Scutellum Protein Bodies

The protein bodies of the aleurone layer and the scutellum have been studied by means of light and electron microscopy. Those of the aleurone layer appear to be similar in structure to ones studied in more detail in other seeds such as castor bean (e.g. Tulley and Beevers [176]). Thus many authors have shown that the aleurone bodies have two types of inclusion [26, 37, 58, 64, 124] although there is apparently some confusion in differentiating and identifying them. Jacobsen *et al.* [58] used differential staining procedures and defined the inclusions as globoids and protein-carbohydrate bodies. The globoid is not retained under many fixation conditions and only a cavity filled with resin is observed (Figure 16a). It is not clear whether this globoid is surrounded by a membrane; Jacobsen *et al.* [58] think not whereas Buttrose [26] and workers with other species (e.g. Tulley and Beevers, [176]) suggest

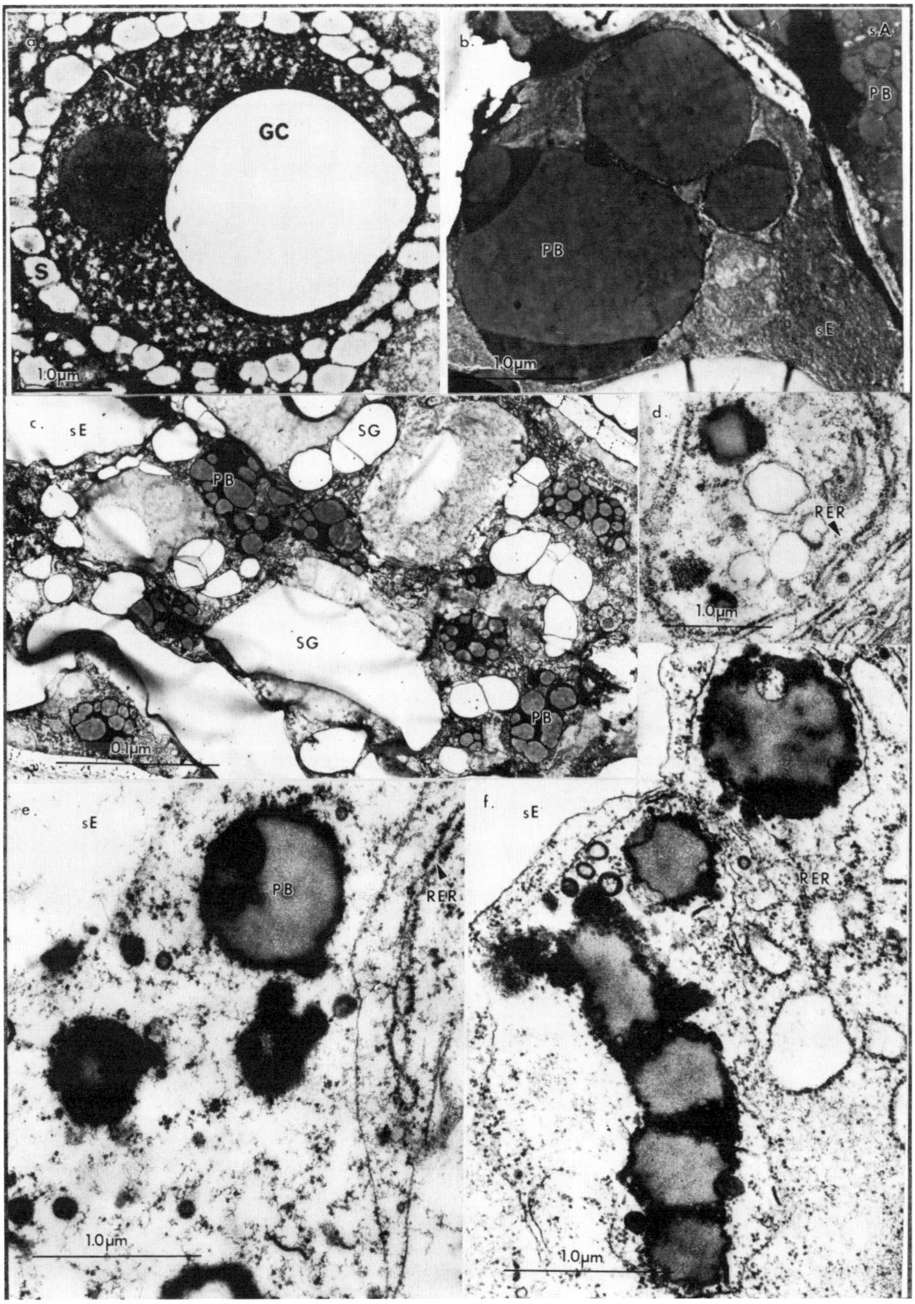

Figure 16. Electron micrographs of protein bodies of barley. a, protein body from mature barley aleurone; GC globoid cavity, PCB protein carbohydrate body, GS ground substance, S spherosome (reproduced with permission from Jacobsen *et al*., [58]. b–f, protein bodies from the developing barley endosperm at about b, 34; c, 37; d and e 24; f, 30 days after anthesis. PB protein bodies, SG starchy grains, RER rough endoplasmic reticulum, SE starch endosperm cells, sA sub-aleurone cell.

that they are. The second inclusion appears directly analogous to the 'protein crystalloid' found in castor bean [176] from which it has been isolated and shown to contain storage globulins. No comparable studies have been done on barley aleurone protein bodies so that little is known of their contents.

Electron microscopy of barley scutellum epithelial cells again shows the presence of protein bodies [117] although in this case they only appear to have one type of inclusion which has the appearance of a globoid cavity. Studies of wheat scutellum tissue also suggest that only globoid inclusions are present [173, 174]. Again nothing is known of the proteins stored in these protein bodies.

Endosperm Protein Bodies

The major storage protein in the endosperm is hordein and it is generally considered that all of the protein bodies in the endosperm contain hordein. This may not be strictly true since certain α and β globulins are located in the endosperm and not solely in the aleurone [30] (Figures 1, 2); if these are present in protein bodies then there may be a minor population of such bodies in the starchy endosperm. However, for the rest of this discussion we will consider only the prolamin-containing bodies. Two types of such protein bodies can be recognized in the starchy-endosperm of the developing seed ([103]; Figure 16b), those of the sub-aleurone layers and those of the more central regions. The sub-aleurone protein bodies are smaller in size and homogeneous under the transmission electron microscope whereas those in cells a few layers nearer the centre are larger and heterogeneous in appearance. As the endosperm develops the protein bodies occupy more and more of the space between the starch grains (Figure 16c) until at maturity they appear to have fused to form a more or less continuous protein matrix which is interspersed with starch grains and remnants of endoplasmic reticulum [8].

Two theories exist as to the nature and origin of barley endosperm protein bodies. The first suggests that, in common with protein bodies of other seeds, they are derived from vacuoles into which the protein is transported and deposited (see Matile, [92]; Ashton, [4] for a general discussion). This theory has been most fully supported by von Wettstein and colleagues [28, 114] and they have produced micrographs to show that the proteins are accumulated within vacuoles. In contrast we have suggested [102, 103] that these protein bodies are derived from the rough endoplasmic reticulum, as has been clearly shown for maize [71, 83, 101]. In our hypothesis the hordeins are considered to be deposited inside the endoplasmic reticulum where they aggregate. As the aggregate increases in size so it breaks away from the endoplasmic reticulum and is deposited within the cytoplasm. It is not considered to be enclosed completely within a membrane. Again electron micrographic evidence can be presented to support this view ([101, 103]; Figure 16d, e, f). Since the developing seed, especially the starchy region, is difficult to fix and section it is to be expected that there may be differences in the interpretation of electron microscope images and we have therefore tried alternative approaches. Attempts to recognize protein bodies as distinct spherical objects in the scanning electron microscope have not been successful in barley [25] or wheat [128], although both sets of authors could identify those present in developing cotyledons of *Pisum* or *Vicia* respectively.

The other approach has been to isolate developing protein bodies and to study their characteristics. Ingversen [54] prepared protein bodies from developing grain on sucrose density gradients and showed that they contained the major hordein polypeptides and we have confirmed this [101, 103]. Our results also show that the 'D' hordein fraction is present in protein bodies. Electron micrographs of the isolated protein bodies [101] do not suggest that they are enclosed within a limiting membrane. To test this we incubated protein bodies in the presence of proteinase k and showed that they were not protected from digestion. In parallel experiments we found that other subcellular organelles of barley and protein bodies from pea cotyledons were protected from digestion [100]. Subcellular separations also show that the hordein polypeptides are associated with the major peak of endoplasmic reticulum and conversely a small amount of endoplasmic reticulum is associated with the protein bodies; we find no evidence for any association of vacuolar enzymes with the protein bodies [101].

In conclusion, therefore, we consider, both from studies on barley and parallel work with other cereals, that there is considerable evidence to suggest that hordein is not deposited in protein bodies of vacuolar origin but is present as aggregates of insoluble protein associated with fragments of the endoplasmic reticulum. As the seed develops these aggregates fuse together to give a proteinaceous matrix surrounding the starch grains (e.g. see Bechtel and Pomeranz, [8]).

In Vitro Synthesis

Convincing evidence has been obtained for the *in vitro* synthesis of hordeins directed by polysomes and mRNA isolated from developing seeds [22, 47, 94]. The *in vitro* synthesis of α-amylase has also been studied in several laboratories but in this case the RNA fractions have been obtained from germinating seeds. Since we are concentrating on those proteins present in the mature seed we will not consider this work further.

The mRNA for hordeins has been shown to be associated with membrane-bound polysomes by using the polysomes themselves, and the poly A^+ RNA fraction from them, to direct cell free protein synthesis in a wheat germ system (Figure 17). These results support the studies with subcellular separations (see above). The polypeptides synthesized by poly A^+ RNA migrate more slowly in SDS-PAGE than do authentic hordeins and this is interpreted as evidence that the poly A^+ fraction leads to the synthesis of precursor proteins which are longer than the authentic proteins. It is considered that the extra protein is a 'signal' which, as it is synthesized, causes

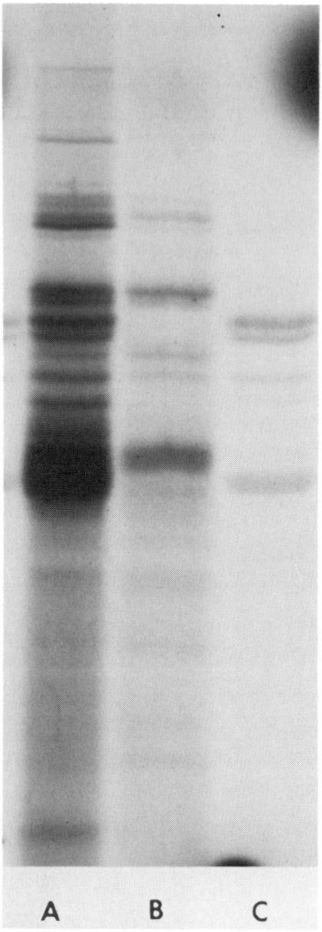

Figure 17. In vitro synthesis of hordeins. Synthesis of radioactive proteins was accomplished using an *in vitro* wheat germ system directed by a, polysomes obtained from a membrane fraction of developing endosperms; b, poly-A^+ RNA derived from such polysomes. c, contains shows hordeins extracted from seeds that had been fed ^{35}S-sulphate during development. Note that polysomes direct the synthesis of products of the same and lower mobility than the *in vivo* synthesized hordein whilst poly A^+ RNA only leads to the formation of the latter (from Matthews [93]).

the polysomes to attach to the endoplasmic reticulum and leads to the co-translational transport of the polypeptide through the membrane into the lumen of the endoplasmic reticulum. During the transport the signal sequence is removed giving rise to authentic proteins. This sequence of events is analagous to that shown to occur for transported proteins in animal systems [17, 18]. Results consistent with this sequence of events have been obtained by Matthews [93] and similar results, except that post-translation transport was observed, were reported by Cameron-Mills and Ingversen [27].

Poly A^+ RNA fractions rich in mRNA for hordein have been used to make cDNA which in turn has been made double stranded, inserted into a plasmid, and the recombinant plasmid cloned in *Eschericia coli* [21, 44]. We have now identified many of these cloned plasmids as containing inserted DNA sequences which are related to the known amino acid sequence of B hordeins ([44]; Figure 10). Consideration of the number and relationships of the clones obtained has led us to conclude that the original poly A^+ RNA fraction must have consisted of a relatively large number of B hordein mRNAs of similar but different sequence [105]. This conclusion is entirely consistent with the hypothesis of a complex multigenic locus (*Hor 2*) coding for the B hordeins (see above). Further study of these clones will lead to more detailed knowledge of the amino acid sequences of the hordeins and will facilitate the isolation and analysis of the genomic DNA for the hordeins. The clones will also allow further studies of the temporal and developmental control of these genes and thus provide a better understanding of the biology of the barley seed.

ACKNOWLEDGEMENTS

We are grateful to our colleagues at Rothamsted (Audrey Faulks, Saroj Parmar, Helen Pratt, Susan Smith, Shirley Burgess, Jayne Matthews, S. Rahman, B.G. Forde, J.M. Field, M.A. Kirkman, and M. Kreis) and to our collaborators at other research institutes (R.A. Finch, R.A. Pickering, I.T. Jones, M.S. Wolfe) for assistance, advice and discussion.

We are also grateful to Dr. H. Doll, Dr. D.D. Kasarda and their colleagues at Risø (Denmark) and Albany (California, U.S.A.), respectively for stimulating discussions.

We gratefully acknowledge receipt of financial assistance from the EEC (grant No. 470).

Finally we are indebted to Sue Wilson for her painstaking preparation of this manuscript.

REFERENCES

1. Åberg, E (1938): *Hordeum agriocrithon*, a wild six-rowed barley. *Ann. Agric. Coll. Swed.* 6: 159–216.
2. Andersen, AJ, and B Køie (1975): N fertilization and yield response of high lysine and normal barley. *Agron. J.* 67: 695–698.
3. Aragoncillo, C, R Sanchez-Monge, and G Salcedo (1981): Two groups of low molecular weight hydrophobic proteins from barley endosperm. *J. Expt. Bot.* 32: 1279–1286.

4. Ashton, F (1976): Mobilization of seed storage proteins. *A. Rev. Pl. Physiol.* 27: 94–117.
5. Autran, J-C, and R Scriban (1977): Recherche sur la pureté variétale d'un malt. *Europ. Brew Conv. Rotterdam*: 47–62.
6. Balaravi, SP, HC Bansal, BO Eggum, and S Bhaskaran (1976): Characterization of induced high protein and high lysine mutants in barley. *J. Sci. Fd. Agric.* 27: 545–552.
7. Bansal, HC, KN Srivastava, BO Eggum, and SL Mehta (1977): Nutritional evaluation of high protein genotypes of barley. *J. Sci. Fd. Agric.* 28: 157–160.
8. Bechtel, DB, and Y Pomeranz (1979): Endosperm structure of barley isogenic lines. *Cereal Chem.* 56: 446–452.
9. Bernard, M, J-C Autran, and P Joudrier (1977): Possibilités d'identification de certains chromosomes de siègle à l'aide de marqueurs biochimiques. *Ann. Amélior: Plantes* 27: 355–362.
10. Bietz, JA (1981): Note on the terminal amino acid sequence of hordein. *Cereal Chem.* 58: 83–85.
11. Bietz, JA, and JS Wall (1973): Isolation and characterization of gliadin-like subunits from glutenin. *Cereal Chem.* 50: 537–547.
12. Biserte, G, and R Scriban (1950): Les protides de l'orge. *Bull. Soc. Chim. Biol.* 32: 959–968.
13. Bishop, LR (1928): The composition and quantitative estimation of barley proteins I. *J. Inst. Brew.* 34: 101–118.
14. Bishop, LR (1929): The composition and quantitative estimation of barley proteins II. *J. Inst. Brew.* 35: 316–322.
15. Bishop, LR (1930): The composition and quantitative estimation of barley proteins III. – The proteins of barley during development and storage and in the mature grain. *J. Inst. Brew.* 36: 336–349.
16. Bishop, LR (1939): The Proteins of Barley Grain, with Special Reference to Hordein. M.Sc. Thesis, University of Birmingham, U.K.
17. Blobel, G (1980): Intracellular protein topogenesis. *Proc. Nat. Acad. Sci. USA* 77: 1496–1500.
18. Blobel, G, and B Dobberstein (1975): Transfer of proteins across membranes. *J. Cell Biol.* 67: 852–862.
19. Brandt, A (1975): *In vivo* incorporation of [^{14}C] lysine into the endosperm proteins of wild type and high-lysine barley. *FEBS Lett.* 52: 288–291.
20. Brandt, A (1976): Endosperm protein formation during kernel development of wild type and high-lysine barley mutant. *Cereal Chem* 53: 890–901.
21. Brandt, A (1979): Cloning of double stranded DNA coding for hordein polypeptides. *Carlsberg Res. Commun.* 44: 255–267.
22. Brandt, A, and J Ingversen (1976): *In vitro* synthesis of barley endosperm proteins on wild type and mutant templates. *Carlsberg Res. Commun.* 41: 312–320.
23. Breidert, D, and W Schön (1974): Biochemische Probleme bei der Qualitätsverbesserung der Getreideproteine. *Göttinger Pflanzenzüchter-Seminar* 1: 89–105.
24. Bright, SWJ, BJ Miflin, and SE Rognes (1981): Threonine accumulation in the seeds of a barley mutant with an altered aspartate kinase. *Biochem. Genet.* 20: 229–243.
25. Burgess, SR, RH Turner, PR Shewry, and BJ Miflin (1982): The structure of normal and high-lysine barley grains. *J. Expt. Bot.* 33: 1–11.
26. Buttrose, MS (1971): Ultrastructure of barley aleurone cells as shown by freeze-etching. *Planta* 96: 13–16.
27. Cameron-Mills, V, and J Ingversen (1978): *In vitro* synthesis and transport of barley endosperm proteins: reconstitution of functional rough microsomes from polyribosomes and stripped polysomes. *Carlsberg Res. Commun.* 43: 471–489.
28. Cameron-Mills, V, and J von Wettstein (1980): Protein body formation in the developing barley endosperm. *Carlsberg Res. Commun.* 45: 577–594.
29. Carbonero, P, F Garcia-Olmedo, and C Hernandez-Lucas (1980): External association of hordothionin with protein bodies in mature barley. *J. Agric. Food Chem.* 28: 399–402.

30. Danielsson, CE (1949): Seed globulins of the Gramineae and Leguminosae. *Biochem. J.* 44: 387–400.
31. Davies, DD (1979): Factors affecting protein turnout in plants. *In*: Hewitt, EJ, and CV Cutting (eds.): Nitrogen Assimilation of Plants: 369–396; Academic Press, London.
32. Davis, BJ (1964): Disc electrophoresis II: method and application to human serum proteins. *Ann. New York Acad. Sci.* 121: 404–427.
33. Djurtoft, R (1961): Salt Soluble Proteins of Barley. Dansk Videnskabs Forlag A/S København.
34. Doll, H, and ADH Brown (1979): Hordein variation in wild (*Hordeum spontaneum*) and cultivated (*H. vulgare*) barley. *Can. J. Genet. Cytol.* 21: 391–404.
35. DuCros, DL, and CW Wrigley (1979): Improved electrophoretic methods for identifying cereal varieties. *J. Sci. Fd. Agric.* 30: 785–794.
36. Duhamel, RC, E Meezan, and K Brendel (1980): Metochromatic staining with Coomassie Brilliant Blue R250 of the proline-rich calf thymus histone, H1. *Biochim. Biophys. Acta* 626: 432–442.
37. Eb, AA van der, and PJ Nieuwdorp (1967): Electron microscopic structure of the aleurone cells of barley during germination. *Acta Bot. Neerl.* 15: 690–699.
38. Einhof, H (1806): Chemische Analyse der kleinen Gerste (*Hordeum vulgare*). *Neues allgem. J. Chem.* 6: 62–98.
39. El Negoumy, AM, CW Newman, and BR Moss (1979): Amino acid composition of total protein and electrophoretic behavior of protein fractions of barley. *Cereal Chem.* 56: 468–473.
40. Ewart, JAD (1980): Isolation of a hordein of low electrophoretic mobility. *J. Sci. Fd. Agric.* 31: 82–85.
41. Ewart, JAD (1980): Isolation of three hordeins. *J. Sci. Fd. Agric.* 31: 1183–1188.
42. Faulks, AJ, PR Shewry, and BJ Miflin (1981): The polymorphism and structural homology of storage polypeptides (hordein) coded by the *Hor-2* locus in barley (*Hordeum vulgare* L.). *Biochem. Genet.* 19: 841–858.
43. Folkes, BF, and EW Yemm (1956): The amino acid content of the proteins of barley grain. *Biochem. J.* 62: 4–11.
44. Forde, BG, M Kreis, MB Bahramian, JA Matthews, BJ Miflin, RD Thompson, D Bartels, and RB Flavell (1981): Molecular cloning and analysis of cDNA sequences derived from poly A^+ RNA from barley endosperm: identification of B hordein-related clones. *Nucleic Acids Research* 9: 6689–6707.
45. Foriers, R, R De Neve, and L Kanarek (1975): Specificity and partial purification of barley-germ agglutinin. *Arch. Int. Physiol. Biochem.* 83: 362.
46. Foriers, R, R De Neve, and L Kanarek (1976): Purification of barley-germ agglutinin. *Arch. Int. Physiol. Biochem.* 84: 617–618.
47. Fox, JE, HM Pratt, PR Shewry, and BJ Miflin (1977): The *in vitro* synthesis of hordeins with polysomes from normal and high lysine varieties of barley. In: *Nucleic Acids and Protein Synthesis in Plants*: 520–524; Centre Nat. Res. Sci. Paris.
48. Fuller, MF, RM Livingstone, BA Baird, and T Atkinson (1979): The optimal amino acid supplementation of barley for the growing pig.1. Response of nitrogen metabolism to progressive supplementation. *Br. J. Nutr.* 41: 321–331.
49. Fuller, MF, J Mennie, and RMJ Crofts (1979): The amino acid supplementation of barley for the growing pig. 2. Optimal additions of lysine and threonine for growth. *Br. J. Nutr.* 41: 333–340.
50. Goldstein, IJ, and CE Hayes (1978): The lectins: carbohydrate binding proteins of plants and animals. *Adv. in Carbohydrate Chem. Biochem.* 35: 127–340.
51. Hejgaard, J, and S Boisen (1980): High lysine proteins in Hiproly barley breeding: identification, nutritional significance and new screening methods. *Hereditas* 93: 311–320.
52. Holder, AA, and J Ingversen (1978): Peptide maping of the major components of *in vivo* synthesised barley hordein: Evidence of structural homology. *Carlsberg Res. Commun.* 43: 177–184.
53. Hoseney, RC, Y Pomeranz, JD Hubbard, and KF Finney (1971): Petroleum ether-soluble lipoprotein of barley flour. *Cereal Chem.* 48: 223–229.

54. Ingversen, J (1975): Structure and composition of protein bodies from wild-type and high-lysine barley endosperm. *Hereditas* 81: 69–76.
55. Ingversen, J, and B Køie (1973): Lysine-rich proteins in high-lysine *Hordeum vulgare* grain. *Phytochem.* 12: 1107–1111.
56. Ingversen, J, B Køie, and H Doll (1973): Induced seed protein mutant of barley. *Experientia* 29: 1151–1152.
57. Ivanov, C, B Mesrob, and Z Prusik (1968): Fractionation of hordein by preparative, continuous carrier-free electrophoresis. *Can. J. Biochem.* 46: 1301–1307.
58. Jacobsen, JV, RB Knox, and NA Pyliotis (1971): The structure and composition of aleurone grains in the barley aleurone layer. *Planta* 101: 189–209.
59. Jensen, J (1981): Coordinators report. Chromosome 5. *Barley Genet. Newsl.* 11: 87–88.
60. Jensen, J, JH Jörgensen, HP Jensen, H Giese, and H Doll (1980): Linkage of the hordein loci *Hor-1* and *Hor-2* with the powdery mildew resistance loci *Mlk* and *Mla* on barley chromosome 5. *Theor. Appl. Genet.* 58: 27–31.
61. Jensen, R, (1952): On isolation and amino acid composition of β globulin extracted from the seeds of barley (*Hordeum vulgare* L). *Acta Chem. Scand.* 6: 771–781.
62. Jonassen, I (1980): Characteristics of Hiproly barley I. Isolation and characterisation of two water-soluble high-lysine proteins. *Carlsberg Res. Commun.* 45: 47–58.
63. Jonassen, I (1980): Characteristics of Hiproly barley II. Quantification of two proteins contributing to its high lysine content. *Carlsberg Res. Commun.* 45: 59–68.
64. Jones, RL (1969): The fine structure of barley aleurone cells. *Planta* 85: 359–375.
65. Kapala, L, I Waitroszak, and J Przybylska (1975): Comparative serological studies in barley (*Hordeum vulgare* L.s.l), *Genet. Polon.* 16: 53–59.
66. Karlsson, KE (1976). Linkage studies on the *lys* gene in relation to some marker genes and translocations. In: Gaul, H (ed.): Barley Genetics III: 536–541; Garching, W. Germany.
67. Kasarda, DD (1980): Structure and properties of α-gliadins. *Ann. Techol. Agric.* 29: 151–173.
68. Kasarda, DD, JC Autran, CC Nimmo, EJL Lew, and PR Shewry (1982): N-terminal amino acid sequences of ω-gliadins and ω-secalins; implications for the evolution of prolamin genes. Manuscript submitted.
69. Kasarda, DD, JE Bernardin, and CO Qualset (1976): Relationship of gliadin protein components to chromosomes in hexaploid wheats. *Proc. Nat. Acad. Sci. USA* 73: 3646–3650.
70. Khan, K, and W Bushuk, (1979): Studies of glutenin XIII, Gel filtration, isoelectric focusing and amino acid composition studies. *Cereal Chem.* 56: 505–512.
71. Khoo, U, and MJ Wolf (1970): Origin and development of protein granules in maize endosperm. *Am. J. Bot.* 57: 1042–1050.
72. Kirkman, MA, PR Shewry, and BJ Miflin (1982): The effect of nitrogen nutrition on the lysine content and protein composition of barley seeds. *J. Sci. Fd. Agric.* 33: 115–127.
73. Kirsi, M, and J Mikola (1971): Occurrence of proteolytic inhibitors in various tissues of barley. *Planta* 96: 281–291.
74. Klaushofer, H, F Mittelbach, and A Szilvinyl (1959): Versuche zur papierelektrophoretischen Trennung von Gerstenhordein. *Mitt. St. Wien* 11/12: 80–85.
75. Kleinschmitt, A (1907): Hydrolyse des Hordeins. *Z. Physiol. Chem.* 54: 110–118.
76. Klemme-Berger, B, A Scheibe, WJ Schön, and M Zoschke (1969): Untersuchungen über die Eiweissfraktionen bei Futtergersten II. *Z. Acker-Pflanzenbau* 130: 86–94.
77. Køie, B, J Ingversen, AJ Andersen, H Doll, and BO Eggum (1976): Composition and nutritional quality of barley proteins. *Evaluation of Seed Protein Alterations by Mutation Breeding*: 55–61; IAEA Vienna.
78. Konarev, VG, JP Gavrilyuk, NK Gubareva, and TI Peneva (1979): Seed proteins in genome analysis, cultivar identification and documentation of cereal genetic resources: a review. *Cereal Chem.* 56: 272–278.

79. Kreft, I, B Javornik, and N Milkovic (1976): Studies on the gene control of electrophoretic protein patterns in barley. In: Gaul, H (ed.): Barley Genetics III: 30–35; Garching, W Germany.
80. Kreis, M (1979): Starch synthesis in barley mutants deficient in storage proteins. PhD Thesis. Catholic University of Louvain, Belgium.
81. Landry, J, and T Moureaux (1970): Hétérogénéité des glutelines du grain de maïs: extraction sélective et composition en acides amines des trois fractions isolées. *Bull Soc. Chim. Biol.* 52: 1021–1037.
82. Landry, J, T Moureaux, and JC Huet (1972): Extractabilité des protéines du grain d'orge: dissolution sélective et composition en acides amines des fraction isolées. *Bios.* 7–8: 281–292.
83. Larkins, BA, and WG Hurkman (1978): Synthesis and deposition of zein proteins in maize endosperm. *Pl. Physiol.* 62: 256–263.
84. Laurière, M, L Charbonnier, and J Mossé (1976): Nature et fractionnement des protéines de l'orge extraites par l'ethanol, l'isopropanol, et le n-propanol à des titres différents. *Biochimie* 58: 1235–1245.
85. Lawrence, G, and KW Shepherd (1980): Variation in glutenin protein subunits of wheat. *Aust. J. Biol. Sci.* 33: 221–233.
86. Lawrence, GJ, and KW Shepherd (1981): Chromosomal location of genes controlling seed proteins in species related to wheat. *Theor. Appl. Genet.* 59: 25–31.
87. Lontie, R, J Rondelet, and J Dulcino (1953): La solubilisation des glutelines de l'orge en présence de réducteurs *Europ. Brew. Conv. Nice:* 33–38.
88. Lontie, R, and T Voets (1959): Contribution à l'étude des prolamines et de l'azote résiduel de l'orge. Europ. Brew. Conv. Rome: 27–36.
89. Mak, AS, and BL Jones (1976): Separation and characterisation of chymotryptic peptides from α and β purothionins of wheat. *J. Sci. Fd. Agric.* 27: 205–213.
90. Mak, AS, and BL Jones (1976): The amino acid sequence of wheat β-purothionin. *Can. J. Biochem.* 54: 835–842.
91. Marchylo, BA, and DE La Berge (1980): Barley cultivar identification by electrophoretic analysis of hordein proteins. *Can. J. Plant Sci.* 60: 1343–1350.
92. Matile, P (1976): Vacuoles. In: Bonner, J, and JE Varner (eds.): Plant Biochemistry, 3rd ed.: 189–224; Academic Press, New York.
93. Matthews, JA (1980): The *in vitro* synthesis of barley storage proteins. PhD Thesis, University of Warwick.
94. Matthews, JA, and BJ Miflin (1980): The *in vitro* synthesis of barley storage proteins. *Planta* 149: 262–268.
95. Melville, JC, and CA Ryan (1972): Chymotrypsin inhibitor I from potatoes. *J. Biol. Chem.* 247: 3445–3453.
96. Mesrob, B, and V Kancheva (1970): Analytical separations of hordein and its fractions by disc electrophoresis on acrylamide gel. *J. Chromatogr.* 46: 94–102.
97. Mesrob, B, M Petrova, and C Ivanov (1968): Fractionation of hordein by gel filtration. *Biochim. Biophys. Acta* 160: 122–125.
98. Mesrob, B, M Petrova, and C Ivanov (1969): Molecular weight determination of the hordein fractions by gel chromatography. *Biochim. Biophys. Acta* 181: 482–484.
99. Mesrob, B, M Petrova, and C Ivanov (1970): A comparative study of hordein fractions. *Biochim. Biophys. Acta* 200: 459–465.
100. Miflin, BJ, and SR Burgess (1982): Protein bodies from developing seeds of barley, maize, wheat, and peas: the effects of protease treatment. *J. Expt. Bot.* 33: 251–260.
101. Miflin, BJ, SR Burgess, and PR Shewry (1981): The development of protein bodies in the storage tissues of seeds: subcellular separations of homogenates of barley, maize and wheat endosperms and of pea cotyledons. *J. Expt. Bot.* 32: 199–219.
102. Miflin, BJ, JM Field, and PR Shewry (1982): Cereal storage proteins and their effects on technolog-

ical properties. In: Mossé, J, and J Daussant (eds.): Seed Proteins. Academic Press, London; in press.

103. Miflin, BJ, and PR Shewry (1979): The synthesis of proteins in normal and high lysine barley seed. In: Laidman, DL, and RG Wyn Jones (eds.): Recent Advances in the Biochemistry of Cereals: 239–273; Academic Press, London.
104. Miflin, BJ, and PR Shewry (1979): The biology and Biochemistry of cereal seed prolamins. Seed Protein Improvement in Cereals and Grain Legumes 1: 137–158; IAEA Vienna.
105. Miflin BJ, PR Shewry, AJ Faulks, BG Forde, M Kreis, J Matthews, MB Bahramian, R Thompson, D Bartels, and R Flavell (1982): The analysis of the *Hor-2* locus of barley. *Barley Genetics IV;* Edinburgh University Press, Edinburgh, UK; in press.
106. Mikola, J, and TM Enari (1970): Changes in the contents of barley proteolytic inhibitors during malting. *J. Inst. Brew.* 76: 182–188.
107. Mikola, J, and M Kirsi (1972): Differences between endospermal and embryonal trypsin inhibitors in barley, wheat and rye. *Acta Chem. Scand.* 26: 787–795.
108. Mikola, J, and EM Suolinna (1959): Purification and properties of a trypsin inhibitor from barley. *Eur. J. Biochem.* 9: 555–560.
109. Moureaux, L, and J Landry (1968): Extraction sélective des protéines du grain de mais et en particulier de la fraction glutelines. *C. R. Hebd. Séanc. Acad. Sci. Paris:* 2302–2305.
110. Mossberg, R, (1969): Evaluation of protein quality and quantity by dye-binding capacity: a tool in plant breeding. New Approaches to Breeding for Improved Plant Proteins: 151–160; IAEA Vienna.
111. Munck, L (1972): Barley Seed Proteins. In: Inglett, GE (ed.): Symposium: Seed Proteins: 144–164; Avi Pub. Co., Westport, Conn.
112. Munck, L (1972): Improvement of nutritional value in cereals. *Hereditas* 72: 1–128.
113. Munck, L, KE Karlsson, A Hagberg, and BO Eggum (1970): Gene for improved nutritional value in barley seed proteins. *Science* 168: 985–987.
114. Munck, L, and D von Wettstein (1976): Effects of genes that change the amino acid composition of barley endosperm. *In:* Genetic Improvement of Seed Protein: 71–82; Nat. Acad. Sci. Washington DC.
115. Murphy, P, J Witcombe, PR Shewry, and BJ Miflin (1982): The origin of six-rowed 'wild' barley from the Western Himalaya. *Euphytica* 31: 183-192.
116. Netsvetaev, VP (1978): *Biologicheskie Osmovy Raseonalnogo Ispozovaniya i Jivotnogo Rastitelnogog Myra*, Riga (USSR): Zinatne (In Russian, translated by VP Netsvetaev) pp. 145–146.
117. Niewdorp, PJ (1964): Electron microscopic structure of the epithelial cells of the scutellum of barley *Acta Bot. Neerl.* 12: 295–301.
118. Nilson, LR, and T Hermelin (1966): Isozyme variations in some barley varieties. *Lantbrukshogsk. Ann.* 32: 297–308.
119. Oram, RN, H Doll, and B Køie (1975): Genetics of two storage protein variants in barley. *Hereditas* 80: 53–58.
120. Osborne, TB (1895): The proteids of barley. *J. Am. Chem. Soc.* 17: 539–567.
121. Osborne, TB (1924): The Vegetable Proteins. Longmans, Green & Co. London; 154 pp.
122. Osborne, TB, and SM Clapp (1907): Hydrolysis of hordein. *Am. J. Physiol.* 19: 117–124.
123. Osborne, TB, and IF Harris (1903): Nitrogen in protein bodies. *J. Am. Chem. Soc.* 22: 323–353.
124. Paleg, LG, and B Hyde (1964): Physiological effects of gibberellic acid VII. Electron microscopy of barley aleurone cells. *Plant Physiol.* 39: 673–680.
125. Partridge, J, L Shannon, and D Gumpf (1976): A barley lectin that binds free amino sugars. I. Purification and characterization. *Biochim. Biophys. Acta* 451: 470–483.
126. Payne, PI, LM Holt, and CN Law (1981): Structural and genetical studies of the high-molecular-weight subunits of wheat glutenin. Part 1. Allelic variation in subunits amongst varieties of wheat (*Triticum aestivum*). *Theor. Appl. Genet.* 60: 229–236.

127. Payne, PI, CN Law, and EE Mudd (1980): Control by homoeologous group 1 chromosomes of the high molecular weight subunits of glutenins, a major protein of wheat endosperm. *Theor. Appl. Genet.* 58: 113–120.
128. Pernollet, JC, and J Mossé (1980): Caractérisation des corpuscules protéiques de l'albumen des caryopses de céréales par micro-analyse élémentaire associée à la microscopie électronique à balayage. *C. R. Acad. Sci. Paris* 290d: 267–270.
129. Peterson, D (1978): Subunit structure and composition of oat seed globulin. *Pl. Physiol.* 62: 506–509.
130. Préaux, G, and R Lontie (1975): The proteins of barley. *In:* Harborne, JB, and CF Van Sumere (eds.): The Chemistry and Biochemistry of Plant Proteins: 89–111; Academic Press, London.
131. Proust, G (1817): De l'orge avant et après sa germination, et conséquences économiques qui en résultent. *Ann. Chim. Phys.* 5: 337–350.
132. Przybylska, J, and A Kapala (1974): Variability of disc electrophoretic patterns of salt-soluble seed proteins in barley (*Hordeum vulgare* L.s.l.) *Genet. Polon.* 15: 231–243.
133. Quensel, O (1942): Untersuchungen über die Gerstenglobuline. Diss; 97 pp. Almqvist and Wiksell, Upsala.
134. Quensel, O, and T Svedberg (1938): Studies on the brewing process by means of ultracentrifugal sedimentation, diffusion and electrophoresis measurements. *Cr. Lab. Carlsberg Ser. Chim.* 22: 441–448.
135. Rahman, S, PR Shewry, and BJ Miflin (1982): Differential protein accumulation during barley grain development. *J. Exp. Bot.* 33: 717–728.
136. Redman, DG, and N Fisher (1968): Fractionation and comparison of purothionin and globulin components of wheat. *J. Sci. Fd. Agric.* 19: 651–655.
137. Redman, DG, and N Fisher (1969): Purothionin analogues from barley flour. *J. Sci. Fd. Agric.* 20: 427–432.
138. Rhodes, AP, and AA Gill (1980): Fractionation and amino acid analysis of the salt-soluble protein fraction of normal and high lysine barleys. *J. Sci. Fd. Agric.* 31: 467–473.
139. Richardson, M (1977): The proteinase inhibitors of plants and micro-organisms. *Phytochem.* 16: 159–169.
140. Ryan, CA (1979): Protease Inhibitors. In: Rosenthal, GA, and DH Janzen (eds.): Herbivores, their Interaction with Secondary Plant Metabolites: 559–618; Academic Press, London.
141. Salcedo, G, R Sanchez-Monge, A Argamentaria, and C Aragoncillo (1980): The A-hordeins as a group of salt-soluble hydrophobic proteins. *Pl. Sci. Letts.* 19: 109–119.
142. Scheibe, A, WJ Schön, M Zoschke, and R Bauer (1968): Untersuchungen über die Eiweissfraktionen bei Futtergersten I. *Z. Acker-Pflanzenbau* 128: 139–150.
143. Schildbach, R, and M Burbridge (1979): Identifizierung von Gerstensorten an Einzelkörnern durch Flachgel-Elektrophorese der Proteine und Aleuronfärbung. *Monatsschrift für Brauerei*, 30th Nov. issue: 470–480.
144. Schjerning, H, (1914): Om byggets proteinstoffer i kornet selv og under brygningsprocesserne. *Medd. Carlsberg Lab. (1914-1917)* 11: 45–98.
145. Schmitt, JM (1979): Purification of hordein polypeptides by column chromatography using volatile solvents. *Carlsberg Res. Commun.* 44: 431–438.
146. Schmitt, JM, and I Svendsen (1980): Amino acid sequences of hordein polypeptides. *Carlsberg Res. Commun.* 45: 143–148.
147. Schmitt, JM, and J Svendsen (1980): Partial amino acid sequence from hordein polypeptide Bl. *Carlsberg Res. Commun.* 45: 549–555.
148. Scriban, R, JC Autran, B Strobbel, and M Nicolaidis (1979): Synthèse des différentes recherches analytiques sur la chimiotaxonomie des orges et des malts. *Europ. Brew Conv. Berlin (West):* 571–586.
149. Shepherd KW (1968): Chromosomal control of endosperm proteins in wheat and rye. *In:* Proc. 3rd Int. Wheat Genet. Symp.: 86–96; Australian Academy of Sciences, Canberra.

150. Shewry, PR, JC Autran, CC Nimmo, EJL Lew, and DD Kasarda (1980): N-terminal amino acid sequence homology of storage protein components from barley and a diploid wheat. *Nature (London)* 286: 520–522.
151. Shewry, PR, JRS Ellis, HM Pratt, and BJ Miflin (1978): A comparison of methods for the extraction and separation of hordein fractions from 29 barley varieties. *J. Sci. Fd. Agr.* 29: 433–441.
152. Shewry, PR, AJ Faulks, and BJ Miflin (1980): Effect of high-lysine mutations on the protein fractions of barley grain. *Biochem. Genet.* 18: 133–151.
153. Shewry, PR, AJ Faulks, S Parmar, and BJ Miflin (1980): Hordein popypeptide pattern in relation to malting quality and the varietal identification of malted barley grain. *J. Inst. Brew.* 86: 138–141.
154. Shewry, PR, AJ Faulks, RA Pickering, JT Jones, RA Finch, and BJ Miflin (1980): The genetic analysis of barley storage proteins. *Heredity.* 44: 383–389.
155. Shewry, PR, JM Field, MA Kirkman, AJ Faulks, and BJ Miflin (1980): The extraction solubility and characterization of two groups of barley storage polypeptides. *J. Expt. Bot.* 31: 393–407.
156. Shewry, PR, JM Hill, HM Pratt, MM Leggatt, and BJ Miflin (1978): An evaluation of techniques for the extraction of hordein and glutelin from barley seed and a comparison of the protein composition of Bomi and Riso 1508. *J. Expt. Bot.* 29: 677–692.
157. Shewry, PR, MA Kirkman, HM Pratt, and BJ Miflin (1978): Storage protein formation in normal and high lysine barley. In: Miflin, BJ, and M Zoschke (eds.): Carbohydrate and Protein Synthesis: 155–170; EEC, Luxembourg.
158. Shewry, PR, EJL Lew, and DD Kasarda (1981): Structural homology of storage proteins coded by the *Hor 1* locus of barley (*Hordeum vulgare* L.). *Planta* 153: 246–253.
158a. Shewry, PR, MS Wolfe, SE Slater, S Parmar, AJ Faulks, and BJ Miflin (1982): Barley storage proteins in relation to varietal identification, malting quality and mildew resistance. *In:* Barley Genetics IV. Edinburgh University Press, Edinburgh, U.K; in press.
159. Shewry, PR, JF March, and BJ Miflin (1980): N-terminal amino acid sequence of C hordein. *Phytochem.* 19: 2113–2115.
160. Shewry, PR, and BJ Miflin (1982): Genes for the storage proteins of barley. *Qualitas Plantarum Plant Foods for Human Nutrition;* in press.
161. Shewry, PR, HM Pratt, MJ Charlton, and BJ Miflin (1977): Two-dimensional separation of the prolamins of normal and high-lysine barley. *J. Expt. Bot.* 28: 597–600.
162. Shewry, PR, HM Pratt, AJ Faulks, S Parmar, and BJ Miflin (1979): The storage protein (hordein) polypeptide pattern of barley (*Hordeum vulgare* L.) in relation to varietal identification and disease resistance. *J. Nat. Inst. Agric. Bot.* 15: 34–50.
163. Shewry PR, HM Pratt, RA Finch, and BJ Miflin (1978): Genetic analysis of hordein polypeptides from single seeds of barley. *Heredity.* 40: 463–466.
164. Shewry. PR, HM Pratt, MM Leggatt, and BJ Miflin (1979): Protein metabolism in developing endosperms of high-lysine and normal barley. *Cereal Chem.* 56: 110–117.
165. Shewry PR, HM Pratt, and BJ Miflin (1978): Varietal identification of single seeds of barley by analysis of hordein polypeptides. *J. Sc. Fd. Agric.* 29: 587–596.
166. Singh, U, and LVS Sastry (1977): Studies on the proteins of the mutants of barley grain I. Extraction and electrophoretic characterization. *Cereal Chem.* 54: 1–12.
167. Sodek, L, and CM Wilson (1971): Amino acid composition of proteins isolated from normal, opaque-2 and floury-2 corn endosperms by a modified Osborne procedure. *J. Agric. Food Chem.* 19: 1144–1150.
168. Solari, RM, and EA Favret (1971): Polymorphism in endosperm proteins of barley and its genetic control. In: Nilan, R. (ed.): Barley Genetics II: 23–31; Washington Univ. Press; Pullman, WA, USA.
169. Sozinov, AA, VP Netsvetaev, EM Grigoryan, and IS Obraztsov (1978): Mapping of Hrd locuses in barley (*Hordeum vulgare* L. emed. Vav et Bacht). *Genetika* USSR 14: 1610–1619 (translated into English in *Soviet Genetics.* 14: 1137–1147 (1979).

170. Sun, SM, RC McLeester, FA Bliss, and TC Hall (1974): Reversible and irreversible dissociation of globulins from *Phaseolus vulgaris* seeds. *J. Biol. Chem.* 249: 2118–2121.
171. Svendsen, I, I Jonassen, J Hejgaard and S Boisen (1980): Amino acid sequence homology between a serine protease inhibitor from barley and potato inhibitor I. *Carlsberg Res. Commun.* 45: 389–395.
172. Svendsen, I, B Martin, and I Jonassen (1980): Characteristics of hiproly barley III. Amino acid sequences of two lysine rich proteins. *Carlsberg Res. Commun.* 45: 79–85.
173. Swift, JG, and MS Buttrose (1972): Freeze-etch studies of protein bodies in wheat scutellum. *J. Ultrastructure Res.* 40: 378–390.
174. Swift, JG, and TP O'Brien (1972): The fine structure of the wheat scutellum during germination. *Aust. J. Biol. Sci.* 25: 469–486.
175. Tallberg, A (1973): Ultrastructure and protein composition in high-lysine barley mutants. *Hereditas* 75: 195–200.
176. Tully, RE, and H Beevers (1976): Protein bodies of castor bean Endosperm. *Plant Physiol.* 58: 710–716.
177. Tuning, B, and W Wilten (1978): Het herkennen van gerstrassen in partijen gerst. *Voedingsmiddelen-technologie* 11: 10–13.
178. Urion, E, and G LeJeune (1940) Fractionnement des constituants azotes de l'orge. *Bull. Soc. Chim. Biol.* 22: 214–220.
179. Van Lonkhuysen, HJ, and JP Marseille (1978): Schnellmethode zur Identifizierung von Weizensorten durch die Stärkegel-Elektrophorese. *Getreide, Mehl und Brot.* 32: 288–291.
180. Viotti, A, E Sala, R Marotta, P Alberi, C Balducci, and C Soave (1979): Genes and mRNAs coding for zein polypeptides in *Zea mays*. *Eur. J. Biochem.* 102: 211–222.
181. Waldschmidt-Leitz, E, and H Brutscheck (1955): Zur Analyse der Gersteneiweisskörper. *Brauwissenschaft* 8: 278–283.
182. Waldschmidt-Leitz, E, and H Brutscheck (1958): Über die Zusammensetzung der elektrophoretisch unterscheidbaren Komponenten des Hordeins. *Hoppe-Seylers Z. Physiol. Chem.* 311: 1–5.
183. Waldschmidt-Leitz, E, and H Kling (1966). Zur Darstellung elektrophoretisch einheitlicher Komponenten von Prolaminen. *Hoppe-Seylers Z. Physiol. Chem.* 346: 17–20.
184. Waldschmidt-Leitz, E, and R Mindemann (1957): Über Zusammensetzung und Eigenart der Glutenine in Getreidemehlen. *Hoppe-Seylers Z. Physiol. Chem.* 308: 257–262.
185. Wendorf, F, R Schild, NE El Hadidi, AE Close, M Kubusiewicz, H Wieckowska, B Issawi, and H Haas (1979): Use of barley in the Egyptian late Paleolithic. *Science* 205: 1341–1347.
186. Wilson, CM, PR Shewry, AJ Faulks, and BJ Miflin (1981): The extraction and separation of barley glutelins and their relationship to other endosperm proteins. *J. Expt. Bot.* 32: 1287–1293.
187. Witcombe, J (1978): Two-rowed and six-rowed wild barley from the Western Himalayas. *Euphytica* 27: 601–604.
188. Wrigley, CW, and KW Shepherd (1973): Electrofocusing of grain proteins from wheat genotypes. *Ann. N.Y. Acad. Sci.* 209: 154–162.

6. Barley Seed Proteins and Possibilities for their Improvement

HANS DOLL

INTRODUCTION

Barley is one of the cereals in which the possibilities for improving the nutritional value of the seed protein have been studied extensively. Detailed knowledge of barley proteins and their genetics is available (for a survey see, e.g. Miflin and Shewry [43], Thomson and Doll [66], and Tallberg [65]). Increases in the content of lysine and other essential amino acids of the protein have been achieved, but unfortunately, the changes in protein composition interfere with the carbohydrate accumulation in the seed. It has not yet been possible to produce a barley variety with improved protein quality and normal grain yield.

This chapter gives a brief introduction to barley seed proteins with special emphasis on the genetic possibilities for improving protein quality. Important characteristics of the mutants influencing protein composition are summarized, and the possibilities for improving the nutritional value of barley protein without reducing the grain yield are evaluated.

Barley seed proteins are traditionally classified into albumin, globulin, prolamin (hordein), and glutelin according to their solubility in water, salt solution, aqueous alcohol, and basic or acid solutions, respectively. This protein fractionation, proposed by Osborne [53], has been used extensively in characterizing barley protein. However, the Osborne fractionation tells little about the physiological functions of the proteins.

A simpler and more useful seed protein classification in the present context utilizes only two groups, viz. functional proteins and storage proteins. The first group includes all proteins having a specific function during seed development or germination. Enzymes, membrane proteins, and histones are examples of such proteins. The second group comprises the storage or reserve proteins that apparently have no function other than nitrogen storage. They are synthesized only in the endosperm, and are further characterized by being deposited in specific organelles denoted protein bodies.

As pointed out by Nelson [49], genetic changes resulting in an improved nutritional value of the seed protein are probably feasible for the storage proteins. Alterations in the storage proteins are not expected to be lethal, as these proteins are not involved in specific metabolic processes, and drastic changes or even complete elimination of a storage protein may be compatible with normal seed development. As most storage proteins are present in relatively large amounts, even single gene changes may have an effect on the overall amino acid composition. Finally, the inferior quality of barley protein is especially due to some of the storage proteins, while the other seed proteins in general have an amino acid composition that fits the requirements of humans and animals. Altogether, these considerations show that attempts to improve the nutritional value of barley seed protein should concentrate on altering the storage proteins.

BARLEY STORAGE PROTEINS

Most of the storage protein in barley is found in the prolamin fraction, which is designated hordein in barley. Hordein constitutes roughly 40% of the seed protein [59], but this amount is strongly dependent on the nitrogen fertilization of the plant [1, 32]. Hordein is synthesized relatively late during seed development in the endosperm [7], where it is deposited in protein bodies [10, 23]. Like the prolamins of other cereals, hordein has a very high content of glutamine and proline. These two amino acids, which are unessential for humans and animals, constitute more than 50% of hordein [59].

The physico-chemically defined hordein, i.e. the protein fraction soluble in aqueous alcohol, contains two genetically defined subfractions, hordein-1 and hordein-2, which are controlled by genes in two corresponding loci, *Hor1* and *Hor2* [18]. The genetics of the two major hordein fractions are illustrated in Figure 1 showing the banding pattern obtained by gel-electrophoresis [17] of the hordein of 6 selected F_2 seeds from a cross between 'Chernomorets-2' and 'Yuzhmyi' that are denoted A and B in the following discussion. The F_2 seeds shown in Figure 1a and b have the same banding pattern as the two parents, respectively A and B, which differ in nearly all the hordein polypeptides.

The next two F_2 seeds (Figure 1c and d) were selected to represent reciprocal F_1 seeds, A(♀) × B(♂) and B(♀) × A(♂), respectively. It is seen that the polypeptides inherited from the female are twice as abundant in amount as those coming from the male parent. This is because the triploid endosperm derives from a fertilization of a diploid central nucleus with a haploid gamete nucleus from the male. The hordein genes are codominant as they are expressed in the heterozygote according to their dose.

Analysis of a number of F_2 seeds shows that recombination may occur among the polypeptides inherited from the two parents (Figure 1e, f). Recombination has been

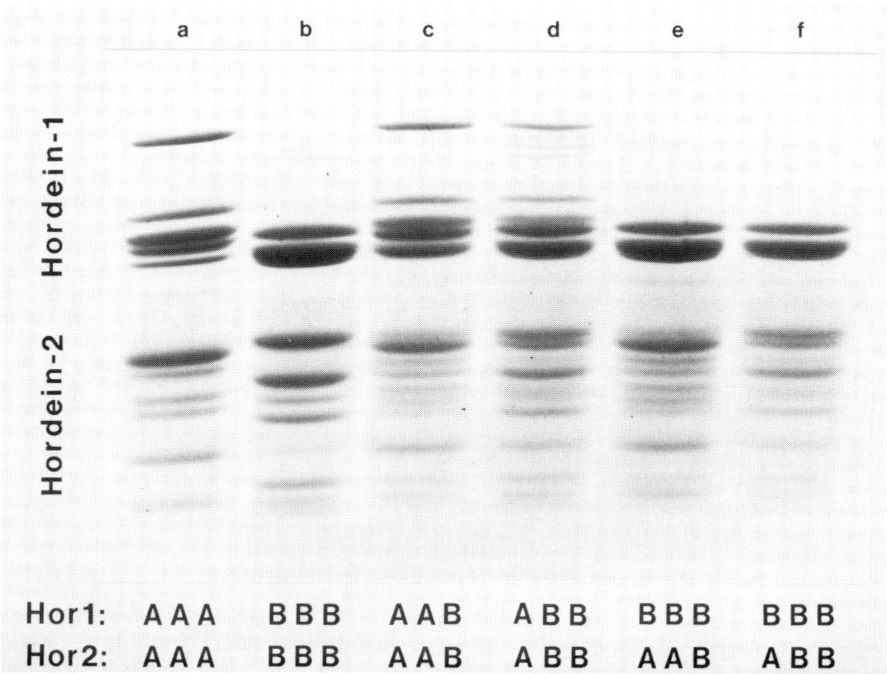

Figure 1. Polypeptide composition as revealed by electrophoresis of hordein and deduced genotype of selected F_2 seeds from a cross between 'Chernomorets' (A) and 'Yuzhmyi' (B). The first 4 F_2 seeds are non-recombinants having the same phenotype as parent A (a), parent B (b), a A (♀) x B (♂) hybrid (c), and a B (♀) x A (♂) hybrid (d), while the last two seeds are recombinants.

observed, however, only between the two groups of polypeptides designated hordein-1 and hordein-2, but not among the polypeptides within these groups. The two recombinants depicted in Figure 1e, f are both homozygous BBB in hordein-1, while they are heterozygous with genotype AAB and ABB, respectively, in hordein-2. As the female gamete contributes two gene doses it can be seen that the two recombinants derive from recombination in the female and the male gamete, respectively.

The two loci controlling the composition of hordein-1 and hordein-2, *Hor1* and *Hor2*, are located on the short arm of chromosome 5 [52, 61]. The linkage between them has been estimated to about 10% [29, 58]. Solari and Favret (1971) reported an endosperm protein locus, *Pr-a*, which probably is the same as *Hor1*. The two hordein loci have also been described as *HrdA* and *HrdB* by Sozinov *et al.* (1979), who further reported three additional loci closely linked with *HrdB*. It is assumed that the *Hor* loci contain the structural genes coding for hordein-1 and hordein-2. The controlling by each of the loci of several polypeptides in the respective protein fraction indicates that the loci contain multigene families. The fine structure of the hordein genes is now studied in detail by DNA cloning techniques [8, 45].

A very large genetic variation has been found in the polypeptide composition of hordein-1 and hordein-2 [41, 60], but the nutritional value has not been reported to be influenced by this variation. The nutritional value of the two hordein fractions is somewhat different. Hordein-1 contains only traces of lysine and sulphur amino acids, while hordein-2 contains about 0.5% lysine and methionine, and 2.5% cysteine [59].

Aside from the two major fractions, hordein-1 and hordein-2, the alcohol-soluble protein fraction of barley also contains several minor, low molecular weight polypeptides. This fraction, the A-hordein [37], has a higher lysine content, ~2%, and a lower content of glutamine and proline [56] than the major hordein subfractions.

Barley contains also a few proteins with storage properties in the water or salt-soluble protein fraction. These proteins are of special interest in breeding for improved nutritional value as they have a much higher lysine content than hordein. Kirsi (1974) studied two proteinase inhibitors and pointed out that they behaved like storage proteins in the sense that they were accumulated in the endosperm during development, and disappeared during germination. The amount of each inhibitor also increased with increased nitrogen fertilization [33]. This typical storage protein feature was also found for β-amylase and Z-protein by Hejgaard & Boisen (1980), who studied the four proteins in normal and 'Hiproly' barley, and denoted the two proteinase inhibitors CI-1 and CI-2.

Another typical storage protein characteristic of CI-1, CI-2 (SP II albumin), and β-amylase (but not Z-protein) is that they are synthesized on the rough endoplasmic reticulum [30] in the same way as hordein [9]. β-amylase, Z-protein, CI-1, and CI-2 contain, respectively, about 5, 7 [22], 9 and 11% [6] lysine in their protein. These proteins are much more lysine rich than hordein, but they constitute only about 4% of the seed protein in normal barley [22]. The four proteins will be referred to as lysine-rich storage proteins in the following though they undoubtedly have other functions in the seed than just nitrogen storage.

PROTEIN MUTANTS

Only mutants having a specific genetic change affecting protein composition will be considered here. Major changes in protein composition usually result in a changed overall amino acid composition, which has been exploited when screening for mutants. Most of the protein mutants found in barley so far have been selected on the basis of their increased lysine content. For convenience these and similar ones have been called high-lysine mutants, and this term will also be used here. However, as the change in overall lysine content derives from a reduced content of hordein (see below) the high-lysine mutants are in principle protein- rather than amino acid mutants.

Mutants in which the changed amino acid content derives from an overproduction of one or several amino acids have also been reported [42]. These are as interesting from a nutritional point of view as the protein mutants. However, they will not be treated here as they represent a completely different class of mutants from a physiological standpoint.

The detection of a high-lysine mutant is somewhat complicated because the genetically conditioned change in lysine content has to be distinguished from the environmentally induced variation in protein and amino acid composition. The latter variation is due mainly to the dependence of the hordein production on the availability of nitrogen [1], which results in a characteristic negative correlation between the protein content of the seed and the lysine content of the protein [67]. A high-lysine mutant was therefore defined [20] as one having a higher lysine content in the protein than other lines/varieties having the same protein content as the mutant. Such a mutant is likely to have an improved nutritional value of the protein. However, the percentage increase in lysine does not prove that the mutant also has a higher absolute amount of lysine. Therefore, it should be required that a high-lysine mutant also has an increased production of lysine per unit area, or at least a higher amount of lysine, e.g. per seed.

Mutant Screenings

The first high-lysine barley, 'Hiproly', was found among about 2500 entries of the World Barley Collection [48]. The designation Hiproly refers to the high content of both protein and lysine of this variety, CI 3947, which comes from Ethiopia. 'Hiproly' was selected on the basis of combined analyses for dye-binding capacity (DBC) and Kjeldahl nitrogen content as suggested by Mossberg (1969), who showed that the DBC is highly correlated with the content of basic amino acids of the sample. Hence, a sample having a higher content of one of the basic amino acids, e.g. lysine, will have an increased DBC while the nitrogen content will be unchanged. Such a sample will, therefore, deviate from the correlation between N% and DBC found in samples of barley with normal lysine content. The combined use of N% and DBC to detect high-lysine mutants has proved very useful and highly reliable, and relatively small changes in lysine content have been detected by this method.

As the natural genetic variation in lysine content in barley apparently was limited, a screening for induced high-lysine mutants was initiated at Risø in 1969. At first we analyzed 92 lines from an EMS treatment and 100 lines derived from a γ-ray treatment. Two of the EMS lines, nos. 29 and 86, were high-lysine mutants [12], and so was one of the γ-lines, no. 56. The reason for the exceptionally high frequency of high-lysine mutants in these materials is not known. The screening was continued by analyzing M_3 seeds of about 15,000 M_2 plants from different mutagenic treatments of Bomi [20]. This gave 7 additional mutants (Table 1), which have increased lysine content in the protein compared to the parent variety from less than 10%, mutant 7,

Table 1. Survey of high-lysine barley mutants and lines (Adapted from Tallberg (1981c)).

Line	Derived from	Mutagen*	Gene	Chromosome
Hiproly	-	-	*lys*	7
Mutant 29	Carlsberg II	EMS	*lys5g*	6
Mutant 86	Carlsberg II	EMS	*lys5h*	6
Mutant 56	Carlsberg II	γ-rays	*Hor2ca*	5
Mutant 7	Bomi	$\underline{n}_{th}$	-	-
Mutant 8	Bomi	EMS	*Lys4d*	5
Mutant 13	Bomi	EMS	*lys5f*	6
Mutant 16	Bomi	$\underline{n}_{f}$	-	1
Mutant 17	Bomi	$\underline{n}_{f}$	-	1
Mutant 527	Bomi	γ-rays	*lys6i*	6
Mutant 1508	Bomi	EI	*lys3a*	7
Mutant 18	Bomi	NaN	*lys3b*	7
Mutant 19	Bomi	NaN	*lys3c*	7
Notch-1	NP 113	EMS	-	-
Notch-2	NP 113	EMS	-	-
lys 95	Perga	EMS	-	-
lys 449	Perga	EMS	-	-
Clipper 500	Clipper	EMS	-	-

* EMS: ethyl methane sulphonate; $\underline{n}_{th}$: thermal neutrons; $\underline{n}_{f}$: fast neutrons; EI: ethyleneimine.

to more than 40%, mutant 1508. These screenings were all done by combined N% and DBC analyses.

In a later screening for protein mutants (unpublished data), 5000 plants derived from a treatment with sodium azide were studied. The first 2000 were analyzed by the DBC and N% method, while the last 3000 were screened for protein mutants by analyzing the hordein from one M_3 seed by SDS-electrophoresis. All seeds having a changed banding pattern were selected. Each of the two screenings gave one mutant, each of which resembled mutant 1508 in phenotype. Crosses with mutant 1508 have shown that the two new mutants, nos. 18 and 19, have mutant genes allelic to that in mutant 1508. Mutants missing one or a few of the hordein bands of the parent variety were not found in the screening by electrophoresis.

In a recent search for low-hordein mutants at Risφ about 50,000 M_2 seeds from treatments with γ-rays, sodium azide, or EMS were analyzed by the turbidity test suggested by Rhodes (1975) [55]. This screening has also yielded mutants resembling mutant 1508 in hordein content, but further testing of selected putative mutants is required to know whether new mutant types have been found (Jens Jensen, personal communication).

A few other laboratories have screened for high-lysine mutants by combined analyses for N% and DBC. Bansal (1974) reported two mutants, Notch-1 and -2, having an increased lysine content of the seed protein. A similar study of EMS-treated barley led to the detection of two other mutants, *lys*95 and *lys*449, which have about 30% more lysine in the protein [11]. Finally, an EMS-induced mutant, Clipper 500, of 'Clipper' has been reported [51].

As most high-lysine mutants have shrunken seeds (see below), Ullrich and Eslick (1978) analyzed 9 spontaneous shrunken endosperm mutants to see if they were high-lysine mutants, All the mutants had a higher lysine content of the protein than the corresponding parent varieties with plump grains. However, only three of them had an increased lysine content of the grain if the mutant kernel weight was adjusted upwards to the normal kernel weight. Hence, the lysine increase may derive from a starch dilution effect in several of these shrunken mutants [69].

Genetics and Protein Alterations

All the high-lysine mutants studied are characterized by having a single gene inheritance of the lysine increase of their protein. This was probably expected because only changes due to single gene mutations are likely to be found in the mutant screenings from which the present mutants derive. Another characteristic of the mutants is that all of them have a more or less reduced hordein content (Figure 2). This is not surprising as a reduction of hordein, which is the only major lysine-poor barley protein, is the simplest way to increase the overall lysine content of the protein. Further, as discussed earlier, genetic changes causing a block in the synthesis of a storage protein are not expected to be lethal.

In most cases the reduced hordein content of the high-lysine mutants is due to mutant genes located outside the *Hor* loci containing the structural hordein genes. The most extreme hordein mutant known is Risφ 1508 [24], which has a very low content of both hordein-1 and hordein-2 [35, 57]. This is also evident from Figure 2c, which shows that the alcohol extract from mutant 1508 contains very little hordein, and it is difficult to see any hordein bands even after a 10-fold concentration of the sample (Figure 2d). This hordein reduction is due to a recessive gene [13], *lys3a*, located on chromosome 7 [26, 31].

Some mutants have a certain reduction in the amount of all the hordein poly-

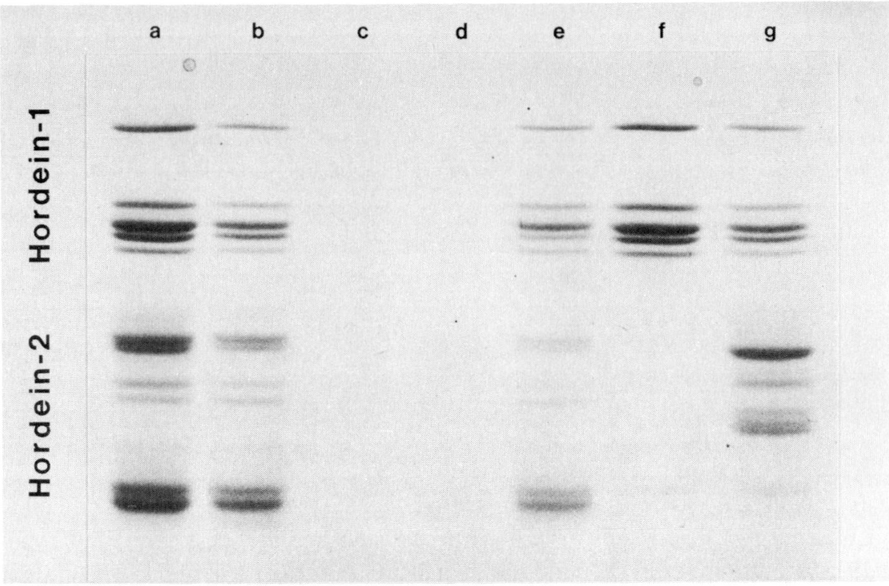

Figure 2. SDS-polyacrylamide gel electrophoresis of hordein extracted from individual seeds (Doll and Andersen, 1981) of a: 'Bomi', b: Mutant 527, c: Mutant 1508, d: as c but concentrated 10 times, e: Mutant 7, f: Mutant 56, and g: 'Carlsberg II'.

peptides seen in gel electrophoresis. Mutant 527 is of this type (Figure 2b), and similar changes were found in mutants 8 [35] and Notch-1 and -2 [44]. Changes of this type are similar to the environmentally induced variation in hordein content, and it is therefore rather difficult to detect these mutant genes by electrophoresis. Other mutants have a more characteristic reduction of some hordein-2 bands that makes them better suited for electrophoretic detection. This is the case for mutant 7 in which the slowest-moving double band in hordein-2 is reduced further than the other hordein bands (Figure 2e) and for the three mutants 13, 29, and 86 [35], which have allelic mutant genes. The effect on hordein of the high-lysine gene *lys* of 'Hiproly' has not been examined in detail, but a summary of several studies [44] showed that 'Hiproly' and lines having the *lys* gene normally have a lower hordein content than corresponding normal-lysine lines.

The available information on the chromosomal location of the genes of the mutants described above is shown in Table 1. All but one of these genes are recessive. The exception is the semidominant *Lys4d* in mutant 8 [27]. The gene symbols used here for the high-lysine genes are those suggested by Munck (1972) and Jensen and Doll (1979). Other symbols have been suggested for some of the mutant genes [65].

So far only one mutant, viz. Risø 56, containing a mutation affecting the structural hordein genes has been found. In mutant 56 the majority of the hordein-2 polypeptides are strongly reduced in amount, while hordein-1 is unchanged in composition but increased in amount ([35]; Figure 2f). This very characteristic hordein change of mutant 56 is due to a mutation at or near the *Hor2* locus [15] containing the structural genes for hordein-2. The mutant gene is therefore considered a *Hor2* allele, *Hor2ca*, originating from a mutation of the *Hor2Ca* allele present in the parent variety 'Carlsberg II' (Figure 2g). *Hor2ca* is recessive in its qualitative effect on the electrophoretic banding pattern of hordein-2 in the sense that only the mutant homozygotes contain very little of the major hordein-2 bands. However, measurements of the amount of hordein-2 in seeds containing different doses of the mutant allele have shown a clear dosage effect [15]. *Hor2ca* is therefore, like other *Hor2* alleles, codominant with respect to the synthesized amount of hordein-2.

Besides the reduction in hordein, several high-lysine mutants have an increased content of other proteins. These changes have especially been studied in 'Hiproly' and lines having the *lys* gene. Hejgaard and Boisen (1980) showed that the content of the lysine-rich proteins β-amylase, Z-protein, and the proteinase inhibitors CI-1 and CI-2 is increased 4 – to 7 – fold in *lys* lines. These increases account for more than 50% of the lysine increase in 'Hiproly'. Similar changes of these four proteins have been found in mutant 56 (Hejgaard, personal communication). Mutant 1508 also contains significantly higher amounts of CI-1 [6], but β-amylase is strongly reduced in this mutant [39].

Grain yield and carbohydrate content

The main problem encountered with the protein/high-lysine mutants is their reduced seed weight and grain yield [14, 20]. The low grain yield of 'Hiproly' was increased considerably by back-crossing to high-yielding varieties, but it was not possible to raise the yield level to more than 85–95% of the recurrent parent [54]. The low grain yield level was due to reduced seed size, and strict selection in the offspring from several crosses did not break the negative correlation between seed size and lysine content. A study of 'Hiproly' and two selected lines containing the *lys* gene showed a clearly reduced content of carbohydrates of the *lys* lines [51]. It is therefore likely that the reduced seed size associated with the *lys* gene is due to an impaired starch accumulation in the seeds. Reduced seed weight and starch content were also reported for the Notch-1 and -2 high-lysine mutants [2], and the *lys*95 and *lys*449 mutants have shrivelled seeds [11].

At Risø we have used chromosome-doubled haploids in studies of the effect of selected high-lysine genes on important agronomic characters. The mutant was crossed with a normal variety, and haploids were produced on the F_1 by the bulbosum technique [25] and chromosome-doubled. The completely homozygous offspring lines derived in this way will segregate into mutant homozygotes and wild-type homozygotes in a 1:1 ratio. All other segregating genes in the cross will be

Table 2. Effect of the high-lysine gene *lys3a* of mutant 1508 on carbohydrate, protein, and grain production in g/m^2.

Yield Component	Wild-type lines	*lys3a* lines	Difference in %
Starch + sugar	197	132	-33
Protein(6.25 x N%)	44.3	38.6	-13
Hordein	20.6	5.3	-74
Lysine	1.66	2.09	+24
Rest dry matter	116	116	0
Total grain	357	287	-20

randomly distributed among the two groups of lines provided these unknown genes are not linked with the high-lysine gene. Hence, the difference between the two groups with respect to a given character is an estimate of the mutant gene effect on that particular character.

A comparison of 10 high-lysine lines with 13 normal lines derived by the haploid technique from a cross between mutant 29 and 'Sultan' showed [36], that the high-lysine gene of mutant 29, *lys5a*, reduced the production of starch + sugar by about 25%. Also hordein production was reduced by *lys5a*, while other proteins and free amino acids were increased. The results of a similar study of the high-lysine gene *lys3a* of mutant 1508 [19] are summarized in Table 2. The 20 lines containing the *lys3a* gene produced 33% less starch + sugar on the average than the 30 lines that were homozygous for the corresponding wild-type gene. While the total protein was affected relatively little, 13%, the largest reduction was found in hordein production, which was reduced 74%. It is seen from Table 2 that the two groups of lines were equal with respect to other dry matter components. Hence, the *lys3a* gene influences only non-structural carbohydrates and hordein, i.e. the main storage components of the seed. There was a large genetic variation in grain yield within the two groups of lines, but the variation in the genetic background had apparently no influence on the negative effect of the mutant gene [19].

Special interest has been devoted to mutant 7 [64] because this mutant apparently has non-shrivelled seeds with normal 1000 kernel weight [14]. However, a recent study (unpublished data) of lines derived from chromosome-doubled haploids made on F_1 of the cross mutant 7 x 'Sultan' showed that this high-lysine gene also affects carbohydrate accumulation and seed weight. The reduction in carbohydrate production per unit area was 6%, and single seed weight was reduced 7%. However, hordein production was much more reduced, viz. 29%. Thus, mutant 7 should also be considered a protein- rather than a carbohydrate mutant.

A study of grain development in mutants 56 and 1508 showed a reduced growth rate and starch accumulation in the mutants compared with the parent varieties, while the number of endosperm cells was unaffected in the mutants [50]. Kreis (1978) studied the accumulation of starch and dry matter in mutant 1508 and 'Bomi' during seed development. The two genotypes were equal in dry weight per seed until 8 days after flowering when starch synthesis commenced. From that time until maturity the mutant was lower in seed weight and starch content per seed than 'Bomi'. The reduction in dry weight was nearly equal to the starch reduction, which was about 25% at maturity. The reduced starch accumulation was not due to sugar deficiency, because mutant 1508 had a higher amount of sugar per seed than 'Bomi' during the period of starch synthesis.

Further studies of starch accumulation and synthesis in mutant 1508 and two other high-lysine mutants, nos. 29 and 527, showed that the shrunken endosperm and smaller single kernel weight were almost entirely a consequence of a reduced starch synthesis throughout the development of the endosperm [39]. The mutant endosperms contained much more sucrose than the parent variety indicating blocks in the biochemical pathways leading from sucrose to starch. However, detailed studies of several enzymes involved in starch metabolism did not reveal any differences between mutant 1508 and 'Bomi' that were likely to be responsible for the reduced starch accumulation in this mutant [39]. The largest enzymatic difference was found in soluble β-amylase, which was strongly reduced in mutant 1508, but this difference is unlikely to affect the rate of starch synthesis during seed development.

The activities of enzymes of starch metabolism have also been studied in the high-lysine mutant Notch-2 and the parent variety NP 113 [4]. Only relatively small differences were found between the mutant and its parent, but it was concluded that the somewhat lower activity of ADPG(UDPG)-starch synthetase could be responsible for the reduced starch accumulation in Notch-2.

BREEDING POSSIBILITIES

Different breeding strategies to improve the nutritional value of barley seed protein have been reviewed recently [16, 65]. The main approach followed so far has been to select for an increased lysine content of the seed protein. This led to substantial and probably also sufficient improvements of the protein quality. However, the reduced grain and starch production of all the selected high-lysine mutants indicated strongly that this breeding strategy is too simple to provide high-quality varieties with acceptable agronomic properties.

In general the elevated lysine content of high-lysine barley is due to a reduced content of the main storage protein hordein. As this reduction in storage protein apparently is always accompanied by a certain reduction in starch it has been hypothesized [21] that the storage protein synthesis is essential for an efficient grain filling. Studies of lines containing two high-lysine genes [40, 63] point in the same direction, as the very low hordein content of such lines is followed by severe reductions in starch. Also studies in maize of the relations between N-fertilization, zein formation, kernel weight, and yield [68] indicated that the kernel N sink capacity is related to the productivity. Although the hypothesis concerning the significance of the storage protein needs further testing, it suggests that the desired protein improvement should be obtained without reducing the total amount of storage protein. If this is true the lysine-poor hordein should not just be decreased or removed, but it should be replaced by improved hordein and/or other storage proteins.

Drastic changes of the amino acid composition of storage proteins can be envisaged by means of genetic manipulations [42]. However, it is too early to comment on this possibility as there probably are several technical and physiological constraints to be solved before such changes are available for practical plant breeding.

A more easily accessible way to improve the nutritional value of hordein was proposed by Blake (1981) who pointed out that the large genetic variation in the polypeptide composition of hordein also may represent some variation in the lysine content of the polypeptides. Based on the use of lysine-specific protein stains, it should be possible to measure the lysine content of the individual hordein polypeptides separated by electrophoresis [5], and thereby select for increased lysine content of the hordein. Two points must be taken into consideration in breeding for improved hordein. Firstly, each *Hor* allele apparently consists of a multigene family coding for several different polypeptides. Secondly, no recombination has been reported among the units coding for the individual polypeptides controlled by a *Hor* allele, and it is probably unfeasible to combine individual high-lysine polypeptides identified in the hordein, coded for by different alleles. Therefore, the lysine content of the hordein can probably be increased only as far as it is possible to find *Hor1* or *Hor2* alleles coding for hordein with a higher average lysine content.

While the prospects for improving the amino acid composition of hordein at present seem rather limited, attempts to replace hordein by more lysine-rich proteins already present in the barley seed may be more promising. Changes in that direction are already present in 'Hiproly' and mutant 56, in which the contents of the lysine-rich proteins β-amylase, Z-protein, and the proteinase inhibitors CI-1 and CI-2 are strongly increased (see above). The increase in these proteins in mutant 56 is probably a secondary effect caused by the hordein reduction in this mutant.

The blocking of the hordein-2 synthesis makes more amino acids available for the synthesis of other storage proteins, and the content of these proteins is therefore increased. The reduced seed weight and starch content of mutant 56 [35] indicates, however, that the total capacity to synthesize storage protein is too low to ensure normal starch accumulation in the mutant. The amount of hordein-1 is increased in mutant 56 [15], but it is not known whether the actual increase in hordein-1 and lysine-rich proteins is as large as the decrease in hordein-2. Until further data are available it is therefore assumed that the increase in lysine-rich storage proteins in mutant 56 is insufficient to compensate for the hordein reduction, and thereby to assure a normal starch accumulation in the mutant grain.

An efficient replacement of a certain part of the hordein by other storage proteins requires probably that the synthesis of these other proteins is increased by genetic means. Such an increase should be possible in principle, e.g. by duplicating the structural genes or by changing the genetic regulation, but it is undoubtedly more difficult to increase than to decrease the content of a protein genetically. Another difficulty could be that there are physiological limitations on the amount of non-hordein storage protein that can be accumulated in the endosperm.

Even if it turns out that an agronomically acceptable improvement in the nutritional value of the protein can be obtained only by replacing hordein by other storage proteins, mutant genes that decrease the hordein content are still very important. This is because the change in protein composition can probably be achieved only by combining genes for increased content of the wanted proteins with genes reducing the hordein content. Such a two-step breeding strategy requires a detailed knowledge of the structural storage protein genes and their regulation, as well as a set of mutant genes governing the desired protein changes. So far we have no gene for which it has been shown that its primary effect is to increase one of the lysine-rich storage proteins, and none of the available low-hordein genes seems to be ideal.

The almost non-functional *Hor2ca* allele of mutant 56 appears to be one of the best low-hordein genes available because it seems to represent a direct change in one of the structural hordein genes. However, an almost complete deletion of hordein-2, which constitutes about 80% of total hordein in normal barley [32], is a drastic hordein reduction that may be difficult to replace completely by other storage proteins. Mutations in the structural genes resulting in a blocking of some hordein-2 polypeptides, or a removal of the very lysine-poor hordein-1 fraction, would probably be more valuable, but such mutants have not been reported. Finally, the mutant 56 gene seems to represent a gross mutation influencing characters other than just hordein. Mutant 56 flowers and matures about one week later than the parent variety 'Carlsberg II'. So far we have been unable to separate the lateness from the hordein-2 reduction in crosses and back-crosses between mutant 56 and other varieties.

It is difficult to appraise the breeding value of the low-hordein genes inherited independently of the structural hordein genes, as the primary effect of these mutant genes is not yet known. However, the almost complete absence of hordein found in mutant 1508 is difficult to utilize at present if the hordein reduction should be completely compensated for by a similar increase in other storage proteins.

Another, probably even more important drawback of mutant 1508 is that the *lys3a* gene apparently not only depresses hordein synthesis, but also influences the synthesis of some of the other proteins with storage properties. Kreis (1979) found a strongly reduced content of β-amylase in mutant 1508, and Hejgaard and Boisen (1980) reported that β-amylase, Z-protein and CI-2 were not increased in mutant 1508. Hence, there are indications that *lys3a* in general interferes with the ability to synthesize storage proteins. Therefore, it may turn out to be very difficult, if not impossible, to utilize this mutant gene and get a normal grain yield.

Although the primary effect of the *lys* gene of 'Hiproly' is unknown, this gene does not restrain the synthesis of the lysine-rich storage proteins. However, more studies are required to know whether this high-lysine gene can be incorporated into lines with a normal grain yield. This is true as well for other mutant genes, e.g. the one in mutant 7, which influence the hordein synthesis in an unknown manner.

So far the progress in the breeding for improved nutritional value of the seed protein of barley and other cereals has been limited in terms of released varieties, and several plant breeders have reduced or terminated their efforts in this field. Nevertheless, the attempts to improve the protein quality have provided a much better knowledge of the seed proteins, and both the more fundamental aspects of storage protein synthesis and genetics are now studied intensively. My intention with this contribution has been to show that the increased knowledge of the seed proteins could lead to new breeding strategies to produce varieties that have both a high protein nutritional value and an acceptable high grain yield. The suggested strategies are certainly not simple, and considerable effort and time will have to be spent before we know how effective they are. However, the expected gain in nutritional value of high-yielding, high-quality varieties is sufficiently large to justify a considerable effort in this breeding area.

REFERENCES

1. Andersen, AJ, and B Køie (1975): N fertilization and yield response of high lysine and normal barley. Agron J 67: 695–698.
2. Balaravi, SP, HC Bansal, BO Eggum, and S Bhaskaran (1976): Characterization of induced high protein and high lysine mutants in barley. J. Sci Food Agric. 27: 545–552.
3. Bansal, HC (1974): Induced variability for protein quantity and quality in barley. Indian J. Genet. Plant Breed. 34A: 657–661.
4. Batra, VIP, and SL Mehta (1981): Enzymes of starch metabolism in developing grains of high lysine barley mutant. Phytochemistry 20: 635–640.

5. Blake, TK (1981): New techniques for evaluating lysine content in hordeins. Barley Genet. Newslett. 11: 79–83.
6. Boisen, S, CY Andersen, and J Hejgaard (1981): Inhibitors of chymotrypsin and microbial serine proteases in barley grains. Isolation, partial characterization and immunochemical relationships of multiple molecular forms. Physiol. Plant. 52: 167–176.
7. Brandt, A (1976): Endosperm protein formation during kernel development of wild type and a high-lysine barley mutant. Cereal Chem. 53: 890–901.
8. Brandt, A (1979): Cloning of double stranded DNA coding for hordein polypeptides. Carlsberg Res. Commun. 44: 255–267.
9. Brandt, A, and J Ingversen (1976): In vitro synthesis of barley endosperm proteins on wild type and mutant templates. Carlsberg Res. Commun. 41: 312–320.
10. Cameron-Mills, V (1980): The structure and composition of protein bodies purified from barley endosperm by silica sol density gradients. Carlsberg Res. Commun. 45: 557–576.
11. Di Fonzo, N, and AM Stanca (1977): EMS derived barley mutants with increased lysine content. Genetica Agraria 31: 401–409.
12. Doll, H (1972): Variation in protein quantity and quality induced in barley by EMS treatment. Induced Mutations and Plant Improvement: 331–341; IAEA Vienna.
13. Doll, H, (1973): Inheritance of the high-lysine character of a barley mutant. Hereditas 74: 293–294.
14. Doll, H (1976): Genetic studies of high-lysine barley mutants. Barley Genetics III: 542–546.
15. Doll, H (1980): A nearly non-functional mutant allele of the storage protein locus *Hor 2* in barley. Hereditas 93: 217–222.
16. Doll, H (1981): Genetic possibilities for improving the nutritional quality of barley protein. Proc. Fourth Intern. Barley Genet. Symp. Edinburgh; in press.
17. Doll, H, and B Andersen (1981): Preparation of barley storage protein, hordein, for analytical sodium dodecyl sulfate-polyacrylamide gel electrophoresis. Analytical Biochemistry 115: 61–66.
18. Doll, H, and AHD Brown (1979): Hordein variation in wild (*Hordeum spontaneum*) and cultivated (*H. vulgare*) barley. Can. J. Genet. Cytol. 21: 391–404.
19. Doll, H, and B Køie (1978): Influence of the high-lysine gene from barley mutant 1508 on grain, carbohydrate and protein yield. Seed Protein Improvement by Nuclear Techniques: 107–114; IAEA Vienna.
20. Doll, H, B Køie, and BO Eggum (1974): Induced high-lysine mutants in barley. Radiation Bot. 14: 73–80.
21. Doll, H, and M Kreis (1979): Significance of storage protein for grain filling in barley. In: Spiertz, JHJ, and Kramer, T (eds.): Crop Physiology and Cereal Breeding: 173–174; Centre for Agricultural Publishing and Documentation, Wageningen.
22. Hejgaard, J, and S Boisen (1980): High-lysine proteins in Hiproly barley breeding: Identification, nutritional significance and new screening methods. Hereditas 93: 311–320.
23. Ingversen, J, (1975): Structure and composition of protein bodies from wild-type and high-lysine barley endosperm. Hereditas 81: 69–76.
24. Ingversen, J, B Køie, and H Doll (1973): Induced seed protein mutant of barley. Experientia 29: 1151–1152.
25. Jensen, CJ (1976): Barley monoploids and doubled monoploids: techniques and experience. Barley Genetics III: 316–345.
26. Jensen, J (1979a): Location of a high-lysine gene and the DDT-resistance gene on barley chromosome 7. Euphytica 28: 47–56.
27. Jensen, J (1979b): Chromosomal location of one dominant and four recessive high-lysine genes in barley mutants. Seed Protein Improvement in Cereals and Grain Legumes I: 89–96; IAEA Vienna.
28. Jensen, J, and H Doll. (1979): Gene symbols for barley high-lysine mutants. Barley Genet. Newslett. 9: 33–37.
29. Jensen, J, JH Jørgensen, HP Jensen, H Giese, and H Doll (1980): Linkage of the hordein loci *Horl*

and *Hor2* with the powdery mildew resistance loci *Ml-k* and *Ml-a* on barley chromosome 5. Theor. Appl. Genet. 58: 27–31.

30. Jonassen, I, I Ingversen, and A Brandt. (1981): Synthesis of SP II albumin, β-amylase and chymotrypsin inhibitor CI-1 on polysomes from the endoplasmic reticulum of barley endosperm. Carlsberg Res. Commun. 46: 175–181.
31. Karlsson, KE (1977): Linkage studies in a gene for high lysine content in Risφ barley mutant 1508. Barley Genet. Newslett. 7: 40–43.
32. Kirkman, MA, PR Shewry, and BJ Miflin. (1981): The effect of nitrogen nutrition on the lysine content and protein compositition of barley seed. J. Sci. Fd. Agric.; in press.
33. Kirsi, M (1973): Formation of proteinase inhibitors in developing barley grain. Physiol. Plant. 29: 141–144.
34. Kirsi, M (1974): Proteinase inhibitors in germinating barley embryos. Physiol. Plant. 32: 89–93.
35. Kφie, B, and H Doll (1979): Protein and carbohydrate components in the Risφ high-lysine barley mutants. Seed Protein Improvement in Cereals and Grain Legumes I: 205–215; IAEA Vienna.
36. Kφie, B, H. Doll, and M Kreis (1976a): Evaluation of a high-lysine barley gene using chromosome-doubled monoploids. Genetika (Beograd) 8: 177–182.
37. Kφie, B, J. Ingversen, AJ Andersen, H Doll, and BO Eggum (1976b): Composition and nutritional quality of barley protein. Evaluation of Seed Protein Alterations by Mutation Breeding: 55–61; IAEA Vienna.
38. Kreis, M (1978): Starch and free sugar during kernel development of Bomi barley and its high-lysine mutant 1508. Seed Protein Inprovement by Nuclear Techniques: 115–120; IAEA Vienna.
39. Kreis, M (1979): Starch synthesis in barley mutants deficient in storage proteins. Thesis, Université Catholique de Louvain, Belgium; 127 pp.
40. Kreis, M, and H. Doll (1980): Starch and prolamin level in single and double high-lysine barley mutants. Physiol. Plant 48: 139–143.
41. Linde-Laursen, I, H Doll, and G Nielsen (1981): Giemsa C-banding patterns and some biochemical markers in a pedigree of European barley. Z. Pflanzenzüchtg. (in press).
42. Miflin, BJ, SWJ Bright, and E Thomas (1981a): Towards the genetic manipulation of barley. Proc. Fourth Intern. Barley Genet. Symp. Edinburgh; in press.
43. Miflin, BJ, and PR Shewry (1979a): The biology and biochemistry of cereal seed prolamins. Seed Protein Improvement in Cereals and Grain Legumes I: 137–157 IAEA Vienna.
44. Miflin, BJ, and PR Shewry (1979b): The synthesis of proteins in normal and high lysine barley seed. In: Laidman, D, and RG Wyn Jones (eds.): Cereals: 239–273; Academic Press, London.
45. Miflin, BJ, PR Shewry, AJ Faulks, BG Forde, M Kreis, J Matthews, MB Bahramian, R Thompson, D Bartels, and R Flavell (1981b): The analysis of the *Hor2* locus of barley. Proc. Fourth Intern. Barley Genet. Symp. Edinburgh; in press.
46. Mossberg, R (1969): Evaluation of protein quality and quantity by dye-binding capacity: a tool in plant breeding. New Approaches to Breeding for Improved Plant Protein: 151–160; IAEA Vienna.
47. Munck, L (1972): High lysine barley – a summary of the present research development in Sweden. Barley Genet. Newslett. 2: 54–59.
48. Munck, L, KE Karlsson, A Hagberg, and BO Eggum (1970): Gene for improved nutritional value in barley seed protein. Science 168: 985–987.
49. Nelson, OE (1969): Genetic modification of protein quality in plants. Adv. Agron. 21: 171–194.
50. Olsen, OA, and T Krekling (1980): Grain development in normal and high lysine barley. Hereditas 93: 147–160.
51. Oram, RN, and H Doll (1981): Yield improvement in high lysine barley. Austr. J. Agric. Res. 32: 425–434.
52. Oram, RN, H Doll, and B Kφie (1975): Genetics of two storage protein variants in barley. Hereditas 80: 53–58.
53. Osborne, TB (1895): The proteids of barley. J. Amer. Chem. Soc. 17: 539–567.

54. Persson, G, and KE Karlsson (1977): Progress in breeding for improved nutritive value in barley. Cereal Res. Commun. 5: 169–179.
55. Rhodes, A (1975): A comparison of two rapid screening methods for the selection of high-lysine barleys. J. Sci. Food Agric. 26: 1703–1710.
56. Salcedo, G, R Sanchez-Monge, A Argamenteria, and C Aragoncillo (1980): The A-hordeins as a group of salt soluble hydrophobic proteins. Plant Science Lett. 19: 109–119.
57. Shewry, PR, AJ Faulks, and BJ Miflin (1980c): Effect of high-lysine mutations on the protein fractions of barley grain. Biochemical Genet. 18: 133–151.
58. Shewry, PR, AJ Faulks, RA Pickering, IT Jones, RA Finch, and BJ Miflin (1980b): The genetic analysis of barley storage proteins. Heredity 44: 383–389.
59. Shewry, PR, JM Field, MA Kirkman, AJ Faulks, and BJ Miflin (1980a): The extraction, solubility, and characterization of two groups of barley storage polypeptides. J. Exper. Bot. 31: 393–407.
60. Shewry, PR, HM Pratt, AJ Faulks, S Parmar, and BJ Miflin (1979): The storage protein (hordein) polypeptide pattern of barley (*Hordeum vulgare* L.) in relation to varietal identification and disease resistance. J. Nat. Inst. Agric. Bot. 15: 34–50.
61. Solari, RM, and EA Favret (1971): Polymorphism in endosperm proteins of barley and its genetic control. Barley Genetics II: 23–31; Washington State Univ. Press.
62. Sozinov, AA, VP Netsvetaev, EM Grigoryan, and IS Obraztsov (1979): Mapping of the *Hrd* loci in barley (*Hordeum vulgare*). Genetika USSR 14: 1137–1147.
63. Tallberg, A (1981a): Protein and lysine content in high-lysine double recessives of barley. I. Combinations between mutant 1508 and a Hiproly back-cross. Hereditas 94: 253–260.
64. Tallberg, A (1981b): Protein and lysine content in high-lysine double-recessives of barley. II. Combinations between mutant 7 and a Hiproly back-cross. Hereditas 94: 261–268.
65. Tallberg, A (1981c): Characterization of high-lysine barley genotypes. Hereditas; in press.
66. Thomson, JA, and H Doll (1979): Genetics and evolution of seed storage proteins. Seed Protein Improvement in Cereals and Grain Legumes I: 109–124; IAEA Vienna.
67. Torp, J (1979): Relations between production of starch and percentage, quality and yield of protein in barley. Z. Acker- Pflanzenbau 148: 367–377.
68. Tsai, CY, DM Huber, and HL Warren (1980): A proposed role of zein and glutelin as N sinks in maize. Plant Physiol. 66: 330–333.
69. Ullrich, SE, and RF Eslick (1978): Lysine and protein characterization of spontaneous shrunken endosperm mutants of barley. Crop. Sci. 18: 809–812.

7. Seed Protein of Rice and Possibilities of its Improvement through Mutant Genes

S. TANAKA

INTRODUCTION

The world production of protein for human beings is estimated at one hundred million tons per year. About seventy per cent of these come from plants because of the higher cost of producing animal protein. Although the animal protein has better quality for human beings, animals are known to have low efficiency of converting protein from plants. It is therefore difficult to expect to increase the production of animal protein in a short term in the near future.

Among the plant protein sources, the seed protein plays an important role owing to its large amount of storage protein. Wheat and rice which are used as staple food provide about 35% of the total plant protein in human nutrition. Although protein content of rice (9%) is lower than that of wheat (12%), rice protein has the highest nutritional value among cereal protein owing to the high content of limiting essential amino acids such as lysine and threonine which are generally deficient in cereals. If protein content of rice is increased through varietal improvement, rice will be a very important protein source. Success in breeding high protein rice would be the most significant achievement to elevate the nutritional standard in Asian countries.

Protein improvement of rice has been carried out through both genetical and physiological methods [1–5, 7, 10, 32, 35, 46, 47, 50–53, 55, 56, 77, 89, 90, 93, 99, 100, 108], as in other cereals and legumes [11, 21, 23, 26, 28, 29, 31, 32, 34, 36, 37, 39, 40, 43, 45, 48, 49, 57, 58, 61–65, 69, 70, 78, 79, 83–85, 103, 107]. It is well-known that there is a close relationship between the genetical background and the environmental conditions in seed protein production. In rice, however, the accumulated results of intensive studies on seed protein point out the possibility of breeding varieties with improved protein through mutations induced artificially [32, 50–53, 77, 99].

Early studies on protein improvement of rice were concentrated on investigating the factors that control the amino acid constituents [47, 92]. Since drastic changes in these constituents can hardly be expected, present researches have been directed towards increased protein content without decreasing grain productivity [88]. Thus, predominantly factors affecting protein productivity of rice and possibilities of breeding high protein varieties through induced mutations are discussed in the present paper.

THE SEED PROTEIN OF RICE

Protein Bodies and their Distribution in Endosperm and Embryo

Protein is accumulated along the cell wall of starch granules as protein bodies in endosperm and embryo [1, 2, 92]. Kasai (1979) found two types of protein bodies. Most bodies show spherical shape with clear margin, but some of them show anomalous spherical shape without clear margin. The formation mechanism of the protein bodies and their activities are still unknown [74].

The study on the distribution of the protein bodies in endosperms and embryos, as measured through a compound or an electron microscope, indicated that the embryo has more protein bodies than the endosperm. In the endosperm, the aleuron layer is richer in protein bodies than inner parts of the tissue. The number of protein bodies decreases with the distance from the surface of the endosperm and there are few bodies in a core of rice kernel. It is known that there are varietal differences of protein body distribution in rice kernel and the proportion of embryo weight to the whole seed weight, though the differences in rice are rather small in comparison to other cereals such as sorghum, maize and rye. Karl (1979) suggested that some Indian varieties have relatively more protein bodies near the core of the kernel which can be used as material for breeding high protein varieties. Omura (1979) also reported that mutants with large embryo can be obtained rather frequently by radiation or chemicals. Both findings have not been utilized for practical breeding programmes [6, 41, 42, 34, 59, 91, 101, 104].

The thickness of the aleuron layer of rice kernel varies with varieties and mutants. In rice, most research aimed at breeding varieties to improve grain quality has succeeded in producing varieties with thick aleuron layer but these could not be released to farmers because of reduced proportion of polished rice to brown rice, even if they have high protein productivity in brown rice. In general, the proportion of polished rice to brown rice is around 90% in weight and the protein content of polished rice (8%) is less than that of brown rice (9%) due to the removal of rice bran and most embryos. Accordingly, about 20% of total protein of rice is lost by polishing.

Although varietal differences of the yield rate exist, they are very small in the commercial varieties, and accordingly there is a close relationship between protein content of the polished rice and that of the brown rice ($\gamma = 0.95$). Therefore, protein content of brown rice has been used as an index for screening high protein plants in practical rice breeding.

Rice protein has the highest quality among cereals in digestability, biological value, protein efficiency and so on. A mixture of extracts from rice and fish meat has been used for baby food as the source of protein and starch, though a few protein bodies from rice cannot be digested by babies.

The amino acids lysine and threonine, which are generally limiting essential amino acids in cereals, are rather rich in rice in comparison to other cereals owing to low prolamine content (5%) in its protein. Accordingly, seed protein of rice has quite different amino acid composition than that of maize, barley and sorghum in which high lysine genes have been found. In addition, varietal and environmental differences in amino acid composition of rice are fewer than those of other cereals [65, 69, 72, 75, 82, 86–88].

Factors Influencing Synthesis and Accumulation of Protein in Rice Kernel

Protein content and amino acid constituent of rice vary with biophysical conditions such as the length of the period from germination to maturity due to the differing growing season, plant type and activities of leaves, stems and roots during zygosis to maturity. Environmental conditions such as daylength, temperature, amount of fertilizer, topdressing time, soil and other factors also affect the seed protein. Furthermore, there exist close interactions between the factors influencing rice protein.

Relationship Between Protein Content and Growing Season. Rice plants comprising several ecotypes show diverse patterns of development depending on their genotypes and environment. The growing period thus determined may affect the accumulation of starch in the kernel. Should a given cultivar be grown in a given paddy, any change of cultivation procedures that affect the development of plants may exert direct or indirect influence on the composition of kernels. Growing season – or planting time – of photosensitive cultivars in the temperate zone, for example, may affect the length of the growing period and influence the protein content. The tendency is recognizable that the earlier the planting, the longer the growing period and, consequently, the larger the starch content per kernel and the lower the protein per starch ratio. Interaction between the growing season and protein content in terms of the ratio with kernel weight is also conceivable under other ecological circumstances which are often complicated with such rigorous components as drought, flood, and low temperature. Generally, the protein content of rice grain decreases with increasing grain yield, although protein yield is the product of the protein content and grain yield. Contrary, protein yield is reduced greatly under abnormal growing conditions, though the protein content of rice is high. For above-mentioned reasons, rice plants grown in a relatively short term tend to have higher protein content in comparison to those grown in a long term, but protein productivity per unit area is lower in the former case than in the latter one. These phenomena in rice are similar to those in other cereals [44].

Relationships Between Protein Yield and Plant Type. Rice varieties with short culm tend to have high protein content and productivity. However, in general, there is no direct relationship between 'high protein genes' and genes controlling plant height. It is well known that plants grown under heavy fertilizer conditions sometimes show abnormal maturity resulting in high protein content and low protein productivity. This phenomenon occurs more frequently in varieties with long culms. Accordingly, lodging resistant varieties under heavy fertilizer conditions should be used as material for increasing protein productivity [55].

Relationships Between Protein Yield and Activities of Plant from Zygosis to Maturity. Accumulation of protein bodies in rice kernel starts just after zygote formation and a great part of stored seed protein is accumulated when the embryo and endosperm begin to develop. However, the process of starch accumulation is rather slow just after zygosis, but gradually increases with seed development. Accordingly, protein content of rice kernel is very high in the beginning of seed development in comparison to that in matured stage. Under normal conditions, large amounts of protein synthesis and accumulation as well as starch accumulation are continuously made during the ripening stage.

As to the accumulation of carbohydrates, about two thirds of the total starch are directly shifted from the leaves, especially from the flag leaf, after the zygotic stage. The substances for the remaining third are transmitted from the stem bases where the carbohydrates have been stored before zygote formation. Although a few data on protein accumulation in rice kernel are available, it is known that the situation on stored protein is somewhat different from that on storage starch: the proteins are mainly shifted from leaves to rice kernel throughout the ripening stage. Therefore, topdressing of nitrogen on rice plants after the heading stage is one of the effective methods to increase the protein content of rice, if the plants have high activities for synthesis of both protein and starch [100].

Relationships Between Protein Productivity, Amount and Time of Fertilization. The amount of fertilizers absorbed by the rice plant from the soil increases with higher temperature ranging from 15 to 30°C. Overabsorption of fertilizers under high temperature, especially of nitrogen, sometimes causes abnormal growth which tends to reduce protein productivity, though the protein content is high. As a preventive measure against the reduction of protein productivity due to the abnormal plant growth, a moderate amount of fertilizer is applied to the plants at the sowing or transplanting period as a basal fertilizer. Additional fertilizer may be applied as topdress. When topdressing is applied, analysis of factors affecting the protein content of rice is difficult because of the complicated interacting factors. However, as the varietal difference of protein productivity is not greatly upset by the amount and time of fertilization, it is recommended that screening high protein varieties should be made in a uniformly cultivated field where the conventional fertilizer has been applied [44, 89, 90, 112].

Relationships Between Protein Productivity, Growing Location and Soil Conditions. It is known that there is a negative correlation between protein content of rice kernels and water content of soil. In general, protein content of upland varieties is considerably higher (10–11%) than that of lowland varieties (9%) although the protein productivity of the former is lower than of the latter due to its low grain productivity. However, protein content of upland varieties grown in paddy field is nearly the same as that of the lowland varieties, when they grow normally [1, 2].

Studies on the effect of soil condition and location on protein productivity are not conclusive due to the complicated interactions between environmental conditions except the water content of soil. Generally speaking, rice plants grown on soil containing much humus or at locations with cold weather tend to have higher protein content than those grown on sandy soil or a place with high temperature [98]. However, since this tendency mainly results from biophysical conditions, the potentialities of high protein varieties being controlled genetically should be confirmed at various locations with various soil types [98, 99].

Relationships Between Protein Yield and Grain Yield. It is well-known that there is a negative correlation between protein content and grain yield [14]. On the other hand, there is a positive correlation between protein yield and grain yield. Coefficient correlation between protein and grain yield is very high among mutants with drastic morphological and physiological changes and among native varieties grown in various out-of-the-way regions but not so high among the presently existing leading varieties.

Among high protein mutants, as shown in Table 1, an increase of protein content does not always result in increased protein yield. 'High protein variety' is defined as a variety with high protein content of the seed meal combined with a correspondingly high grain productivity under favorable growth conditions.

Relationships Between Protein Content and Amino Acid Compositions of the Proteins. As previously mentioned, lysine and threonine are the limiting essential amino acids in cereals. Although rice protein contains higher amounts of the essential amino acids than other cereals, making it higher in nutritive value, one of the important objectives of breeding rice varieties should consist in increasing these amino acids [95].

It is well-known that there is a negative correlation between protein content and

Table I. Relationships between protein content and protein productivity, grain yield and protein productivity and protein content and grain yield among seven high protein mutant lines and original variety 'nihonbare' grown under various environmental conditions.

CONDITIONS		PROTEIN CONTENT and PROTEIN PRODUCTIVITY	GRAIN YIELD and PROTEIN PRODUCTIVITY	PROTEIN CONTENT and GRAIN YIELD
CULTIVATION PERIOD	FERTILIZER LEVEL N, P_2O_5, K_2O (kg/ha)			
EARLY CULTIVATION (April – July) 145 days	70, 90, 70	0.68^{**}	0.98^{**}	0.80^{**}
STANDARD CULTIVATION (June – October) 130 days	60, 90, 60	0.35^{ns}	0.96^{**}	0.59^{*}
STANDARD CULTIVATION	90,120, 90	0.25^{ns}	0.93^{**}	0.53^{*}

S. Tanaka (1975, unpublished).

lysine content in cereals [15, 16, 39, 40, 105, 106, 109]. Since individual species have a specific amino acid composition, close relationships between genes controlling amino acid composition and protein content cannot be expected. Therefore, many studies have been focused to the improvement of both protein quality and quantity at the same time by using genes controlling amino acid composition and other ones controlling protein content of the seed meal. As pointed out previously, the amino acid composition of rice protein shows specific characteristics being different from crops with high amounts of prolamine in protein such as corn, barley and sorghum in which high lysine genes have been found [60, 68, 71, 81]. Since it could hardly be expected to change the amino acid composition of rice drastically, breeding high protein rice results in a high lysine product, though the negative correlation exists in rice protein.

POSSIBILITIES OF RICE PROTEIN IMPROVEMENT THROUGH MUTANT GENES

The historical study on protein and oil contents of maize which has been carried out since 1876 is a common knowledge. Hopkins and co-workers made the successive mass screening for protein and oil contents of maize for more than 70 generations using an autogamous variety, Burr-white. At present, lines with four times as much protein content and fifteen times as much oil content as the original material are available [17, 110]. Although these lines are not usable as parents of high protein and/or high oil varieties, they play an important role in genetic studies on storage components in plant seeds. The results obtained in maize could also be expected in rice though the maximum content of the two crops might differ. However, in the case of maize, the results obtained strongly suggest that protein and oil content are controlled by polygenes.

After the discovery of mutant major genes that control the lysine content in maize endosperm [60], intensive work was carried out in cereals to breed varieties with a higher content of high quality protein which brought interesting results [60, 69–71]. At the same time, since the protein gap in the world was emphasized by the FAO, WHO, IAEA and related agencies of United Nations, intensive national and international research programmes were carried out throughout the world with the aim to improve protein quantity and quality of many crops [76, 80]. Under these circumstances, breeding rice varieties with high protein productivity became one of the most important breeding objectives in this crop although protein improvement was a minor aim of rice breeding for a long time.

Varietal Differences in Protein Content

Before starting the improvement of rice protein, one should study the varietal differences with regard to this trait and the possibilities of altering it by means of induced mutations. Many studies on rice protein have been carried out throughout the world. Some recently obtained results in this field are discussed related to the progress in improving rice varieties.

Commercial Varieties in Japan. Commercial varieties grown in each prefectual experimental station which were distributed as seeds to farmers were determined for their protein content to screen the high protein sources, because all these varieties have clear genealogy. Among eight hundred commercial varieties, recommended by the prefectual governments, there is no extremely high or low protein variety, though four varieties in some prefectures had more than 10% protein. However, it is clearly concluded that the variety 'Fukunishiki' having high protein content (9–10%) has contributed as a donor to high protein commercial varieties (8.5–9.5%), suggesting the presence of genes controlling seed protein content [98].

Japanese Native Varieties. Native varieties, collected from many remote country areas in Japan and maintained at Kyushu University, were determined for their protein content for the purpose of screening high protein genes. Protein content of more than seven hundred native varieties varied widely ranging from 5% to 12%. Among them, six varieties with more than 10% protein content and modernized plant types were screened as high protein gene source.

Relationships between high protein native and high protein commercial varieties have not yet been delineated, but some native high protein varieties should be good sources for high protein commercial varieties. It is possible that spontaneously mutated genes responsible for high protein content may have been eliminated at a time when breeding was carried out without the aim of improving seed protein. Although negative correlation exists between starch and protein accumulation in rice, the relationship should be considered for breeding high protein varieties, if the increase of protein productivity results from increase of protein content without decreasing grain yield [98].

Varieties of the World Collection. Varieties of the world collection of rice in the International Rice Research Institute in the Philippines were determined for their protein content. A large variation of protein content, ranging from 6% to 14%, was observed. Some of these varieties have high protein content which might have resulted from abnormal growth under the specific environmental conditions in the Philippines, because considerably large numbers of Japanese varieties including upland varieties have extremely high protein content due to insufficient starch synthesis. Accordingly, some Indian varieties with high protein content, screened among these varieties, have been used as material for cross breeding programmes in IRRI with promising results for developing high protein varieties [46, 47].

Mutants. More than five hundred mutant lines with changes of visible characters,

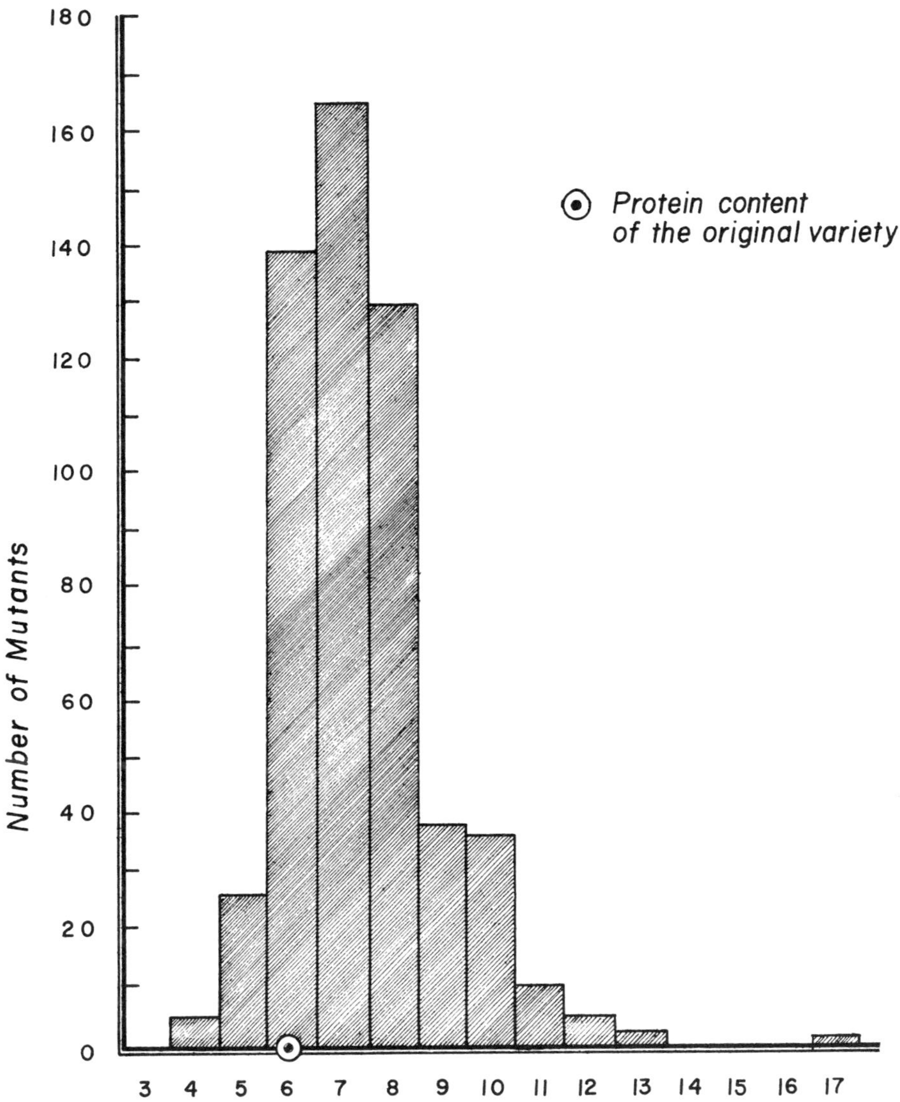

Figure 1. Protein content (%).

obtained from the Japanese variety Norin 8 by gamma irradiation at the Institute of Radiation Breeding, were examined for their protein content. Results obtained are given in Figure I. As shown in the figure, great differences in protein content of the mutant lines were found ranging from 4–16%. Most extremely high or low protein mutants showed drastic changes in some agronomic or morphological traits such as extreme earliness or dwarfness suggesting pleiotropic action of the genes involved.

However, mutants with similar phenotypes had different protein content suggesting the occurrence of mutations in specific genes controlling protein content. As recommended by the International Atomic Energy Agency, extremely high protein mutants resulting from insufficient carbohydrate synthesis ability due to abnormal growth should not be used as a gene source for breeding high protein varieties; on the contrary, normal looking high protein mutants should be screened [95, 97, 99].

Effectiveness of Mutation Breeding for Rice Protein Improvement

It is well-known that mutation and cross breeding are the matchless twin stars for varietal improvement of crops at present. Although the two breeding methods are closely interdependent on each other, merits and demerits of mutation breeding are sometimes compared with those of cross breeding because of the differences in breeding efficiency for utilization of useful genes. The merit of mutation breeding is that desirable mutant varieties without any changes of other superior agronomic characters could be isolated in early generations; thus, breeding term and cost are shortened [8, 9, 22, 27, 29, 30, 38, 73, 94, 96, 102, 111]. On the other hand, accumulation of useful genes into a commercial variety is hardly expected by induction of mutations. Accordingly, newly bred leading varieties should be used as material to induce mutations which can be used directly as released varieties; moreover, useful mutant genes can be used as genetic sources in cross breeding.

Mutations occur spontaneously or are induced artificially by radiations and/or chemical mutagenic treatments in genes, chromosomes and cytoplasm. Among these, gene mutations are frequently used for breeding varieties, directly or indirectly. Most gene mutations are recessive to their normal type and they act as major genes. Although many reports on dominant mutations are available, the finding might have resulted from recessive mutations which occurred in supplemental genes of the major genes. From this point of view, the efficiency of mutation breeding depends on the screening and utilization of mutated recessive major gene.

As mutation is an event in a single cell, we often encounter with the chimeric structure that sorts out sectors of mutated and original genotypes in a plant and make the screening of mutants tedious. Purification of mutated lines is rather easy in sexually reproducible plants; for the embryogenesis from a single zygote would bring about non-chimeric mutant plant. Most plants with mutated characteristics in M2 population of such sexually-reproducing species would be isogenic, although simultaneous mutations at nearby loci as well as chimeric structures might be recognizable at a very low frequency.

Since rice originated during the ancient times, seed components are somewhat different from those of other cereals such as maize, sorghum, and so on and the amino acid composition of *Oryza sativa* is rather specific among the family [88, 90]. This holds particularly true with regard to the prolamine content which is considerably lower in comparison to that of other cereals. Accordingly, it is hardly expected to discover major genes affecting the amino acid composition of seed proteins and so far, no mutant gene of this category has been found in rice.

The protein content of rice is affected by genetic elements as well as by environmental factors. However, the large differences in protein content of different rice varieties, grown under the same environmental conditions, indicate that varieties with high protein productivity can be bred. Moreover, heritability of the protein productivity is rather high in rice, suggesting the effectiveness of rice protein improvement through mutant genes [95].

Mutation Breeding Procedure for High Protein Varieties. Mutations influencing protein content are rather frequently induced, because it is thought that genes on many loci affect protein synthesis of rice. Although the mutant genes have no monogenic action, the heritability of the altered traits is considerably high. However, analysis of the respective genes is not conclusive due to their complicated behavior in successive generations. Therefore, in M_1 generation, sufficient seeds of a panicle or plant are harvested to raise M_1 lines. Highly sterile plants must be eliminated, because correlation between M_1 fertility and M_2 mutation is not always significant.

Mutants with beneficial characteristics other than protein content might be visible in the population for high protein breeding. Theoretically, the mutation ratio of a given locus is not affected by the simultaneous mutation in an independent locus. Therefore we can include the visible mutants in the population for screening without any adverse effect on the efficiency of screening. Our experience tells us, however, that many drastic mutations are accompanied with the alteration of protein content, high or low, that is tightly associated with unfavorable traits, such as very small grain, inability of starch accumulation, partial sterility, etc. Practically, it is desirable to eliminate drastic visible mutants from the selected lines so as not to disturb the breeding procedures as well as statistics.

M_3 lines fixed for visible characters do not always breed true for their protein content, but also other complicated facts resulting from heterogeneity, chimerism, three nuclei in endosperm cells and so on should be taken into consideration. Therefore, screening for high protein mutants must be made on a single plant basis in each M_3 line. Investigations on sterility of plants should also be made carefully because small reduction of fertility sometimes results in high protein content of fertile seeds. Even in M_4 generation, some lines segregate for morphological traits though their frequency is very low for reasons not yet known. As it is conceivable that high protein lines segregating for visible characters might reach the same protein level as the mother variety, lines segregating for visible characters should be eliminated in early generations. In successive generations after M_4, yield trials should be made under various environmental conditions, because protein content is strongly influenced by many factors.

Achievements Up to Date

Kataoka [52] reported that mutation frequency for protein in M_2 generation (number of lines segregating for high protein variants/number of tested lines × 100) was 7–14% after ethylenemine treatment with concentrations ranging from 0.01% to 0.03%. Several high protein lines were obtained but most of them had simultaneous morphological changes [50–53]. Tanaka (1974) confirmed Kataoka's findings that mutation frequency of protein productivity (grain yield × protein content) in M_2 generation was about 0.01% after gamma irradiation of dormant seeds. The large difference of mutation frequency in both the experiments might have resulted from the different screening methods, namely, Kataoka isolated high protein variants with morphological changes while Tanaka isolated high protein variants from normal looking M_2 lines not segregating for any visible characters. Mutant 83–5–1, isolated by Tanaka, and the original variety were investigated with regard to their nitrogen absorption into various organs under various environmental conditions, in order to discern the effect of the mutant gene on seed protein production during the pre-heading to the maturing period. The amount of nitrogen absorbed in the leaf blade of the mother variety decreased greatly in its growing stages, but leaves of the mutant maintained the activity up to the harvesting period. This tendency was clearer under standard than under heavy fertilization conditions, resulting in a marked difference in protein content between the 2 genotypes grown under standard fertilization conditions (Figure 2). The nitrogen content of panicles of both the mutant and the original parent was nearly constant in their later growing stages, despite a rapid accumulation of carbohydrates, though the nitrogen content of the mutant was higher than that of its parent. This fact indicates that the nitrogen is still accumulated into panicles from other organs during later growing stages. The grain yield of the mutant was slightly higher than that of its mother variety due to the better tillering ability. The protein yield of the mutant was considerably increased owing to an increased grain yield and protein content as shown in Table II.

From these results it is concluded that the high protein mutants in general have higher nitrogen absorption ability than the original cultivar. High concentration of nitrogen in organs other than panicles at late growing stages may contribute to the high seed protein production. Nitrogen absorbed from the soil at late growing stages was also shown to contribute to the seed protein production in these plants with a vigorous absorption system lasting to maturity. Seeds of high protein mutants were analysed for any change of amino acid constitution in hopes of finding lines with improved amino acid composition. However, none of the high protein mutations induced to date have been shown to be accompanied with the alteration of amino acid constitution. Hagu obtained a high protein mutant Habiganj–11–43 with opaque endosperm, but mutations for high lysine content like opaque-2 in maize were not found in this mutant, either.

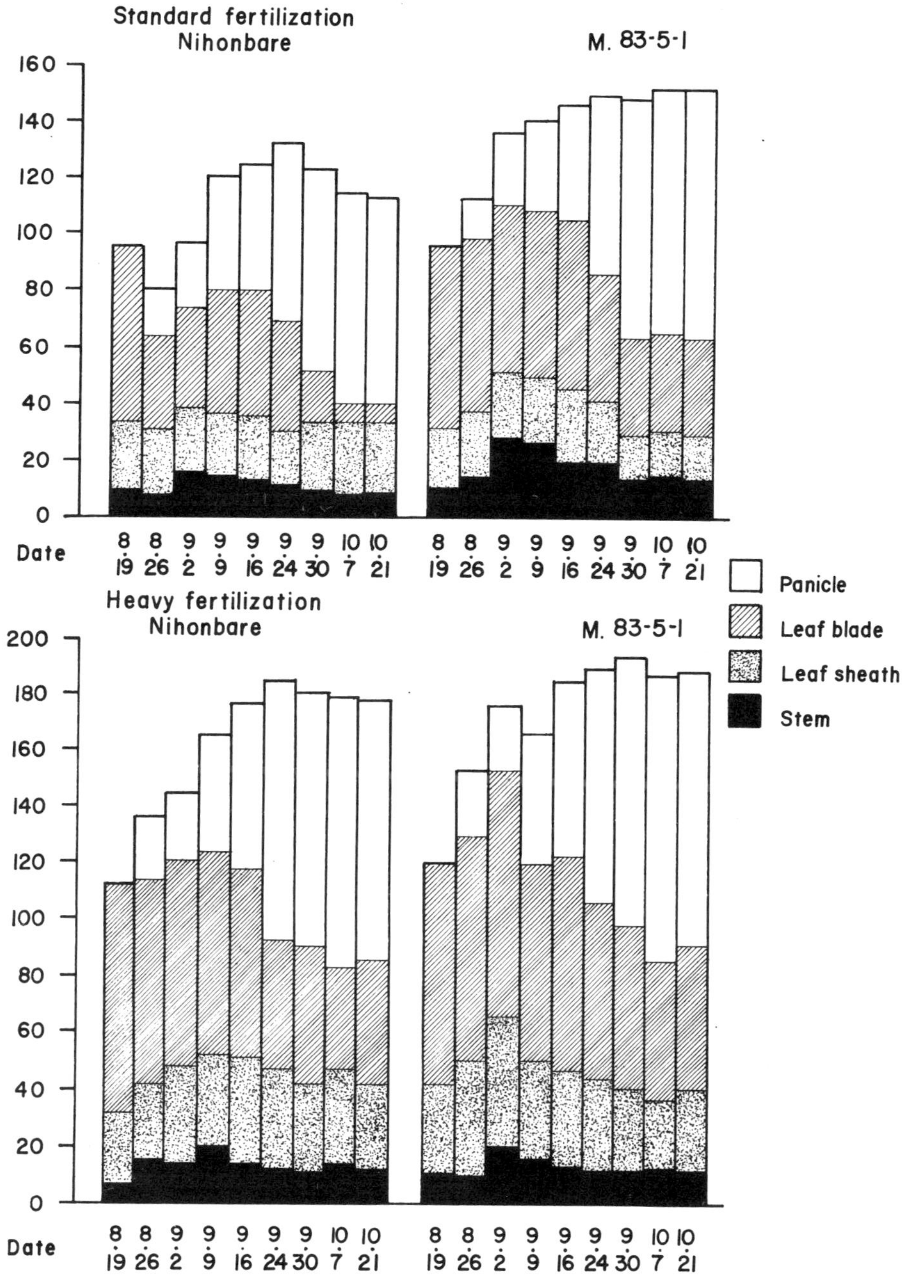

Figure 2. Nitrogen absorption invarius parts of the rice plant.

Table II. Agronomic traits of mutant 83-5-1 and its original variety nihonbare grown under standard and heavy nitrogen fertilizations (field experiment).

	HEADING DATE	MATU-RING DATE	STEM LENGTH (cm)	PANICLE LENGTH (cm)	PANICLE NUMBER per m^2	GRAIN WEIGHT (t/ha)	BROWN RICE WEIGHT (t/ha)	PROTEIN CONTENT (%)	PROTEIN YIELD (kg/ha)
STANDARD NITROGEN FERTILIZATION									
Nihonbare	27 Aug.	8 Oct.	84	18.4	256	5.48	4.55	7.9	359.4
M. 83-5-1	27 Aug.	8 Oct.	83	18.2	274	5.50	4.57	8.7	397.2*
HEAVY NITROGEN FERTILIZATION									
Nihonbare	30 Aug.	10 Oct.	88	18.9	261	5.77	4.73	9.4	444.6
M. 83-5-1	30 Aug.	10 Oct.	88	19.0	272	5.86	4.84	9.9	479.1*

CONCLUSIONS

As the amino acid composition of rice protein is rather specific among cereal proteins and as it is hardly expected to breed a variety with drastic genetical changes of lysine content due to the limited prolamine component, intensive efforts for rice protein improvement have been focused on breeding varieties with high protein productivity. Since it is clear that the genes controlling the protein content of rice kernel have higher heritability among those of cereals, mutation breeding has been employed as well as cross breeding. Radiation breeding has an advantage over other breeding methods in that it is possible to improve a single characteristic without affecting other desirable agronomic characters in the parent. The fact that most of the high protein mutations obtained in rice are of fairly high heritabilities renders the mutation breeding for this purpose especially advantageous. High protein genes discovered in rice cultivars are diverse in heritability; some are as high as mutant genes and some are very low. Therefore, it is hardly possible to select high protein lines with the least change of other characteristics from the hybrid of cultivars, and virtually no high protein lines can be selected if the progenies of the hybrids were first selected for characteristics other than protein content. Although time for determinating protein content of rice seeds has been reduced markedly, the time for the screening is still the major factor limiting the efficiency of the breeding. Recurrent selection for high protein characters may also bias the population for poor behavior in agronomic characters. We can circumvent these problems by adopting mutation breeding.

For these reasons, it is concluded that conventional mutation breeding methods by which many crop varieties have been bred for recent forty (40) years should be employed for breeding varieties with high protein productivity. These methods should be supplemented by utilizing mutations in major genes which are certainly more effective than those in polygenes although protein production is a quantitative character. As mentioned above, it is possible to improve rice protein through mutant genes. However, it is important to screen high protein content mutants without reduction of grain yield due to simultaneous changes of other agronomic traits. At the same time, especially for screening protein mutants, the material should be grown normally under uniform environmental conditions.

REFERENCES

1. Ann. Rep. Int. Rice Res. Inst. (1968): Improvement of the Protein Content of Rice.
2. Ann. Rep. Int. Rice Res. Inst. (1968): Screening of Rice Varieties for Protein.
3. Ann. Rep. Int. Rice Res. Inst. (1969): Improvement of the Protein Content of Rice.
4. Ann. Rep. Int. Rice Res. Inst. (1972): Improvement of the Protein Content of Rice.
5. Beachell, HN, GS Kush, and BO Juliano (1972): In Rice Breeding; Int. Rice Res. Inst. Los Baños, Philippines: 419.

6. Bloemendel, H (1967): High resolution techniques, Electrophoresis II. Academic Press: 739.
7. Britten, RJ, and EH Davidson (1969): Gene regulation for higher cells: A theory. Science 165: 349–357.
8. Brock, RD (1972): The role of induced mutations in plant improvement. Rad. Bot. 11: 181–196.
9. Brock, RD, and HF Shaw (1969): Response to a second cycle of mutagenic treatment in *Arabidopsis thaliana*. Induced Mutations in Plants: 457–467; IAEA Vienna.
10. Cagampang, GB, LJ Cruz, G Espiritu, RG Santiago, and BO Juliano (1966): Studies on the extraction and composition of rice protein. Cereal Chem. 43: 145–155.
11. Campbell, AR, and RC Pickett (1968): Effect of nitrogen fertilization on protein quality and quantity and certain other characteristics of 19 strains of *Sorghum bicolor* (L.) Moench. Crop Sci. 8: 545–547.
12. Clark, JA (1926): Breeding wheat for high protein content. J. Am. Soc. Agron. 18: 648–661.
13. Clark, JA, and RW Smith (1928): Inheritance in Nodak and Kahla *durum* wheat crosses for rust resistance, yield, and quality at Dickinson, North Dakota. J. Am. Soc. Agron. 20: 1297–1304.
14. Clark, JA, and KS Quisenberry (1929): Inheritance of yield and protein content in crosses of Marquis and Kota spring wheats grown in Montana. J. Agric. Res. 38: 205–217.
15. Doll, H (1971): Variation in protein quantity and quality induced in barley by EMS treatment. Induced Mutations and Plant Improvement: 331–342; IAEA Vienna.
16. Doll, H, B Køie, and BO Eggum (1974): Induced high lysine mutants in barley. Rad. Bot. 14: 73–80.
17. Dudley, JW, and RJ Lambert (1969): Genetic variability after 65 generations of selection in Illinois high oil, low oil, high protein, and low protein strains of *Zea mays* L. Crop Sci. 9: 179–181.
18. Dumanović, J, L Ehrenberg, and M Denić (1970): Induced variation of protein content and composition in hexaploid wheat. Improving Protein by Nuclear Techniques: 107–120; IAEA Vienna.
19. Favret, EA, L Manghers, R Solari, A Avila, and JC Monesiglio (1970): Gene control of protein production in cereal seeds. Improving Plant Protein by Nuclear Techniques: 87–97; IAEA Vienna.
20. Favret, EA, R Solari, L Manghers and A Avila (1969): Genetic control of the qualitative and quantitative production of endosperm proteins in wheat and barley. New Approaches to Breeding for Improved Plant Protein: 87–107; IAEA Vienna.
21. Frey, KJ (1949): The inheritance of protein and certain of its components in maize. Agron. J. 41: 113–117.
22. Futsuhara, Y, K Toriyama, and K Tsunoda (1967): Breeding of a rice variety 'reimei' by gamma-ray irradiation. Jap. J. Breed. 17: 13.
23. Golovchenko, VI et al. (1970): Amino-acid composition of protein in radiation-induced variety of lupin for fodder and in economically valuable mutants of spring wheat. Improving Plant Protein by Nuclear Techniques: 149–162; IAEA Vienna.
24. Gottschalk, W (1960): Über züchterisch verwendbare strahleninduzierte Mutanten von *Pisum sativum*. Züchter 30: 33–42.
25. Gottschalk, W (1965): Der Einfluß der Penetranzverhältnisse mutierter Gene auf die Leistungsfähigkeit von Positivmutanten. Publ. Europ. Atom. En. Comm. EUR 2510, d: 1–14.
26. Gottschalk, W (1966): The yield capacity of useful mutants. A critical review of a collection of mutant types of *Pisum*. Mutations in Plant Breeding: 85–101; IAEA Vienna.
27. Gottschalk, W (1970): The productivity of some mutants of the pea (*Pisum sativum* L.) and their hybrids. A contribution to the heterosis problem in self-fertilizing species. Euphytica 19: 91–96.
28. Gottschalk, W, and HP Müller (1970): Monogenic alteration of seed protein content and protein pattern in X-ray induced *Pisum* mutants. Improving Plant Protein by Nuclear Techniques: 201–215; IAEA Vienna.
29. Gustafsson, Å, and D v. Wettstein (1958): Mutationen und Mutationszüchtung. In: H Kappert and W Rudorf: Handb. d. Pflanzenzüchtung I: 612–699; Parey Berlin, Hamburg.

30. Hagberg, A, and KE Karlsson (1969): Breeding for high protein content and quality in barley. New Approaches to Breeding for Improved Plant Protein: 17–21; IAEA Vienna.
31. Haq, MS, MM Rahman, and MH Chowdhury (1973): Studies of the quality of induced mutants of rice. Nuclear Techniques for Seed Protein Improvement: 139–144; IAEA Vienna.
32. Hartwig, EE (1969): Breeding soybeans for high protein content and quality. New Approaches to Breeding for Improved Plant Protein: 67–70; IAEA Vienna.
33. Haunold, A, VA Johnson, and JW Schmidt (1962): Genetic measurements of protein in the grain of *Triticum aestivum* L. Agron. J. 54: 203–206.
34. Hillerislambers, D, JN Rutger, CO Qualset, and WJ Wiser (1973): Genetic and environmental variation in protein content of rice (*Oryza sativa* L.) Euphytica 22: 246–273.
35. Hiraiwa, S, S Tanaka, and S Nakamura (1976): Induction of mutants with higher protein content in soy bean. Evaluation of Seed Protein Alterations by Mutation Breeding: 185–196; IAEA Vienna.
36. Huebner, FR, and JA Rothfus (1968): Gliadin proteins from different varieties of wheats. Cereal Chem. 45: 242–253.
37. Ichijima, K (1934): On the artificially induced mutations and polyploid plants of rice occuring subsequent generations. Proc. Imp. Acad. Japan (Tokyo) 10: 368–381.
38. Ingversen, J, and B Køie (1971): Protein patterns of some high lysine barley lines. Hereditas 69: 319–323.
39. Ingversen, J, B Køie, and H Doll (1970): Protein studies of barley mutants with increased lysine content. Barley Newsletter 14: 32–33.
40. Jacobsen, JV, RB Knox, and NA Pyliotis (1971): The structure and composition of aleurone grains in the barley aleurone layer. Planta 101: 189–209.
41. Johnson, BL (1967): Tetraploid wheats: Seed protein electrophoretic patterns of the Emmer and the Timopheevi groups. Science 158: 131–132.
42. Johnson, VA, PJ Mattern, and JW Schmidt (1967): Nitrogen relations during spring growth in varieties of *Triticum aestivum* L. differing in grain protein content. Crop. Sci. 7: 664–667.
43. Johnson, VA, PJ Mattern, and JW Schmidt (1976): The breeding of wheat and maize with improved nutritional value. Proc. Nutr. Sci. 29: 20–31.
44. Johnson, VA, JW Schmidt, and PJ Mattern (1968): Cereal breeding for better protein impact. Econ. Bot. 22: 16–25.
45. Juliano, BO (1964): Variability in protein content, amylose content and alkali digestability of rice varieties in Asia. Philippine Agriculture 48: 234–241.
46. Juliano, BO, CC Ignacio, VM Panganiban, and CM Perez (1968): Screening for high protein rice varieties. Cereal Sci. Today 13: 299.
47. Kaizuma, H, and J Fukui (1974): Specific characteristics and varietal differences for seed protein percentage and sulfur-containing amino acids contents in Japanese wild soybean (*Glyzine soja*), and its significance on the soybean (*G. max*) breeding program. Japan J. Breed. 24: 65–72.
48. Kaizuma, N, H Taira, H Taira and J Fukui (1974): On the varietal differences and heritabilities for seed protein percentage and sulfur-containing amino acid contents in cultivated soybeans. Japan J. Breed. 24: 81–87.
49. Kataoka, K (1969): Studies on the chemical quality of rice kernel varietal difference of protein content in rice. Bull. Fac. Agric. Tamagawa Univ. 9.
50. Kataoka, K (1971): High protein content mutants induced by ethylenimine treatment. Bull. Fac. Agric. Tamagawa Univ. 11: 29–36.
51. Kataoka, K (1974): Induction of high protein mutants in rice with ethylenimine treatment. Gamma Field Symp. 13: 49–59.
52 Kaul, AK, RD Dhar, and MS Swaminathan (1970): Microscopic and other dye-binding techniques of screening for proteins in cereals. Improving Plant Protein by Nuclear Techniques: 253–263; IAEA Vienna.

53. Kido, M, and S Yanatori (1965): Histochemical studies of protein accumulating process in rice. Proc. Crop Sci. Soc. Japan 34: 204.
54. Kido, M, and S Yanatori (1968): Studies on influence of cultural conditions of rice quality, especially amount of protein content in rice kernel. Proc. Crop. Sci. Soc. Japan 37: 32–36.
55. Knipfel, JE (1969): Comparative protein quality of *Triticale*, wheat, and rye. Cereal Chem. 46: 313–317.
56. Larsen, AL, and BE Caldwell (1968): Inheritance of certain proteins in soybean seed. Crop Sci. 8: 474–476.
57. Lott, JNA (1975): Protein body composition in *Cucurbita maxima* cotyledons as determined by energy dispersive X-ray analysis. Plant Physiol. 55: 913–916.
58. Mertz, ET, LS Bates, and OE Nelson (1964): Mutant gene that changes protein composition and increases lysine content of maize endosperm. Science 145: 279–280.
59. Mikaelsen, K (1972): An *erectoides* barley mutant with increased seed protein content. Hereditas 72: 201–204.
60. Mikaelsen, K (1973): Studies on the inheritance of the high seed protein content of an *erectoides* mutant (H-14) in barley. Nuclear Techniques for Seed Protein Improvement: 217; IAEA Vienna.
61. Misra, PS, R Jambunathan, ET Mertz, DV Glover, HM Barbosa, and KS McWhirter (1972): Endosperm protein synthesis in maize mutants with increased lysine content. Science 176: 1425–1426.
62. Mossberg, R (1970): Evaluation of protein product, IBP. Symp. Pergamon Press: 41.
63. Mossé, J (1966): Alcohol-soluble proteins of cereal grains. Fedn. Proc. 25: 1663–1669.
64. Mossé, J, J Baudet, and T Moureaux (1966): Etude sur les protéines du mais. I. Composition en acides amines des fractions azotées du grain. Ann. Physiol. Veg. 8: 321.
65. Munck, L (1972): Improvement of nutritional value in cereals. Hereditas 72: 1–128.
66. Munck, L, KE Karlsson, A Hagberg, and BO Eggum (1970): Gene for improved nutritional value in barley seed protein. Science 168: 985–987.
67. Nagl, K (1973): Mutation breeding for improved protein in *durum* wheat. Nuclear Techniques for Seed Protein Improvement: 181–192; IAEA Vienna.
68. Nelson, OE (1969): The modification by mutation of protein quality in maize. New Approaches to Breeding for Improved Plant Protein: 41–54; IAEA Vienna.
69. Nelson, OE, ET Mertz, and LS Bates (1965): Second mutant gene affecting the amino acid pattern of maize endosperm proteins. Science 150: 1469–1470.
70. Nishimura, Y, and H Kurakami (1952): Mutations in rice induced by X-rays. Japan. J. Breed. 2: 65–71.
71. Ogawa, M, K Tanaka, and Z Kasai (1975): Isolation of high phytin containing particles from rice grains using an aqueous polymer two phase system. Agr. Biol. Chem. (Tokyo) 39: 695–700.
72. Ory, RL, and KW Henningsen (1969): Enzymes associated with protein bodies isolated from ungerminated barley seeds. Plant Physiol. 44: 1488–1498.
73. Parpia, HAB (1970): Bridging the protein-calorie gap. Improving Plant Protein by Nuclear Techniques: 3–12; IAEA Vienna.
74. Sampath, S, S Patnaik, and GN Mitra (1968): The breeding of high protein rice. Curr. Sci. 9: 248.
75. Sandhu, SS, WF Keim, HF Hodges, and WE Nyquist (1974): Inheritance of protein and sulfur content in seed of chickpeas. Crop Sci. 14: 649–652.
76. Schweizer, CJ, and SK Ries (1969): Protein content of seed: Increase improves growth and yield. Science 165: 73–75.
77. Sigurbjörnsson, B (1970): The role of plant breeding in bridging the protein gap. Improving Plant Protein by Nuclear Techniques: 13–17; IAEA Vienna.
78 Singh, R, and JD Axtell (1973): High lysine mutant gene (*hl*) that improves protein quality and biological value of grain sorghum. Crop Sci. 13: 535–539.
79. Sodek, L, and CM Wilson (1970): Incorporation of leucine-14C and lysine-14C into protein in the developing endosperm of normal and *opaque*-2 corn. Arch. Biochem. Biophys. 140: 29–38.

80. Stuber, CW, VA Johnson, and JW Schmidt (1962): Grain protein content and its relationship to other plant and seed characters in the parent and progeny of a cross of *Triticum aestivum* L. Crop Sci. 2: 506–508.
81. Swaminathan, MS, A Austin, AK Kaul, and MS Naik (1969): Genetic and agronomic enrichment of the quantity and quality of proteins in cereals and pulses. New Approaches to Breeding for Improved Plant Protein: 71–86; IAEA Vienna.
82. Swaminathan, MS, MS Naik, AK Kaul, and A Austin (1970): Choice of strategy for the genetic upgrading of protein properties in cereals, millets and pulses. Improving Plant Protein by Nuclear Techniques: 165–183; IAEA Vienna.
83. Swift, JG, and MS Buttrose (1972): Freeze-etch studies of protein bodies in wheat scutellum. J. Ultrastruct. Res. 40: 378–390.
84. Swift, JG, and TP O'Brien (1972): The fine structure of the wheat scutellum before germination. Aust. J. Biol. Sci. 25: 9.
85. Taira, H, S Matushima, and A Matsuzaki (1970): Analysis of yield-determing process and its application to yield-prediction and culture improvement of lowland rice, XCII. Possibility of increasing yield and nutritional value of rice protein by nitrogen dressing. Proc. Crop. Sci. Soc. Japan 39: 33–40.
86. Taira, H, and H Taira (1965): Studies on amino acids contents in food crops (part 6) – Effect of the difference of transplanting and ripening stage on total and free amino acids in rice. Bull. Food Res. M.A.F. 19.
87. Taira, H, H Taira, and A Matsuzaki (1974): Effect of nitrogen fertilizer application on chemical composition of lowland brown rice. Proc. Crop. Sci. Soc. Japan 43: 144–150.
88. Tanaka, K, M Ogawa, and Z Kasai (1976): The rice scutellum: Studies by scanning electron microscopy and electron microprobe X-ray analysis. Cereal Chem. 53: 643–649.
89. Tanaka, S, (1968): Radiation-induced mutations in rice. Rice Breeding with Induced Mutations: 53–64; IAEA Vienna.
90. Tanaka, S (1969): Some useful mutations induced by gamma irradiation in rice. Induced Mutations in Plants: 517–527; IAEA Vienna.
91. Tanaka, S (1973): Varietal differences in protein content of rice. Nuclear Techniques for Seed Protein Improvement: 107–113; IAEA Vienna.
92. Tanaka, S. (1976): Induction of mutations in protein content of rice. Evaluation of Seed Protein Alterations by Mutation Breeding: 139–140; IAEA Vienna.
93. Tanaka, S, and S Hiraiwa (1978): Induction of high-protein mutants in rice. Seed Protein Improvement by Nuclear Techniques: 191–198; IAEA Vienna.
94. Tanaka, S, and S Murata (1969): Mutations in protein content of rice. Nogyo-Gijutsu 24: 80–81.
95. Tanaka, S and F Sekiguchi (1966): Studies on effective techniques to induce mutations in rice. Japan J. Breed. 16: 184–190.
96. Tanaka, S, and Y Takagi (1970): Protein content of rice mutants. Improving Plant Protein by Nuclear Techniques: 55–62; IAEA Vienna.
97. Tanaka, S, and S Tamura (1969): A short report on gamma-ray induced mutants having high protein content. Japan Agric. Res. Quart. 3: 1.
98. Thomas, H, and DIH Jones (1968): Electrophoretic studies of proteins in *Avena* in relation to genome homology. Nature 220: 825–826.
99. Toda, M (1970): Breeding of new rice varieties by gamma-rays. Gamma Field Symp. 18: 73–82.
100. Vaughan, JG, and A Waite (1967): Comparative electrophoretic studies of the seed proteins of certain amphidiploid species of *Brassica*. J. Exper. Bot. 18: 269–276.
101. Villegas, E, CE McDonald, and KA Gilles (1970): Variability in the lysine content of wheat, rye and *Triticale* proteins. Cereal Chem. 47: 746–757.
102. Virupaksha, TK, and LVS Sastry (1968): Studies on the protein content and amino acid composition of some varieties of grain sorghum. J. Agric. Food Chem. 16: 199–203.

103. Viuf, TB (1969): Breeding of barley varieties with high protein content with respect to quality. New Approaches to Breeding for Improved Plant Protein: 23–28; IAEA Vienna.
104. Viuf, TB (1972): Varietal differences in nitrogen content and protein quality in barley. Roy. Vet. Agric. Univ. Copenhagen, Yearbook 1972: 37–61.
105. Webb, BD, CN Bollich, CR Adair, and T Johnston (1968): Characteristics of rice varieties in the U.S. Department of Agriculture Collection. Crop. Sci. 8: 361–365.
106. Wolf, JW, GE Babcock, and AK Smith (1961): Ultracentrifugal differences in soybean protein composition. Nature 191: 1395–1396.
107. Woodworth, CM, ER Leng, and RW Jugenheimer (1952): Fifty generations of selection for protein and oil in corn. Agron. J. 44: 60–65.
108. Yamagata, H (1964): Mutations induced with radiations in heading date of rice. Gamma-Field Symp. 3.
109. Zoschke, M (1970): Effect of additional nitrogen nutrition at later growth stages on protein content and quality in barley. Improving Plant Protein by Nuclear Techniques: 345–356; IAEA Vienna.

8. Genetic Basis of Storage Protein Synthesis in Maize

M. DENIĆ

INTRODUCTION

There are several reasons to review the work covering the title of this article. First of all storage proteins, situated mainly in the endosperm, are a group of proteins without metabolic activity and therefore can be exposed to severe changes without interfering with the viability of the plant. Because of this role, and due to the fact that they are synthesized in one generation and used in the next generation the possibility is suggested that these types of proteins are under a loose type of control, rather than a stringent one, which is the case for enzymes. Furthermore in ordinary maize zeins contribute half of the total proteins, which is important for the study of gene products in in vitro systems.

Gene mutations were found to modify the amino acid pattern of seed proteins in maize [58, 73]. Thus twice as much lysine and tryptophan was found in the endosperm of the *opaque*-2 mutant of maize. Besides its practical importance for breeding for improvement of the nutritional value of maize proteins, this finding is of fundamental interest in the studies of gene action in protein synthesis. Indeed soon after this discovery many investigations were started first in breeding and then in a study of the mechanism of gene action. As a result of such work numerous papers have been published and reports presented. In addition to breeding and genetics, the work was naturally expanded to related fields such as physiology, biochemistry, etc. According to Maize Quality Protein Abstracts [53] where papers and reports on the above mentioned fields have been reviewed, it can be seen that more than 900 abstracts have been listed during the last 10 years. Although most of the papers reviewed are of importance, to cover the title of this article it is selfevident that all those references cannot be quoted. Therefore, essential and first publications on the important findings are included. Finally it should be mentioned that work in connection with this article is being continued including the use of modern techniques such as gene cloning [7].

INHERITANCE AND THE INFLUENCE OF SPECIFIC GENES ON PROTEIN CONTENT AND COMPOSITION IN MAIZE ENDOSPERM

Protein content as a quantitative trait is generally controlled by many genes. Therefore, it is difficult to describe the influence of any particular gene which is controlling the level of protein accumulation.

The first studies showed that a lower protein percentage was a partial to complete dominance character [34, 35]. Later studies, however, showed that additive genetic variance plays a predominant role in protein inheritance [3, 83] or partial dominance of lower protein content [24, 27]. Detailed experimental work [27] showed that when crossing parental forms with different protein content a complete dominance of the character 'lower protein content' over the alternative characteristic 'higher protein content' was obtained in the F_1 generation in most cases. The parental predominance of the characteristics 'lower protein content' or intermediate inheritance, was only manifested in a few cases. Thus, out of the 22 combinations tested, there were only two combinations with partial dominance of the parent with 'lower protein content' and one with intermediate protein content, respectively. There was not a single case of either complete or partial dominance of the parent with higher protein content. And more than that, no tendency for such occurrence of protein content was observed.

Very remarkable findings were obtained at Purdue University, namely, that *opaque*-2 and *floury*-2 genes change the protein composition [58, 73]. This brought scientists in many countries to study the influence of these genes on other agronomic traits. Our studies [84] with 7 inbred lines have shown that protein content was lower in all *opaque*-2 versions than in their normal counterparts (Table 1). The extent of reduced protein content was different for different *opaque*-2 versions ranging from 4.5 to 25.7 percent.

Some of the changes of protein levels are caused by certain genes which primarily alter the amounts of protein fractions. Thus, it has been found [58] that the *opaque*-2 (o_2) gene changed the protein composition and increased the lysine and tryptophan content. Further work along this line indicated a second mutant gene of maize *floury*-2 (fl_2) with elevated lysine concentration [73]. In 1971 [56] a third *opaque*-7 (o_7) mutant was discovered with a similar effect to the previous one. The genes o_2, fl_2, and o_7 are not linked and they are located in chromosomes 7, 4 and 10, respectively.

Some of the genes are known to affect starch synthesis in maize endosperm, but at the same time they also alter the amino acid patterns. Thus, it has been reported that *waxy* (*wx*) [89], *sugary*-1 (su_1) [88], *shrunken*-2 (sh_2) [52], *brittle*-1 (bt_1) and *brittle*-2 (bt_2) [11] and *dull* (*du*) [10] genes affect starch synthesis in maize endosperm. Comparative studies performed at Purdue University in Lafayette (Indiana) [64], showed that the mutant endosperm of *opaque*-2, *opaque*-7, *floury*-2, and the *brittle*-2 had higher concentrations of albumins, globulins and glutelins, but lower concen-

Table 1. Influence of opaque-2 gene on the amino acid composition[1], and protein content in endosperm of seven inbred lines.

Amino acid	Normal		Opaque 2		Opaque
	Average	Range	Average	Range	Normal
Lys	1.77	1.5 - 2.0	3.29	3.0 - 4.0	1.90
His	2.80	2.5 - 3.3	3.34	3.0 - 3.5	1.20
Arg	3.30	3.0 - 3.9	5.0	4.7 - 5.7	1.52
Asx[2]	6.59	6.2 - 7.4	9.57	8.6 -11.1	1.46
Thr	3.39	3.3 - 3.6	3.91	3.8 - 4.2	1.16
Ser	4.40	4.2 - 4.6	4.56	4.4 - 4.9	1.03
Glx3	21.54	20.3 - 22.3	18.46	16.6 - 20.6	0.85
Pzo	9.91	9.5 -10.2	9.23	8.4 - 9.6	0.93
Cys	1.27	1.1 - 1.5	1.77	1.5 - 2.1	1.40
Gly	3.03	2.7 - 3.5	4.41	3.9 - 4.9	1.46
Ala	8.11	7.8 - 8.4	6.79	6.5 - 6.9	0.84
Val	4.47	4.2 - 4.6	4.97	4.5 - 5.2	1.11
Met	1.80	1.5 - 2.4	1.67	1.4 - 2.3	0.93
Ile	3.54	3.4 - 3.6	3.59	3.5 - 3.7	1.01
Leu	15.23	14.3 -16.2	11.24	9.4 -10.1	0.74
Tyr	3.83	3.6 - 4.2	3.59	3.3 - 3.8	0.93
Phe	5.19	4.7 - 5.7	4.73	4.4 - 5.1	0.91
Protein content (%)	12.3	11.1 -12.8	10.6	9.5 -11.2	0.87

1) Amino acids are expressed in % of protein

2) Includes contents of aspartic acid and asparagine

3) Includes contents of glutamic acid and glutamine

trations of prolamins than their normal counterparts. The double mutant *opaque*-2, *brittle*-2 even enhanced these differences. Further studies [36, 64] have shown that other starch-modifying genes provoked a distribution of endosperm proteins resembling that found in the *floury* types of maize with high lysine content (o_2, o_7, fl_2). The same data [36, 64] showed that protein content (as % of dry matter) was reduced in endosperms of near isogenic Oh 43 sublines in the presence of o_2, wx, fl_2

du and o_7 in a W 22 genetic background. The genes bt_1, bt_2, sh_1, sh_2, however, increased the concentration of proteins. The rest of the genes (fl_2, su_2, *ae*) considerably changed the level of protein content in the endosperm of inbred Oh 43. When fl_2, su_2, *wx*, *ae*, *du* and bt_1 were combined with the o_2 gene in an inbred background (Oh 43) the concentration of protein was reduced, whereas in the case of combinations with bt_2, su_2, su_1 and sh_2 the amount of protein was increased.

In the experiments with hybrid backgrounds [36] the concentration of protein in the o_2 single-mutant and all the o_2 double-mutant combinations was significantly below the mean of the comparable single non-*opaque* hybrids. *Opaque*-2 single-mutant and most of the double-mutant combination hybrids showed decreased protein per endosperm as well. *Opaque*-2 double-mutant combinations in hybrids did not show significantly lower concentrations of endosperm protein as compared to the o_2 single-mutant hybrids. Thus the estimates obtained in the well-known long-term experiment which started in 1896 and is still being carried on [24] suggest that the number of effective genes controlling protein percentage is 122. This number is larger than the number of genes controlling oil percentage (54 genes).

On the basis of literature data [39, 85], it seems that linkage has no effect on protein content before 10 generations. Analysis of diallel crosses showed that 7% of the entry sum of squares were accounted for by heterosis for protein percentage [26]. The same studies [24] showed that after 76 generations of selection for high protein content (percent) in maize genetic variation for this trait has not been exhausted even though progress of 20% was made in the direction of high protein content. The same data indicated that gene frequencies for favourable alleles in the original population ('Burr's White') were below 0.37 for protein percentage. Due to the low quality of maize proteins (see Chapter by Eggum in the present book) much attention is paid to the limiting essential amino acids, primarily to the lysine and tryptophan content. The comparative studies of Glover *et al.* [36] show that all mutants exhibited higher concentrations of lysine and tryptophan in comparison with their normal counterparts. These studies showed that when combined with the *opaque*-2 gene (double mutants) all genes with the exception of fl_2 increased the concentration of lysine and tryptophan more than the o_2 gene alone.

Using Landry and Moureaux's [49] modification of the classical Osborne procedure [74] the same investigators [36] found a reduced concentration of the zein fraction in all mutant endosperms in comparison with the normal counterpart. The concentration of the glutelin fraction, however, was increased in all mutant endosperms. In this respect the biggest difference between mutant endosperm and normal counterpart was found in the case of the o_2 gene, and the smallest with the *wx* gene. The concentration of zein was further decreased when the mutant genes were combined with the o_2 gene. This effect was most pronounced in the mutants sh_2o_2, bt_1o_2, bt_2o_2 and su_1o_2.

Besides this group of genes with synergistic effects there are genes which modify or suppress the effect of the o_2 gene. The first publication on the modified *opaque*-2

Table 2. Influence of modifying gene(s) on protein and lysine content in maize endosperm.

Endosperm type	Protein content		Lysine in dry matter		Lysine in protein	
	%	modified/Opaque-2	%	modified/Opaque-2	%	modified/Opaque-2
Normal	10.8	1.17	0.20	0.63	1.84	0.53
Near normal	10.5	1.13	0.26	0.81	2.50	0.71
Normal cca 75%	10.4	1.12	0.27	0.84	2.66	0.76
Normal cca 50%	9.4	1.01	0.28	0.88	3.04	0.87
Normal cca 75%	9.4	1.01	0.29	0.91	3.11	0.89
Traces of modification	9.5	1.03	0.32	1.00	3.43	0.98
O_2 standard	9.3	1.00	0.32	1.00	3.50	1.00

endosperm appeared in 1969 [75]. Different data were obtained with respect to the influence of modifying genes on lysine content. The first report [75] stated that the lysine content of modified kernels did not differ from standard *opaque*-2 kernels, whereas some other authors found modified biochemical traits in modified *opaque*-2 kernels [21, 29, 31, 84, 91]. The data presented in Table 2 suggest that with the increase in the degree of normalization total protein content is also increased. It is interesting from these results that the total average protein content undergoes hardly any change up to a normalization of 50 percent of the endosperm. At this level of normalization, however, the lysine content in dry matter and protein fell by 12 and 13 percent, respectively. On the basis of these results and those obtained at CIMMYT [91] the following can be concluded:

1. The total endosperm protein content of modified o_2 types of endosperm increases, with an increase in degree of normalization.
2. The content of lysine and tryptophan of modified o_2 endosperm, as a rule, decreases with increase in degree of normalization.
3. The vitreous fraction of normalized *opaque*-2 endosperm differs from the soft fraction in having more protein and less tryptophan, lysine and ribonuclease activity.
4. Decrease in lysine and tryptophan found in modified o_2 endosperm as compared to standard *opaque*-2 types seems to be due to the increased zein fraction. When

comparing the standard o_2 endosperm to modified o_2 endosperm the following changes were obeserved: as a rule, besides a reduction in tryptophan and lysine content, the content of histidine, arginine, aspartic acid and glycine slightly decreased, and that of glutamic acid, leucine, alanine, tyrosine and phenylalanine slightly increased. These findings would suggest that the mode of action of the modifier gene (s) is opposite to that of the *opaque*-2 gene.

5. The lysine content of modified o_2 types is in general satisfactory up to the normalization of about 50 per cent of the endosperm.
6. Increased normalization due to modifying gene (s) shows no regular association with increased test weight of the kernels.
7. Embryo size seems to be influenced by a modifying gene (s).

Great variation was observed with respect to the extent and type of normalization in kernel appearance [29, 91]. This would suggest that the mode of inheritance of gene modifiers is complex and under polygenic control. An inheritance pattern of kernel vitreousness controlled by a modifying gene complex was studied by Vasal [91]. The data obtained indicated partial dominance in the expression of kernel vitreousness. The estimates of different components of genetic variance showed that the additive variance seems to be more important than dominance variance. The same studies showed that in reciprocal crosses, when modified parents were used as female parents, ears with a modified and segregating type were obtained more frequently. The difference between reciprocal crosses suggests maternal influence of the expression of kernel vitreousness. Using a solution of nitrosomethyl urea, a mutant was obtained with corneous tissue mainly located on both sides of the endosperm [55]. The same studies showed that this trait is controlled by a dominant factor not linked with the *opaque*-2 gene. Besides these two types of gene interactions, the interaction of complementary cfl_2 with fl_2 results in a slight decrease of seed weight and increases of lysine and methionine contents in the seed proteins [76].

BIOCHEMICAL PATHWAYS OF GENE ACTION ON PROTEIN SYNTHESIS IN MAIZE ENDOSPERM

Studies in vivo

The main component of maize endosperm proteins is the alcohol-soluble fraction zein. This fraction constitutes about 60 percent of the total protein in the endosperm of normal maize. The work conducted at Purdue University [90] showed that zein may account for as much as 10 percent of the total protein in the mature embryo of normal maize. The synthesis of the zein component in the embryo was controlled by the *opaque*-2 and *floury*-2 genes. Polypeptide patterns on SDS-PAGE[1] were similar to that of the endosperm, but there were differences in the amino acid composition between the zein components from these two tissues. Because of this the zein

[1] Sodium dodecyl sulfate-polyacrylamide gel electrophoresis.

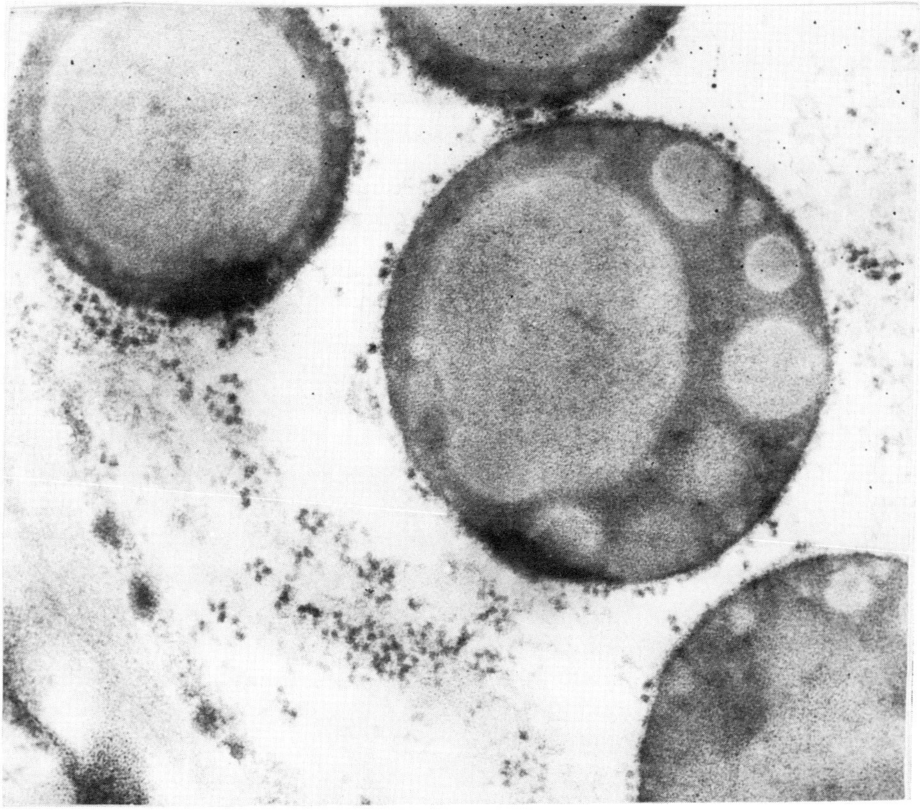

Figure 1 General picture of protein granules with associated ribosomes and polysomes on the surface. Protein granules are from normal inbred V-312; age of endosperm 26 days after pollination; magnification 72,000 fold.

fraction became a kind of a model system for the study of the synthesis of storage proteins in cereals.

The great discovery that the *opaque*-2 gene increases lysine content in maize [58] was later extended by the finding that this gene is responsible for a decreased amount of alcohol-soluble proteins and for an increased quantity of non-zein proteins (particularly glutelins) in mature endosperm [43, 67]. It was also found that the total amount of free amino acids [13, 68, 86, 87] and ribonuclease activity [14, 17, 94] was higher in the *opaque*-2 mutant than in the normal genotype.

Following endosperm development differences between normal and mutant maize were observed with respect to endosperm weight, nitrogen accumulation and amounts of different Osborne-protein fractions. Our studies [86] showed that the fresh weight of the *opaque*-2 endosperm was higher than in the normal genotype up to 35 days after pollination and then became lower. However, the weight of dry

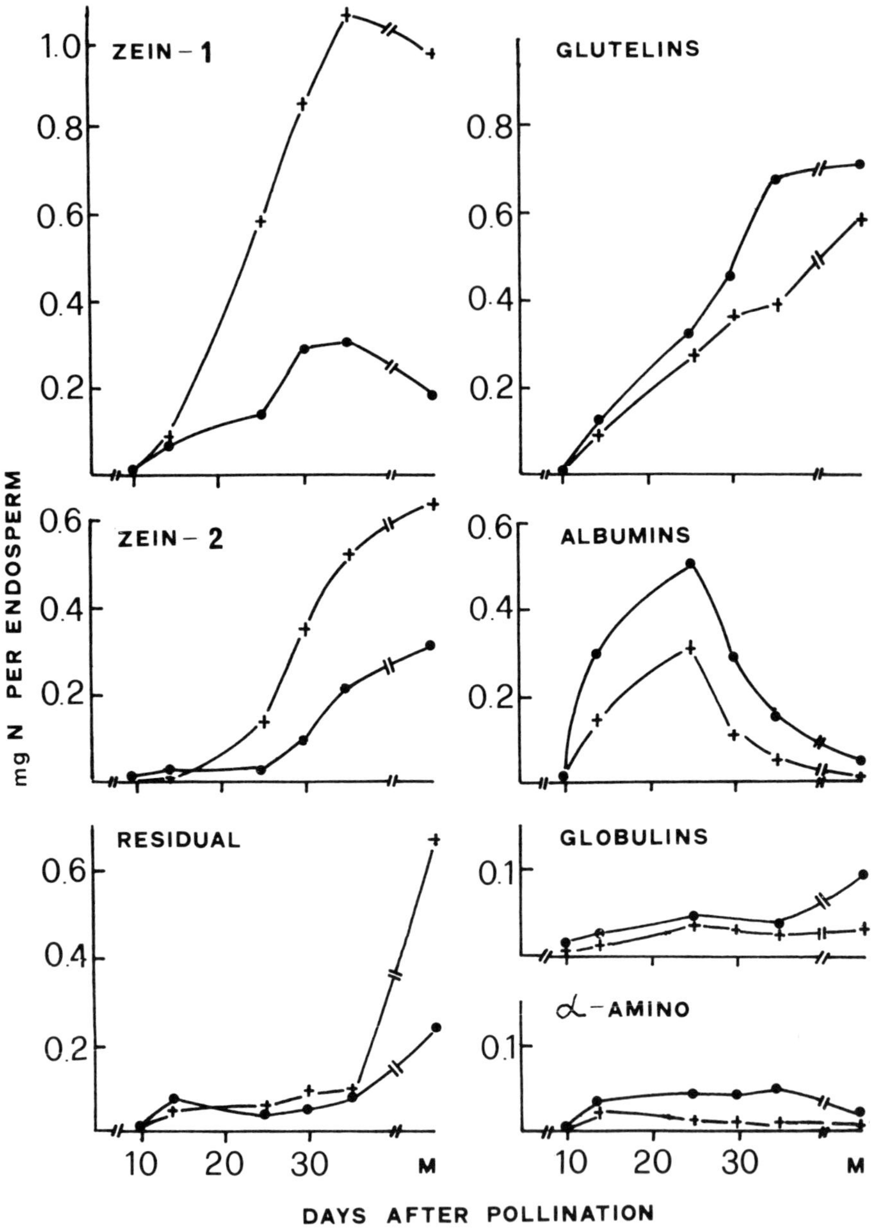

Figure 2 Amount of different protein fractions during endosperm development of *opaque*-2 mutant (o) and normal maize (+).

endosperm was higher in normal genotypes from 25 days after pollination to the mature stage. These differences between the two genotypes are thus due to the differences in water content in the endosperm. The much higher water content in *opaque*-2 endosperm than in the normal genotype is the main reason for the greater weight of fresh endosperm. The data of Dalby [13] and our data [86] showed that the amount of total nitrogen was very similar in both genotypes until day 22. After this period there was a much greater accumulation of nitrogen in the normal genotype. Investigations on the contribution of each protein fraction to the total nitrogen in the endosperm, with the exception of the butanol extract and residual nitrogen, showed apparent differences between normal and *opaque*-2 maize (Figure 2 and references [68, 86]). Initially, water-soluble proteins-albumins [86] or saline extracts-globulins, together with water-soluble proteins [68] represented the largest nitrogenous fractions in both genotypes. These fractions and free amino acids were reduced during endosperm development. The extent of reduction was higher in normal endosperm thus leading to a higher amount of these fractions in mature *opaque*-2 endosperm. The ethanol extract (prolamins, zeins) increases in *opaque*-2 maize up to 30 days and then remains at about 15% of total nitrogen to the mature stage. In the case of the normal genotype a much more rapid accumulation and prolonged synthesis of the zein-1 fraction was observed. (Zein-1 fraction in Figure 2 and reference [86] is equal to the ethanol fraction in Dalby's papers).

From the data in Figure 2 it is evident that the amounts of zein-2 (ethanol soluble protein in the presence of 2-mercaptoethanol) are reduced in the endosperm of the mutant, whereas the amounts of the sodium hydroxide-soluble fraction (glutelins) are reduced. It is interesting to compare the same differences in two major ethanol-soluble, alkali-soluble fractions, which resulted from the application of different methods. Introduction of 2-mercaptoethanol in ethanol after the separation of zein-1 leads to the separation of the new protein fraction which was designated by Sodek and Wilson [87] as zein-2. The rate of synthesis of this fraction was almost the same in both genotypes to day 22. After this stage there was greater accumulation of this fraction in the normal genotype. The amount of zein-2 at the mature stage was 0.31 and 0.63 mg per endosperm of mutant and normal genotype, respectively.

Since Dalby used a procedure without introduction of 2-mercaptoethanol, the glutelin fraction contained some zein-2 fraction. This author indicated that no differences occurred between normal and mutant maize in the amounts of alkali-soluble proteins per endosperm [13]. However, our results showed that the amount of the glutelin fraction is higher in the mutant even at 14 days after pollination. The final concentration of this fraction was 0.72 and 0.58 mg per mature endosperm of mutant and normal maize, respectively.

Taking into consideration the better separation of these two major protein fractions (zeins and glutelins) after application of a modified Osborne procedure, the differences between these two genotypes are more pronounced. Thus, the rate of accumulation of ethanol-soluble proteins (zein-1 and zein-2) is even higher in the endosperm of the normal genotype than was found earlier. The amount of ethanol-soluble proteins at maturity was 0.49 and 1.59 mg per endosperm of mutant and normal genotype, respectively, and this fraction formed 30.6 and 55.2% of the total protein content of the endosperm from mutant and normal genotype, respectively.

Our data [86] suggested a kind of negative correlation between the amount of total nitrogen and the amount of free amino acids. This tendency was valid for both genotypes. In addition, the larger amounts of total nitrogen in the normal genotype were accompanied by a smaller amount of free amino acids and in the mutant endosperm *vice versa*. When the relative amounts (mol per cent) of free amino acids in developing endosperm are considered, differences between the genotypes were not marked as in the case of absolute amounts (mol per endosperm). The proportions of free amino acids were changed mainly due to the increased amounts of ALA, ASP, ASN, GLU, GLN, in the mutant endosperm. On the basis of these data it seems that most differences between the genotypes at the level of free amino acids are a secondary effect of *opaque*-2 gene action.

Studies in vitro

It has been suggested [32, 33] that zeins are accumulated in protein granules (protein bodies, proteoplasts, proteinoplasts). In 1964 the existence of two independent systems for the incorporation of amino acids into storage and soluble proteins was reported [65]. In 1971 we showed that protein granules (proteinoplasts) isolated from developing maize endosperm incorporated labelled amino acids into proteins [20]. The same studies showed that the amino acid incorporating activity in the cell-free system was increased with the time of endosperm development. About 20 days after pollination a much higher incorporation of leucine was found in comparison with lysine incorporation. Leucine incorporating activity *in vitro* was in agreement with the rate of zein accumulation *in vivo*. A positive correlation was also found between the amino acid incorporating activity and the size of protein granules [23].

Our previous results reported in 1970, which were obtained in a cell-free protein synthesizing system with a microsomal fraction showed a relatively higher incorporation of lysine in comparison with the system with protein granules [17, 23]. These findings suggested that protein granules are able to synthesize the proteins that are the sites of synthesis of zein-like proteins (leucine-rich and lysine-poor proteins). Later studies performed in 1975 at Purdue [50] and in Milan [93] showed that protein granules are surrounded by polysomes. In further work with *in vitro* translation of membrane polyribosomal [51] and protein body-polyribosomal [8] m-RNAs two proteins differing in size (but quite similar to *in vivo* major zein group proteins) were obtained. Some discrepancies that appeared in different papers with respect to the size and number of m-RNAs coding for zeins and zeins polypeptides were reviewed and discussed by Mossé and Landry [66]. Therefore, only the results

Table 3. Influence of the microsomes from maize endosperm on the incorporation of lysine and leucine into proteins[a].

Days after pollination	Genotype	Lysine (DPM per assay)[b]	Leucine	Lysine/Leucine	Mutant/Normal
15	Normal.	393	971	0.41	1.57
	Mutant	550	854	0.64	
20	Normal	553	1937	0.29	1.70
	Mutant	697	1437	0.49	
25	Normal	587	2435	0.24	2.12
	Mutant	738	1442	0.51	
30	Normal	460	2699	0.17	1.66
	Mutant	568	2018	0.28	
Average	Normal	500	2011	0.25	1.79
	Mutant	638	1437	0.44	

a) S-100 fraction from endosperm of normal W64A inbred line.

b) Mean of two experiments

obtained on the biochemical pathways of gene action in storage protein synthesis are reviewed here. It has been suggested [69] that *opaque*-2 is a regulatory gene and not a structural gene for zein. Thus besides its practical importance in breeding work for improvement of protein quality the *opaque*-2 gene is a good model system for the study of the mechanism of gene action in storage protein synthesis.

The first report along this line appeared in 1968 [16]. When the amino acids bound to t-RNA *in situ* in endosperms were compared, a 19% higher lysine content was found in the deacylated t-RNA from *opaque*-2 mutant than in the normal genotype.

To study the possible influence of the microsomal fraction on the changed amino acid composition of the *opaque*-2 mutant endosperm, an investigation was made of amino acid incorporation into protein in the presence of microsomes from both normal and mutant endosperms [17]. Since the most remarkable changes were found in the contents of lysine and leucine [58], these two amino acids were chosen for studies of amino acid incorporation. The data presented in Table 3 show a higher incorporation of lysine in the presence of the polysomes from mutant endosperm in comparison with the polysomes from a normal genotype. The leucine incorporation was opposite to that with lysine. The higher lysine/leucine incorporation ratios with

Table 4. Amino acid incorporation with microsomes and protein granules from opaque-2 and normal maize endosperm[a].

Cell component	Genotype	Lysine (DPM per assay)[b)]	Leucine	Lysine/Leucine	Mutant/Normal
Microsomes	Normal	495	1827	0.27	1.56
	Mutant	655	1556	0.42	
Protein granules	Normal	328	2158	0.15	4.80
	Mutant	373	446	0.72	

a) S-100 fraction from normal endosperm

b) Average of two experiments

mutant endosperm polysomes were compared with the corresponding ratios of the normal systems; higher genotype (mutant/normal) ratios were obtained. An experiment on protein synthesis *in vitro* with protein granules and microsomes from normal and *opaque*-2 endosperm was performed [20]. The results obtained show a relatively higher incorporation of lysine in the presence of protein granules from the mutant endosperm than with the protein granules from a normal genotype. When the lysine/leucine ratios of incorporations were compared within the same cell component and between mutant and normal genotypes, a higher genotype ratio was found for protein granules than for microsomes (Table 4). Thus, if the lysine/leucine ratio of incorporation into protein is taken as the criterion, then the differences between protein granules from mutant and normal maize are higher than the differences between the microsomes.

Comparative studies on the size and incorporating activity of protein granules indicated that granules from both *opaque*-2 and normal genotypes showed differences in the time of formation, size and amino acid incorporating activity (Table 5, reference [23]). In the *opaque*-2 mutant formation starts later, the size is smaller and protein synthetic activity is lower than with the normal inbred line. The same data show that regarding endosperm development the size and protein synthesizing activity of protein granules are considerably lower in the *opaque*-2 than in the normal inbred line. It should be pointed out that the size of protein granules in mature endosperm remained smaller in *opaque*-2 than in normal endosperm [95].

Experiments performed at Milan [93] showed that qualitatively the profiles of the protein body and total polysomes are unchanged in *opaque*-2 mutant in comparison with the normal genotype. Quantitatively, however, at day 21 protein-body ribosome represented 50 and 40% of the total in normal and *opaque*-2 mutants,

Table 5. Size and in vitro activity[a] of protein granules during endosperm development of opaque-2 and normal maize.

Days after pollination	Size (nm) Normal	Size (nm) Mutant	Activity Normal	Activity Mutant
16	395	absent	170	120
20	535	205	290	125
26	760	235	705	335

a) Incorporation of ^{14}C-1-leucine in DPM

respectively, while at day 35 they reached approximately 85% in the normal and 20% in the mutant.

Similar work conducted later at Purdue [42] showed that ethanol-soluble protein was synthesized primarily by the membrane-bound polysomes of both normal and *opaque*-2 endosperm. It was found that approximately 33 and 50% of the label incorporated into TCA-insoluble protein was soluble in hot ethanol in the probes with membrane-bound polysomes from mutant and normal genotypes, respectively. Both *opaque*-2 and normal genotypes had similar patterns of free polysomes. However, the very large polysome size classes in the group of membrane-bound polysomes were almost completely absent in *opaque*-2 kernels. Extended work on the same group [43] showed that r-RNA concentration in the mutant at day 16 after pollination was reduced by 30% from that of the normal, and the levels increased more slowly and plateauned several days earlier in the mutant. When total RNA extracts were analyzed, a distinct 16 S peak was detected in both normal and mutant genotypes [44]. When these two m-RNAs were translated in the wheat-germ system, and the products analyzed on SDS-PAGE, both the 21 and 22 components were synthesized. These results suggested that *opaque*-2 mutation is acting at the level of translation. When purified zein m-RNA, prepared from normal and mutant genotypes, was incubated with wheat S-100 fraction and free- or membrane-bound ribosomes from normal or mutant kernels, comparable levels of ^{14}C-leucine were obtained. Free ribosomes from *opaque*-2 maize incorporated nearly equal radioactivity with either source of zein m-RNA. The mutant membrane-bound ribosomes, however, significantly reduced the incorporation of ^{14}C-leucine in the presence of m-RNA of either origin. The same studies showed that reduction of elongation rate or premature termination of nascent chains in the mutant did not occur [43]. The studies on the initiation rate were not successful and thus it is not clear where the primary point of o_2 gene action is.

Our studies in cell-free RNA synthesizing systems showed a higher incorporation of purine than pyrimidine bases in the probes with chromatin from *opaque*-2 mutants than in the probe with chromatin from normal endosperm [48]. The same studies showed that non-histone proteins from normal and mutant endosperm differ mainly quantitatively. Characteristic differences between the two genotypes were found in the alcohol-soluble proteins synthesized on chromatin as a template. These proteins corresponded to native zein isolated from normal and mutant endosperm.

Besides the template characteristics of chromatin from maize endosperm [47] the studies with a DNA-dependent protein synthesizing system (coupled transcription and translation) showed that in the presence of chromatin from the mutant endosperm 25 days after pollination there was a relatively higher incorporation of lysine as compared to the probe with chromatin from a normal genotype [46]. It should be pointed out, however, that in this coupled system much higher incorporation of lysine than of leucine was found irrespective of the origin of chromatin. The reason for this is not explained.

Besides the large amount of additional work done the statement of Nelson [72] that 'In spite of reasonably complete data covering the net effect of genes such as *opaque*-2, *opaque*-7 and *floury*-2 on endosperm proteins we do not yet understand in any instance the primary biochemical lesion responsible for the effect', is still valid after a period of 6 years.

GENETIC VARIABILITY OF MAIZE SEED PROTEINS

Taking into consideration that maize is an outcross pollinating plant and that there is a principal difference in the propagation of this crop through hybrid production, where a vigour effect is explored, genetic variability of maize seed proteins will be considered separately in inbred lines, varieties and hybrids.

Genetic Variability of Seed Proteins in Inbreds

In studies of the possibilities of improving maize seed proteins genetic variability, either induced or spontaneous, is of great importance. For this purpose a group of 45 genetically different inbred lines and 76 mutant lines (in the X_5 generation after mutagenic treatment of maize line V-312) was studied with respect to the content and composition of endosperm proteins [18]. Sixteen amino acids were determined by a semi-nondestructive procedure [15] involving excision for analysis of a part of the endosperm without interfering with the viability of the kernels. The kernels were sown after analysis for a test of heritability by intergeneration regression analysis.

Since variations in lysine content are of major interest in maize breeding the lysine values are presented diagrammatically as a function of protein content (Figure 3). Lysine in protein exhibits a negative regression similar to that for wheat [19]. The figure demonstrates clearly that the total protein content exhibits the large varia-

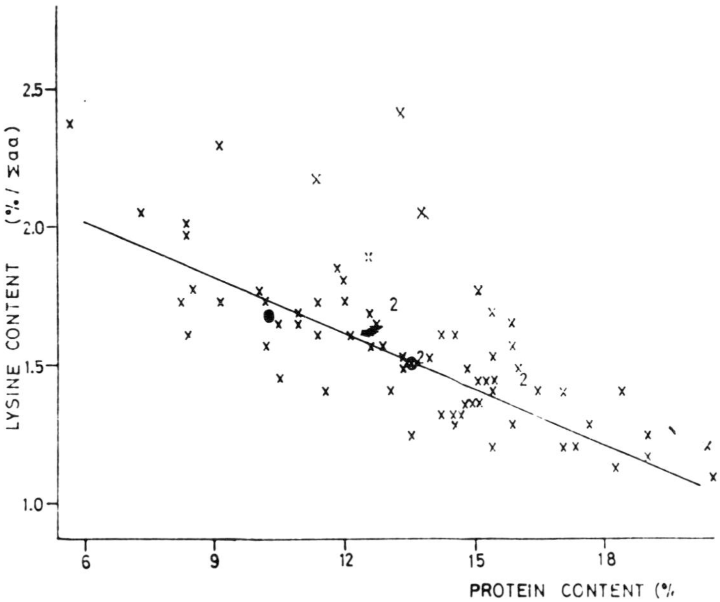

Figure 3 Regression of the relative abundance of lysine (% lys/εaa) on protein content. The mean value of the group is encircled; '2' means two points with identical values.

tion, mainly towards a higher content of the bulk protein. A few lines show lysine contents about 50 percent higher than that of the mother V-312 line. When lysine production is considered, *i.e* lysine in the sample (% lys/w), a positive regression is observed (Figure 4) which means that the induced variation in protein content concerns lysine-rich as well as lysine-poor protein fractions, although the relative abundance of the latter increases slightly with increased protein content. A similar diagram was obtained for genetically different inbred lines but the points in this case were more scattered [28]. Visualizing now the total variation of the contents of the sixteen amino acids (% / εaa) in a sixteen dimensional space, we found that nine dimensions are needed to explain 96 percent of the total variation in the mutant group, in contrast to only six dimensions in the case of the group of genetically different lines. The total variation in the case of mutant lines was considerably smaller (variance = 8.0) than with the different inbred lines (variance = 16.2).

The inter-year regression found in the mutation material corresponds to a heritability of about 65% of the total variation. Since all amino acids are positively correlated with the protein content in the mutant as well as with the inbred line group any simple correlation coefficient of two amino acids (in % / w) is positive and significant. For certain amino acids (*e.g.* lysine, histidine and glycine, especially in the mutant group) the relative abundance in the protein (% / εaa) is negatively correlated with the protein content. It is observed that in the mutant as well as in the

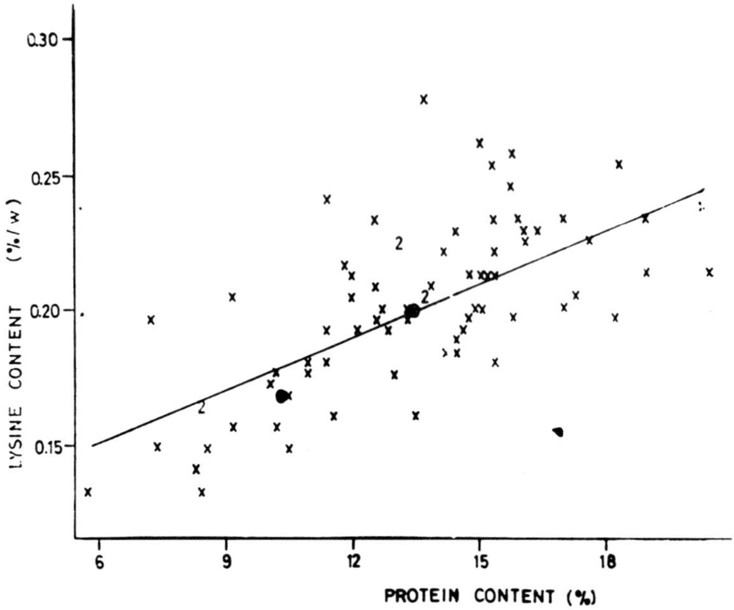

Figure 4 Regression of the production of lysine (% lys/weight of sample) on protein content. The mean value of the group is encircled; '2' means two points with identical values.

line group the lysine content of the protein is positively correlated with histidine, arginine and asx, and negatively correlated with alanine and leucine and probably also with glx which are characteristic of zein. Data in the partial correlation coefficients (excluding the influence of all other amino acids) showed that in the mutant group lysine is correlated positively with histidine, arginine and isoleucine, negatively with leucine in the inbred line group, positively with arginine and threonine, negatively with tyrosine and possibly also with proline and leucine. The mode of intercorrelation of the three basic amino acids is, however, in principle the same in the two groups, where $r_{lys, his}$ and $r_{lys, arg}$ are weakened by the formation of $r_{lys, his, arg, his}$, *i.e* by subtraction of the influence of arginine and histidine, respectively, from the correlations. The fact that the weakening of $r_{lys, his, arg}$ is stronger shows that lysine and arginine have a covariation which is partly independent of histidine, whereas the whole of the covariation of lysine and histidine is dependent on the variation of arginine. Similar results were obtained with isoleucine and leucine which are positively correlated.

Variability of Grain Proteins in Maize Varieties and Hybrids

Variability in protein content and composition of different varieties is of great importance for breeding work concerning maize in these traits. The data obtained

Table 6. Protein, lysine and tryptophan content[1] of some local varieties in three European countries.

Maize types	Protein (%)	Lysine (g/16g N)	Tryptophane (g/16g N)
1. Classified maize varieties in Yugoslavia /38,39,78-82/			
Flints:			
Montenegrin flints	9.4-16.1	1.23-2.92	0.32-0.49
Kosovien flints	11.7-14.7	1.60-2.81	0.27-0.50
Macedonien flints	10.9-14.7	0.74-2.69	0.35-0.52
Mediterranean flints	12.4-14.9	0.79-2.35	0.31-0.39
Eight-rowed type N.E. America	10.7-16.1	0.84-3.52	0.27-0.40
Derived	10.9-16.9	0.74-3.73	0.23-0.47
Related types	10.8-16.3	0.67-2.81	0.23-0.36
Semiflints-semidents:			
Bosnian semiflint	8.4-14.7	1.27-3.13	0.31-0.54
Morava semiflint	13.1-16.1	0.51-1.31	0.33-0.41
Flinty dents	9.4-15.0	1.25-2.89	0.33-0.55
Denty flints	9.6-14.5	0.60-3.80	0.22-0.49
Dent types:			
Similar to USA corn belt dents	9.7-14.7	0.32-2.46	0.32-0.49
Similar to southern USA area dents	9.6-14.4	0.97-2.19	0.27-0.36
Serbian dents	10.4-14.2	1.07-2.39	0.22-0.30
Derived dents	10.6-14.8	0.82-2.79	0.25-0.46
Eight-rowed soft dents	11.5-15.2	0.56-1.96	0.33-0.41
	Protein (%)	Lysine (%)	Tryptophane(%)
2. Romanian flints /5/	10.3"16.5	0.13-0.33	0.12-0.18
Romanian dents /5/	9.7-15.3	0.16-0.34	0.13-0.17
3. Italian flints /5/	10.1-14.9	0.18-0.42	0.12-0.23
Italian dents /5/	10.5-14.8	0.08-0.45	0.12-0.22

1) Data for Yugoslav varieties are refered to endosperm, and for Romanian and Italian to whole grains

on Yugoslav local varieties [81] showed that out of 1,187 varieties there were 453 flints, 470 dents and 264 semiflints (semidents). The results for protein, lysine and tryptophan content are shown in Table 6. In Yugoslav local varieties protein content in the endosperm of flint types ranged from 9.4 to 16.9%, in semiflints from

8.4 to 16.1%, and in dents from 9.6 to 15.2%. In Rumanian and Italian flints protein contents in whole kernel varied from 12.7 to 15.3 and 10 to 12.2%, respectively; whereas in the case of dents Rumanian varieties showed 12.1–14.2 and Italian 9.7–11.4%.

Regarding the lysine and tryptophan contents in the endosperm the range in Yugoslav flint types was between 0.23 and 0.52 g/16 g N respectively; in semiflints lysine was from 0.51 to 3.80 and tryptophan from 0.22 to 0.55 g/16 g N, and in dents lysine content varied from 0.32 to 2.79 and tryptophan from 0.22 to 0.49 g/16 g N. Both flint and dent types from Rumania and Italy showed a similar range of variation.

According to Bressani *et al.* [6] among ten varieties of maize planted in Mexico, Guatemala, Nicaragua and Panama the average values were: nitrogen, 1.55%, lysine 172 mg and tryptophan 26 mg per g of nitrogen. Tryptophan and lysine showed significant varietal differences in each of the four localities.

The data on the variation of protein content in grains of some American and Yugoslav hybrids realised and grown in two different periods are shown in Table 7. The largest variation in the 1960-62 group was found in three-way cross hybrids and the smallest in single cross hybrids. However, the highest mean value for protein content was found in single cross types. The 1973–75 group of hybrids showed a smaller range of variation, but mean values for all types of hybrids were higher in comparison with corresponding hybrids of the 1960–62 group. With respect to lysine and tryptophan content in endosperm the range was from 1.41 to 2.74 and from 0.35 to 1.15 g/16 g N respectively, whereas in the case of whole kernels the range of lysine content was from 1.90 to 3.54, and for tryptophan from 0.48 to 1.17 g/16 g N. This range of variation of lysine and tryptophan was enlarged because

Table 7. Protein content (%) in kernels in the groups of some American and Yugoslav hybrids grown in two different periods.

Types of hybrids	Period 1960-1962[1)]			Period 1973-1975[2)]		
	Range	Mean	n[3)]	Range	Mean	n[3)]
Single cross	9.8-13.1	11.7	9	10.3-13.6	12.1	11
Three-way cross	9.0-14.8	11.3	11	10.4-12.4	11.6	6
Double cross	9.3-14.1	11.2	9	11.2-12.9	12.1	8

1) Ref. 65

2) Calculated values from ref. 62-64

3) Number of tested hybrids

Table 8. Protein, lysine and tryptophane content of some hybrids grown in Yugoslavia /61/.

Hybrid	Endosperm			Kernel		
	Protein (%)	Lysine (g/16g N)	Trypto-phane (g/16g N)	Protein %	Lysine (g/16g N)	Trypto-phane (g/16g N)
ZP 448	11.9	1.43	0.42	12.6	2.22	0.47
SC-546A	11.4	1.49	0.43	12.1	1.39	0.50
SC-58C	11.3	1.51	0.44	11.9	2.11	0.51
SC-6	11.5	1.47	0.35	12.1	2.07	0.50
SC-71C	11.6	1.47	0.43	12.4	1.93	0.48
SC-4	12.1	1.41	0.41	12.6	1.90	0.48
SC-74 o_2	7.7	2.74	1.15	9.6	3.54	1.17

of the presence of one *opaque*-2 hybrid (Table 8). The work performed in South Africa [97] in the last decade showed a range of protein content from 8.6 to 10.0%.

BREEDING FOR IMPROVED PROTEIN QUALITY IN MAIZE

This topic is a broad one and useful results are just arriving. However, some problems that have appeared in this work suggest that the difficulties will be overcome, some of the results achieved and essential references are listed.

When all aspects of the world's food situation are considered the following should be underlined in the context of this paper: a) the possibilities of improving the quality of the present food output, most notably characterized by a shortage of protein [30], b) 70% of edible proteins are contributed by plants, c) more than 2/3 of total plant protein production is cereal protein [40], d) about 1/4 of the world's cereal protein is contributed by maize, namely, about 50% in North and Latin America, about 30% in Africa and 20% in Europe [40], e) the maize kernel is a very valuable chemical entity with 80% carbohydrates, 10% proteins and the remaining 10% oil, fiber and minerals [37]; about 50–60% of total grain protein consists of low quality prolamins or zeins, and f) the discovery of the *opaque*-2 mutant with a reduced amount of zein and elevated quantities of other protein fractions and hence increased amounts of lysine and tryptophan [58]. Taking into consideration all these facts the *opaque*-2 mutant was not only a fortunate discovery but also a very logical source for genetic improvement of the nutritional value of poor maize proteins. Great enthusiasm in many breeding programmes was quite evident. The results obtained have been presented at important scientific symposia, some of them especially dedicated to this topic, e.g. High lysine corn conference 1966, Lafayette U.S.A.; CIMMYT – Purdue Symp. on protein quality maize, 1972, El Batán, Mexico. However, soon after this enthusiasm many disappointments have appeared because of the low performance of *opaque*-2 hybrids and varieties. The problems that are associated with breeding for improved quality of maize protein have been discussed many times by various scientists [1, 2, 12, 38, 70, 71]. Recently these problems were discussed and summarized by Vasal *et al.* [92] as follows: (1) reduced grain yield; (2) unacceptable kernel appearance; (3) greater vulnerability to ear rot; (4) more infestation by weevils during storage; and (5) slower drying of grain following physiological maturity. It was also stated that the degree and magnitude of these problems vary considerably in different parts of the world. In order to overcome or to reduce the extent of some of the above mentioned problems the following approaches were suggested:

1. Population improvement through the introduction of *opaque*-2 and *floury*-2 alleles into two maize synthetics (Iowa super stiff stolk, SSS and Synthetic Disease-Oil, Do) in order to increase genetic diversity of the starting material [25].
2. Introduction of modifying genes in the genotypes with the *opaque*-2 allele [25, 91, 92].
3. Introduction of the su_2 gene in genetic backgrounds with the *opaque*-2 allele [22, 92].
4. Improving protein quality in maize without the use of specific mutants [96], where the endosperm/embryo ratio or the aleurone layer is changed.

Following these suggestions, with necessary breeding procedures, including those especially designed to increase protein quality with the *opaque*-2 gene in floury maize in both American [91] and other hybrids, successful results were reported for *opaque*-2 hybrids and varieties with comparable yields [25, 38, 55, 92]. As would be expected this work is still in progress.

REFERENCES

1. Alexander, DE (1966): Problems associated with breeding *opaque*-2 corns, and some proposed solutions. In: Proc. High Lysine Corn Conf.: 143; Corn Res. Found., Washington.
2. Alexander, DE (1975): Breeding for protein quality in maize: Current issues and problems. In: High Quality Protein Maize; Proc. Symp. El Batán, Mexico: 83; Dowden, Hutchinson and Ross, Stroudsburg.
3. Bjarnason, M, WG Pollmer, and D Klein (1977): Inheritance of modified endosperm structure and lysine content in *opaque*-2 maize. II. Lysine content. Cereal Res. Commun. 5: 49.
4. Brandolini, A (1970): Razze Europee Di Mais. Maydica XV: 5.

5. Brandolini, A (1975): Research on high quality protein maize in Southern Europe. In: High Quality Protein Maize; Proc. Symp. El Batán, Mexico: 443; Dowden, Hutchinson and Ross, Stroudsburg.
6. Bressani, R, LG Elias, NS Scrimshaw, and MA Guzman (1962): Nutritive value of Central American corns. VI. Varietal and environmental influence on the nitrogen, essential amino acid and fat content of ten varieties. Cereal Chem. 39: 59–67.
7. Burr, B (1979): Identification of zein structural genes in the maize genome. Seed Protein Improvement in Cereals and Grain Legumes I: 175–178; IAEA Vienna.
8. Burr, B, and FA Burr (1976): Zein synthesis in maize endosperm by polyribosomes attached to protein bodies. Proc. Nat. Acad. Sci. U.S.A. 73: 515–519.
9. Burr, FA (1979): Zein synthesis and processing on zein protein body membranes. Seed Protein Improvement in Cereals and Grain Legumes I: 159–164; IAEA Vienna.
10. Cameron, JW (1947). Chemico-genetic bases for the reserve carbohydrates in maize endosperm. Genetics 32: 459.
11. Cameron, JW and HJ Teas (1954): Carbohydrate relationships in developing and mature endosperms of *brittle* and related genotypes. Am. J. Bot. 41: 50–55.
12. Carangal, VR (1975): Breeding for protein quality in maize. In: High Quality Protein Maize; Proc. Symp. El Batán, Mexico: 136; Dowden, Hutchinson and Ross, Stroudsburg.
13. Dalby, A (1966): Protein synthesis in maize endosperm. In: Proc. High Lysine Corn Conf.: 80; Corn Ind. Res. Found., Washington.
14. Dalby, ADC, and II Davies (1967): Ribonuclease activity in the developing seeds of normal and *opaque*-2 maize. Science 155: 1573–1575.
15. Denić, M (1968): An improved method for determination of amino acid composition in bulk protein of individual maize kernels. Acta Chem. Scand. 22: 1809.
16. Denić, M (1969): On the role of some components involved in the synthesis of storage protein in maize. Agrochimica (Pisa) 13: 143–150.
17. Denić, M (1970): Role of the microsomal fraction in the regulation of protein synthesis in maize endosperm. Improving Plant Protein by Nuclear Techniques: 381–389; IAEA Vienna.
18. Denić, M, J Dumanović, KG Bergstrand, and L Ehrenberg (1969): On principal components of variation in amino acid composition of maize endosperm proteins after γ-irradiation. Genetika (Bgd) 1: 25.
19. Denić, M, J Dumanović, and L Ehrenberg (1969): Induced variation of protein content and composition in wheat. Contemp. Agr. (Novi Sad) 17 (11, 12): 85–93.
20. Denić, M, M Džamić, and J Dumanović (1971): On the role of proteinoplasts in genetic regulation of amino acid composition of maize endosperm proteins. Agrochimica (Pisa) 15: 564–568.
21. Denić, M, K Konstantinov, and J Dumanović (1971): Influence of the *opaque*-2 gene on some characters of maize endosperm. Genetika (Bgd) 3:55.
22. Denić, M, K Konstantinov, and R Petrović (1981): Influence of some genes on protein content and composition in maize endosperm. II. Cong. Yugoslav Geneticists 1981: Vrnjačka Banja, Yugoslavia.
23. Denić, M, and D Milivojević (1978): Study of activity of protein granules in leucine incorporation into endosperm proteins of normal and *opaque*-2 maize. J. Sci. Agr. Res. (Bgd) XXXI, No. 116: 3.
24. Dudley, JW (1977): 76 Generations of selection for oil and protein concentration in maize kernel. In: 13th Ann. Illinois Corn Breeding School: University of Illinois, Urbana.
25. Dudley, JW, DE Alexander, and RJ Lambert (1975): Genetic improvement of modified protein maize. In: High Quality Protein Maize; Proc. Symp. El Batán, Mexico: 120; Dowden, Hutchinson and Ross, Stroudsburg.
26. Dudley, JW, RJ Lambert, and IA de la Roche (1977): Genetic analysis of crosses among corn strains divergently selected for percent oil and protein. Crop. Sci. 17: 111–117.
27. Dumanović, J. (1960): Effect of heterosis on the oil and protein content in the corn kernel of varietal and line crosses of F_1 and F_2 generations. Thesis, University of Belgrade.

28. Dumanović, J, and M Denić (1969): Variation and heritability of lysine content in maize. New Approaches to Breeding for Improved Plant Protein: 109–122; IAEA Vienna.
29. Dumanović, J, M Denić, and K Konstantinov (1974): Variation in kernel phenotype and some biochemical properties in *opaque*-2 from different genetic backgrounds. Genetika (Bgd) 6: 211.
30. Dumanović, J, and L Ehrenberg (1968): Modern breeding methods for improving protein quality and quantity in plants. In: Proc. Symp. on Isotope Studies on the Nitrogen Chain: 325; FAO, IAEA, ICSU, Vienna.
31. Dumanović, J, M Mišović, Č Jovanović, and M Denić (1979): Changes in protein content and quality in modified *opaque*-2 endosperm types of maize. In: Proc. IX. Meet. EUCARPIA, Maize and Sorghum Section: 441; Krasnodar Agric. Res. Inst., Krasnodar 1977.
32. Duvick, DN (1955): Cytoplasmic inclusions of the developing and mature maize endosperm. Am. J. Bot. 42: 717–725.
33. Duvick, DN (1961): Protein granules of maize endosperm cells. Cereal Chem. 38: 374–385.
34. East, EM, and DF Jones (1920): Genetic studies on the protein content of maize. Genetics 5: 543.
35. Frey, KJ (1949): The inheritance of protein and certain of its components in maize. Agron. J. 41: 113–117.
36. Glover, DV, PL Crane, SP Misra, and ET Mertz (1975): Genetics of endosperm mutants in maize as related to protein quality and quantity. In: High Quality Protein Maize; Proc. Symp. El Batán, Mexico: 228–240; Dowden, Hutchinson and Ross, Stroudsburg.
37. Greech, RG, and DE Alexander (1978): Breeding for industrial and nutritional quality in maize. In: Intern. Maize Symp.: Genetics and Breeding: 249; Urbana 1975; John Wiley and Sons, New York.
38. Hadjinov, MI, and KI Zima (1979): Breeding maize for improved protein quality. In: Proc. IX. Meet. EUCARPIA, Maize and Sorghum Section: 373; Krasnodar Agric. Res. Inst., Krasnodar 1977.
39. Hanson, WD (1959): Theoretical distribution of the initial linkage block lengths intact in the gametes of a population intermated for n generations. Genetics 44: 839–846.
40. Jalil, ME, and WM Tahir (1970): Review of the world's plant protein resources. Improving Plant Protein by Nuclear Techniques: 21–32; IAEA Vienna.
41. Jelenić, D, J Pavličić, G Radović, and V Hadži-Tašković-Šukalović (1976): L'étude de certains qualités des types de maïs Yougoslave classifiés: les types de maïs denté derive. Contemp. Agr. (Novi Sad) 24 (7, 8): 5–13.
42. Jimenez, JB (1966): Protein fractionation studies of high lysine corn. In: Proc. High Lysine Corn. Conf. Purdue Univ.: 74–79; Res. Found. Washington.
43. Jones, RA (1976): Genetic regulation of storage protein biosynthesis in developing maize endosperm. Thesis Purdue University, West Lafayette.
44. Jones, RA, BA Larkins, and CY Tsai (1977): Storage protein synthesis in maize. II. Reduced synthesis of a major zein component by the *opaque*-2 mutant of maize. Plant Physiol. 59: 525–529.
45. Jones, RA, BA Larkins, and CY Tsai (1977): Storage protein synthesis in maize. III. Developmental changes in membrane-bound polyribosome composition and *in vitro* protein synthesis of normal and *opaque*-2 maize. Plant Physiol. 59: 733–737.
46. Konstantinov, K (1978): Contribution to the study of the role of chromatin in the synthesis of protein in maize endosperm. Genetika (Bgd) 10: 115–121.
47. Konstantinov, K. and M Denić (1975): Isolation and some characteristics of chromatin from developing maize endosperm. Breeding for Seed Protein Improvement using Nuclear Techniques: 99–103; IAEA Vienna.
48. Konstantinov, K, and M Denić (1979): A study of genetic control of RNA and protein synthesis in maize endosperm. Genetika (Bgd) 11: 121–134.
49. Landry, J, and T Moureaux, (1970): Heterogeneity of the glutelins of the grain of corn: selective extraction and composition in amino acids of the three isolated fractions. Bull. Soc. Chim. Biol. 52: 1021–1037.

50. Larkins, BA, and A Dalby (1975): *In vitro* synthesis of zein-like protein by maize polyribosomes. Biochem. Biophys. Res. Commun. 66: 1048–1054.
51. Larkins, BA, RA Jones, and CY Tsai (1976): Isolation and *in vitro* translation of zein messenger ribonucleic acid. Biochem. 15: 5506–5511.
52. Laughnan, JR (1953): The effect of sh_2 factor on carbohydrate reserves in the mature endosperm of maize. Genetics 38: 485.
53. Maize Quality Protein Abstracts; CIMMYT, Commonwealth Agricultural Bureaux, Cambridge.
54. Maize Research Institute, Zemun (1980): ZP hybrids, Ed. Ass. Agric. Eng. Techn. of S.R. Serbia, Belgrade.
55. Maschenkov, AS, and MI Khadzinov (1979): Dominant mutation partially suppressing expression of *opaque*-2 allele. In: Proc. IX. Meet. EUCARPIA, Maize and Sorghum Section: 447; Krasnodar Agric. Res. Inst., Krasnodar 1977.
56. McWhirter, KS (1971): A floury endosperm, high lysine locus on chromosome 10. Maize Genet. Coop. Newslett. 45: 184.
57. Mehta, SL, ML Lodha, MS Naik, and J Singh (1975): RNA polymerase from *opaque*-2 and normal *Zea mays* Endosperm. Phytochemistry 14: 2145–2146.
58. Mertz, ET, LS Bates, and OE Nelson (1964): Mutant gene that changes protein composition and increases lysine content of maize endosperm. Science 145: 279–280.
59. Mihajlović, M (1977): Contribution to the study of chemical composition of some domestic maize hybrids especially on amino acid and faty acid composition. Thesis University of Skopje.
60. Mihajlović, M (1978): Some physical and chemical properties of hybrid maize grain. Hrana i ishrana XIX (1–2): 29.
61. Mihajlović, M. (1979): Some physical and chemical kernel properties of ZP hybrids. Proc. 5th Yugoslav Livestock Conf. Ohrid.
62. Mihajlović, M (1980): Maize grain as the source of high quality properties for human food. Hrana i ishrana *XXI* (1–2): 35.
63. Mišović, M, M Mihajlović, and V Trifunović (1966); Yield, oil and protein content of some American and domestic experimental high oil hybrids of maize. J. Sci. Agr. Res. (Bgd) 19, No. 671 78 pp.
64. Misra, PS, SR Jambunathan, ET Mertz, DV Glover, HM Barbosa, and KS McWhirter (1972): Endosperm protein synthesis in maize mutants with increased lysine content. Science 176: 1425–1427.
65. Morton, RK, and JK Raison (1964): The separate incorporation of amino acids into storage and soluble proteins catalyzed by two independent systems isolated from developing wheat endosperm. Biochem. J. 91: 528–539.
66. Mossé, J, and J Landry (1980): Recent research on major maize proteins: Zeins and glutelins. In: Inglet, GE, and L Munck (eds.): Cereals for Food and Beverages: 255–272; Acad. Press, New York.
67. Mossé, T (1966): Alcohol soluble proteins of cereal grains. Federation Proc. 25: 1663–1669.
68. Murphy, JJ, and A Dalby (1971): Changes in the protein fractions of developing normal and *opaque*-2 maize endosperm. Cereal Chem. 48: 336–349.
69. Nelson, OE (1969): Genetic modification of protein quality in plants. Adv. Agron. 21: 171–194.
70. Nelson, OE (1970): Improvement of plant protein quality. Improving Plant Protein by Nuclear Techniques: 43–51; IAEA Vienna.
71. Nelson, OE, (1975): Breeding for protein quality in maize. Current issue and problems. In: High Quality Protein Maize; Proc. Symp. El Batán, Mexico: 193; Dowden, Hutchinson and Ross, Stroudsburg.
72. Nelson, OE (1978): Gene action end endosperm development in maize. In: Intern. Maize Symp.: Genetics and Breeding: 389; Urbana 1975; John Wiley and Sons, New York.
73. Nelson, OE, ET Mertz, and LS Bates (1965): Second mutant gene affecting the amino acid pattern of maize endosperm proteins. Science 150: 1469–1470.

74. Osborne, TB, and LB Mendel (1914): Nutritive properties of proteins of the maize kernel. J. Biol. Chem. 18: 1–16.
75. Paez, AV, JL Helm, and MS Zuber (1969): Lysine content of *opaque*-2 maize kernels having different phenotypes. Crop. Sci. 9: 251–254.
76. Paliy, AF, and AI Rotar (1979): The effect of complementary gene cfl_2 in relation with recessive allele fl_2 on the seed weight and protein quality in maize. Genetika (USSR) 15: 478–481.
77. Pavličić, J, and D Jelenić (1977): Maize types classified in Yugoslavia and their significance in breeding. Contemp. Agr. (Novi Sad) 25 (9, 10): 31–42.
78. Pavličić, J, D Jelenić, and G Radović (1976): L'étude de certaines charactères des types de maïs classifiés en Yougoslavie (le maïs denté – type du maïs denté de principale région des Etats Unies et les maïs dentés blancs et jaunes lui semblable. Contemp. Agr. (Novi Sad) 24 (1, 2): 65–74.
79. Pavličić, J, D Jelenić, and G Radović (1976): L'étude de certaines caractères des types de maïs classifiés en Yougoslavie: Les variétés de type de maïs denté des régions du sud des Etats Unies et les dentés de Serbie. Contemp. Agr. (Novi Sad) 24 (9, 10): 5–13.
80. Pavličić, J, D Jelenić, and G Radović (1977): L'étude de certaines caractères des types de maïs Yougoslaves classifiés: Le maïs corné tendre. Contemp. Agr. (Novi Sad) 25 (1, 2): 21–30.
81. Pavličić, J, and V Trifunović (1966): A study of some important ecologic corn types in Yugoslavia and their classification. J. Sci. Agr. Res. (Bgd) 19: 44.
82. Pieterse, MJ (1976): Protein content in South African commercial maize. Proc. Second South African Maize Breed. Symp.: 42; Pietermaritzburg.
83. Pollmer, WG, D Eberhard, and D Klein (1978): Inheritance of protein and yield of grain and stover in maize. Crop. Sci. 18: 757–759.
84. Popović, V, M Denić and J Dumanović (1974): Influence of *opaque*-2 gene on proteins in maize of different genetic backgrounds. Genetika (Bgd) 6: 7.
85. Robertson, A (1970): A theory of limits in artificial selection with many linked loci. In: KI Kojima (ed.): Mathematical Topics in Population Genetics. Biomathematics 1: 246–288; Springer, New York, Heidelberg, Berlin.
86. Simić, R, and M Denić (1975): Relationship between protein composition and the content of free amino acids in the endosperm of a normal genotype and an *opaque*-2 mutant of maize. Genetika (Bgd) 7: 25.
87. Sodek, L, and CM Wilson (1971): Amino acid composition of proteins isolated from normal, *opaque*-2, and *floury*-2 corn endosperm by a modified Osborne procedure. J. Agr. Food Chem. 19: 1144–1150.
88. Sommer, JB, and GF Somers (1944): The water soluble polysaccharides of sweet corn. Arch. Biochem. Biophys. 4: 7.
89. Sprague, GF, B Brimhall, and RM Hixon (1943): Some effects of the waxy gene in corn on properties of the endosperm starch. J. Amer. Soc. Agron. 35: 817–822.
90. Tsai, CY (1979): Tissue-specific zein synthesis in maize kernel. Biochem. Genet. 17: 1109–1120.
91. Vasal, SK (1975): Use of genetic modifiers to obtain normal type kernels with the *opaque*-2 gene. In: High Quality Protein Maize; Proc. Symp. El Batán, Mexico: 197–216; Dowden, Hutchinson and Ross, Stroudsburg.
92. Vasal, SK, E Villegas, and R Bauer (1979): Present status of breeding quality protein maize. Seed Protein Improvement in Cereals and Legumes II: 127–150; IAEA Vienna.
93. Viotti, A, E Sala, P Alberti, and C Soave (1975): RNA metabolism and polysome profiles during seed development in normal and *opaque*-2 maize endosperms. Maydica 20: 111.
94. Wilson, CM, and DE Alexander (1967): Ribonuclease activity in normal and *opaque*-2 mutant endosperm of maize. Science 155: 1575–1576.
95. Wolf, MJ, V Khoo and HL Seckinger (1967): Subcellular structure of endosperm protein in high lysine and normal corn. Science 157: 556–557.
96. Zuber, MS, and JL Helm (1975): Approaches to improving protein quality in maize without the use

of specific mutants. In: High Quality Protein Maize; Proc. Symp. El Batán, Mexico: 241; Dowden, Hutchinson and Ross, Stroudsburg.

97. Zuber, MS, WH Skrdla, and BH Choe (1975): Survey of maize selections for endosperm lysine content. Crop. Sci. 15: 93–94.

9. Seed Protein Fractions of Maize, Sorghum, and Related Cereals

CURTIS M. WILSON

INTRODUCTION

More and more studies are being made on seeds as the final storage site for the products of photosynthesis and nitrogen metabolism. The cereals are especially noted for storing starch, yet cereal proteins make up a large proportion of the protein in the diets of many societies around the world. Maize is the highest yielding cereal, while sorghum and other more distant maize relatives are either high-yielding species or capable of relatively high yields in harsh environments. This may be related to their possession of the C4 photosynthetic system. Unfortunately, the major storage proteins in most cereals are prolamins which are low or lacking in the essential amino acids lysine and tryptophan. Table 1 shows that the average content of essential amino acids found in 379 species of plants is adequate, except for the sulfur amino acids, and that the non-zein proteins of maize endosperm are also adequate. Zein, however, possesses little or no lysine and tryptophan, less than adequate levels of threonine and valine, and has an imbalance in the leucine/isoleucine ratio which is also deleterious. Zein makes up half or more of the total endosperm protein in normal genotypes, thus producing the well-known amino acid deficiency in maize protein. Mutants have been found in maize, barley, and sorghum which contain elevated levels of lysine, or more precisely, lowered levels of prolamins. This has stimulated much research on the proteins of these species and the search has intensified for other mutants in these and other species with the desired properties of high-lysine, high-protein, and high-yield. Given this interest, it is deemed important that we have a good understanding of the many proteins which are found in cereal seeds.

No report on seed protein would be complete without a reference to the pioneering work of Osborne, whose careful methodology still forms the basis for the classification of seed proteins. Although he recognized that the use of solubility was unsatisfactory from many standpoints [81, 82], his system is still useful. Osborne

Table 1. Essential amino acids.

Amino Acid	1973 FAO Provisional Pattern[1]	Seeds of 379 species[2]	Maize Non-Zein Proteins[3]	Zein[4]
		mol %		
Lysine	4.3	4.6	4.7	0.1
Threonine	3.9	4.3	4.9	3.0
Valine	4.2	6.0	7.1	3.6
Cysteine + Methionine	3.0	1.6[5]	4.5	1.9
Isoleucine	3.5	4.2	4.0	3.8
Leucine	6.2	7.1	9.0	18.7
Phenylalanine + Tyrosine	4.0	6.1	6.2	8.7
Tryptophan	0.5	-	0.6	0
Total	29.6	33.9	41.0	39.8

[1] Calculated from 1973 FAO/WHO pattern in FAO Nutrition Report Series No. 52.

[2] From Reference [123].

[3] Weighted average for RS-protein (Table III), albumin, globulin and glutelin (Table VII).

[4] From Table III.

[5] Methionine only.

reported that variations in techniques led to variations in the proteins classed as albumins and globulins, and that techniques suitable for animal proteins (especially serum proteins) could not be applied to vegetable proteins without causing problems [82]. He also noted that some globulins could be extracted from seeds into an aqueous solution and that proteins extracted by water did not form a distinct class which should be termed albumin. He recommended that proteins which are soluble in saline solution, but which are precipitated by dilution or dialysis be classed as globulins, while the albumins are the proteins which remain soluble. A classical Osborne extraction starts by extracting the albumins and globulins together with a saline solution. It has often been noted that extractability is not the same as

Table 2. Dry matter and nitrogen distribution in whole grain of maize, its parts, and in protein extracts.

	% of grain weight or Nitrogen in parts		
	Whole	Endosperm	Germ
Dry weight	100[1]	80	13
Nitrogen	100[1]	78	18
	% of Nitrogen recovered in each extract		
Osborne and Mendel [83]:			
N soluble in 10% KCl	22.0	7.8	77.2
N soluble in 90% alcohol	41.0	50.0	2.0
N soluble in 0.2% KOH	31.0	38.2	0.6
N insoluble and loss	6.0	4.0	20.2
Consensus from literature[2]:			
Nonprotein N	6	3	20
Albumin	7	3	35
Globulin	5	3	18
Salt-soluble protein	(12)[3]	(6)	(53)
I.[4] Salt-soluble N	(18)	(9)	(73)
II. Zein-1 (Alcohol)	42	48	4
III. Zein-2 (Alcohol + ME)	10	12	1
Total Zein=II + III	(52)	(60)	(5)
IV. Glutelin-2 (pH 10 buffer + ME)	8	9	3
V. Glutelin-3 (buffer + ME + SDS)	17	17	15
Osborne Glutelin≃III + IV + V	(35)	(38)	(19)
Residue	5	5	4

[1]The hull, tip cap, etc. make up about 7% of the dry weight and contain about 4% of the N.

[2]References: 7,50,51,56,68,84,85,87,107,114,133.

[3]Subtotals for fractions which may be extracted simultaneously.

[4]Numbering for Landry-Moureaux fractions [50].

solubility, and that seeds contain non-protein substances, such as salts and phytate, which affect both properties [19]. New techniques, have, of course, made it possible to obtain many highly purified proteins, but the use of the specific property of prolamin alcohol solubility is perhaps the simplest single technique known for isolating large quantities of any protein. However, recent studies have shown that prolamins are not single polypeptides, and new extraction solvent mixtures have led to some minor differences over the classification of cereal seed proteins. The glutelins were classed by Osborne as those proteins which were not soluble in the previously used solvents, but which were soluble in alkali – thus they were in general poorly characterized [82]. Newer techniques are now making it possible to more readily extract and characterize these proteins, and thus advance our understanding of the mixture of proteins originally included in the glutelin fraction. I will concentrate on the 80% of the seed found in the endosperm, especially because the proteins in this tissue are subject to genetic and environmental manipulations.

Table 2 is basically an outline of this review, for it presents a consensus collection of data on the distribution of dry matter and nitrogen in the maize seed. After completing the consensus figures, I found that they were remarkably similar to the data presented decades ago by Osborne and Mendel. The present system is based on the procedure put forth by Landry and Moureaux in 1970 [50], which has proven useful for the complete extraction of zein and for a beginning of the fractionation of the glutelins. The five basic extraction steps are identified by Roman numerals to avoid the possible confusion which could result from the use of different descriptive names for the fractions.

There have been a number of books and review articles which have covered the field of maize seed proteins, and which contain additional information [18, 63, 64, 72, 127, 131, 76]. One report of recent research has a good, though brief, review of the field along with original experimental data on the protein fractions [51]. In this review, I will be using previous reviews and the original literature to prepare an outline of the protein fractions of maize and related cereals as we know them today. In addition to maize, the near maize relatives teosinte and *Tripsacum*, and the other major coarse grains, sorghum and the millets, will be considered. The properties of the prolamins of these species indicate that they are much more closely related to each other than to the other cereals (the small grains) reviewed in other chapters. The variety of techniques employed on different genotypes grown under different environments requires that I make a number of assumptions and extrapolations of the data to make them fit into some kind of pattern – I hope that I will not have done great violence to the ideas of the original authors, and that this review will provoke critics to do the experiments which alone can fill the gaps in our knowledge.

ZEIN AND OTHER ALCOHOL-SOLUBLE PROTEINS

Zein is the predominant and characteristic protein of maize endosperms, and perhaps is the only protein which functions as a true storage protein. Miflin and Shewry [64] concluded that storage proteins probably have certain characteristics:

1. They have no metabolic function other than to provide N for the germinating seedlings.
2. They are formed relatively late in seed development.
3. They are increased preferentially if the N supply to the plant is increased.
4. They may be stored in a separate package, usually known as a protein body.
5. They may be composed of a limited number of similar polypeptides.

The specialized storage proteins of cereals were defined by Osborne [82]: 'The group of proteins soluble in relatively strong alcohol deserves a definite name, for it is one of the best characterised groups yet found in either plants or animals . . . The writer has proposed calling this group "prolamins", since all its members which have thus far been hydrolysed yield a relatively large quantity of both proline and amide nitrogen. The prolamins are characterised by their solubility in alcohol of from 70 to 90 per cent. They are nearly of wholly insoluble in water, but their salts with acids or alkalis dissolve freely therein. They yield much glutaminic (sic) acid, proline and ammonia, small amounts of arginine and histidine and little or no lysine.' This definition is still satisfactory, when broaded to include alcohols other than ethanol, usually at lower concentrations.

Zein, the maize prolamin, is at once a simple protein, easily prepared, and a complex protein difficult to study with many of the common methods of protein chemistry. The distinctive solubility in alcohol makes it easy to prepare zein uncontaminated by other proteins, but its insolubility in most aqueous solutions and its tendency to form aggregates hinders further study. The properties of solubility and extractability must be clearly distinguished. Proteins may not always be completely extracted by solvents in which they are soluble. Also, solvents carried over from one extraction to the next may affect extractability [51].

A protein which I will term zein ('true' zein, classical zein, or zein-1) is easily prepared by extracting ground maize seeds or endosperms with 70% (v/v) ethanol or 55% (w/w) isopropanol (after extracting with a saline solution). The resulting solution contains proteins which may be precipitated by dialysis against water, or by dilution with water or an aqueous salt solution. The amino acid composition of zein is shown in the first column of Table 3, giving the average of the values obtained in six laboratories over the past decade. The range of values is remarkably small (column 2), considering the variable sources, and it agrees with the composition determined before the advent of the automatic amino acid analyzer (column 3) [113a]. The samples included U.S. commercial maize, several identified hybrids and inbreds, and several low-zein *opaque*-2 lines. Two laboratories used whole kernels and four used isolated endosperms, four first defatted the meal, all first extracted

Table 3. Amino Acid Composition of Zein and other Alcohol Soluble Proteins from Maize.

	Zein[1] Ave.	Zein[1] Range	Zein[2]	Zein-2[3]	Small Prolamins[4]	Reduced Soluble Protein[5]
				mol %		
Lysine	0.1	0.1- 0.3	0	0.2	0.3	0.4
Histidine	1.0	0.8- 1.1	1.0	1.7	1.2	6.8
Arginine	1.2	1.1- 1.3	1.1	1.6	2.4	2.4
Aspartic Acid	5.1	4.8- 5.5	4.0	3.8	2.4	0.7
Threonine	3.0	2.9- 3.1	3.3	3.4	3.7	4.3
Serine	6.3	6.0- 6.4	7.7	6.0	5.6	4.5
Glutamic Acid	21.4	19.7-23.6	21.0	19.6	20.3	14.8
Proline	10.7	10.3-12.0	10.4	13.4	12.5	24.8
Glycine	2.2	1.9- 2.6	0	4.8	8.6	7.5
Alanine	13.3	12.7-13.8	13.5	12.0	11.1	5.9
Cysteine[6]	(1.0)	0 - 1.0	0.8	0-2.1	0-1.4	3.7
Valine	3.6	2.7- 4.0	3.9	3.9	3.5	7.5
Methionine[6]	0.9	0.1- 1.5	1.8	4.9	3.9-7.1	1.0
Isoleucine	3.8	3.6- 4.0	4.4	2.8	1.2	2.4
Leucine	18.7	16.9-19.8	18.5	15.0	10.8	10.1
Tyrosine	3.5	3.2- 3.7	3.3	4.0	3.4-6.7	2.9
Phenylalanine	5.2	4.8- 5.8	5.1	4.1	2.3	1.8

[1] Average, references: 32, 49, 68, 84, 107, 133.
[2] Reference: 113a
[3] Average, references: 32, 50, 59, 107.
[4] Average, references: 32, 59.
[5] Average, references: 88, 137.
[6] Values for cysteine and methionine are uncertain, because they may be oxidized during hydrolysis, and cysteine is not always reported.

salt-soluble proteins with approximately 3% sodium chloride (with phosphate buffer in one case), two employed the traditional 70% (v/v) ethanol while four used iso-propanol. The salt extracts were performed in the cold, while the alcohol extractions were performed at room temperature. One preparation was specially purified [49]. The yields of zein ranged from 37 to 59 per cent of the total N when reported. Zein has a characteristic amino acid composition distinctly different from that of most other common cereal prolamins (see Table 9), and from the other endosperm proteins (see Table 7). Among the prolamins, the zein family is low in glutamic acid and proline and high in alanine and leucine. Because half of the protein in maize endosperm is zein, its amino acid composition is of primary importance in determining the whole seed composition. The distinctive amino acid pattern may also be noted when zein has not been completely separated from other protein fractions. In particular, the lack of lysine is important. Tryptophan is also

Table 4. Relative Solubility or Extractability of Zein in different Alcohols

Sample	Ranking	Reference
Purified zein	40% n-Propanol>22% t-Butanol>50% Ethanol>70% Methanol (mol/mol)[1]	Dill, in [71]
Gluten meal	55% iso-Propanol>60% Ethanol (w/w)	Swallen, in [71]
Seed powder	70% Ethanol>Water-saturated n-Butanol (v/v)	[91]
Industrial zein	65% iso-Propanol = 70% Ethanol>72% n-Butanol = 80% Methanol (v/v?)	Reiners, in [131]
Endosperm meal	70% iso-Propanol>70% Ethanol (v/v)	Table V, [116]

[1]Conversion Table for Expression of Alcohol Percentages

	v/v	w/w	mol/mol
Ethanol	70	63	40
Iso-Propanol	61	55	27
n-Propanol	77	69	40
t-Butanol	63	54	22

Calculations by author from [71] and density data in handbooks.

Table 5. Effects of alcohol, mercaptoethanol, and temperature on extraction of prolamins from maize (abridged from Ref. 116), courtesy American Association of Cereal Chemists.

Mercaptoethanol		Extraction Conditions				
%	mM	70% Ethanol		70% Isopropanol		55% Isopropanol
		25°C	60°C	25°C	60°C	60°C
0	0	5.4[1]	7.1	6.4	7.0	7.0
0.008	1	5.8	8.1	6.5	8.0	8.1
0.08	10	6.5	8.1	7.3	8.1	8.2
0.79	100	7.6	8.1	8.0	8.0	8.2

[1]Protein extracted as percent of sample.

absent from zein, but is seldom considered because it is not recovered after the usual acid hydrolysis procedure. Amides are not determined for the same reason, but it has been estimated that as much as 90% of the aspartate and glutamate residues in zein may be present as amides [49, 93]. The amide composition will be known when zein has been completely sequenced.

The complexities of zein begin to appear once an investigator moves beyond the stage of performing a simple extraction with alcohol and determining the amino acid composition. The potential variations in techniques are considerable.

Several alcohols have been used in studies of zein extractability and solubility (Table 4). Iso- or n-propanol appear to be more effective than ethanol, but comparisons are difficult because of the different types of samples and the variable alcohol concentrations used. A conversion table for alcohol percentages is included in Table 4 so that the original data can be presented. Expression of alcohol percentages by weight is preferable, for it is temperature independent. Aqueous alcohols do not extract prolamins completely. Various procedures to improve prolamin extractions have been reviewed [92]. Although the older literature contains examples of improved zein extraction techniques, none were widely adopted until Landry and Moureaux introduced the use of 55% (w/w) iso-propanol with 0.6% ME [50]. A more detailed study of the effects of different alcohols, added ME, and temperature on zein extraction is shown in Table 5 [116]. The combination of 60° and only 1 mM ME is fully effective, while the use of a higher temperature can be avoided by using isopropanol and 100 mM ME.

Some of the variations used for the complete extraction of zein are given in Table 6, along with differing maize genotypes, to illustrate the range of zein contents which may be encountered. Alcohol alone may extract from 36 to 59% of the total protein of normal varieties, with the higher values being found for endosperm samples. The zein content of *opaque*-2 samples is lower than that of dent corn, but it varies rather

Table 6. Prolamin recovery as affected by genotype, alcohol, mercaptoethanol, sodium acetate and heat.

Genotype	Alcohol	Additive	% of seed (endosperm) N extracted: Alcohol	Second Extraction with Additive	Total[1]	Reference
Normal hybrid (K)[2]	55% Isopropanol (w/w)	0.6% ME	38	11.5	49.5	50
Normal hybrid (E)	55% Isopropanol (w/w)	0.6% ME	42.9	18.8	60.7	107
o2 hybrid (E)	55% Isopropanol (w/w)	0.6% ME	17.8	13.9	31.7	107
o2 R802 (E)	55% Isopropanol (w/w)	0.6% ME	2.9	13.6	16.5	107
Normal Oh43 (E)	70% Isopropanol (v/v)	0.6% ME	59	6	65	65
Normal W22 (E)	70% Isopropanol (v/v)	0.6% ME	40.6	15.3	55.9	65
o2 Oh43 (E)	70% Isopropanol (v/v)	0.6% ME	26.9	8.4	35.3	65
fl2 Oh43 (E)	70% Isopropanol (v/v)	0.6% ME	49.1	9.0	58.1	65
o2bt2 Oh43 (E)	70% Isopropanol (v/v)	0.6% ME	3	5	8	65
Normal hybrid (K)[2]	70% Ethanol (v/v)	60°	36	7.25	43.25	7
Normal hybrid (E)	70% Ethanol (v/v)	0.5% sodium acetate	37.6		45.8	84
o2 hybrid (E)	70% Ethanol (v/v)	0.5% sodium acetate	14.6		24.4	84
Normal hybrid (K)	70% Ethanol (v/v) + 0.5% sodium acetate	0.1 M ME	41.4	12.6	54	87

[1] Total of extractions without and with additive if done sequentially, or of a separate extraction with additive and without previous alcohol extraction.

[2] K - whole kernels, E - endosperms.

widely depending upon genotype. For example, it may range from 3% to 27% of the endosperm protein. In contrast, zein is high in the *floury*-2 mutant [117]. The lowest total zein content was in the double mutant of *opaque*-2, *brittle*-2. Various additives – ME, sodium acetate, heat – removed up to an additional 19% of the seed protein. The size of this fraction will depend upon the efficiency of the first, alcohol-only, extraction, and may also be affected by genotype. Normal endosperms released

Table 6a. Amino acid residues per zein chain.

	A30 Clone[1]	Average[2]
Lysine	0	0
Histidine	2	2
Arginine	2	2
Aspartic Acid	0	11
Threonine	5	6
Serine	15	13
Glutamic Acid	1	45
Proline	23	23
Glycine	5	5
Alanine	29	28
Cysteine	2	2
Valine	5	8
Methionine	0	2
Isoleucine	9	8
Leucine	43	40
Tyrosine	8	7
Phenylalanine	13	11
Tryptophan	0	-
Asparagine	10	-
Glutamine	41	-
Total	213	213

[1]From [31a]

[2]Calculated from column 1, Table III, using the closest whole residue numb totaling 213, and ignoring chain heterogeneity.

from 45 to 65% of their N to the two alcoholic solutions. Whether this range is a true representation of variable zein contents in the different samples, or results from the different techniques cannot be determined from the available data.

The zein extracted by alcohol alone [Landry-Moureaux (L-M) Fraction II] will be termed "zein-1" to distinguish it from the prolamins and other polypeptides which can be subsequently extracted by the use of alcoholic ME (L-M Fraction III) or other more effective extractants. The latter has been termed glutelin-1 of G_1 [50], zein-2 [107], or alcohol-soluble glutelin [88]. I prefer the term zein-2 because many such preparations contain a large proportion of polypeptides which are soluble in alcohol (though not extractable), and many of these polypeptides appear identical with those extracted by alcohol alone. The zein-2 fraction is always a mixture of zein, small prolamins, and some unrelated alcohol-soluble polypeptides. The amino acid pattern of the zein-2 mixture is dominated by the zein pattern (Table 3). The most striking difference from that of zein-1 preparations is the high level of methionine. The data in this table came from four laboratories, and there was considerably more variation than found for the zein-1 composition. This is to be expected from a mixture of at least three different types of polypeptides.

A single solvent including alcohol and a reducing agent will extract a prolamin fraction which may be termed 'zein 1 + 2', in which the major polypeptides would be zein-1 along with varying amounts of the other components. When the more vigorous extractants are used, zein polypeptides should not appear in the subsequent glutelin fractions [116, 137]. At present there seems to be no extraction system which will completely remove zein-1 from the endosperm meal without also removing some proteins of a slightly different nature. Many different systems have been used, and each will produce a mixture of proteins.

Sodium dodecyl sulfate-polyacrylamide gel electrophoresis (SDS-PAGE) is very useful for resolving the mixed polypeptides in alcohol extracts. Two major zein bands are usually identified by estimated molecular weight, but there is no agreement on the true molecular weights of these polypeptides, other than that they lie between 18,000 and 25,000 daltons. SDS-PAGE shows discontinuities in the size range of the zeins [27, 112]. Inspection of the literature will show that several small polypeptides used as standards, including soybean trypsin inhibitor [33], RNase A and cytochrome c, show deviations from expected mobilities. Deviations from expected molecular weights have been noted for other small, high-proline polypeptides such as collagen [78]. The high-proline proteins may be more difficult to stain than typical proteins, and they may give metachromatic staining with Coomassie Blue R [26]. Similar staining effects have been noted for zein [23, 28], but these reactions are difficult to reproduce or to quantitate. The unusual solubility properties of zein make molecular weight determinations by other methods more difficult also.

Until reliable molecular weight can be determined by SDS-PAGE, it would be

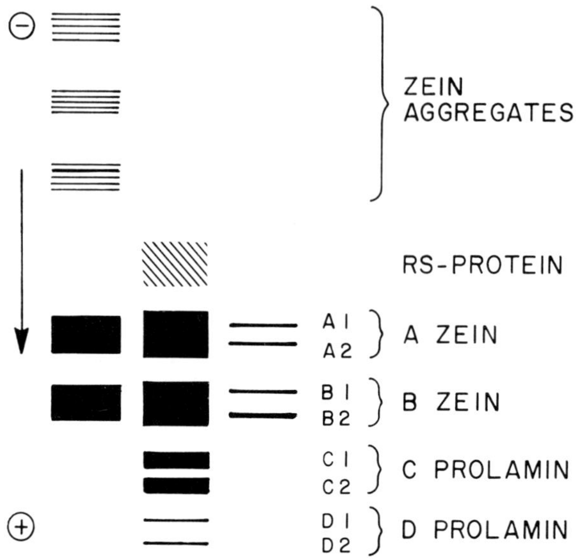

Figure 1. Outline of SDS-PAGE patterns by alcohol-soluble proteins from maize. 1, a zein extract prepared for SDS-PAGE without ME in sample buffer. Slow moving bands are apparent aggregates of zein. 2, a zein extract reduced and alkylated before SDS-PAGE. 3, same as 2, but with a small sample size.

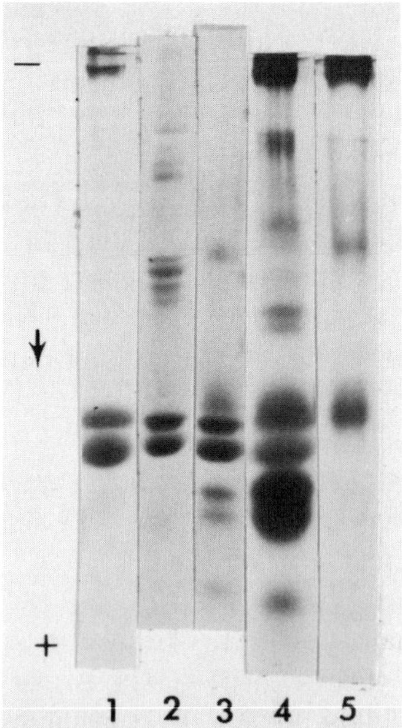

Figure 2. Examples of different zein preparations on SDS-PAGE. 1, alkylated zein-1; 2, zein prepared for SDS-PAGE without ME; 3, alkylated total zein; 4, alkylated zein-2; 5, alkylated reduced-soluble protein, recovered from a zein extract. From [137].

better to identify the bands by relative mobilities only. Such a system is shown in Figure 1. Zein-1 preparations of normal endosperms always give the bands termed A zein and B zein, in order of decreasing molecular weight (Figure 1, Figure 2, lane 1). These bands have been reported to consist of two or sometimes three bands each, depending upon genotype and upon the amount of protein applied to the gels [103, 125]. These differences may be characteristic of certain genotypes, but SDS-PAGE is not the system of choice for more complete resolution of zein polypeptides.

Few other polypeptides are noted in most zein-1 preparations, but if the samples are prepared for SDS-PAGE without using ME, two or more groups of slow-moving multiple bands appear (Figure 1 and Figure 2, lane b). These appear to be aggregates of A and B zeins, sometimes being made up of more A than B [49]. Because they are found in extracts made without ME, it was suggested that these aggregates are native polymers of zein held together by disulfide bonds, and that they exist as such in the endosperm [49]. Apparently identical aggregates are seen in samples which were extracted with ME, but then subsequently prepared for SDS-PAGE without ME [116, 137]. This suggests that the aggregates could be artifacts formed in solution.

Major components of zein-2 preparations, (Figure 1 and Figure 2, lane 4) and minor components of zein 1 + 2 preparations (Figure 2, lane 3) are small polypeptides here called C and D prolamins. These also may be subdivided into at least two bands under certain conditions [54, 116, 137]. Sharper patterns may be obtained with alkylated polypeptides [137]. C and D prolamins have been isolated by SDS-PAGE [32] and by cryoprecipitation [59]. The amino acid compositions were sufficiently similar that I averaged them in Table 3. The amino acid composition differs from that of zein-1, having higher levels of methionine and glycine, and lower levels of aspartic acid, leucine, and phenylalanine. The differences are enough to warrant using the general term of prolamin rather than zein. Tsai [114] isolated similar small prolamins from SDS-PAGE, and found a different amino acid pattern, while Paulis and Wall [88] reported on a water-insoluble, alcohol-soluble glutelin sub-fraction with an amino acid composition much like zein, but having low glycine and high methionine. The high methionine content of the C prolamins is also shown by the high degree of labelling when protein body polysome or zein mRNA translation products are synthesized in vitro in the presence of [H]methionine [52, 58, 124]. The small prolamins are the only maize polypeptides identified so far with the high methionine content noted in all zein-2 fractions. The A and B zeins are the major prolamins of normal, *opaque-2*, and *floury-2* endosperms, but the C prolamins appear to be the major polypeptide recovered from an alcoholic extract of the *sh2o2* double mutant (Figure 3) [120]. The A and B zeins can be partially separated on a phosphocellulose column, and some fractions can be enriched in C and D prolamins (Figure 4) [29]. Additional studies are needed to confirm the properties of the small prolamins as summarized here.

Another component of zein-2 or zein 1 + 2 preparations produces a somewhat

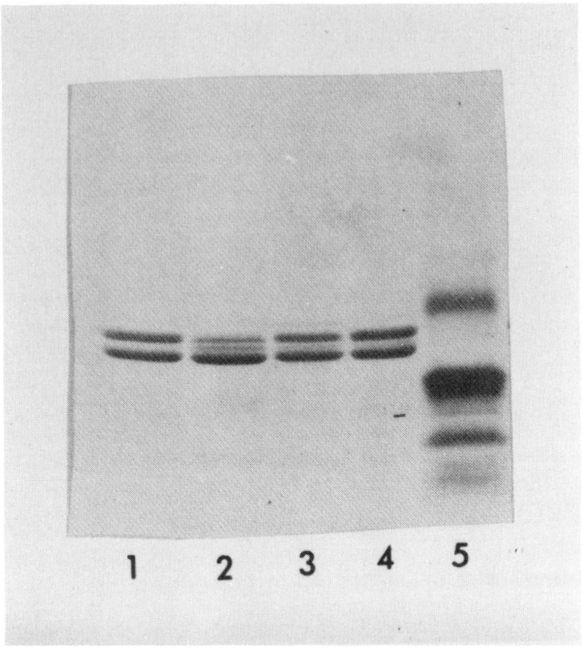

Figure 3. SDS-PAGE patterns of zein extracts from the mature endosperms of different maize genotypes. 1, normal; 2, *o2*; 3, *fl2*; 4, *sh2*; 5, *sh2o2*. From [120], courtesy of Plenum Publishing Corp.

diffuse band which runs slightly slower on SDS-PAGE than A zein (Figure 1, Figure 2, lane 5). Here it will be called reduced soluble protein (RS-protein) [137], but it has also been isolated and named water-soluble, alcohol-soluble glutelin [88]. It is readily isolated from the other components of an alcohol-ME extract because it remains in solution upon dialysis or addition of aqueous salt solutions, which causes zeins and prolamins to precipitate [88, 137]. According to Osborne's definition, it is not a prolamin. The RS-protein is not found in protein bodies [62]. The amino acid pattern is quite distinctive (Table 3). It is about one-fourth proline, and is high in histidine, cysteine, glycine, and valine, and is lower in aspartate, glutamate, and leucine than zein. It is low in lysine also. It may be found in the salt-soluble fraction if reducing agents are used [50, 137]. Its presence in the L–M G_2 (IV) fraction is shown by the amino acid pattern [50] and it can be seen in SDS-PAGE patterns of L–M fractions III, IV, and V [23]. The high cysteine content suggests that it may readily form disulfide bonds with other proteins, which would affect its extractability and cause it to appear in various fractions.

Zein-1 may be still further sub-divided by isoelectric focusing (IEF), where 10 to 15 major bands have been reported for most genotypes, and at least 26 bands have been detected among the maize lines examined so far [93, 122]. About 10 bands may

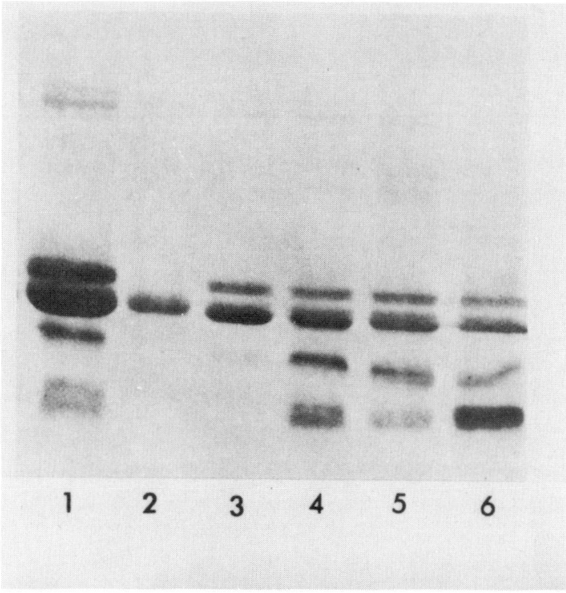

Figure 4. SDS-PAGE of zein fractions from a phosphocellulose column. 1, whole zein; 2, shoulder of first peak; 3, main peak; 4, second peak; 5, third peak; 6, fifth peak. From [29], courtesy of the American Association of Cereal Chemists.

be present or absent, or may vary considerably in quantity, depending upon genotype. These characteristics of IEF patterns can be used to classify maize inbreds [79]. Three examples, two inbreds and the hybrid, are shown in Figure 5. IEF has been used to find the chromosomal locations of zein genes. Some IEF bands have been found to contain both A and B zein, when separated by SDS-PAGE as well [125]. Two-dimensional separation with IEF + SDS-PAGE reveals a large number of polypeptides (Figure 6) [64, 137], with differences among genotypes [35]. The small C prolamins also separate into several polypeptides, as do the more basic RS-protein polypeptides (Figure 6B).

This area of research is moving rapidly, but is not well standardized. IEF patterns obtained by different workers differ, probably the result of using different reagents and amphoteric substances [63]. Zein has shown apparent pI values 7 to 8.5 [93], 6.3 to 7.4 [90], or 5.7 to 6.3 [35]. Standard proteins for calibration of IEF gels are becoming available – their use may make comparisons easier.

Zein may also be separated into multiple bands by polyacrylamide or starch gel electrophoresis in urea at acid pH. These separations depend primarily upon charge, and thus may be related to the banding patterns found with IEF. The differences among genotypes are greater than those reported for SDS-PAGE [49, 87].

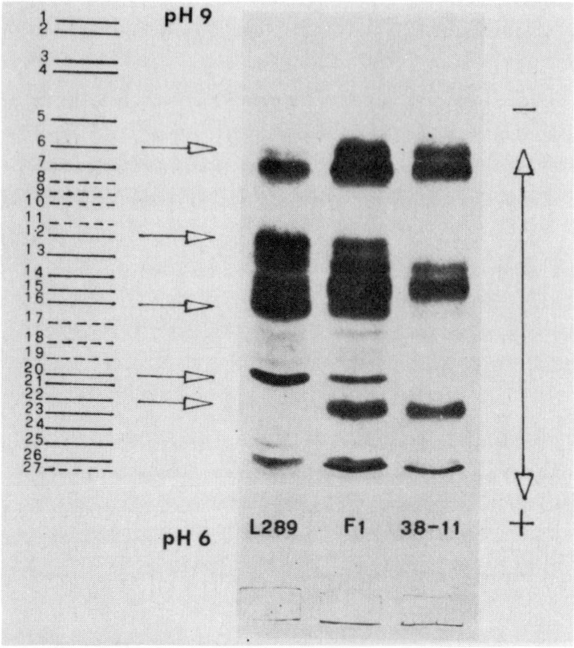

Figure 5. IEF patterns of zein from two inbreds amd a hybrid. 1, L289; 2, Fl; 3, 38–11. On the left are indicated the band positions which have been identified on one or more genotypes. The arrows indicate the bands which differ between the two parents. From [122], courtesy of the Genetical Society of Great Britain.

The terminal amino acid sequence of zein has been determined for 33 residues from the amino end [11]. The sequences from two hybrids were very similar, and the sequence of zein from an *opaque-2* hybrid showed only relatively minor differences. These results imply that there is a very large degree of homology among the different zein polypeptides detected by IEF and SDS-PAGE, in spite of a number of differences in charge and some small differences in size. The larger zein chains may have insertions in the middle or at the C-terminal. Shorter sequences determined in other laboratories were similar for the most part [25, 53]. A complete amino acid sequence for one zein polypeptide has been derived recently from the nucleotide sequence of a zein cDNA clone [31a]. It agrees quite well with the partial sequences derived from the protein itself. The amino acid composition for the complete peptide, with 213 amino acids, is given in Table 6a. Although the sequence represents only a single zein polypeptide, the amino acid composition is similar to that calculated from the zein data of Table 3. Notable is the almost complete amidation of the acidic amino acids. The molecular weight calculated from the sequence is 23,329 [31a], and it is thought to be one of the smaller (or B zein) polypeptides. This suggests that SDS-PAGE often underestimates the true molecular weight of zein.

When sequences for other polypeptides are obtained, we should learn the nature of the differences which produce the large number of fractions obtained by IEF. It seems unlikely that these differences will be large.

The zein sequence showed evidence for a seven or eight tandem repetition of a 20 amino acid repeating unit, which was highly conserved [31a]. The RS-protein (water-soluble, alcohol-soluble glutelin) has been partially sequenced [10a, 29a]. This protein also has a repeating unit, in this case the hexapeptide -Pro-Pro-Pro-Val-His-Leu-, which occurred up to eight times. The proteins of maize endosperm have some unusual properties, and the detailed data is just beginning to appear.

SALT-SOLUBLE NITROGEN

The first fraction obtained when seeds are extracted by the classical Osborne-Mendel technique (the L-M Fraction I) includes the albumins, globulins, and a non-protein nitrogen (NPN) fraction which includes free amino acids and small peptides. The amount of N in fraction I varies considerably among different reports, even when only normal genotypes are considered. It may be particularly susceptible to the history of the sample (i.e., harvest date, storage conditions, etc.) and to the specific techniques of drying [130], grinding, homogenizing, etc.

Most of the germ proteins occur in this fraction, so that there will be differences in quantity and quality between whole seed and endosperm extracts. In addition, the germ is rich in phytic acid [80], one of the factors affecting solubility of proteins after extraction [19]. The proteins in the salt-soluble fraction are usually considered to be metabolic and perhaps structural proteins, but in the cereals are not storage proteins. Mutations which affect this fraction probably adversely affect the N metabolism of the seed, and thus changes of a magnitude which change the amino acid composition are unlikely.

The free amino acid fraction is the least known of the nitrogen fractions, and a search of the literature provided a rather variable pattern of amino acids from normal genotypes (Table 7). Only a few generalizations may be made. Aspartic acid and glutamic acid (or their amides) are present in relatively large amounts, as expected from their role as transport amino acids. High levels of proline and alanine occur in some samples. Among the non-transport amino acids, lysine is present in relatively large amounts, while leucine is found in small amounts, espectially in relation to the concentration found in the seed protein. The developing endosperm appears to be adequately supplied with lysine from the vascular sap [2,100,106], and during development the level of free lysine in the endosperm exceeds that of leucine [3,74]. There is also evidence that the developing endosperm has enzymes which can carry out extensive N metabolism [104,105]. The data do not support the idea that zein might be a preferred storage protein because the input of lysine is limiting – relative to the rates of incorporation into protein leucine appears to be the limiting amino acid in the maize endosperm.

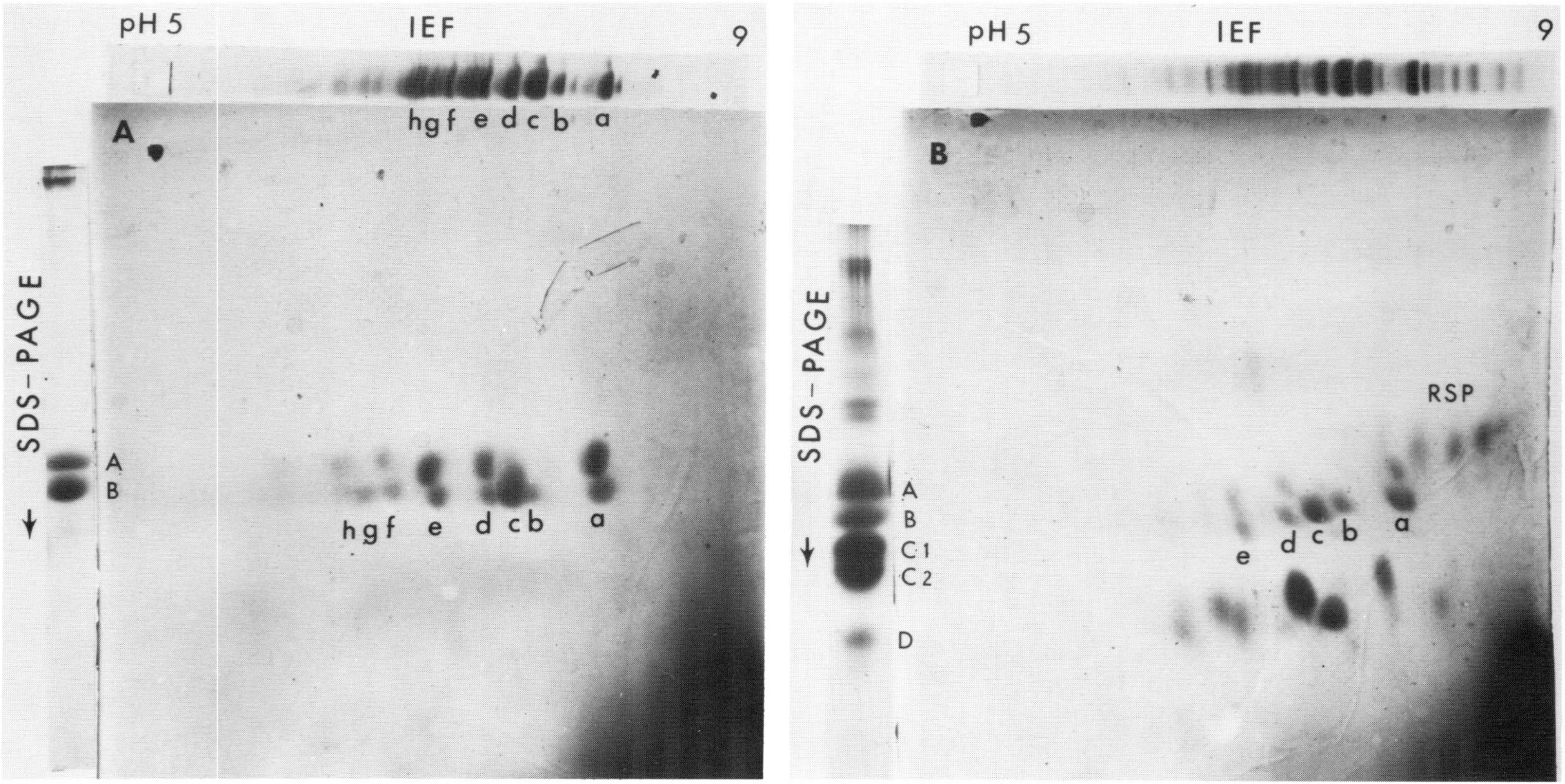

Figure 6. Two-dimensional separation of alkylated zein-1 (A) and zein-2 (B). The first dimension (IEF) is matched with a similar gel from another run, while the second dimension (SDS-PAGE) is matched with a similar gel, also shown as 1 and 4 in Fig. 2. From [137].

Table 7. Amino acid compositions of non-prolamin maize endosperm proteins and free amino acids, and the average of seed proteins from 379 species.

Amino Acid	Albumin[1]	Globulin Sequential[2]	Globulin Dialysis[3]	Glutelin[4]	Free Amino Acids[5], Range	Mean, seed from 379 species[6]
			mol %			
Lysine	5.9	5.0	5.5	5.4	2.5 - 10	4.6
Histidine	2.1	2.2	3.3	2.7	0.6 - 2.4	2.3
Arginine	5.6	5.4	9.5	4.7	1.1 - 7.3	7.6
Aspartic Acid	9.2	9.1	7.8	8.3	24 - 33	9.8
Threonine	5.6	5.0	3.9	5.0	1.0 - 3.6	4.3
Serine	6.2	6.4	7.4	6.0	3.5 - 6.3	6.1
Glutamic Acid	11.5	11.9	14.6	12.2	12.5 - 37	17.6
Proline	6.1	7.3	4.4	6.8	5.2 - 19.6	5.8
Glycine	11.2	10.5	9.9	8.6	1.2 - 8.9	10.0
Alanine	11.0	10.5	8.5	9.8	2.2 - 22.5	6.9
Cysteine	0-4.6	0-2.9	0-2.0	0-2.9	0 - 3.5	-
Valine	6.9	6.3	6.6	7.2	1.1 - 2.3	6.0
Methionine	1.3	1.4	1.3	2.2	0 - 0.6	1.6
Isoleucine	3.9	3.8	3.4	4.4	0.5 - 1.8	4.2
Leucine	7.1	7.7	6.6	9.7	0.6 - 1.5	7.1
Tyrosine	2.5	2.7	2.5	2.5	1.3 - 2.3	2.5
Phenylalanine	2.9	3.5	4.1	4.0	0.7 - 1.0	3.6
Asparagine					(7.2 - 9.1)	
Glutamine					(4.9 - 15.9)	
% of total N	1.3-6	1.2 - 5.7		15-27		

[1]Average from references: 51,73,85,87,107.

[2]Globulins extracted after albumins had been extracted by water, average from references: 51,107,133.

[3]Globulins separated from albumins by dialysis against water, after extraction by a saline solution. Average of references: 73,85,87.

[4]Fraction V (G_3) isolated by the Landry-Moureaux procedure, average from references: 51,87,68,97.

[5]The values for aspartic acid and glutamic acid include the amides, because they were not always reported separately. References: 3,7,74,107.

[6]Reference: 123.

The albumins and globulins were defined by Osborne [82] as 'soluble in water and coagulable by heat' and 'proteins which are insoluble in water but readily soluble in dilute saline solutions from which they are precipitated by dilution or dialysis', respectively. Osborne did not employ sequential extractions with water and then saline solutions, and commented that a water extract will include some globulins [82]. Reproducible distinctions between these two fractions are difficult to make. The amino acid compositions of albumins, taken from several sources, do not differ much according to the method of preparation, but the globulin amino acid compositions are different for samples prepared by sequential extraction compared to those prepared by dialysis of a single extract (Table 7). Albumin and the 'sequential' globulins are much alike. The 'dialysis' globulins contain appreciably more arginine, but also differ in histidine, aspartic acid, threonine, and proline. An example of the overlapping of these two fractions occurred during preliminary work on the extraction of maize RNase I (an albumin) from lyophylized mature endosperms (Wilson, unpublished). Water extracted between 30 and 90% of the total activity, depending upon the genotype. The same amount of RNase was obtained with a single extraction using 0.5M KCl as with water extractions followed by saline extractions.

As expected, gel electrophoresis reveals that these fractions consist of a large number of different polypeptide chains. When examined separately, the albumins and globulins show differences, which can be either major [85] or minor [84]. As shown in Figure 7, the two fractions differ in contents of some polypeptides, some being concentrated in the albumin, others in the globulin, but still others occurring about equally in both. There appears to be some overlap with polypeptides found in the glutelin fractions as well [135].

Only minor differences have been found for SDS-PAGE [137, Wilson unpublished] and starch gel electrophoresis [87] patterns of the salt-soluble proteins of various lines and mutants. The *o2* version of W64A lacked one SDS-PAGE band found in the normal and *fl2* versions [23]. Patterns characteristic for certain lines and mutants are clearly shown after Davis-type PAGE (Figure 8) [136]. The three normal inbreds differed in the relative amounts of the three bands labelled B1, B2, and B3. The other major band, labelled A', has been identified as sucrose synthase [17, 136]. The introduction of the *opaque-2* mutation into these inbreds caused different changes in each. New bands appeared in Oh45B, the B2 band was greatly reduced in Oh7N, and the B2 and B3 bands were reduced in N28. Only one minor difference, not readily detected here, was found in all *o2* lines. This suggests that changes in N metabolism differ according to the background in which the mutant gene is placed. The four *floury-2* lines, however, all showed a decrease in the relative intensity of the B1 band. These patterns are not unique for most lines, but may be of use when combined with other markers in a system for identification of maize lines by seed characteristics.

A few of the many enzymes and other specific proteins of the salt soluble fraction have been isolated and analyzed. RNase I makes up only 0.025% of the total endosperm protein, and cannot be detected if stained for protein after PAGE of a

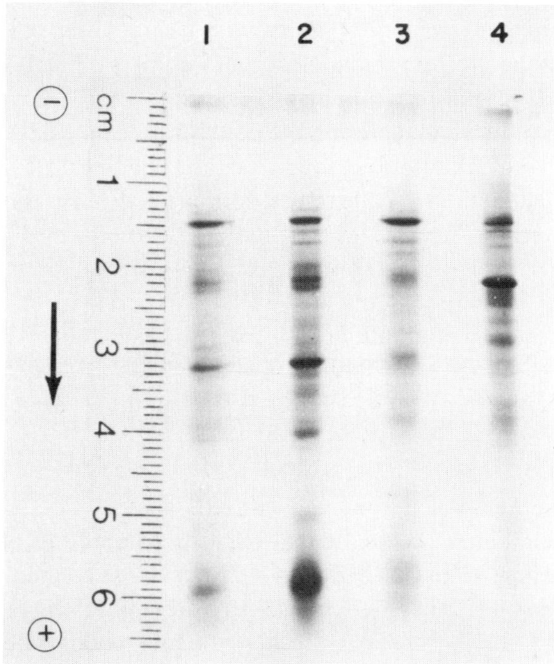

Figure 7. SDS-PAGE of salt-soluble proteins and glutelins. 1, total salt-soluble proteins; 2, albumins; 3, globulins; 4, alkylated glutelins [137].

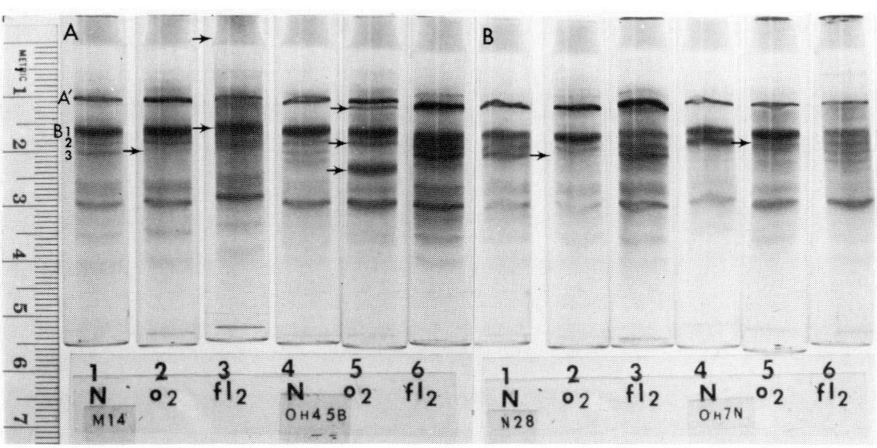

Figure 8. PAGE of soluble proteins from normal, *opaque*-2, and *floury*-2 endosperms of inbreds harvested 50 days after pollination [136].

whole endosperm homogenate [134]. Sucrose synthase, at the other extreme, may make up 2.8% of the soluble protein in the endosperm [110], but it can be detected by PAGE only at or before physiological maturity [136]. The amino acid compositions of RNase [134] and sucrose synthase [110] are similar to the average values for albumins. Trypsin inhibitors are found in maize endosperm [113], and the levels are higher in *o2* lines [60]. It is difficult to determine how much protein is involved, but may be between 0.1 and 1% of the total endosperm protein [my calculations from 109 and 113]. The reported amino acid compositons are similar to the average values for endosperm albumins, but are low in lysine and the aromatic amino acids [109, 113]. The trypsin inhibitor has no effect on corn borer larvae [109], but fortunately also has no effect on the growth of rats [69]. The isolated maize inhibitor is a much less effective inhibitor on a molar basis than the soybean Kunitz inhibitor [109].

Some very unusual polypeptides, thought to be similar to the purothionins found in wheat flour, were extracted from maize using 0.05 M sulfuric acid [41]. They are also soluble in ammonium acetate, so they might be components of the salt-soluble fraction. Six fractions were isolated from a CM-cellulose column, and had amino acid compositions with up to 13% lysine and cystine and 25% arginine. They made up only 0.1% of the total protein, and seem not to have been noticed in other maize extracts.

Somewhat different pictures of the developmental changes in the salt-soluble proteins are obtained when sequential albumin-globulin fractions are compared to a single saline extract. In several reports, the albumin fraction reached a definite peak 20–25 days after pollination, then declined to much lower levels [13, 115, 118]. If radioactive amino acids were provided, loss of activity from the albumins occurred, further confirmation that turnover and net loss of protein was occurring [108]. The globulin fraction showed only a slow increase throughout development. The magnitude of the albumin changes was such that it should be detectable in a single saline extract, but most such studies showed increases later in development than occurred with the albumins alone, and little or no decline at maturity [55, 66, 74, 102]. A drop was noted in one experiment [23]. The rise to a peak followed by a decline for the albumin fraction follows the course often seen for endosperm enzymes [105, 121]. However, the proteins may still be present, but may have become difficult to extract as the endosperms dried down, as was RNase I. These proteins might well be recovered in the glutelin fraction if they are not extracted by a saline solution [75].

GLUTELINS

The glutelin fraction is difficult to work with, almost by definition – starting with Osborne's definition which 'includes those proteins which are not dissolved by neutral aqueous solutions, by saline solutions, or by alcohol' [82]. Classical studies examined those proteins which could be extracted by dilute alkaline solutions (which would cause protein degradation) or considered the residue left after alcohol extractions as the glutelins (an unsatisfactory state for most studies). Before reducing agents or sodium acetate were added to the alcohol used for extracting the prolamins, the glutelin fraction often contained sizeable amounts of contaminating zein. Now most of the residue proteins can be made soluble through the use of reducing agents such as ME and DTT and detergents such as SDS. The glutelins are thus much more accessible for further study. This fraction has a strong tendency to form disulfide bonds in solution, if they did not already exist *in vivo*. Some of the problems and some solutions have been recently reviewed [131]. Some of the reactions involving disulfide bonds in cereal proteins have been outlined in more detail [129]. Alkylation of the sulfhydryl groups will assist studies of glutelin polypeptides.

It is the author's viewpoint that as proteins are isolated from Osborne's glutelin fraction and then characterized, they should be identified by these characteristics, and no longer considered to be glutelins. Thus the mixture of proteins in the L-M fraction III (zein-2 or G_1) was found to consist of zein-1, several small prolamins, and the RS-protein. The RS-protein, with its high cysteine content and its unusual solubility requirements, is difficult to classify. As noted above, it has been identified or recovered from saline extracts, alcohol extracts, and high pH extracts, all with reducing agent, and it also can be detected in the insoluble residue after most of the glutelin has been extracted [86]. It has been called 'glutelin-like', as part of the L-M fraction IV, but this term is not informative. If the RS-protein is active in forming disulfide cross-links with other proteins, the stage of fractionation at which it is removed could have considerable effect on the components which are finally recovered as glutelin or L-M Fraction V (G_3).

The fraction V protein, extracted with SDS, makes up about 1/6 of the total protein of the endosperm (Table 2), so that its composition is of importance to the overall amino acid pattern. The glutelin pattern (Table 7) is similar to those of the albumins and globulins, and on the average is of good nutritional quality (cf. Table 1). Research on glutelins has been reviewed recently [131, 137]. There are a number of polypeptides in the glutelin fraction but these have not been thoroughly investigated on a quantitative basis. Only one polypeptide, solubilized by urea, was present in appreciable quantity after SDS-PAGE (at 21 mm, in Figure 7), but it was not considered to be a storage protein [137]. Some of the bands revealed on SDS-PAGE appear to coincide with those of the salt-soluble fraction. Many bands are also seen after IEF [137]. This is not surprising, considering that some enzymes become insoluble during the final stages of seed maturation, and may be displaced to the glutelin fraction. No specific proteins have been traced from one fraction to another during development, however. It would be interesting to determine the fate of sucrose synthase, which disappears from the salt-soluble fraction at maturation

[136]. Are the tetrameric polypeptide chains hydrolyzed to release amino acids for incorporation into storage proteins, or are the intact chains made insoluble and perhaps enzymatically inactive, either in the form of monomers bound to some cell structure or as aggregates?

THE EFFECTS OF GENOTYPE AND FERTILITY ON PROTEIN COMPOSITION

The *opaque*-2 mutation causes a decrease in the amount of zein accumulated in maize endosperm with an increase in the amounts, at least in relative proportions, of the other proteins, and an increase in the lysine content expressed either as a percentage of the total amino acids or as a percentage of the dry weight. Osborne and Mendel, in pioneering studies on essential amino acids [83], noted that zein, lacking in lysine and tryptophan, did not permit maintenance of body weight by rats. Preparations of the other maize proteins did allow for growth. Although the *o2* mutants possess different relative amounts of the different protein fractions, the average amino acid compositions of the fractions are unchanged [68, 84, 94, 107]. However, the response of the maize endosperm to the introduction of the mutation is variable, as shown by the examples in Table 6. Thus zein may make up 3% or 36% of the total nitrogen, perhaps a one-third reduction from the normal level. Differences in reported zein levels may be due to extraction techniques or to the genotypes examined. *Opaque*-2 lines usually cease the accumulation of zein at an early stage, 30 to 35 days after pollination [20, 23, 102, 115, 119], but mutant hybrids continue zein accumulation for a longer period than do inbreds [20, 102]. As noted earlier (Figure 8), there are no changes in soluble proteins which are characteristic of the mutation – different inbreds have different changes in PAGE patterns. The *o7* mutation also represses zein synthesis at an early stage [24].

The level of free amino acids in higher in *o2* endosperms, up to five times the normal level [3, 67, 107]. A high free amino acid content is a potential screening test for detection of high-lysine varieties [61], but the levels of total free amino acids do not increase in proportion to the total endosperm lysine content [67]. There is a potential drawback to the use of the free amino acid levels as a screen for high-lysine varieties. The free amino acids accumulate when the endosperm loses the ability to synthesize normal quantities of zein and the synthesis of other proteins does not fully utilize the available amino acids. The high free amino acid content may inhibit transport of nitrogen compounds into the endosperm, reducing the yield of nitrogen in the mature grain. It has been postulated that this could inhibit sucrose transport and thus starch yield [119]. Further, considering subsequent storage, processing, cooking, etc., of maize before it is used for food or feed, protein is a better storage product than are the free amino acids. The most desirable high-lysine genotype is one which has replaced the zein by accumulating higher levels of the other proteins, which on the average have adequate lysine (Table 1). A screen for increased free amino acids would miss such a genotype. If the meal is first extracted with trichloroacetic acid to remove free amino acids, the ninhydrin color reaction can be used to determine the lysine content of the protein [10].

When zeins from mutants are compared by SDS-PAGE, it is found that *o2* reduces the relative level of A-zein [35, 42, 49, 54, 103, 120], *o7* reduces the B-zein [24] while *fl2* reduces both major zein bands [42, 54, 103, 120]. Double mutants of *o2* and various starch mutants have variable effects, but when zein synthesis is most strongly inhibited, the alcohol-soluble proteins include relatively large amounts of the small C- and D-prolamins and the RS-protein (Figure 3) [120]. Zein synthesized *in vitro* by polysomes from *opaque*-2 endosperms was deficient in A-zein as compared to that using normal polysomes [43].

More detailed studies of the mutant endosperms are reported in the chapter by Denić.

Protein and the zein-containing protein bodies are not uniformly distributed throughout the endosperm [6, 16, 48, 94, 95, 138]. Different formations of protein bodies ranging from complete absence through small, clumped protein bodies mixed with matrix protein, to full-sized protein bodies, are reported in a series of genotypes from *opaque*-2 through modified *opaque*-2 to normal [31].

The zein content rises at a more rapid rate than any other protein fraction as the available N is increased [13, 21, 118, 119, 139]. The lysine content of the endosperm increases in absolute amount, and as a percentage of the dry weight, but decreases as a percentage of protein [118, 139]. Thus the gains in nutritive value are less than the gains in total protein content in normal maize lines. The proportion of zein does not increase in *o2* lines [21, 119, 140], so the nutritive value of the product will be in proportion to the total protein content. Unfortunately, the *o2* inbreds and hybrids do not respond to fertilizer treatments as efficiently as do normal inbreds and hybrids, and the relative yield penalty increases as yields increase [22, 119].

RNase activity may be as much as five times higher in *opaque*-2 inbred endosperms than in the corresponding normal lines [135]. Although this enzyme has the potential for disruption of normal RNA metabolism, and thus of protein synthesis, the mode of action is unknown, and it may not be directly related to the reduction in zein synthesis. The level of RNase activity of some normal inbreds is as high as that in some *o2* lines [135]. The wide ranging levels of RNase during endosperm development raises questions about the relationships between measured enzyme activities and the rates of metabolism *in vivo*. *Opaque*-2 endosperms have been reported to have higher, lower, or unchanged levels of several enzymes involved in nitrogen [115] or carbohydrate metabolism [30, 45]. Comparisons involving only one inbred may not be meaningful, when the normal range of enzyme activities are not known. However, *opaque*-2 mutants are characterized by a premature cessation of the accumulation of starch as well as of zein, so that changes in the enzymes of starch synthesis might well be involved.

The reduced yield in *o2* endosperms is usually considered to be the result of impaired metabolism within the endosperm. When *o2* plants are pollinated by a mixture of normal and *o2* pollen, the heterozygous normal endosperms have higher dry weights and protein contents than the mutant endosperm [106, 119]. Lysine metabolism also differs [106]. The normal and mutant endosperms on mixed ears have equal nutrient supplies, so that differences must be related directly to differences in endosperm metabolism. However, yield studies on *o2* hybrids, pollinated with only normal pollen, show that the average yields on the *o2* plants are less than on homozygous normal plants, even though kernel sizes are the same [57]. Yields are equal for some lines. Photosynthetic rates may also be lower for *o2* inbreds than for normal inbreds [70]. These results suggest that the *o2* mutation may have an affect on the vegetative plant. These results need to be confirmed on a wider variety of inbreds and hybrids, for they pose a potential limit in the search for high-yielding, high-lysine maize varieties.

PROTEINS OF RELATED SPECIES

The wild ancestor of maize is probably the tropical grass *Zea mays* spp *mexicana*, commonly called teosinte. The genus *Tripsacum* includes other wild grasses which are more distantly related, but both teosinte and *Tripsacum* form hybrids with maize. Sorghum and the millets are still more distantly related, but are much closer than the other cereals. These relationships are shown in Table 8.

Comparisons were recently made between the L-M fractions of maize, teosinte, and *Tripsacum* [87]. The amino acid composition (Table 9) and the mobilities in SDS-PAGE of the prolamin fraction of teosinte closely resemble those of maize zein, and therefore this prolamin should be termed zein. Even the small molecular weight teosinte prolamins appeared to have a similar mobility to those of maize. The *Tripsacum* prolamin showed slight differences on PAGE at pH 3.1 as well as on SDS-PAGE, and there were appreciable differences among the small molecular weight prolamins of Fraction III polypeptides. These results are consistent with the close relationship between teosinte and maize, and the slightly more distant relationship of *Tripsacum*.

Sorghum, the nearest relative to maize of the major cereal crops, is grown widely around the world. Several recent reviews and articles present much useful data on sorghum and sorghum proteins [34, 37a, 96, 131]. There are five cultivated races (and ten intermediate races) which can be characterized by grain spiklet and head types, all with variable vegetative characters [36]. There has been little comparative work on proteins of the differing races. A preliminary study suggested that the PAGE patterns of the soluble proteins and of some isoenzymes could be used to

Table 8. Names and relationships of maize, maize relatives, and other cereals.

Family - <u>Gramineae</u>

Subfamily - <u>Pooideae</u>: wheat, rye, barley, oats

Subfamily - <u>Oryzoideae</u>: rice

Subfamily - <u>Panicoideae</u>

Tribe - <u>Paniceae</u>

Genus and species - <u>Pennisetum americanum</u> (L.) <u>Leecke</u>: pearl millet

- <u>Panicum miliaceum</u> L.: proso millet

- <u>Setaria italica</u> (L.) <u>Beauv</u>.: foxtail millet

Tribe - <u>Andropogoneae</u>

Genus and species - <u>Sorghum bicolor</u> (L.) Moench: sorghum

- <u>Zea mays</u> L.: maize, corn

- <u>Zea mays</u> ssp <u>mexicana</u> (Schrad.) Iltis: teosinte

- <u>Tripsacum dactyloides</u> L.

identify the races [99]. The highly productive sorghums grown in the United States are almost all hybrids.

The first isolated sorghum prolamin was obtained from the race known as Kafir, thus the name kafirin [40], which has been used regardless of race. As with zein, kafirin can be isolated relatively free of other proteins. The amino acid composition of kafirin (Table 9) is much like that of zein, with slightly less leucine and slightly more alanine [9, 34, 89]. From the first isolation [40], there have been reports that kafirin forms gels in alcoholic solutions, but most recent papers do not mention this – whether this is due to the kafirin itself or to some contaminant perhaps found only in some races is not known. SDS-PAGE patterns for kafirin are similar to those of zein, but it is not established whether kafirin has one or two major weight classes [34, 89]. The alcohol-ME extract contains minor amounts of small molecular weight polypeptides [89]. Urea-lactic acid PAGE separates kafirin into a number of bands [89], again similar to the zein pattern [87], and in this case varietal differences are apparent. Kafirin is best extracted with t-butanol, or hot ethanol, [9, 89, 98] but the yield of kafirin is at least doubled by the use of ME [34, 38, 89].

Colored sorghum lines contain high levels of polyphenol compounds (usually called tannins). These lines may have been selected for resistance to damage by birds, but they have a much lower nutritional quality [38]. The tannins have a high potential for binding to proteins, and drastically change the proportions of N

Table 9. Amino acid composition of zein and zein family prolamins.

	Zein Maize[1]	Zein Teosinte[2]	Tripsacum[2]	Kafirin Sorghum[3]	Pearl Millet[4]	Proso Millet[5]
			mol%			
Lysine	0.1	0.1	0.1	0.2	0.0	0.1
Histidine	1.0	1.1	1.3	0.7	1.3	1.6
Arginine	1.2	1.2	1.2	0.9	0.8	1.0
Aspartic Acid	5.1	5.0	4.7	6.2	7.0	4.0
Threonine	3.0	3.0	3.1	2.7	3.7	2.8
Serine	6.3	6.4	5.8	4.8	6.1	7.9
Glutamic Acid	21.4	21.2	21.2	21.2	22.4	21.2
Proline	10.7	10.1	9.9	10.5	8.1	8.4
Glycine	2.2	2.1	2.0	2.0	1.5	2.0
Alanine	13.3	13.7	15.9	16.4	13.9	17.2
Cysteine	(1.0)				1.1	1.3
Valine	3.6	4.3	4.1	5.2	6.1	5.2
Methionine	0.9	1.4	2.2	0.7	1.8	2.8
Isoleucine	3.8	3.8	3.8	4.2	5.0	4.2
Leucine	18.7	18.0	16.8	16.7	13.8	12.6
Tyrosine	3.5	3.4	3.6	3.4	2.2	3.0
Phenylalanine	5.2	5.2	4.4	4.5	5.1	4.9

[1]Table III.

[2]Reference: 87.

[3]Average from references: 9,34,89.

[4]Reference: 133.

[5]Average from references: 44,128.

recovered from different protein classes [14, 34, 38, 77]. The differences produced by the variable genetic background, differing growth environments, and the extractability problems caused by the tannins make it difficult to interpret results from different laboratories. The saline-soluble protein and kafirin extractions are most affected by the tannins. The tannin-protein complexes are recovered in the L-M fraction V [34, 38], so this glutelin fraction is especially contaminated by prolamin. Glutelin fractions extracted from high tannin lines are low in lysine and high in glutamic acid, alanine, and leucine, signs that kafirin is present [38, 77, 89]. Removal of tannins by dehulling whole seeds increases the recovery of saline-soluble proteins and kafirin from the dehulled portion of the seed, [14, 38], but this may involve the loss of a large fraction of the total seed protein [14]. The yield of kafirin as a percentage of the total protein increases when only the endosperm is extracted, but the increase is more than would be expected when high tannin lines are involved [14]. Only 5% of the total protein was recovered in the salt extract of a high tannin line, while 13% was recovered from a low tannin line [14]. Others have recovered 20–30% of the total N in the same fraction [77]. Kafirin recovery was as low as 11%, but increased to 44% after dehulling [14, 38] while recoveries of 50% have been reported when ME is included [89]. There are serious problems in extracting sorghum proteins. The various fractions may be similar to those of maize in some cases, but different in others, largely because of the tannin complexes.

Protein bodies occur in sorghum endosperms [1, 37, 98, 111], they dissolve in t-butanol or hot ethanol [37, 98], and thus appear to contain kafirin. The number, size, and packing of the protein bodies differs among different lines [37]. Whole sorghum apparently has a fairly good tryptophan level, about 1.3 g/100 g protein [38], higher than that reported from maize.

The discovery of the high lysine mutants of maize triggered a search for and the discovery of opaque sorghum lines which were also high in lysine [34, 101, 111]. A chemically induced mutant, P-721, also has high lysine, but the seed weight is reduced about 14% [4]. The high lysine lines also have increased tryptophan, and the increase in feeding efficiency paralleled that of *opaque*-2 maize [38, 101]. The high lysine lines typically have higher levels of salt-soluble proteins and lower levels of kafirin, but the magnitude of the reported changes differs in two reports [34, 39, 89]. Shrunken endosperms, reduced seed weight, and loss of protein bodies occur in the high lysine sorghums [4, 101, 111], but these changes are not well documented.

The parallel to maize extends even to the effect of nitrogen fertilizer – as the protein content increases the lysine content increases as percent of sample, but decreases as percent of protein [126]. The amino acid composition changes showed that kafirin increased more rapidly than the other proteins, at least in normal varieties. The response of high lysine lines has not been tested.

The millets are small-seeded grasses cultivated around the world, representing at least five genera: *Panicum*, *Pennisetum*, *Setaria*, *Eleusine*, and *Echinochloa*, with many common names. Only the first two are widely grown outside of Africa and Asia. Protein bodies occur in *Pennisetum* [1, 5], *Panicum* [44], and *Eleusine* [1]. The published data on millet proteins has been collected recently [37a]. Proso and pearl millet prolamins, at least, have amino acid compositions similar to that of zein (Table 9), with the characteristic high levels of leucine and alanine and low levels of lysine. However, the lysine content of whole grain pearl millet is higher than that of sorghum [5] or maize [133].

The prolamins of the major cereals are compared in Table 10. The prolamins of maize, teosinte, *Tripsacum*, sorghum, proso millet, and pearl millet are all similar, forming a 'zein family' with moderately high levels of glutamic acid and proline, and four to seven per cent aspartic acid. The small maize prolamins differ from zein more than do the major prolamins from the maize relatives, while the reduced-soluble protein is distinctly different. Rice and oat prolamins differ but are intermediate between the zein family and the other cereal prolamins. Wheat, rye, and barley are characterized by high levels of the amino acids which gave the prolamin fraction its name, glutamic acid (glutamine) and proline, and much lower amounts of leucine plus alanine and also aspartic acid. The zein family tends to have a higher content of the large non-polar amino acids, but the differences are not large, and barley, with high proline, has a high content of non-polar amino acids. The highest level is reached by the reduced-soluble protein. The frequency of hydrophobic side chains is appreciably higher than that in many common proteins [12]. The average hydrophobicity of zein is well above that of most proteins, including gliadin [132], but this is not necessarily related to the ability of a protein to take part in hydrophobic interactions [47].

Table 10. Prolamin comparisons among gramineae.

	Hydrophobic Amino Acids[1]	Glutamic Acid	Proline	Leucine + Alanine	Aspartic Acid
			mol %		
Maize, Zea mays[2] (zein)	45.8	21.4	10.7	32.0	5.1
Teosinte, Zea mexicana[3]	44.8	21.2	10.1	31.7	5.0
Tripsacum dactyloides[3]	42.9	21.2	9.9	32.7	4.7
Sorghum bicolor[3]	44.5	21.2	10.5	33.1	6.2
Proso millet, Panicum miliaceum[3]	38.3	21.2	8.4	29.8	4.0
Pearl millet, Pennisetum typhoides[3]	40.3	22.5	8.1	27.7	7.0
Maize, small prolamins[2]	35.3	20.3	12.5	21.9	2.4
Maize, reduced-soluble protein[2]	49.5	14.8	24.8	16.0	0.7
Rice, Oryza sativa [133]	39.4	20	5.2	21.4	7.5
Oats, Avena sativa [133]	38.6	34.6	10.4	16.4	2.3
Wheat, Triticum aestivum [133]	38.5	37.7	16.9	9.9	2.7
Rye, Secale cereale [133]	37.9	36.0	18.7	9.0	2.4
Barley, Hordeum vulgare [133]	44.9	35.9	23.4	8.6	1.7

[1]Larger nonpolar side chains: Isoleucine, tyrosine, phenylalanine, proline, leucine, valine [132].

[2]From Table III.

[3]From Table IX.

It is interesting that the amino acid compositions of the prolamins fall into two distinct classes for those species with high levels of prolamin, while the prolamins of the two species, oats, and rice, with low levels of prolamin have intermediate amino acid compositions. A comparative study on wheat, barley, and maize protein bodies suggests that the prolamins are synthesized on rough endoplasmic reticulum, but that maize differs from the other two in that the deposited protein is completely surrounded by endoplasmic reticulum [62]. Glutelins function as storage proteins in rice [46]. Three types of protein bodies have been identified in rice [8] – it is tempting to speculate that they store different proteins. Pea storage proteins may also be synthesized on the rough endoplasmic reticulum, but they may be deposited within vacuoles [62]. These differences suggest that there may be a relationship between the properties of the storage protein and the nature of the protein bodies in which they are stored. Solubility, glycosylation, the effect of proline on the three dimension tertiary structure of the proteins, etc., may determine the ability of the cell to move the storage protein from the site of synthesis to the site of storage. Different storage sites may thus be adapted to particular storage polypeptides.

ADDENDUM

I am indebted to Dr. B.A. Larkins for information on three recent papers which extend our knowledge of zein, as derived from the nucleotide sequences of cDNA clones. The first two papers (Marks, M.D., and Larkins, B.A. 1982. Analysis of Sequence Microheterogeneity among Zein Messenger RNAs. J. Biol. Chem. 257: 9976–9983; and Pedersen, K., Devereux, J., Wilson, D.R., Sheldon, E., and Larkins, B.A. 1982. Cloning and Sequence Analysis Reveal Structural Variation Among Related Zein Genes in Maize. Cell 29: 1015–1026) show that there is about 60% homology for three mRNAs for each of the two molecular weight classes of zein. The molecular weights determined from the amino acid sequences are about 23,500 for the smaller zein polypeptides and 26,700 for the larger. There are tandem repeats of approximately 20 amino acids in both size classes. From these data, a speculative structural model was derived (Argos, P., Pedersen, K., Marks, M.D., and Larkins, B.A. 1982. A Structural Model for Maize Zein Proteins. J. Biol Chem. 257: 9984–9990). Zein may have a structure with nine adjacent helices clustered within a distorted cylinder, with glutamine-rich turns and caps, which favor stacking. Specific structures of this type could limit the allowable amino acid variations among the zein polypeptides. Perhaps the ‘zein family’ of prolamins (Table 10) shares a common basic structure which accounts for the similarities of amino acid composition.

REFERENCES

1. Adams, CA, L Novellie, and N v.d.W Liebenberg (1976): Biochemical properties and ultrastructure of protein bodies isolated from selected cereals. Cereal Chem. 53: 1–12.
2. Arruda, P and WJ Da Silva (1979): Amino acid composition of vascular sap of maize ear peduncle. Phytochem. 18: 409–410.
3. Arruda, P, WJ Da Silva, JPF Teixeira (1978): Protein and free amino acids in a high lysine maize double mutant. Phytochem. 17: 1217–1218.
4. Axtell, JD, SW Van Scoyoc, PJ Christensen, and G Ejeta (1979): Current status of protein quality improvement in grain sorghum. Seed Protein Improvement in Cereals and Grain Legumes II: 357–366; IAEA Vienna.
5. Badi, SM, RC Hoseney, and AJ Casady (1976): Pearl millet. I. Characterization by SEM, amino acid analysis, lipid composition, and prolamine solubility. Cereal Chem. 53: 478–487.
6. Baenzinger, PL, and DV Glover (1977): Protein body size and distribution and protein matrix morphology in various endosperm mutants of *Zea mays* L. Crop Sci. 17: 415–421.
7. Baudet, J, J Mossé, J Landry, and T Moureaux (1966): Étude sur les protéines du maïs. I. Composition en acides aminés des fractions azotées du grain. Ann. Physiol. Veg. 8: 321–329.
8. Bechtel, DB, and BO Juliano (1980): Formation of protein bodies in the starchy endosperm of rice (*Oryza sativa* L.) A Re-investigation. Ann. Bot. 45: 503.
9. Beckwith, AC, and RW Jones (1972): Physical chemical characterization of grain sorghum prolamine fractions and components. J. Agric. Food Chem. 20: 259–261.
10. Beckwith, AC, JW Paulis, and JS Wall (1975): Direct estimation of lysine in corn meals by the ninhydrin color reaction. J. Agr. Food Chem. 23: 194–196.

10a. Bietz, JA, A Esen, JW Paulis, and JS Wall (1981): Amino Acid Sequence Investigations of Isolated Corn Endosperm Protein. Cereal Foods World 26: 500.

11. Bietz, JA, JW Paulis, and JS Wall (1979): Zein subunit homology revealed through amino-terminal sequence analysis. Cereal Chem. 56: 327–332.
12. Bigelow, CC (1967): On the average hydrophobicity of proteins and the relation between it and protein structure. J. Theor. Biol. 16: 187–211.
13. Carter, JN (1950): The effects of balanced and unbalanced conditions of nitrogen and phosphorus of the soil on the quantity and quality of protein in corn grain. Thesis Univ. of Illinois, Urbana; 324 pp.
14. Chibber, AK, ET Mertz, and JD Axtell (1978): Effects of dehulling on tannin content, protein distribution, and quality of high and low tannin sorghum. J. Agric. Food Chem. 26: 679–683.
16. Choe, Bong-Ho, BG Cumbie, and MS Zuber (1974): Association of zein body classification with lysine content of corn (*Zea mays* L.) endosperm. Crop Sci. 14: 187–190.
17. Chourey, PS, and OE Nelson (1976): The enzymatic deficiency conditioned by the shrunken-1 mutations in maize. Biochem. Genet. 14: 1041–1055.
18. CIMMYT (1975): High Quality Protein Maize. Dowden, Hutchinson and Ross, Stroudsberg, PA; 524 pp.
19. Craine, EM, and KE Fahrenholtz (1958): The proteins in water extracts of corn. Cereal Chem. 35: 245–259.
20. Dalby, A, and CY Tsai (1974): Zein accumulation in phenotypically modified lines of *opaque-2* maize. Cereal Chem. 51: 821–825.
21. Decau, J, and B Pujol (1969): Influence de l'alimentation hydrique et azotée sur la production et la qualité comparcé des matières protéiques de grains de maïs portant ou non le gène *opaque-2* (*o2*). C. R. Acad. Sc. Paris 268: 2343–2346.
22. Decau, J, and B Pujol (1976): Influence de reductions artificielles de la fécondation de l'épi sur la photosynthèse nette et le rendement du mais (*Zea mays* L.). Etude comparée d'un maïs normal et d'un mutant '*opaque 2*' (*o2*). C. R. Acad. Sci. Paris 283: 923–926.

23. Di Fonzo, N, E Fornasari, F, Salamini, and C Soave (1977): SDS-protein subunits in normal, *opaque-2* and *floury-2* maize endosperms. Maydica 22: 77–88.
24. Di Fonzo, N, E Gentinetta, and F Salamini (1979): Action of the *opaque-7* mutation on the accumulation of storage products in maize endosperm. Plant Sci. Lett. 14: 345–354.
25. Drenska, AI, KD Ganchev, and CP Ivanov (1980): Amino acid sequence of zein fractions in the N-terminal region. C. R. Acad. Bulg. Sci. 33: 67–70.
26. Duhamel, RC, Meezan, and K Brendel (1980): Metachromatic staining with Coomassie Brilliant Blue R-250 of the proline-rich calf thymus histone, Hl. Biochim. Biophys. Acta 626: 432–442.
27. Dunker, AK, and RR Rueckert (1969): Observations on mulecular weight determinations on polyacrylamide gels. J. Biol. Chem. 244: 5074–5080.
28. Esen, A (1978): A simple method for quantitative semiquantitative and qualitative assay of protein. Anal. Biochem. 89: 264–273.
29. Esen, A (1980): Fractionation of zein by ion-exchange chromatography on phosphocellulose. Cereal Chem. 57: 75–76.
29a. Esen, A, JA Bietz, JW Paulis, and JS Wall (1982): Tandem repeats in the N-terminal sequence of a proline-rich protein from corn endosperm. Nature 296: 678–679.
30. Fullerton, SG, and LV Svec (1976): Acid invertase activity in kernels of normal and *opaque-2* corn at different growth stages through maturity. Crop. Sci. 16: 419–422.
31. Gentinetta, E, T Maggiore, F Salamini, C Lorenzoni, F Pioli, and C Soave (1975): Protein studies in 46 *opaque-2* strains with modified endosperm texture. Maydica 20: 145–164.
31a. Geraghty, D, MA Peifer, I Rubenstein, and J Messing (1981): The primary structure of a plant storage protein: zein. Nucleic Acids Res. 9: 5163–5174.
32. Gianazza, E, V Viglienghi, PC Righetti, F Salamini, and C Soave (1977): Amino acid composition of zein molecular components. Phytochem. 16: 315–317.
33. Gomes, JC, U Koch, and JR Brunner (1979): Isolation of a trypsin inhibitor from navy beans by affinity chromatography. Cereal Chem. 56: 525–529.
34. Guiragossian, V, AAK Chibber, S Van Scoyoc, R Jambunathan, ET Mertz, and JD Axtell (1978): Characteristics of proteins from normal, high lysine and high tannin sorghums. J. Agric. Food Chem. 26: 219–223.
35. Hagen, G, and I Rubenstein (1980): Two-dimensional gel analysis of the zein proteins in maize. Plant Sci. Lett. 19: 217–223.
36. Harlan, JR, and JMJ deWet (1972): A simplified classification of cultivated sorghum. Crop Sci. 12: 172–176.
37. Hoseney, RC, AB Davis, and LH Harbers (1974): Pericarp and endosperm structure of sorghum grain shown by scanning electron microscopy. Cereal Chem. 51: 552–558.
37a. Hulse, JH, EM Laing, and OE Pearson (1980): Sorghum and the Millets: Their Composition and Nutritive Value. Academic Press, New York; 997 pp.
38. Jambunathan, R, and ET Mertz (1973): Relationship between tannin levels, rat growth, and distribution of proteins in sorghum. J. Agric. Food Chem. 21: 692–696.
39. Jambunathan, R, and ET Mertz (1975): Fractionation of soluble proteins of high-lysine and normal sorghum grain. Cereal Chem. 52: 119–121.
40. Johns, CO, and JF Brester (1916): Kafirin, an alcohol-soluble protein from Kafir, *Andropogon sorghum*. J. Biol. Chem. 28: 59–65.
41. Jones, BL, and DB Cooper (1980): Purification and characterization of a corn (*Zea mays*) protein similar to purothionins. J. Agric. Food Chem. 28: 904–908.
42. Jones, RA (1978): Effects of *floury-2* locus on zein accumulation and RNA metabolism during maize endosperm development. Biochem. Genet. 16: 27–38.
43. Jones, RA, BA Larkins, and CY Tsai (1976): Reduced synthesis of zein in vitro by a high lysine mutant of maize. BBRC 69: 404–410.
44. Jones, RW, AC Beckwith, U Khoo, and GE Inglett (1970): Protein composition of proso millet. J. Agric. Food Chem. 18: 37–39.

45. Joshi, S, ML Lodha, and SL Mehta (1980): Regulation of starch biosynthesis in normal and *opaque-2* maize during endosperm development. Phytochem. 19: 2305–2310.
46. Juliano, BO, and D Boulter (1976): Extraction and composition of rice endosperm glutelin. Phytochem. 15: 1601–1606.
47. Keshavarz, E, and S Nakai (1979): The relationship between hydrophobicity and interfacial tension of proteins. Biochim. Biophys. Acta 576: 269–279.
48. Khoo, U, and MJ Wolf (1970): Origin and development of protein granules in maize endosperm. Amer. J. Bot. 57: 1042–50.
49. Landry, J (1979): La zéine du grain de maïs. Préparation et caractérisation. Biochimie 61: 549–558.
50. Landry, J, and T Moureaux (1970): Hétérogénéité des glutélines du grain de maïs: extraction sélective et composition en acides aminés des trois fractions isolées. Bull. Soc. Chim. Biol. 52: 1021–1037.
51. Landry, J, and T Moureaux (1980): Distribution and amino acid composition of protein groups located in different histological parts of maize grain. J. Agr. Food Chem. 28: 1186–1191.
52. Larkins, BA, K Pedersen, AK Handa, WJ Hurkman, and LD Smith (1979): Synthesis and processing of maize storage proteins in *Xenopus laevis* oocytes. PNAS 76: 6448–6452.
53. Larkins, BA, K Pedersen, WJ Hurkman, AK Handa, AC Mason, CY Tsai, and MA Hermodson (1980): Maize storage proteins: characterization and biosynthesis. In: Leaver, CJ (ed.): Genome Organization and Expression in Plants: 203–218; Plenum Press, New York.
54. Lee, KH, RA Jones, A Dalby, and CY Tsai (1976): Genetic regulation of storage protein content in maize endosperm. Biochem. 14: 641–650.
55. Lodha, ML, KN Srivastava, PC Ram, and SL Mehta (1978): Developmental changes in endosperm proteins, lysine and tryptophan of normal and *opaque*-2 *Zea mays*. Physiol. Pflanz. 173: 123–128.
56. Ma, Y, and OE Nelson (1975): Amino acid composition and storage proteins in two new high-lysine mutants in maize. Cereal Chem. 52: 412–419.
57. Makonnen, D (1977): Effect of the *opaque-2* gene and outcrossing on characteristics of the kernel protein and lysine in maize. Econ. Bot. 31: 61–65.
58. Melcher, U (1979): *In vitro* synthesis of a precursor to the methionine-rich polypeptide of the zein fraction of corn. Plant Physiol. 63: 354–358.
59. Melcher, U, and B Fraij (1980): Methionine-rich protein fraction prepared by cryoprecipitation from extracts of corn meal. J. Agric. Food Chem. 28: 1334–1336.
60. Mertz, ET (1972): Recent improvements in corn protein. In: Inglett, GE (ed.): Symposium: Seed Proteins, Westport, CN: 136–143; Avi Publishing Co.
61. Mertz, ET, PS Misra, and R Jambunathan (1974): Rapid ninhydrin color test for screening high-lysine mutants of maize, sorghum, barley, and other cereal grains. Cereal Chem. 51: 304–307.
62. Miflin, BJ, SR Burgess, and PR Shewry (1981): The development of protein bodies in the storage tissues of seeds: Subcellular separations of homogenates of barley, maize, and wheat endosperms and of pea cotyledons. J. Exp. Bot. 32: 199–219.
63. Miflin, BJ, and PR Shewry (eds.) (1977): Techniques for the Separation of Barley and Maize Seed Protein. Comm. Europ. Comm., Luxembourg; 114 pp.
64. Miflin, BJ, and PR Shewry (1979): The biology and biochemistry of cereal seed prolamins. Seed Protein Improvement in Cereals and Grain Legumes I: 137–158; IAEA Vienna.
65. Misra, PS, ET Mertz, and DV Glover (1975): Studies on corn proteins. VI. Endosperm protein changes in single and double endosperm mutants of maize. Cereal Chem 52: 161–166.
66. Misra, PS, ET Mertz, and DV Glover (1975): Studies on corn proteins. VII Developmental changes in endosperm proteins of high-lysine mutants. Cereal Chem. 52: 734–739.
67. Misra, PS, ET Mertz, and DV Glover (1975): Studies on corn proteins. VIII. Free amino acid content of *opaque-2* double mutants. Cereal Chem. 52: 844–848.
68. Misra, PS, ET Mertz, and DV Glover (1976): Studies on corn proteins. IX. Comparison of the

amino acid composition of Landry-Moureaux and Paulis-Wall endosperm fractions. Cereal Chem. 53: 699–704.

69. Mitchell, HL, DB Parrish, M Cormey, and CE Wasson (1976): Effect of corn trypsin inhibitor on growth of rats. J. Agric. Food Chem. 24: 1254–1255.
70. Morot-Gaudry, JF, J Farineau, and E Jolivet (1979): Effect of leaf position and plant age on photosynthetic carbon metabolism in leaves of 8 and 16 day old maize (*Zea mays*) cultivar Wisconsin seedlings W-64A with and without the gene *opaque-2*. Photosynthetica 13: 365–375.
71. Mossé, J (1961): Monographie sur une protéine du maïs: la zéine. Ann. Physiol. Veg. 3: 105–139.
72. Mossé, J, and J Landry (1980): Recent research on major maize proteins: zeins and glutelins. In: Inglett, GE, and L Munck (eds.): Cereals for Food and Beverages: 255–272; Academic Press, New York.
73. Moureaux, T, J Baudet, and J Mossé (1966): Fractionnement des albumines du Mais par chromatographie sur Sephadex. C. R. Acad. Sc. Paris 262: 1710–1713.
74. Moureaux, T, and J Landry (1972): La maturation du grain de Maïs. Évolution qualitative et quantitative des différentes formes azotées. Physiol. Veg. 10: 1–18.
75. Moureaux, T, and J Landry (1972): Effets du gène opaque 2 sur protéogenèse du grain de Maïs au cours de la maturation. C. R. Acad. Sci. Paris 274: 3309–3312.
76. Nelson, OE (1980): Genetic control of polysaccharide and storage protein synthesis in the endosperms of barley, maize, and sorghum. In: Pomeranz, Y (ed.): Advances in Cereal Science and Technology III: 41–71; Am. Assoc. Cereal Chemists; St. Paul, MN.
77. Neucere, NJ, and G Sumrell (1979): Protein fractions from five varieties of grain sorghum: amino acid composition and solubility properties. J. Agric. Food Chem. 27: 809–812.
78. Noelken, ME, BJ Wisdom Jr., and BG Hudson (1981): Estimation of the size of collagenous polypeptides by sodium dodecyl sulfate-polyacrylamide gel electrophoresis. Anal. Biochem. 110: 131–136.
79. Nucca, R, C Soave, M Motto, and F Salamini (1978): Taxonomic significance of the zein isoelectric focusing pattern. Maydica 23: 239–249.
80. O'Dell, BL, AR De Boland, and SR Koirtyohann (1972): Distribution of phytate and nutritionally important elements among the morphological components of cereal grains. J. Agric. Food Chem. 20: 718–720.
81. Osborne, TB (1908): Our present knowledge of plant proteins. Science 28: 417–427.
82. Osborne, TB (1924): The Vegetable Proteins. 2nd Ed. Longmans, Green and Co., London; 154 pp.
83. Osborne, TB, and LB Mendel (1914): Nutritive properties of proteins of the maize kernel. J. Biol. Chem. 18: 1–16.
84. Paulis, JW, C James, and JS Wall (1969): Comparison of glutelin proteins in normal and high-lysine corn endosperms. J. Agric. and Food Chem. 17: 1301–1305.
85. Paulis, JW, and JS Wall (1969): Albumins and globulins in extracts of corn grain parts. Cereal Chem. 46: 263–273.
86. Paulis, JW, and JS Wall (1971): Fractionation and properties of alkylated-reduced corn glutelin proteins. Biochim. Biophys. Acta 251: 57–69.
87. Paulis, JW, and JS Wall (1977): Comparison of the protein compositions of selected corns and their wild relatives, teosinte and *Tripsacum*. J. Agr. Food Chem. 25: 265–270.
88. Paulis, JW, and JS Wall (1977): Fractionation and characterization of alcohol-soluble reduced corn endosperm glutelin proteins. Cereal Chem. 54: 1223–1228.
89. Paulis, JW and JS Wall (1979): Distribution and electrophoretic properties of alcohol-soluble proteins in normal and high-lysine sorghums. Cereal Chem. 56: 20–23.
90. Paulis, JW and JS Wall (1979): Note on mill for pulverizing single kernels of cereals for isoelectric focusing. Cereal Chem. 56: 497–498.
91. Pollmer, WG, and HK Fromberg (1973): Improved lysine in maize (*Zea mays*, L.) with alcoholic extraction at high temperature. Cereal Res. Commun. 1 (2): 45–53.

92. Preaux, G, and R Lontie (1975): The proteins of barley. In: Harborne, JB, and CF van Sumere (eds.): The Chemistry and Biochemistry of Plant Proteins: 89–111; Academic Press, New York.
93. Righetti, PG, E Gianazza, A Viotti, and C Soave (1977): Heterogeneity of storage proteins in maize. Planta 136: 115–124.
94. Robutti, JL, RC Hoseney, and CW Deyoe (1974): Modified *opaque*-2 corn endosperm. I. Protein distribution and amino acid composition. Cereal Chem. 51: 163–172.
95. Robutti, JS, RC Hoseney, and CE Wassom (1974): Modified *opaque*-2 corn endosperms. II. Structure viewed with a scanning electron microscope. Cereal Chem. 51: 173–180.
96. Rooney, LW, MN Khan, and CF Earp (1980): The technology of sorghum products. In: Inglett, GE, and L Munck (eds.): Cereals for Food and Beverages: 513–554; Academic Press, New York.
97. Ryadchikov, VG, VP Neudachin, TB Filipas, and AV Lebedev (1979): Fractional composition of proteins of corn endosperm and amino-acid composition of isolated fractions. Prikl. Biokhim. Mikrobiol. 15: 923–292. (Transl.) Appl. Biochem. Microbiol. 15: 688–692.
98. Seckinger, HL, and MJ Wolf (1973): Sorghum protein ultrastructure as it relates to composition. Cereal Chem. 50: 455–465.
99. Shechter, Yaakov, and JMJ de Wet (1975): Comparative electrophoresis and isozyme analysis of seed proteins from cultivated races of sorghum. Amer. J. Bot. 62: 254–261.
100. Silva, WJ da, and P Arruda (1979): Evidence for the genetic control of lysine catabolism in maize endosperm. Phytochem. 18: 1803–1806.
101. Singh, R and JD Axtell (1973): High lysince mutant gene (*hl*) that improves protein quality and biological value of grain sorghum. Crop Sci. 13: 535–539.
102. Soave, C, F Pioli, and A Viotti (1975): Synthesis and heterogeneity of endosperm proteins in normal and *o2* maize. Maydica 20: 83–94.
103. Soave, C, PG Righetti, C Lorenzoni, E Gentinetta and F Salamini (1976): Expressivity of the *opaque*-2 gene at the level of zein molecular components. Maydica 21: 61–76.
104. Sodek, L (1976): Biosynthesis of lysine and other amino acids in the developing maize endosperm. Phytochem. 15: 1903–6.
105. Sodek, L, and WJ da Silva (1977): Glutamate synthase: A possible role in nitrogen metabolism of the developing maize endosperm. Plant Physiol. 60: 602–605.
106. Sodek, L, and CM Wilson (1970): Incorporation of leucine-^{14}C and lysine-^{14}C into protein in the developing endosperm of normal and *opaque*-2 corn. Arch. Biochem. Biophys. 140: 29–38.
107. Sodek, L, and CM Wilson (1971): Amino acid composition of proteins isolated from normal, *opaque*-2, and *floury*-2 corn endosperms by a modified Osborne procedure. J. Agric. Food Chem. 19: 1144–50.
108. Sodek, L, and CM Wilson (1971): Metabolism of ^{14}C-amino acids in developing endosperm of corn. Plant Cell Physiol. 12: 889–893.
109. Steffens, R, FR Fox and B. Kassell (1978): Effect of trypsin inhibitors on growth and metamorphosis of corn borer larvae *Ostrinia nubilalis* (Hubner). J. Agric. Food Chem. 26: 170–174.
110. Su, JC, and J Preiss (1978): Purification and properties of sucrose synthase from maize kernels. Plant Physiol. 61: 389–393.
111. Sullins, RD, LW Rooney and DT Rosenow (1975): Endosperm structure of high lysine sorghum. Crop Sci. 15: 599–600.
112. Swank, RT, and KD Munkres (1971): Molecular weight analysis of oligopeptides by electrophoresis in polyacrylamide gel with sodium dodecyl sulfate. Anal. Biochem. 39: 462–477.
113. Swartz, MJ, HL Mitchell, DJ Cox, and GR Reeck (1977): Isolation and characterization of trypsin inhibitor from *opaque*-2 corn seeds. J. Biol. Chem. 252: 8105–8107.
113a. Tristram, GR, and RH Smith (1963): The amino acid composition of some purified proteins. Adv. Protein Chem. 18: 227–318.
114. Tsai, CY (1979): Tissue-specific zein synthesis in maize kernel. Biochem. Genetics 17: 1109–1120.
115. Tsai, CY (1979): Early termination of zein accumulation in *opaque*-2 maize mutant. Maydica 24: 129–140.

116. Tsai, CY (1980): Note on the effect of reducing agent on zein preparation. Cereal Chem. 57: 288–290.
117. Tsai, CY, and A Dalby (1974): Comparison of the effect of *shrunken*-4, *opaque*-2, *opaque*-7, and *floury*-2 genes on the zein content of maize during endosperm development. Cereal Chem. 51: 825–829.
118. Tsai, CY, DM Huber, and HL Warren (1978): Relationship of the kernel sink for N to maize productivity. Crop Sci. 18: 399–404.
119. Tsai, CY, DM Huber, and HL Warren (1980): A proposed role of zein and glutelin as N sinks in maize. Plant Physiol. 66: 330–333.
120. Tsai, CY, BA Larkins, and DY Glover (1978): Interaction of the *opaque*-2 gene with starch-forming mutant genes on the synthesis of zein in maize endosperm. Biochem. Gen. 16: 889–896.
121. Tsai, CY, F Salamini, and OE Nelson (1970): Enzymes of carbohydrate metabolism in the developing endosperm of maize. Plant Physiol. 46: 299-306.
122. Valentini, G, C Soave, and E Ottaviano (1979): Chromosomal location of zein genes in *Zea mays*. Heredity 42: 33–40.
123. Van Etten, CH, WF Kwolek, JE Peters, and AS Barclay (1967): Plant seeds as protein sources for food or feed. Evaluation based on amino acid composition of 379 species. J. Agric. Food Chem. 15: 1077–89.
124. Viotti, A, E Sala, P. Alberi, and C Soave (1978): Heterogeneity of zein synthesized in vitro. Plant Sci. Lett. 13: 365–375.
125. Vitale, A, C Soave and E Galante (1980): Peptide mapping of IEF zein components from maize. Plant Sci. Lett. 18: 57–64.
126. Waggle, DH, CW Deyoe, and FW Smith (1967): Effect of nitrogen fertilization on the amino acid composition and distribution in sorghum grain. Crop. Sci. 7: 367–368.
127. Walden, DB (ed.) (1978): Maize Breeding and Genetics. Wiley New York; 794 pp.
128. Waldschmidt-Leitz, E, and H Kling (1967): Die Prolamine. Fortschr. Chem. Org. Naturst. 25: 251–268.
129. Wall, JS (1971): Disulfide bonds: Determination, location and influence on molecular properties of proteins. J. Agric. Food Chem. 19: 619–25.
130. Wall, JS, C James, and GL Donaldson (1975): Corn proteins: Chemical and physical changes during drying of grain. Cereal Chem. 52: 779–790.
131. Wall, JS, and JW Paulis (1978): Corn and sorghum grain proteins. In: Pomeranz, Y (ed.): Advances in Cereal Science and Technology II: 135–219; Am. Assoc. Cereal Chemists; St. Paul, MN.
132. Waugh, DF, (1954): Protein-protein interactions. Adv. Protein Chem. 9: 325–437.
133. Wieser, H, W Seilmeier and HD Belitz (1980): Vergleichende Untersuchungen über partielle Aminosäuresequenzen von Prolaminen und Glutelinen verschiedener Getreidearten. I. Proteinfraktionierung nach Osborne. Z. Lebensm. Unters. Forsch. 170: 17–26.
134. Wilson, CM (1967): Purification of a corn ribonuclease. J. Biol. Chem. 242: 2260–2263.
135. Wilson, CM (1973): Plant nucleases. IV. Genetic control of ribonuclease activity in corn endosperm. Biochem. Genet. 9: 53–62.
136. Wilson, CM (1981): Variations in soluble endosperm proteins of corn (*Zea mays* L.) inbreds as detected by disc gel electrophoresis. Cereal Chem. 58: 401–408.
137. Wilson, CM, PR Shewry and BJ Miflin (1981): Maize endosperm proteins compared by sodium dodecyl sulfate gel electrophoresis and isoelectric focusing. Cereal Chem. 58: 275–281.
138. Wolf, MJ, U Khoo, and HL Seckinger (1969): Distribution and subcellular structure of endosperm protein in varieties of ordinary and high-lysine maize. Cereal Chem. 46: 253–263.
139. Zink, F (1979): Effect of late application of nitrogen on the grain protein of three high yielding maize hybrids with different protein or lysine content. Seed Protein Improvement in Cereals and Grain Legumes I: 273–281; IAEA Vienna.
140. Zink, F, and E Wilberg (1976): Einfluss einer Stickstoffspätdüngung auf das Korneiweiss einer lysinreichen Maishybride. Z. Pflanzenern. Bodenk. 2: 229–238.

10. The Genetic Control of Seed Protein Production in Legumes

H.P. MÜLLER

INTRODUCTION

Data which could be discussed under this broad heading are numerous and diverse. They include the ascertainment of the genetic diversity in legume seed composition and the manifold attempts to throw light on the biochemical basis of the different constituents, the trials to utilize mutant genes as experimental tools for analysing the inheritance of distinct biochemical traits, and the breeding experiments for elucidating the regulation of gene activities which are influenced by divergent external factors. The diversity of facts, however, necessitates selection and emphasizes on research fields where considerable progress has been made within the last few years and those which need further development. Not to be forgotten are those areas connected with the author's particular interests which, perhaps, may act as a basic and prerequisite selection factor. In the light of this eclectic presentation, I refer also to several of the articles available in this book.

In preparing the manuscript a great volume of literature including research papers, reviews and general works on seed proteins was consulted. Most of these are cited as references. Nevertheless, it embodies only a part of the literature. In attempting to compare the facts concerning the qualitative and quantitative composition of legume seeds in a more uniform way, certain difficulties arose. They are mainly based on the comparisons of analytical data from different laboratories in which various species or different cultivars of the same species are used by different workers often without indicating precisely environmentally conditioned variations in seed composition. Frequently, the information concerning a particular species is incomplete, involving also a variety of different experimental conditions used by independent laboratories investigating similar phenomena. These findings make biochemical and genetical comparisons often difficult. The above-mentioned difficulties are pointed out in order to draw attention to a crucial point with regard to genetic considerations, rather than as an excuse for deficiencies in the contribution.

The examples chosen to elucidate the genetic background for seed protein sythesis and acculumation in *Leguminosae* refer mainly to economically important members of this family [6, 15, 20, 22, 130, 132, 154, 202, 238, 239]. This is due to the fact that legume seed proteins as characters have been used extensively for determining their ranges of variation, with the aim of improving seed proteins for human nutrition [2, 28, 71, 130, 132, 149, 154, 160, 202, 221]. This work was mainly performed within the framework of international research programmes (see chapters 1 and 2). Generally speaking, the data obtained in this way reveal numerous and highly interesting findings that contribute to the understanding of the basic mechanisms responsible for seed protein biosynthesis and accumulation. Furthermore, these investigations revealed that numerous other nitrogenous compounds are available in seeds which are not storage proteins but which are physiologically active in different ways [125, 132].

The seeds are differentiated organs and constitute the endpoint of an integrated developmental process. The genetic information for e.g. storage proteins appears to be repressed in all tissues except for certain developmental stages in the cotyledons, when it is actively expressed [101, 159, 162, 178]. Studies concerning the kinetics of protein incorporation in the cotyledons suggest that numerous biological processes contribute to differences in protein content and quality among genotypes within species and greater taxonomic groups. The prevention or reduction of the synthesis of a single protein can modify seed protein content and quality. Furthermore, the latter may be affected by the time of onset and termination and the rate of synthesis of the component proteins. Environment markedly influences the composition of mature seeds with site-to-site and year-to-year variations being well documented while the effects of changes in single environmental factors have not been extensively studied [89].

The study of seed protein genetics in legumes must necessarily comprise quantitative and qualitative aspects. When comparing seed proteins of species and cultivars, the information available is highly dependent upon the resolving power of the methods applied. Thus, total protein content which is the sum of different individual protein components, largely includes total nitrogen content, multiplied by a specific factor. The values obtained do not reflect the protein composition. On the other hand, the overall amino acid composition of seed proteins, determined quantitatively after hydrolysis also reflects qualitative characters of the seed proteins. But there are no indications with regard to the identification of distinct individual proteins showing high or low levels in distinct amino acids. Furthermore, the natural non-protein amino acids, constituting a significant part of nitrogenous compounds, largely escape consideration. The fractionation of native seed proteins by means of different extraction techniques, gel electrophoresis and isozyme analysis reveals a quite different pattern of seed protein structures. Furthermore, a qualitative aspect in investigations on seed proteins involves a group of likewise tissue-specific genetic characters expressed preferentially in cotyledons of leguminous species exerting distinct psysiological functions, these being the lectins and protease inhibitors. The informations obtained from studies of seed proteins on the different levels indicated are represented in the following sections. They will contribute to the characterization of the genetically determined variability on the seed protein composition.

I. THE TOTAL PROTEIN CONTENT OF SEEDS

There is no doubt that apart from considerable influences of different environmental factors, the protein content is strongly influenced by several genes. Nevertheless, the question concerning the number of genes involved in seed protein synthesis and their possible role in seed protein synthesis is presently unknown [168]. The mature seed showing its unique genetically determined developmental and physiological composition is a complex assembly of tissues. Their developmental sequence is distinct and temporarily defined and genetically controlled by a number of tissue-specific genes [59, 111, 146; see section III]. It is to note that a wide spectrum of biochemical alterations is possible without influencing the seeds function of supplying nutrients for the growing embryo.

The first step with regard to the analysis of the genetic basis of seed protein composition consists of screening for total nitrogen content, mainly by the classical Kjeldahl-method. It is a convenient method for indicating relevant relationships between different genotypes. Within the family of *Leguminosae* several genera were subject of numerous investigations. Some species of food legumes including *Cajanus*, *Glycine*, *Phaseolus*, *Vigna*, *Vicia*, *Pisum*, *Lathyrus*, *Lens*, *Arachis*, *Cicer* and *Lupinus* which are known to be sufficiently important for human nutrition are subject of intensive investigations since they justify substantial efforts on genetic improvement. The main interests are focused on the seed protein quantity and quality in order to assess the extent of variability between species and genotypes.

Figure 1 illustrates the variability in seed protein contents of 9 *Leguminosae* genera. The range of the character seed protein is indicated by a horizontal line. The classification of the genera follows that of [191]. In order to assess the extent of environmental-induced variability within the genera, the values for the different genera are further grouped according to the areas in which they are preferentially grown. It should be emphasized that the number of samples analysed is inadequate. The presentation for the variation has been chosen in order to demonstrate the whole ranges. For many of the genotypes included detailed analyses concerning the sources of variability resulting in representative graphs for the percentages of proteins exist which support the general picture [e.g. 20, 22, 65, 91, 160, 238, 256]. The phenotypic expression of the character seed protein shows a considerable variability, depending on the genotypes and environmental factors. On the other hand, comparing the variation within different genera, the coincidence of the ranges in seed protein contents is astonishing. Seed protein content is a component of total

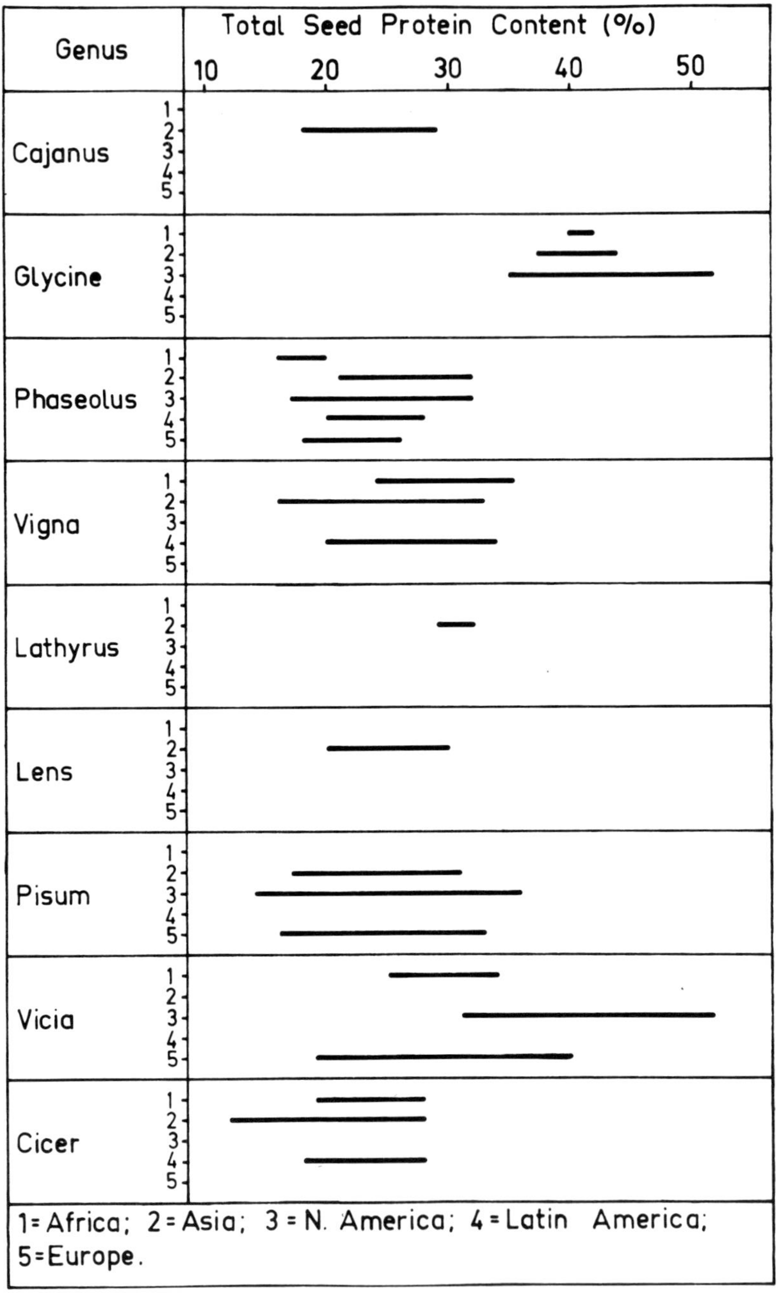

(References: *Cajanus* [6, 15, 132, 154, 187, 204, 238]; *Glycine* [110, 116, 130, 132, 149, 187]; *Phaseolus* [34, 47–49, 100, 104, 132, 135, 187, 255]; *Vigna* [15, 20, 31, 122, 132, 154, 187, 221]; *Lathyrus* [221]; *Lens* [221, 238]; *Pisum* [8, 22, 90–95, 166, 171–173, 187, 231, 238]; *Vicia* [120, 138, 149, 180, 184]; *Cicer* [132, 154, 221, 238].

Figure 1. Variability in total seed protein content of *Leguminosae* genera.

yield and is composed of various protein fractions, each with a different pattern of amino acids. The protein traits will resemble quantitative characters with a continuous type of variation. They are under the control of multiple genes. As one of the mechanisms responsible for the continued release of genetic variation recombination, assisted by mutations and some retaining heterozygosity, is discussed [22, 48, 91, 116, 120]. The genetic background of the variability in protein content has been usually described and explained through different terms, as 'major genes', 'modifier genes', 'multiple factors' and 'polygenes', contrasting to the 'oligogenes' controlling the discontinuous or qualitative characters. It could be helpful, perhaps, to refer only to the terms 'polygene', indicating a gene which individually exerts a slight effect but which controls a number of other equivalent genes of a character, and 'modifier-gene' which interacts with other genes and affects the expression at other loci and modifies their gene action. Until now, our knowledge of the molecular basis for their functions is scarce [162]. With regard to practical breeding purposes the possibility exists that positively extreme genotypes showing only increases in protein content or protein content in combination with yield and other agronomically important traits can be detected and isolated from the observed continuous variability. This becomes obvious from experimental demonstrations of induction of new variation [15, 22, 47, 90–95, 120, 171, 184, 187, 221]. Selection within the variation of quantitative characters is difficult because quantitative steps are generally small and not easy to detect because of the genotype-environmental interactions which often mask the genetic variation. The possibility of inducing inheritable changes of protein content in a 'positive direction' can be regarded as a consequence of the fact that this character was not subjected to intensive selection during the development of the respective species. On the other hand, the protein spectra shown (figure 1) may reflect the fact that selection work on pulses has led to a considerable conservation of the genetic information for storage proteins. Such a conservation also indicates that it will not be easy to achieve large qualitative or quantitative alterations of seed proteins [162, 238].

The graphic presentation becomes more meaningful if one considers the number of genotypes which have been analysed (table 1). Even if one takes into account that some values used in figure 1 derive from review papers without indicating the number of genotypes studied the data compiled in table 1 represent a suitable basis for comparisons. The classification into varieties and mutants was chosen because for most of the improved varieties of crop plants the term 'variety' or 'cultivar' is often without precise definition. It may involve clones or pure lines, selected hybrid lines from crosses between established varieties showing specific characters, land races and primitive forms. The range of variability within leguminous genera found in protein characters indicates their narrow genetic basis. Even the induction of mutations as a less disruptive method as compared with hybridization did not show large advantages in breeding for increased protein content [22]. On the other hand, it was recognized that many of the currently cultivated leguminous species still possess

Table 1. Number of genotypes analysed for total seed protein content in different genera of *Leguminosae*.

Genus	Number of varieties	mutants	References
Cajanus	2274	5	6,15,132,154,187,204,238
Glycine	10	25	110,116,130,132,149,187
Phaseolus	329	1	34,47,48,49,100,104,132,135,187,255
Vigna	99	83	15,20,31,122,132,154,187,221
Lathyrus	5	-	221
Lens	11	25	221,238
Pisum	2218	249	8,22,90–95,166,171–173,187,231,238
Vicia	352	479	120,138,149,180,184
Cicer	17	30	132,154,221,238

a broad number of 'wild-type' characteristics [221, 238] that were obviously beneficial during natural evolution so that the genetic variability seems not to be fully exploited [71, 110, 221, 238].

AMINO ACID BALANCE

Protein Amino Acids

A more outstanding qualitative feature of the storage proteins produced is their amino acid composition. The intensive search for an appropriate amino acid balance in pulse seeds revealed the character 'total seed protein variations' in amino acid balance as being largely attributable to the genetic differences of the respective leguminous genotypes.

Figure 2 represents an attempt to obtain an understanding of the variation in the amino acid composition and content in seeds of different genera of the *Leguminosae*. Being aware of the fact that the values are derived from results of different laboratories working with different species and varieties and the experimental difficulties which exist in evaluating amino acids, the ranges of the amino acid contents may be indicative of the genetical potential existing within different genera. The range of the essential amino acid content is presented by horizontal bars with single values being shown as points. For comparison the values for the respective amino acids deviated from a 'standard protein' [232] are indicated by the vertical lines. The numerical values are shown in the lower part of the diagram. There is a considerable variation in single amino acids between different genera depending in part of the number of genotypes analysed. Generally, the seed of peas and beans contain more sulphur amino acids than other genera. When comparing the range of values of essential amino acids found in leguminous seeds to those of the standard

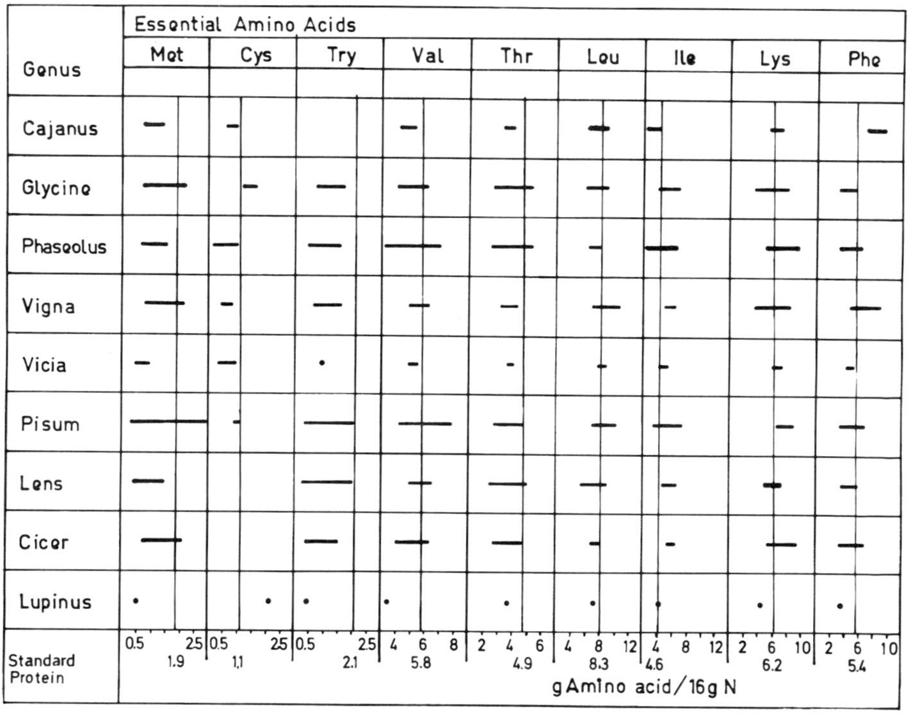

(References: *Cajanus* [15, 204, 208]; *Glycine* [5, 130, 225]; *Phaseolus* [2, 5, 130, 133, 134, 135, 155, 198]; *Vigna* [5, 15, 20, 30]; *Vicia* [180, 189]; *Pisum* [5, 71, 130]; *Lens* [5]; *Cicer* [5]; *Lupinus* [68].

Figure 2. Variation in essential amino acid contents in seed proteins of *Leguminosae* genera.

protein, the limiting character of the sulphur-containing amino acids is obvious. But with a small increase in one of these two amino acids, tryptophane would become the next limiting amino acid. This sequence can be extended to the following amino acids according to the plant species under investigation. From these findings it can be deduced that in breeding programs for protein quality, the whole spectrum of amino acids must be considered for application of the appropriate selection techniques. The biological value of the seed proteins depends on a balanced amino acid composition [20, 28, 35, 133–135, 138, 154, 225].

Generally, the data reveal that the qualitative composition of leguminous seed proteins, as with protein in general, is governed by a conservative manner of biosynthesis. This conservation of the genetic information for seed proteins could be the cause of the difficulties with regard to an enlargement of the genetic variability in the above-mentioned traits [26, 162, 180]. Furthermore, there is only scarce information with regard to the genetic background of the control mechanisms responsible for amino acid composition of seeds. On the other hand, a wide range of values, especially for the amino acid methionine in the genus *Phaseolus* has been detected. In assaying the available methionine in the germplasm collection of *Phaseolus*, the frequency distribution among 4115 bean varieties was found (figure 3). The data indicate that a sufficient variability exists in this individual constituent of legume seed protein. In that investigation, 63 lines exhibited an increase in the methionine content of more than 30% compared to the standard variety Sanilac [130]. Similar results on other *Phaseolus* lines have been reported [65] as well as an enhancement of the essential amino acid variability [91, 94, 130]. Obviously, the genetic basis for these traits can be exploited to change the seed protein quality. The sulphur-containing amino acids play an important role in breeding strategies for protein quality improvement. A prerequisite for this aim is the knowledge of the inheritance of the required characters based on their genetic variation [2,8, 15, 20, 28, 31, 48, 49, 58, 73, 130, 134, 160, 168, 169, 180, 184, 213, 255].

In this connection it should be emphasized that changes in the seed protein composition cannot be regarded without considering the correlation between percentage available amino acids on the one hand and their relationships to the yielding

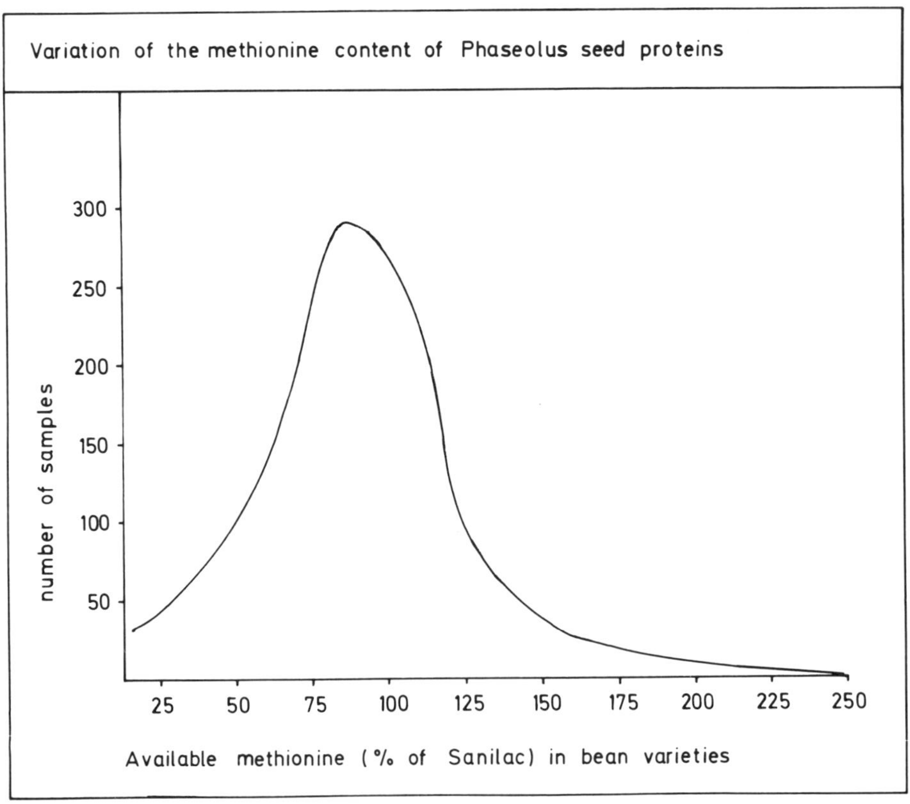

Figure 3. Frequency distribution for available methionine among 4115 common bean varieties (*Phaseolus vulgaris* L.), Schematically redrawn from [130].

capacity of the respective genotypes on the other. It is not the aim of this review to discuss the manifold interactions between these three factors [21, 22, 25, 26, 30, 31, 34, 65, 71, 100, 107, 116, 134, 143, 149, 154, 155, 172, 185, 186, 211, 221, 227, 228, 230, 239]. Nevertheless, we must keep in mind that the protein levels as well as those of the amino acids are strongly influenced by environmental factors [92, 133, 143, 172] which influence the regulatory processes responsible for seed protein incorporation. The relationships between yield and percentage protein of seeds are mainly characterized by negative correlations: *Phaseolus*: −0.45 [148], −0.64 [239]; *Vigna*: −0.14 (phenotypic correlation [21]), −0.38 (genotypic correlation within 11 varieties) 0.23 to −0.29 (phenotypic correlation) [25]; *Pisum*: 0.02 to 0.49 [185, 186]. But as can be seen, most correlations are low, sometimes absent [95] or even positive [22, 133]. According to these data, it seems possible to improve both traits. Similar findings concerning the relationships between seed weight and protein content are available. Within different genera, no consistent correlation was found: none in peas [94], a negative one in chick peas [213], beans [133] and a positive one in mungbeans and soybeans [110, 156]. Since seed size is an important component of yield, it should be possible to select and to develop lines which simultaneously show high yield and high protein content. The succesful selection of lines with improvement in both characters in *Glycine* [110] indicates that similar results may be possible in other legumes. The genetic variability between homozygous varieties, mutants and recombinants can also be used effectively for isolating progenies showing higher seed protein contents than the parent genotypes [48, 49, 91–94]. With regard to the trait protein content, the effects of the genes involved have been defined in terms of low or high additive variance and narrow-sense or broad-sense heritability. In selected pea lines, the inheritance of the protein content shows an additive dominance character [107, 238a] and in beans in relatively low additive genetic variance [148, 213]. The broad sense heritability estimates range from 30.7–63% for beans [148], 29% for cowpea [21] and 78–79% for peas [238a]. The heritability coefficients in the narrow sense were 5–12% for beans, in peas only 7.6%. No generalizing rules can be established with regard to the character protein content. Even the relationships between protein percentage and methionine percentage of protein vary between a negative correlation for *Vicia* [184] to a partly negative, partly positive one in *Cicer* [213] and to a positive one in *Phaseolus* [133]. Thus, more knowledge of the biochemical diversity of the seed composition in different leguminous genotypes is needed in order to obtain more complete informations of the genetic background of the phenomenon observed.

Non-protein amino acids

Among higher plants legumes are extraordinarily rich in uncommon or non-protein amino acids [13, 79]. In contrast to the protein amino acids, the non-protein amino acids mainly play a role in taxonomic studies because of their great number and their inter- and intraspecific variability [14, 72]. More than 200 non-protein amino acids have been identified. They occur in free state or as condensation products in the plants [72, 79]. The occurrence and distribution of different structural analogs of the known protein amino acids is well established. They include neutral aliphatic; sulphur-containing; imino, acidic and basic; heterocyclic and aromatic amino acids and non-analogs of protein amino acids [13, 79]. Until now, data concerning the biosynthesis and the accumulation of these compounds are scarce [13, 72, 233–236]. The same is valid for the question how the often high concentrations of distinct non-protein amino acids are stored in the free amino acid pool of seeds [233]. Even less is known about the genetic basis of the species-specific compounds which disappear or accumulate during seed germination [13]. The demonstration of the quantitative isolation [233], determination and the structural characterization [234, 236] of γ-hydroxyornithine and γ-hydroxyarginin in *Lens culinaris* seeds is of great interest. The content in γ-hydroxyarginin accounts for 76–91% of the free basic amino acids and 0.76–1.70% of the dry matter, while the content of γ-hydroxyornithine was only 3–7% of the free basic amino acids and 0.04–0.08% of the dry matter. The proportions varied depending on the genotypes analysed. Both amino acids show quantitative differences according to the respective lentil varieties derived from different locations and can be regarded as species-specific compounds [235]. The occurrence of γ-hydroxyarginine in seeds of different *Vicia* lines has been reported as characteristic for the genus *Vicia* [14, 235]. This indicates that the occurrence of certain non-protein amino acids in different genera of the legumes should receive more attention. No reliable data for their physiological compatibility in animal and human nutrition are yet available. It is however known that the consumption of certain non-protein amino acids like 2-aminobutyric acid, 3-cyanoalanine and N2-oxalyldiaminopropionic acid may represent a causative factor of human neurolathyrism [235], but little is known about the physiological action of the hydroxyamino acids in relation to the disturbances in the natural balance between the essential amino acids [233]. Therefore, the presence of large amounts of rare amino acids in legume seeds, which, perhaps, are toxic and only partially destroyed by cooking [233], require more intensive investigations of their biochemical, nutritional and organoleptic properties as well as their genetic basis.

LEGUME SEED PROTEINS

Structurally, the mature seed represents a mosaic of several differentiated tissue types and at least two genetically distinct clones of cells which are involved in seed formation. The embryo has a diploid hybrid genome, the nucellus, integuments and ovary wall a diploid maternal genotype. Consequently, the deposition of food reserves of seeds is closely related with seed structures. In leguminous seeds generally two main types of protein characterized on the basis of their solubility are found. The water-soluble proteins are referred to as albumins and the salt-soluble proteins as globulins [53, 54].

The albumins are mainly metabolic proteins involved in cellular activities, including the synthesis and degradation of the second group [9, 12] which constitute the bulk of storage proteins. The latter serve as nitrogen source for the germinating embryo [54, 55]. The degradation of proteins following imbibition of seeds is also known for albumins [9, 176]. The presence and their variable amounts in seeds of different leguminous species [91, 99, 114, 115, 173, 176, 200, 201] indicate that the albumins must have more functions than presently attributed to them. Therefore, in defining the storage proteins it has been provisionally proposed to consider all extracted proteins exceeding 5% of the extractable total seed proteins as storage proteins [63]. Including the albumins in the group of storage proteins undoubtedly needs more interest not only with regard to the possible alterations of the nutritional value of seed proteins.

The function of storage proteins does not necessitate a particular amino acid composition since any given amino acid may serve as nitrogen source. The large variation of the composition of seed storage proteins from species to species is taken as evidence for that function. The seed protein phenotypes as identified by highly resolving analytical techniques (see chapter 3) revealed a sufficient heterogeneity between distinct protein species encoded in the genetic material. But also high degrees of homology exist. The latter allow comparisons concerning the inheritable variations of the subunit composition.

ALBUMINS AS STORAGE PROTEINS

Variation of the albumin contents

Because albumins are deposited in leguminous seeds as are other proteins during seed development [12, 104, 114, 115, 164, 173, 188, 217], it is conceivable that they serve as storage proteins. Therefore, it is expected that variability exists with regard to their quantitative proportions in different genotypes. The situation found in different species shall be exemplified for *Phaseolus* and *Pisum*. Similar results are available for *Lupinus* [68]. Even if we realize that no definite values for that particular fraction can be given because of the manifold genotype-environment-interactions [94, 203, 242], the variation seems large enough not to be ignored. Furthermore, the proportion of the different protein fractions remains stable within one variety and is independent of the biological or mineral nitrogen supply [131]. The quantitative but not the qualitative variations [58] are of interest since this fraction shows an amino acid profile with a relatively high content of methionine and other essential amino acids [8, 58, 119, 176, 217], compared to the globulin fraction. This becomes obvious if one compares the distribution of the essential

Table 2. Variation in the albumin contents of *Phaseolus*- and *Pisum* seeds.

Species	Number of genotypes analysed	Characteristics	Albumin content (% total protein content)	Ref.
Phaseolus vulgaris	5	standard lines	12.4 - 15.0	155
	10	F_5-lines from crosses between genotypes showing different percentages in protein and methionine	11.1 - 19.8	155
	5	lines with high seed protein content	11.2 - 15.6	104
	5	lines with low seed protein content	14.3 - 19.8	104
	2	standard lines	12.5 - 15.5	104
Pisum sativum	2	varieties	17.0	200,201
	1	variety, dwarf	33.0	12
	2	varieties	13.0 - 14.0	99
	8	varieties of different origin	15.6 - 38.2	58
	1	variety, different nitrogen supplementation	18.0 - 30.0	131
	9	1 variety, 8 mutants	25.0 - 35.0	173
	1	variety	42.0	176
	45	wild forms, field peas, garden peas, round-wrinkled-seeded types	20.0 - 35.0	217

amino acids in different leguminous genera to a standard protein [232] as shown in table 3. The data indicate that the amino acid profiles of the albumin fractions show in some amino acids similar or in some exceeding values compared with those of the standard protein. The values for the sulphur-containing amino acids reach in some cases ‘optimal’ amounts. As the albumin fraction is one third of the total storage proteins this fraction needs more conceptual consideration. Its composition is complex with regard to the species and number of proteins involved [91, 94, 101, 151, 168, 222, 244, 255] and their relations to the other groups of storage proteins. These relationships are of great interest in connection with attempts to improve nutritionally the seed proteins (see chapters 1, 2, 15). The proportion of the albumin and legumin fractions, the latter belonging to the seed globulins, within the mono-specific genus *Pisum* is of interest. A negative correlation exists between the two quantitative characters. In the wild forms of *Pisum* a high legumin content is associated with a low albumin content. This relation is adversely changed in field peas which at the same time show the least variation in legumin but the widest range for albumin contents. Round-seeded garden peas exhibit the greatest variation in legumin content while wrinkled-seeded garden peas have the least variation for both these components [217]. The close relationships between the amounts of legumin and seed shape controlled by the r_a locus [59] indicate a possibility for approaching the problem of seed protein genetics more directly and in a defined way.

Protein Profiles of the Albumin Fraction

The multiplicity of polypeptides in the albumin fraction can be revealed by electrophoretic methods. The banding patterns obtained, not influenced by external factors [3, 83, 84] are convenient for taxonomic purposes [18, 29, 140–142, 246] for characterizing the genetically determined variability in seed protein composition on species, variety and mutants levels [46, 64, 101, 102, 119, 123, 127, 144, 167, 175, 192] and for providing a basis for identification of distinct genotypes. Furthermore, they are useful for identifying developmental steps in seed protein incorporation [101, 104, 136, 237] and degradation [9, 101]. More information concerning the variability of the polypeptide composition of the albumin fraction is obtained by SDS-PAGE analysis. The data available concerning the polypeptide structure of different genotypes are scarce. In *Pisum* the albumin fraction contains components of 78.000, 47.700 and 26.000 Daltons. These are made up of subunits of 25.500 and 15.500 [99], and additional six minor components showing molecular weights of 200.000, 110.000, 48.000, 32.000, 21.500 and 18.000. In different pea species, four fractions were obtained after gel filtration with molecular weights of 80.000, 40.000, 18.000 and 7.000 Daltons while the gel patterns of total albumin extracts showed three distinct bands of 76.000, 23.000 and 17.000 Daltons [127]. Similar findings were obtained in other laboratories [58, 131]. Of the pea species analysed one of the four chromatographically-separated fractions shows a considerably higher amount

Table 3. Distribution of essential amino acids in the albumin fractions of different legume genera.

Amino acids (g amino acid/ 16 g N)	Standard protein (Ref.: 232)	Phaseolus (10 F_5-lines) (Ref.: 155)	Pisum (5 species, 14 cultivars) (Ref.: 8,99,127,176)	Lupinus (1 variety) (Ref.: 68)
Met	1.90	1.01	0.96 - 1.88	0.51
Cys	1.10	0.23	0.23 - 2.60	1.70
Try	2.10	-	1.20	-
Val	5.80	4.90	3.62 - 6.73	3.62
Thr	4.89	7.44	5.20 - 7.50	4.96
Leu	8.30	6.62	3.70 - 8.30	7.31
Ile	4.60	4.32	3.28 - 6.10	4.26
Lys	6.21	10.85	7.37 - 10.60	5.40
Phe	5.40	4.04	3.15 - 7.26	3.15

of cystine than the other fractions [127], indicating qualitative differences which may be genotype-specific. A major albumin subunit with a molecular weight of 105.000 and two subunits of 32.000 and 22.500 with other minor components were detected in *Vigna unguiculata* [136]. The few data cited also indicate that the composition and structure of the proteins of the albumin fraction contribute essentially to the understanding of the seed protein characteristics.

Variation of the Banding Patterns and Genetic Control of Isozymes

Numerous data prove that the genetic variability of seed protein phenotypes can also be expressed in terms of altered enzyme activities [129, 173]. Studies on various enzymes in different plant organs have revealed that most have electrophoretically detectable isozymes [214, 215]. Since the variant alleles are generally codominant [37, 44, 97, 248] it is possible to identify positively heterozygotes and homozygotes. Their expression proved not to be influenced by environmental factors [3, 46, 83–85, 131, 167, 168]. Considering these facts, the identification of seed protein phenotypes by means of isozyme techniques is potentially useful in genetic research. The isozymes, as genetic markers, directly express a part of the genetic makeup of the plants and their patterns are representative. For comparisons it is appropriate to analyse physiologically dormant tissues like cotyledons, because their genotype-specific patterns are not confused by ontogenetic changes. In this way the polymorphism present indicates the genetic differences within and between genotypes. Since isozymes represent specific gene products, variants are more likely to represent single gene lesions than are complex morphological traits.

The electrophoretic data obtained for seed of different leguminous genotypes include many enzymes, assayed with artificial substrates. Their in-vivo substrate and subcellular locations are unknown mostly. The enzymes analysed include esterases, phosphatases, peptidases and peroxidases. Therefore, differences in the number of bands can be caused by several factors, including changes in substrate specifity, reaction requirements and changes in the genomatic structure of the organism. The relative importance of these factors cannot be evaluated with the available electrophoretic findings. Nevertheless, intensive studies on a large number of isozyme systems in seeds of numerous legumes have shown that isozymes are effective in efforts to answer basic genetic questions. In seeds of different legume species, at least 14 isozyme systems were analysed for identification of varietal characteristics. A survey of the enzyme systems used in characterizing leguminous genotypes is given in table 4 while in tables 5 and 6 the data available for *Glycine*, *Phaseolus*, *Pisum* and *Vicia* are arranged for specifying their characteristics.

The number of isozymes in the different enzyme systems analysed varies between genera. The unequal distribution of isozyme systems analysed in different leguminous genera may indicate that either the respective enzymes are not available in the seed or not detectable by the usual methods. Apart from a functional meaning, the variability at enzyme loci has proven suitable for characterizing species [170, 226] and varieties on a large scale [43, 44, 96, 97, 111, 112, 113, 114–146, 258] and mutants and recombinants of one species [165, 168, 169, 173]. Especially the intravarietal analysis of isozymes shows both qualitative and quantitative differences [44], although the quantitative aspect is still unclear in most cases. But it is possible to measure not only the genetic distances between cultivars of closely related species but also to determine the genotype-specific isozyme patterns characterizing the level of inbreeding [84, 251].

According to the data, isozymes show Mendelian inheritance while maternal inheritance could be excluded [43]. Codominance has been generally stated in studies of enzyme inheritance [37, 44, 97, 248]. Thus, isozymes can be used as genetic markers at the molecular level. They represent specific gene products, while variants are more likely to represent single lesions [137]. Structural variations can be caused either by altered mobility or missing isozymes [97] both of which indicate changes in the amino acid sequence of the respective enzymes. When considering the genetic control of any enzyme system, it should be kept in mind that the banding pattern may be influenced by numerous experimental factors as well as by protein-protein interactions. In higher plants, complex regulation mechanisms govern gene expression, so that it is also conceivable that enzyme loci underlie these mechanisms resulting in altered patterns.

The genetic analysis performed in breeding tests in which the segregation of the progeny isozyme patterns is examined, can be supplemented by comparing the electrophoretic pattern of pollen extracts with those of leaf tissue from the same organisms [251]. Since the haploid pollen contains only one allele for each gene locus, the nature of distinct enzymes or hybrid enzymes can be evaluated. This method can also be used for analysing external factors influencing the leaf isozyme pattern [85].

With regard to genetic and biochemical data of the different enzyme systems different models are proposed for the observed polymorphism [97]. Alcohol dehydrogenase (ADH) exists as functional dimer with the subunits coded for by different genes form intra- and interlocus at least two stable heterodimers. Two of the ADH-loci are closely linked (ADH-1, ADH-4). In the case of the enzyme TO four genes are responsible for a homoheterodimer relationship, while the AM-system shows three independent amylase loci [97]. Two of the bands (AM-1 and AM-2) are invariant and at the Am-3 locus two codominant alleles (F and S) and two null variants were detected [137]. The null variants are characterized as recessive [97] and show an undetectable or markedly reduced activity for the variant AM-3 band. The occurrence of AM null 1 and 2 is specific for distinct varieties [97]. The variant AM-3 band was identified as beta-amylase [111], the null 1 and null 2 variants were related to the Sp_1^a- and Sp_1^b-protein fractions of *Glycine* seed proteins [111, 112, 146], and the Sp1 codes for beta-amylase [112]. AM null 1 was shown to be recessive to the codominant alleles Sp_1^a and Sp_1^b for both β-amylase activity and the seed protein

Table 4. Isozyme systems investigated in legume seeds.

	Genera							
	Glycine	Ref.	Phaseolus	Ref.	Pisum	Ref.	Vicia	Ref.
Alcoholdehydrogenase (ADH)	+	96,97	+	248	-	-	-	-
Amylase (AM)	+	96,97,111 112,137	-	-	-	-	-	-
Esterase (EST)	-		+	248	+	81,165,167 168-170, 254	+	84,85,254
Glutamate dehydrogenase (GDH)	-		-		-		-	
Glutamate oxaloacetic transaminase (GOT)	-		-		-		+	84,85
INT-oxidase	+	145	-		-		-	
Lactate dehydrogenase (LDH)	-		-		+	165	-	
Leucine aminopeptidase (LAP)	-		-		+	165,170 173,215	-	
Lipoxygenase of *Pisum* (PL)	-		-		+	258	-	
Malate dehydrogenase (MDH)	-		+	252	+	165,166, 168	-	
Peroxidase (PE)	+	43	-		+	224	-	
Phosphatase (acid) (AP)	+	96,97	-		-		-	
Tetrazolium oxidase (TO)	+	96,97	-		-		-	
Urease	+	44	-		-		-	

band. AM null 2 was shown to be recessive to Sp_1^a and Sp_1^b for β-amylase activity but codominant to Sp_1^b for the seed protein band. Therefore, the symbols Sp_1 are proposed for the AM null 1 allele and Sp_1^{an} for the Am null 2 allele [112]. Another example for a genetically conditioned, close relationship between seed protein fractions synthesized very early in seed development because they are contained in mature seeds independently from environmental conditions, is the genetically conditioned linkage of an acid phosphatase with the Kunitz Trypsin Inhibitor (Ti) [113]. In this case, the linkage between two chemical components of the soybean seeds controlled by loci with codominant alleles was reported for the first time.

Table 5. Isozyme systems investigated in seeds of *Glycine*.

Isozyme system	Number of isozymes	Genotypes analysed	Genetic basic	Remarks	Ref.
ADH	4 - 7	113 varieties	stable varietal characteristic; 4 genes: ADH-1,2,3,4; ADH-4 independently segregating character; ADH-1 and ADH-4 closely linked	classification of varieties	96,97
AM	3 - 4	113 varieties	stable varietal characteristic; 3 genes: AM-1,2,3; AM-1,AM-2 invariant; AM-3 independently segregating	classification of varieties	96,97
		116 varieties F_2-seeds	2 AM-3 variants	Seed protein locus, Sp_1, codes for β-amylase	111
	2	4 varieties F_2,F_3 progenies from reciprocal crosses	AM-null 1; recessive allel AM-null 2; recessive/codominant allel	Inheritance of β-amylase nulls	112
	3 - 4	176 varieties 105 accessions of wild soybeans and their relatives	AM-3 locus with 4 variants: F,S,S^w,n_1 allelic in their effects on AM-3; null variants	Comparison of polymorphism between cultivars and wild soybeans and their relatives	137
INT-oxidase	2 - 5	42 varieties	stable varietal characteristic	iso-dehydrogenases (see also TO)	145
PE	3	28 varieties 20 parental lines	stable varietal characteristic; 1 gene	classification of varieties	43
AP	3	113 varieties	stable varietal characteristic	classification of varieties	96,97
TO-oxidase	9 - 11	113 varieties	stable varietal characteristic ; 4 genes: TO-1, 2,3,4; TO-3 independently segregating marker	classification of varieties (see also INT)	96,97
Urease	1	50 varieties	stable varietal characteristic ; 1 gene (Eu/eu)	classification of varieties	44

Comparison of the isozyme patterns between cultivars or cultivated varieties shows that the development of cultivars has undoubtedly led to genetic uniformity. This becomes obvious from genetic analyses of amylase AM-1, AM-2 and AM-3 variants in cultivated and wild soybeans [137]. The cultivar purity proved to be very high except in one cultivar in which a recent mutation had caused the null variant. Only 0.42% of the cultivar seeds showed heterozygosity at the AM-3 locus, which probably represents products of natural outcrossing. Quite another picture is given

Table 6. Isozyme system investigated in *Phaseolus*, *Pisum*, and *Vicia*.

Genus	Isozyme system	Number of isozymes	Genotypes analysed	Genetic basis	Remarks	Ref.
Phaseolus	ADH	3 - 6	38 collections of wild and cultivated Ph. vulgaris and Ph. coccineus	ADH-1: one locus, ADH-3: two codominant alleles; ADH-2: interaction products	evolutionary problems of Ph. vulgaris - Ph. coccineus complex	248
	EST	2		EST-1, EST-2: variants controlled by codominant alleles		
	MDH	4	1 genotype	stable varietal characteristic		252
Pisum	EST	4 - 6	12 varieties	stable varietal characteristic; genotype-specific banding patterns	classification of genotypes	81
		4 - 8	29 species, varieties, mutants, recombinants			165,167,168-170
		6 - 14	13 species, subspecies varieties, 7 mutants			254
	LDH	3 - 4	1 variety, 4 mutants	genotype-specific banding patterns		165
	LAP	2	20 inbred strains	AMP1: 2 codominant alleles; AMP2: one gene; stable genotype-specific characteristic		215
		2	18 species			170
		2 - 4	6 mutants			165,173
	PL	4	1 variety	stable varietal characteristic		258
	MDH	2 - 8	23 species, varieties, and mutants	genotype-specific banding patterns		165,166,168
	PE	6	2 varieties	stable varietal characteristic		224
Vicia	EST	6 - 13	2 inbred lines, 8 species, 3 subspecies	genotype-specific banding patterns	identification of inbred lines, classification of genotypes	84,85 254
	GOT	3 - 4	26 inbred lines	genotype-specific banding patterns	identification of inbred lines, classification of genotypes	84,85

by the wild forms of *Glycine*. Here, a relatively high genetic variation was found which can potentially be used to soybean breeding [137]. The average heterozygosity among a distinct collection was 7.4% to 8.1%. Thus, investigations concerning the genetic control of the isozymes provide a valuable source of genetic information of the seed storage proteins.

Protease Inhibitors

Quite another group of water-soluble seed proteins in leguminous seeds are the protease inhibitors. They constitute up to 6% of the total seed protein [181, 218] and are characterized by low molecular weights [207, 218, 219, 249]. Trypsin inhibitor of soybeans (Kunitz Inhibitor), the chymotrypsin inhibitor and the subtilisin inhibitor [88, 218] have been well characterized. The protein inhibitors are widespread within the legumes [33, 88, 132, 151, 161, 174, 205, 207, 219, 249] and exhibit as a general trait a broad specifity towards animal and plant proteases [1, 35, 125]. These characteristics indicate a possible protective effect for plants. They have been shown to consist of several molecular species (isoinhibitors) [16, 207, 218] and some additional low molecular weight proteins. The amino acid composition of *Lathyrus sativus* trypsin inhibitor is characterized by the absence of methionine which is also observed for other legumes [207]. But there are also opposite examples. The genetic control of the synthesis of trypsin inhibitors in soybeans has been studied [113, 121, 181, 229]. The trypsin inhibitor in *Glycine* seeds is controlled at a single locus with two, codominant alleles [121]. This control is different for the Kunitz trypsin inhibitor [113].

Lectins

The seeds of numerous leguminous species and varieties contain a group of natural products, lectins (phytohemagglutinins) which show some apparently unrelated characteristics. Their common feature is that they are proteins or glycoproteins. Their specifity and effect are measured in agglutination tests with treated or untreated erythrocytes [67, 152]. When acting on lymphocytes, they may be mitogenic [179]. They are water-soluble in a purified state [158]. This group of compounds shows remarkable biological properties insofar as these glycoproteins are able to bind to carbohydrates and to interact with animal and plant cells. This reaction mechanism is comparable to that of human antibodies, but lectins are not induced or the result of an immune response, they are inborn. Many aspects of the chemical and biological properties of the lectins are intensively reviewed by various authors [152, 153, 176, 196, 210, 244]. In the family *Leguminosae* more than 600 species and varieties contain lectins [152]. Samples for biological activity assays are generally obtained by extracting the seed material with physiological saline solutions or buffers (PBS), thus they generally cannot be associated with a distinct protein fraction [76, 106, 117, 126, 182, 196, 199, 209]. In the genus *Phaseolus* the G2/albumin fraction [39, 40, 196] and the albumin and globulin fractions contain lectins [158, 198]. In leguminous seeds lectins constitute between 2% and 10% [152] and more [198] of the total seed protein. But for other species the following data are available: *Canavalia ensiformis* (Con A) 2% – 3% [153]; *Glycine max* (SBA) 1% – 1.5% [153], and 0.25% – 1.2% and 3% [182] for defatted seed meal [194]; *Phaseolus vulgaris* (PHA) up to 10% [196]. Although numerous lectins have been isolated and characterized from different legume species [4, 39, 40, 41, 105, 106, 117, 118, 126, 195, 196, 199, 210], their biological role is a matter of speculation [152, 210, 245].

Based on their sugar-binding specifity, the lectins can be classified into several groups. Lectins binding D-mannose and D-glucose are found in seeds of *Pisum*, *Vicia*, *Lens* and *Canavalia* and are mitogenic towards lymphocytes [69, 70, 117, 118, 223]. N-acetyl-D-galactoseamine-binding lectins isolated from *Glycine* and *Phaseolus lunatus* are specific for blood group A and in the case of lima beans it is mitogenic. *Arachis* exhibits D-galactose-binding lectins and agglutinates type B erythrocytes. *Phaseolus vulgaris* produces lectins with complex carbohydrate-binding sites [4, 196, 199, 223]. Thus, there are obviously no recognizable relationships between the chemical properties of the lectin and the taxonomic distribution within the *Leguminosae* family. Lectins with the same specifity are found in different genera, whereas within one genus, specific and non-specific compounds are found [210]. In one species more than one lectin may be present. For example, seed extracts of *Vicia cracca* contain a glucose-specific lectin which nonspecifically agglutinates erythrocytes and a blood group A-specific lectin [10, 11, 209, 210]. There is a great variability in the biological effects of lectins especially in the genus *Phaseolus* which has been intensively studied in this respect [39, 40, 41, 76, 126]. These biological effects are associated with different polypeptides [40] which may explain the observed variability in agglutinating and mitogenic activities since the polypeptide composition may vary in protein preparations from the same source. The ability to agglutinate erythrocytes necessitates polyvalent binding sites. Most of the analysed lectins are tetramers having molecular weights of 120.000, built up of four subunits [74, 106, 126, 147, 196, 209], although other subunit structures have been reported [11, 75, 199, 209]. Numerous hemagglutinins are likewise mitogenic. In *Phaseolus* lectins, this is due to the presence of different subunits in the lectin molecule [126]. The hybrid tetramer consists of erythrocyte-reactive subunits (E), and lymphocyte-reactive subunits (L) [75]. The five possible isolectins can have the following tetrameric structures: L4 – L3E1 – L2E2 – L1E3 – E4 [210]. The five isolectins can be separated by ion change chromatography and recovered in high yield for further analyses of their properties [147].

In comparative analyses of 62 cultivars of *Phaseolus vulgaris* derived from different locations, the presence of the five isolectins in most of the cultivars could be confirmed [76] and the same is valid for German cultivars of *Phaseolus vulgaris*, as indicated in figure 4 which shows a typical elution profile of *Phaseolus* isolectins [24]. The same holds true for the isolectin composition of seeds of several varieties of *Arachis hypogaea* and *Arachis villosulicarpa*, two related species, which is quite similar and can be used for characterizing the species and varieties [193]. The coincidence of the findings concerning the structure of the isolectins in genotypes of different origin within the *Leguminosae* family suggests that many of these proteins belong to few classes of homologous proteins which were, perhaps, conserved as storage proteins.

This view is supported by electrophoresis [76, 147, 196] and primary structure comparison of the lectins of *Arachis*, *Glycine*, *Lens*, *Pisum* and *Vicia* [11, 78]. The amino acid sequences reveal homologies between the N-terminal regions of the α- and β-chains of lectins of different genotypes of agglutinins and the E and L subunits. This does not exclude alterations in the carbohydrate specifity and biological properties. Therefore, one cannot expect an unequivocal answer regarding the biological properties of the lectins. Focusing on the diversity shown by lectins, it is often not easy to discern to what extent the differences observed are genetically conditioned or due to the testing methods used. The genus *Phaseolus* exhibits a large variability in hemagglutinating and mitogenic activity. Thus, in four of nine wildgrowing populations of *Phaseolus aborigineus* in South America the seed extracts gave positive and negative agglutination reactions, while five were uniformly positive [41]. The positively reacting extracts had the same specifity as the common bean (PHA). Genetic analyses for the inheritance of PHA suggest a single

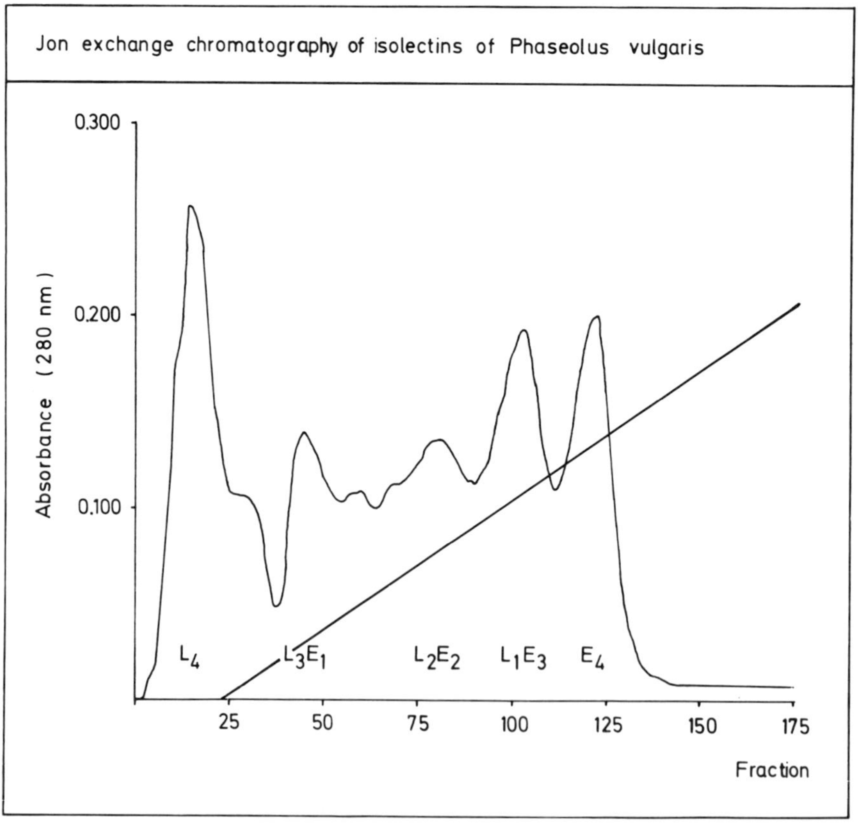

(Reference: [24]).

Figure 4. Elution profile of *Phaseolus* isolectins. (The hybrid tetrameric molecules are indicated.)

dominant gene, responsible for this trait. The analysis of 21 bean cultivars revealed the same types of specifity [126]. The great range of the existing variability has been amply demonstrated in investigations on 107 different bean cultivars and plant introduction lines [39]. These cultivars were screened for variation in the polypeptide composition of the lectin-containing G2/albumin protein group and their agglutination activities. Eight different G2/albumin electrophoretic patterns could be identified by which the genotypes could be characterized and classified. But the occurrence of deviating genotypes indicates that variability may be even greater. Similar results were obtained from analyses of other 62 cultivars of *Phaseolus vulgaris* [76]. Considering the inheritance of the lectins in seeds of *Glycine max* the presence of the major lectin has been shown to be controlled by a single dominant gene named Le. The same is valid for bean lectins [198]. The homozygous recessive genotypes (le/le) produce no lectin [198]. On the other hand in the case of *Vicia cracca* two genes are responsible for coding lectins [11]. In this contect, the question concerning the increased number of lectins in one diploid species is discussed in terms of duplication of structural genes, similar to that of isozymes. But the explanation of the loss of the ability to produce lectin offers at least two possible answers considering the molecular structure of genes: whether it is an example of an inherited point mutation or whether an insertion element within the gene prevents the transcription of the lectin gene or the processing of its products. The available informations of these genetic aspects are scarce. No generalization can be made at present as more investigations are still necessary.

Regarding the biological functions of lectins, the available data are mostly contradictory. Some aspects may be indicative. The ability of lectins to bind to characteristic carbohydrate structures on one side on the cell surface and the observed variability in lectin production in different genotypes of leguminous plants on the other side have initiated a number of systematic investigations concerning the function of lectins as important determinants of host range specifity in *Rhizobium* – legume symbiosis [23, 52, 190, 253]. Thus, the lectin of *Glycine* binds only to symbiontic strains of *Rhizobium* [16, 17, 23]. But the hypothesis that lectins are responsible for recognizing the symbiontic *Rhizobia* is too simple to explain the relationships between bacteria and legume roots [60, 61, 62]. This view is supported by investigations on 102 lines of *Glycine* on the presence of the soybean lectin (molecular weight 120.000) [194]. Five lines lacked the lectin although they were effectively nodulated by several strains of *Rhizobium japonicum*. These results reveal that the lectin of soybean is probably not essential for initiating the symbiontic relationship. That may be partly due to the fact that the distribution of the lectins within the plant during development changes. It could be shown that 9% of the *Arachis* lectins were located in the cotyledons in early seedling development, while two weeks later lectin was found in all tissues but not in the roots [193]. Its amounts in different tissues indicate that either they have been synthesized or transported from the cotyledons into the tissues. We don't know much concerning the functions of lectins within the plant. One interesting aspect is discussed in connection with the characterization of a lectin extracted from seeds of *Vigna radiata* [106] which possesses a strong enzymatic activity (α-galactosidase). By comparing data concerning synthesis, distribution and function within plants it is postulated that legumes lectins may, in general, be plant enzymes.

Quite another biological function as insecticides is ascribed to the lectins [128]. This function is interpreted in terms of an adaptative significance of the lectins in *Phaseolus vulgaris* for protecting seeds from attack by insect seed predators. Futhermore, the nutritional toxicity of *Phaseolus vulgaris* for monogastric animals is, at least partly, due to the presence of high amounts of lectins in the seeds [197, 198]. In contrast to these findings the lectins of *Glycine* seem to play a minor role with regard to deleterious effects of unheated soybean flour [245]. The lack of available information on different aspects of the biological function of lectins raises more questions than the answers provided.

Albumins and Allergens

For human nutrition, the protein quality which depends on the composition of the available protein fractions is decisive. In cotton seeds the 2S fraction has proven to be allergenic. This fraction constitutes one third of the total seed protein. The protein is chemically very similar to that isolated from castor beans [57], as evaluated on the basis of their amino acid composition. Thus, functionally similar proteins are closer related in taxonomically unrelated species than the 2S fraction as compared with other globulin fractions within the same species [259, 260]. But albumin storage proteins with allergenic properties are not restricted to both species mentioned. They are also found in leguminous seeds. Thus, aqueous extracts of seeds from green peas, soybeans and alfalfa showed allergenic activities [157]. The albumin fractions, obviously, represent a common or familiy-allergenic determinant among legumes. In view of the abundance and the ubiquitous occurrence of the albumins in leguminous seeds and their multiple functions more informations are needed for an adequate characterization of this seed protein fraction precisely.

GLOBULINS AS STORAGE PROTEINS

The globulins as seed reserve components, deposited in the cotyledons, constitute for each plant taxon definable sets of organ- and tissue-specific gene products [101, 104, 159, 162, 178, 240, 243]. Comprising of a limited number of protein species, they offer the possibility for analysing the control exerted by the genome and also the physiological influence of the seed-bearing plant. Most of the storage proteins are coded for by the DNA of the embryo [163] and their synthesis is under normal growing conditions essentially influenced by environmental factors [19, 30, 203].

They are quantitatively inherited [177]. Their genetically controlled variation is well established [93, 113, 144, 146, 241]. In mature seeds the globulins account for different percentages of the total protein content in different species: *Glycine* 90% [139]; *Phaseolus* 50–75% [104, 123, 155, 206]; *Pisum* 60–80% [173, 217, 242, 243]; *Vicia* 60–90% [77, 159, 257]; *Lupinus* 80% [68, 220]. Therefore, the globulins proved to be useful indicators for the quantitative and qualitative variability within the family of *Leguminosae* [18, 19, 27, 29, 32, 38, 45, 46, 50, 53, 54, 56, 63, 66, 82, 87, 98, 103, 104, 114, 115, 124, 136, 139, 144, 159, 162, 163, 164, 175, 177, 178, 212, 217, 240, 242, 243, 250]. The electrophoretic analysis is one of the most reliable methods for revealing genetic relationships in this group of proteins. Its advantages and limitations are explained in chapter 3. It serves for elucidating structural problems and inter- and intravarietal relationships (see chapter 11, 14, 16) as well as provides solutions to questions concerning the inheritance of distinct seed proteins components [38, 56, 103, 113, 124, 140–146]. The protein profiles of the major protein fraction (G1) of *Phaseolus* lines which are high and low in seed methionine content showed variations with regard to the banding patterns of the lines. There were three bands in the first-mentioned and two in the other group of genotypes. One of the 3 polypeptides is controlled by a single gene. Thus, the banding pattern indicates the availability of methionine in seeds of unknown strains [206]. The examples available reveal that for the understanding of the genetic basis of the quantitative and qualitative seed protein composition, more basic knowledge about the relations between protein structure and function and their genetic regulation on a broader genotype scale is needed. Structurally, the globulins consist mainly of two types of oligomeric proteins: the 7–8S proteins, generally termed vicilin and the 11–12S proteins, named legumin [63]. In some leguminous species a group of smaller globulin molecules is present which show a sedimentation coefficient of approximately 2. They are found in *Glycine* [114, 115, 139] in *Lupinus* [220] and in some other legumes [53, 54]. These proteins are often found as dissociation products and in early seed developmental stages [63, 114, 115]. But in mature seeds they could correspond to the lectins and/or protease inhibitors, according to their molecular weights. Recently a further storage protein distinct from both the 7S and 11S proteins was isolated from *Pisum* seeds [51]. This protein was named 'convicilin' because it shows a cross-reactivity with vicilin.

Legumin-like and vicilin-like proteins are found in a wide variety of leguminous seeds as shown in table 7. The data indicate collectively numerous similarities for these proteins isolated from different sources, identified according to their sedimentation coefficients and their molecular weights. But before making appropriate genetic comparisons with regard to genotype-specific traits, the homology of the characters has to be proved. So far, only in few cases the homology of globulin storage proteins could be ascertained. It is mainly based on identical serological reactions [50, 66] and similarities in primary structure. In this way the legumin-like proteins occurring in *Pisum sativum* (legumin), *Vicia faba* (legumin), *Arachis hypo-*

gaea (arachin) and *Glycine max* (glycinin) have been shown to be homologous proteins [87]. These proteins are possibly homologous with the legumin-like proteins from *Phaseolus* and *Vigna.* Provided that the genetic relations between legumin-like proteins can be proved for other leguminous species these proteins would have the same genetic origin. This would signify that few structural genes would be responsible for producing a class of main storage proteins within the *Leguminosae* family.

The second class of probably homologous proteins are the vicilin-like proteins found in *Vicia* (vicilin), *Pisum* (vicilin) and *Arachis* (α-conarachin). Unfortunately, only little is known on their distribution in other members of the *Leguminosae* family. Furthermore, the lack of a methodical system of naming the different protein classes often makes appropriate comparisons difficult. The storage proteins of *Phaseolus* obviously occupy an exceptional position within the leguminous seed proteins [159, 195]. For further characterization of the homology of legume seed proteins it will be necessary to collect more data concerning the amino acid sequences of the respective proteins.

Considering the quantitative distribution of the two main globulin fractions the legumin : vicilin ratio reveals genotype-specific differences. The following ratios were found: *Vicia* (2.30–4) : 1 [26, 45, 175, 257]; *Pisum* (1–3) : 1 [12, 173, 243]; *Phaseolus* G1 : G2 : 6 : 1 [104], 1 : 9 [45]. In *Vigna unguiculata* the 7S globulins represent the major protein fraction [45]. The available data indicate that the legumin : vicilin ratio is not a constant trait. It can vary not only between species but also within a species and varieties [63, 174, 241, 243, 257]. These findings are of great interest with regard to the seed protein quality, because both fractions differ in their essential amino acid composition. This is shown in table 8. With regard to the amounts of the limiting amino acids methionine and cystine the legumin fraction is superior to the vicilin fraction. As will be discussed later, both fractions are independently genetically controlled, so that changes in their respective proportions can be expected. Variations in the proportion of legumin are present in different genotypes as indicated by the L : V-ratios. Moreover, it has been shown that the amount of legumin in different genotypes of *Pisum* is controlled by genes which are closely linked to genes responsible for distinct morphological characters [59, 217]. The effect exerted by the r_a locus on the composition of storage proteins of the wrinkled seed mutant in *Pisum* thus indicates that also in legumes ‘regulatory mutants’ similar to those found in maize are available [59]. Alterations in the seed protein quality by changes of the proportions of both globulin fractions will only be successful if both components are available in nearly equal amounts. Furthermore, the data available from *Phaseolus* and *Vigna* suggest that the 11S fraction seems to be nonessential during seed protein development. The question of qualitative improvement of legume seeds is complicated by the fact that the relatively large amounts of protease inhibitors and lectins in legume seeds influence the nutritional value of the seed proteins [197, 198]. In connection with seed protein improvement it

Table 7. Legumin-like and vicilin-like proteins of legume seeds (data from 63).

Species	Legumin-like proteins		Vicilin-like proteins	
	Sedimentation coefficient (S)	Molecular weight	Sedimentation coefficient (S)	Molecular weight
Glycine max	12.2 - 14.0	309000 - 380000	6.7 - 8.0	105000 - 330000
Phaseolus aureus	11.3	-	8.0	-
Phaseolus coccineus	12.16	-	7.4	-
Phaseolus nanus	10.10	-	6.6	-
Phaseolus vulgaris	11.02 - 11.6	340000	6.5 - 7.6	140000 - 151000
Vigna unguiculata	11.2	320000	7.3	-
Lathyrus clymenum	13.0	-	7.6	-
Lathyrus sativus	13.04	-	7.5	-
Lathyrus silvestris	12.97	-	7.5	-
Lathyrus odoratus	12.0	-	7.6	-
Arachis hypogaea	12.0 - 14.7	330000 - 396000	7.8 - 8.7	142000 - 190000
Pisum sativum	12.1 - 13.7	330000 - 410000	7.1 - 8.1	-
Vicia faba	11.4 - 11.5	328000	6.8 - 7.1	150000
Vicia sativa	11.48 - 12.9	208000 - 360000	7.1 - 7.5	193000
Lupinus albus	12.29 - 12.6	393000	8.2 - 8.3	204000
Lupinus angustifolius	11.6 - 13.05	363000	7.8 - 8.2	181000
Lupinus luteus	11.4 - 11.50	-	7.4 - 8.3	-
Lupinus polyphyllus	12.20	-	8.7	-

Table 8. Essential amino acid contents in 7S and 11S seed protein fractions of different legumes (g/16 gN).

Amino acids	Standard protein	Glycine max		Phaseolus vulgaris		Vigna unguiculata	Pisum sativum		Vicia faba	
	Ref.:232	Ref.:139		Ref.:104	Ref.:63,195	Ref.:45	Ref.:63,174		Ref.:26,63,163,175	
Fraction		7S	11S	G1	11S	7S	7S	11S	7S	11S
Met	1.90	0.3	1.3-1.8	0.57-1.36	1.5	0.98	0.22	0.65	0.2 -0.7	0.3 -0.7
Cys	1.10	0.3	1.4-1.7	0.0 -0.13	0.6	0.55	0.35	0.71	0.13-0.4	0.74-1.26
Try	2.10	0.3	1.5-1.6	-	2.9	2.23	0.09	1.06	-	1.13
Val	5.80	5.1	4.9-5.1	4.10-5.89	7.0	4.96	4.6	4.6	4.3 -5.6	4.4 -5.1
Thr	4.89	2.8	3.4-4.1	2.56-3.41	4.9	2.99	3.4	2.9	2.6 -3.8	3.4 -3.9
Leu	8.30	10.3	7.2-8.1	9.42-11.82	8.7	6.36	9.2	8.1	8.9 -9.7	7.3 -8.5
Ile	4.60	6.4	4.7-4.9	4.66-5.47	4.9	3.86	5.1	4.0	5.1- 5.7	4.0 -4.3
Lys	6.21	7.0	4.9-5.7	7.38-8.87	7.8	6.75	7.9	4.9	6.5- 8.2	4.2 -5.3
Phe	5.40	7.4	5.5-5.7	7.78-10.82	3.6	6.66	6.2	4.9	4.0- 6.8	3.2 -4.8

is discussed to alter the relative proportions of embryo axis to cotyledons because the amino acid composition of the first one is superior to that of the cotyledons [183]. But the embryo accounts for only 1% while the cotyledons represent approximately 90% of the seed proteins. These considerations for seed protein improvement remain speculations.

Both the legumin-like and vicilin-like proteins are heterozygous. They can be split in subunits by electrophoretic methods. This important subject of protein structures is extensively discussed in chapters 11 and 14. The different levels of seed protein incorporation as indicated by the different legumin : vicilin ratios raise the question which regulatory mechanisms are involved in seed development that are responsible for the final protein composition. A convenient answer includes investigations on seed development in order to understand how distinct processes are regulated, which processes are limiting in terms of seed maturation and which processes underlie genetic control. Intensive experimental morphogenetic studies on the embryogeny in the genus *Phaseolus* indicate an experimental approach [247]. Seed development in *Phaseolus* passes through nine distinct stages from fertilisation to maturation. Although until maturation all morphogenetic stages are nearly of equal duration the maturation time differs significantly between two species analysed. The maturation time for *Phaseolus vulgaris* is of 16 days, that of *Phaseolus coccineus* needed additional 30 days for bringing to end the maturation. Thus, genotype-specific differences in maturation time should be taken into consideration while analysing the accumulation of proteins in seeds. The metabolic activities decline towards the end of the maturation period but in terms of differential gene activity the different duration would mean that in both species deviating regulatory mechanisms could be effective. The measurement of biochemical parameters in accordance with morphogenetic and physiological parameters offers the possibility to study an additional aspect of the regulation of seed protein accumulation. This view is supported by the fact that the rates of dry matter and protein accumulation are strongly influenced by the time of flowering in *Glycine* [86]. Investigations involving developmental mutants, e.g. early flowering mutants, could contribute to understanding in which way different processes are regulated and to what extent their genetic control may be modified [92, 172].

In order to follow the globulin accumulation in developing seeds electrophoretic and diverse serological methods, density centrifugation and quantitative analyses of the incorporation of radioactive labelled amino acids into the globulin were used. The findings of several authors reveal that the synthesis of globulins starts some days after flowering and that the pattern of the synthesis of legumin and vicilin changes during development, resulting in different legumin : vicilin ratios [45, 104, 136, 163, 164, 174, 178, 237, 242, 247, 257]. The genotype-dependent accumulation of four protein families related to vicilin and legumin during seed development has been shown in different lines of *Pisum*. Although the overall protein sequence appeared similar in the three genotypes during development, the cross immunoelectrophoretic profiles revealed distinct quantitative differences mainly in the legumin fraction. The results indicate that the relative proportions of the various proteins are under genetic control. One of the antigens found in mature cotyledons has been shown to be a hemagglutinin [164, 241, 242]. Similar quantitative differences could be found in mutants of one *Pisum* variety [93, 169]. Thus, for some legume species the pattern of seed protein accumulation seems to be clear. The first crucial point is the duration of the seed development after flowering which is a genetically controlled and modifiable character. The second comprises the onset of protein synthesis, its kinetics and the termination which accompany seed development. This is shown in table 9, in which data concerning the protein accumulation in different genotypes are compiled and compared on the basis of developmental time-tables.

The morphogenetic changes in seed development as well as the onset, the duration and the termination of protein synthesis vary not only between species but also between the varieties of one species. The variation in accumulation of the main storage protein of *Phaseolus* (G1) becomes obvious by comparing the onset and the termination of the protein synthesis. Two of the three cultivars start at the same time, but the termination is different while the third starts later, resulting in different G1-contents at maturity (38, 38 and 47% G1/protein) [178]. In lines of *Pisum* the low legumin content is due to the later onset of legumin synthesis [242]. The accumulation rates of the different protein components are not continuous. There are genotype-specific pauses in synthesis of different lengths during seed development. Similar results were obtained in *Pisum* [164, 242] and in *Vigna* [45]. These findings demonstrate that onset, kinetics and termination times of the synthesis of seed protein components are genetically controlled and genotype-specific. The discontinuous synthesis of different proteins indicates possibly differential gene expression and activity [178]. Furthermore, the findings underline that the regulation of seed protein synthesis at that level must be considered in selection work for altered protein composition.

Based on cytological and cell-physiological criteria, the ontogeny of seed generally can be divided in three [174] or four phases [45]. Generally, stage 1 starts with zygote formation. The cells of the storage tissue exhibit a high mitotic activity and growth is mainly due to cell divisions. Storage components are not synthesized. Stage 2 begins concomitantly with cessation of mitotic activity. Cell elongation causes a rapid increase of the storage tissues and consequently of the seeds. Mainly in this stage the synthesis of storage compounds and their accumulation occurs. The other stages are characterized by successive desiccation of the seed until the stage of a dormant seed is reached. Therefore, the seed length is often used as an index for the respective developmental stages as the reference parameter. But it is not suitable for very early and late stages of seed development. The parameter 'days after flowering' is needed additionally [237]. Generally, at least both physical parameters combined with other physiological data should be available for appropriate and reliable comparisons.

The rapid synthesis of large amounts of a small number of gene products after the embryo has reached its final cell number very early in ontogeny raises the question of the regulatory processes involved. Studies on developing cotyledons in *Vicia* [163] and *Pisum* [216] show that the cotyledonous cells are highly polyploid, due to endoreduplication. They exhibit DNA contents of 16C in *Vicia* [163] and up to 64C in *Pisum* [216]. The polyploidization is accompanied by a high rate of RNA-synthesis per cell, suggesting that the formation of polyploid nuclei is closely related to the intensity of protein synthesis (gene dosage effect) [216]. Comparisons of the legumin accumulation in developing seeds and the proportions of legume-specific polypeptides after in-vitro synthesis show that the transcription of globulin-specific mRNA is another decisive level for the regulation of the biosynthesis of storage proteins of legumes [174]. The incomplete knowledge on the regulatory processes involved in seed development, especially in the onset of the synthesis of the different protein classes, would be essentially enlarged by analysing a wide range of genetic variants. These should include developmental mutants carrying genes for early flowering and photoperiod insensitivity, with altered flowering duration, alterations in seed size and seed coat formation in the spectrum of proteins. According to the rule of parallel variation, such mutants can be expected because striking differences in these characters are apparent between species and varieties.

MAINTENANCE OF GENETIC RESOURCES

The desired enhancement and enlargement of the genetic variability in legume seeds in view of the analysis of the genetic basis of regulation of seed development includes an additional aspect, namely the genetic resources. The techniques of plant breeding have not only increased the productivity of developed varieties but also narrowed the genetic base of many varieties as compared with the land races from which they are derived and their wild relatives [80]. Concomitantly, a disease-vulnerable genetic pool in numerous plants was produced [46]. The manifold aspects concerning the conservation of plant genetic resources are extensively discussed in several reviews [37, 80, 108, 109]. Primitive and wild gene pools will be the main sources for genes for disease resistance, for adaptive gene complexes and for a broadened genetic base. Though induced mutations represent a very efficient source of variability [22, 36, 89, 92, 94, 120, 171, 173], exploration of the genetic variation in wild and cultivated plant populations will become more and more important. One method for obtaining population genetic data consists of analysing electrophoretically the protein variation [37]. The application of the isozyme technique has proved to be a suitable tool for comparing levels of genic variations [29, 32, 42, 46, 64, 84, 93, 137, 165, 170, 214, 215, 246, 254]. Isozymes are excellent genetic markers because they fulfill all the criteria for genetic markers (see above). The study of the levels of genetic variability in natural populations as measured by allele

Table 9. Developmental time-tables and protein accumulation in seeds of *Phaseolus*, *Pisum*, and *Vicia*.

			Ref.
Phaseolus vulgaris	0 10 20 30 36 0.7 4.7 15 16.5 15.1 0 7 11 20 36 globular cotyledon maturation Zygote heart dormant	days after flowering seed length (mm) days after flowering embryo stage	247
Phaseolus coccineus	0 10 20 30 40 50 0.75 5 22 22 22 22 0 7 12 18 36 globular cotyledon maturation Zygote heart dormant	days after flowering seed length (mm) days after flowering embryo stage	247
Phaseolus vulgaris cv. Tender-green	0 10 20 30 40 50 6 15 21 21 13	days after flowering seed length (mm) protein accumulation G_1 accumulation	104,237
Phaseolus vulgaris cv.WI 74-2047 cv.Sanilac cv.Endogava Z.N.	0 10 20 30 40 50	days after flowering G_1 accumulation	178

Table 9 continued.

			Ref.
Pisum sativum	0 10 20 30 40 50	days after flowering	164
line 034 (high legumin			242
line 076 (high legumin)	maturation	legumin accumulation	
line 086 (low legumin)			
	maturation	vicilin accumulation	
Vicia faba	25 30 40 50 60 70 80 90	days after flowering	257
		legumin accumulation	
		vicilin accumulation	
Vicia faba var.minor		legumin accumulation	174
		vicilin accumulation	

frequency per locus, employing electrophoretic techniques [42, 150], revealed quantitative estimates of the genetic diversity in populations of two *Lupinus* species concerning genetic variability and ecological amplitude [7] even though the adaptive or selective significance of the allozymes are poorly understood. Although there are no difficulties with regard to the applicable strategies [37, 46, 80, 108, 109], in the *Leguminosae* family only a few results concerning comparisons of the genetic diversity in gene pools are available so far [137, 170, 248]. Generally, the attempts for comparing the genetic diversity in some genera of legumes indicate that the 'ancestors' of cultivars exhibit a broader genetic variation. This needs to be investigated more intensively. A prerequisite for understanding differences between genotypes and populations is the precise knowledge of the genetic systems responsible for the variability observed. The evaluation of the genetic structure of natural populations of crop relatives could be useful with regard to possible breeding strategies for reduced genetic vulnerability for diseases [46] and for genotypes adapted to non-optimal environments [7]. Generally, it must be underlined that the genetic resources of the wild relatives of our crops should be analysed systematically [37]. Population genetic studies as indicated will substantially contribute to the exploration, the conservation and the use of genetic resources, that we need urgently.

REFERENCES

1. Abbey, BW, G Norton, and RJ Neale (1979): Effects of dietary proteinase inhibitors from field bean (*Vicia faba* L.) and field bean meal on pancreatic function in the rat. Br. J. Nutr. 41: 39–45.
2. Adams, WM (1973): On the quest for quality in the field bean. Nutritional Improvement of Food Legumes by Breeding (Milner, M. ed.): 143–149; PAG New York.
3. Adriaanse, A, W Klop, and JE Robbers (1969): Characterization of *Phaseolus vulgaris* cultivars by their electrophoretic patterns. J. Sci. Food Agric. 20: 647–650.
4. Allen, HJ, and EAZ Johnson (1976): Isolation and partial characterization of a lectin from *Vicia faba*. Biochim. Biophys. Acta 444: 374–385.
5. Altman, PL, and DS Dittmer (eds.) (1968): Metabolism. Fed. Amer. Soc. Exp. Biology, Bethesda, Maryland, U.S.A.
6. Aykroyd, WR, and J Doughty (1964): Legumes in Human Nutrition. FAO Rome.
7. Babbel, GR, and RK Selander (1974): Genetic variability in edaphically restricted and widespread plant species. Evolution 28: 619–630.
8. Bajaj, S (1973): Biological value of legume proteins as influenced by genetic variation. Nutritional Improvement of Food Legumes by Breeding (Milner, M ed.): 223–232; PAG New York.
9. Basha, SMM, and L Beevers (1975): The development of proteolytic activity and protein degradation during the germination of *Pisum sativum* L. Planta 124: 77–87.
10. Baumann, CM, and H Rüdiger (1981): Interactions between the two lectins from *Vicia cracca*. FEBS Lett. 136: 279–283.
11. Baumann, C, H Rüdiger, and AD Strosberg (1979): A comparison of the two lectins from *Vicia cracca*. FEBS Lett. 102: 216–218.
12. Beevers, L, and R Poulson (1972): Protein synthesis in cotyledons of *Pisum sativum* L. I. Changes in cell-free amino acid incorporation capacity during seed development and maturation. Plant Physiol. 49: 476–481.

13. Bell, EA (1981): Non-protein amino acids in the *Leguminosae*. In: Polhill, RM, and PH Ravens (eds.): Advances in Legume Systematics, II: 489–499; Royal Botanic Gardens, Kew; Ministry of Agriculture, Fisheries and Food, London.
14. Bell, EA, and ASL Tirimanna (1963): Occurrence of γ-hydroxyarginine in plants. Nature 197: 901–902.
15. Bhagwat, SG, CR Bhatia, T Gopalakrishna, DC Joshua, RK Mitra, P Narahari, SE Pawar, and RG Thakare (1979): Increasing protein production in cereals and grain legumes. Seed Protein Improvement in Cereals and Grain Legumes, II: 225–236; IAEA Vienna.
16. Bhuvaneswari, TV, and WD Bauer (1978): Role of lectins in plant-microorganism interactions. III. Influence of rhizospere/rhizoplane culture conditions on the soybean lectin-binding properties of *Rhizobia*. Plant Physiol. 62: 71–74.
17. Bhuvaneswari, TV, SG Pueppke, and WD Bauer (1977): Role of lectins in plant-microorganism interactions. I. Binding of soybean lectins to *Rhizobia*. Plant Physiol. 60: 486–491.
18. Bingham, ET, and KJ Yeh (1971): Electrophoretic patterns among alfalfa seed proteins from selected varieties, experimental stocks, and species accessions. Crop Sci. 11: 58–61.
19. Blagrove, RJ, JM Gillespie, and PJ Randall (1976): Effect of sulphur supply on the seed globulin composition of *Lupinus angustifolius*. Aust. J. Plant Physiol. 3: 173–184.
20. Bliss, FA (1973): Cowpeas in Nigeria. Nutritional Improvement of Food Legumes by Breeding (Milner, M ed.): 151–158; PAG New York.
21. Bliss, FA, LN Barker, JD Franckowiak, and TC Hall (1973): Genetic and environmental variation of seed yield, yield components, and seed protein quantity and quality of cowpeas. Crop Sci. 13: 656–660.
22. Blixt, S (1979): Natural and induced variability for seed protein in temperate legumes. Seed Protein Improvement in Cereals and Grain Legumes, II: 3–21; IAEA Vienna.
23. Bohlool, BB, and EL Schmidt (1974): Lectins: a possible basis for specifity in the *Rhizobium* – legume root nodule symbiosis. Science 185: 269–271.
24. Bollmann, J (1982): unpublished.
25. Bond, DA (1966): Yield and components of yield in diallel crosses between inbred lines of winter beans (*Vicia faba*). J. Agric. Sci. 67: 325–336.
26. Boulter, D (1974): Biosynthese, Struktur und Zusammensetzung der Samenproteine von *Vicia faba* im Hinblick auf die Auslese zur Verbesserung der Proteinqualität. Göttinger Pflanzenzüchter-Seminar 2: 106–120.
27. Boulter, D (1979): Structure and biosynthesis of legume storage proteins. Seed Protein Improvement in Cereals and Grain Legumes, I: 125–136; IAEA Vienna.
28. Boulter, D (1980): Ontogeny and development of biochemical and nutritional attributes in legumes seeds. In: Summerfield, RJ, and AH Bunting (eds.): Advances in Legume Science, I: 127–134; Ministry of Agriculture, Fisheries and Food, London.
29. Boulter, D, and E Derbyshire (1971): Taxonomic aspects of the structure of legume proteins. In: Harborne, JB, D Boulter, and BL Turner (eds.): Chemotaxonomy of the *Leguminosae*: 285–308; Academic Press, London and New York.
30. Boulter, D, IM Evans, and E Derbyshire (1973): Proteins of some legumes with reference to environmental factors and nutritional value. Qual. Plant. 23: 239–250.
31. Boulter, D, IM Evans, A Thomson, and A Yarwood (1973): The amino-acid composition of *Vigna unguiculata* (cow pea) meal in relation to nutrition. Nutritional Improvement of Food Legumes by Breeding (Milner, M ed.): 205–215; PAG New York.
32. Boulter, D, DA Thurman, and E Derbyshire (1967): A disc electrophoretic study of globulin proteins of legume seeds with reference to their systematics. New Phytol. 66: 27–36.
33. Bowman, DE (1971): Isolation and properties of a proteinase inhibitor of navy bean. Arch. Biochem. Biophys. 144: 541–548.
34. Bressani, R (1973): Legumes in human diets and how they might be improved. Nutritional Improvement of Food Legumes by Breeding (Milner, M ed.): 15–42; PAG New York.

35. Bressani, RL, G Elias, and AT Valiente (1963): Effect of cooking and of amino acid supplementation on the nutritive value of black beans (*Phaseolus vulgaris* L.). Br. J. Nutr. 17: 69–78.
36. Brock, RD (1971): The role of induced mutations in plant improvement. Rad. Bot. 11: 181–196.
37. Brown, AHD (1978): Isozymes, plant population genetic structure and genetic conservation. Theor. Appl. Genet. 52: 145–157.
38. Brown, JWS, Y Ma, FA Bliss, and TC Hall (1981): Genetic variation in the subunits of globulin-1 storage protein of French bean. Theor. Appl. Genet. 59: 83–88.
39. Brown, JWS, TC Osborn, FA Bliss, and TC Hall (1982a): Bean Lectins. Part 1: Relationships between agglutinating activity and electrophoretic variation in the lectin-containing G2/albumin seed proteins of French bean (*Phaseolus vulgaris* L.). Theor. Appl. Genet. 62: 263–271.
40. Brown, JWS, TC Osborn, FA Bliss, and TC Hall (1982b): Bean Lectins. Part 2: Relationship between qualitative lectin variation in *Phaseolus vulgaris* L. and previous observations on purified bean lectins. Theor. Appl. Genet. 62: 361–367.
41. Brücher, O, M Wecksler, A Levy, A Palozzo, and WG Jaffé (1969): Comparison of phytohaemagglutinins in wild beans (*Phaseolus aborigineus*) and in common beans (*Phaseolus vulgaris*) and their inheritance. Phytochem. 8: 1739–1743.
42. Bulmer, MG (1971): Protein polymorphism. Nature 234: 410–411.
43. Buttery, BR, and RI Buzzell (1968): Peroxidase activity in seeds of soybean varieties. Crop Sci 8: 722–725.
44. Buttery, BR, and RI Buzzell (1971): Properties and inheritance of urease isoezymes in soybean seeds. Can. J. Bot. 49: 1101–1105.
45. Carasco, JF, R Croy, E Derbyshire, and D Boulter (1978): The isolation and characterization of the major polypeptides of the seed globulin of cowpea (*Vigna unguiculata* L. Walp) and their sequential synthesis in developing seeds. J. Exp. Bot. 29: 309–323.
46. Cherry, JP, RY Mayne, and RL Ory (1974): Proteins and enzymes from seeds of *Arachis hypogaea* L. IX. Electrophoretically detected changes in 15 peanut cultivars grown in different areas after inoculation with *Aspergillus parasiticus*. Physiol. Plant Pathol. 4: 425–434.
47. Crocomo, OJ, TS Gerald Lee, E Derbyshire, and D Boulter (1979): Biochemical investigations on the seed proteins of a Brazilian variety and mutant of *Phaseolus vulgaris*. Seed Protein Improvement in Cereals and Grain legumes, I: 217–229; IAEA Vienna.
48. Crocomo, OJ, AT Neto, A Ando, S Blixt, and D Boulter (1978): Breeding for improved protein content and quality in the bean (*Phaseolus vulgaris*): II. Further work in selections from spontaneous variation; new work on mutagenic treatments and the influence of added nitrogen levels. Seed Protein Improvement by Nuclear Techniques: 207–222; IAEA Vienna.
49. Crocomo, OJ, AT Neto, S Blixt, and K Mikaelsen (1975): Breeding for protein in the bean (*Phaseolus vulgaris* L.): I. Inventory of some Brazilian varieties and a number of lines of differing origin. Evaluation of Seed Protein Alteration by Mutation Breeding: 197–212; IAEA Vienna.
50. Croy, RRD, E Derbyshire, TG Krishna, and D Boulter (1979): Legumin of *Pisum sativum* and *Vicia faba*. New Phytol. 83: 29–35.
51. Croy, RRD, JA Gatehouse, M Tyler, and D Boulter (1980): The purification and characterization of a third storage protein (convicilin) from the seeds of pea (*Pisum sativum* L.). Biochem. J. 191: 509–516.
52. Currier, WW, and GA Strobel (1977): Chemotaxis of *Rhizobium* ssp. to a glycoprotein produced by birdsfoot trefoil roots. Science 196: 434–436.
53. Danielsson, CE (1949): Investigations of vicilin and legumin. Acta chem. Scand. 3: 41–49.
54. Danielsson, CE (1949): Seed globulins of the *Gramineae* and *Leguminosae*. Biochem. J. 44: 387–400.
55. Daussant, J, NJ Neucere, and EJ Couberton (1969b): Immunochemical studies on *Arachis hypogaea* proteins with particular reference to the reserve proteins. II. Protein modification during germination. Plant Physiol. 44: 480–484.

56. Daussant, J, NJ Neucère, and LY Yatsu (1969a): Immunochemical studies on *Arachis hypogaea* proteins with particular reference to the reserve proteins. I. Characterization, distribution, and properties of α-arachin and α-conarachin. Plant Physiol. 44: 471–479.
57. Daussant, J, RL Ory, and LL Layton (1976): Characterization of proteins and allergens in germinating castor seeds by immunochemical techniques. J. Agric. Food Chem. 24: 103–107.
58. Davies, DR (1976): Variation in the storage proteins of peas in relation to sulphur amino-acid content. Euphytica 25: 717–724.
59. Davies, DR (1980): The ra locus and legumin synthesis in *Pisum sativum*. Biochem. Genet. 18: 1207–1219.
60. Dazzo, FB, and WJ Brill (1978): Regulation by fixed nitrogen of host-symbiont recognition in the *Rhizobium*-clover symbiosis. Plant Physiol. 62: 18–21.
61. Dazzo, FB, and DH Hubbell (1975): Cross-reactive antigens and lectin as determinants of symbiotic specificity in the *Rhizobium*-clover association. Appl. Microbiol. 30: 1017–1033.
62. Dazzo, FB, WE Yanke, and WJ Brill (1978): Trifolin: a *Rhizobium* recognition protein from white clover. Biochim. Biophys. Acta 539: 276–286.
63. Derbyshire, E, DJ Wright, and D Boulter (1976): Legumin and vicilin, storage proteins of legume seeds. Phytochem. 15: 3–24.
64. Derbyshire, E, HP Müller, MTV Carvalho, and OJ Crocomo (1981): Protein profiles of Brazilian beans (*Phaseolus vulgaris*) obtained by electrophoresis in slabs of polyacrylamide gel. Energ. Nucl. Agric. 3: 100–109.
65. Dickson, MH, and LR Hackler (1973): Protein quantity and quality in high-yielding beans. Nutritional Improvement of Food Legumes by Breeding (Milner, M ed.): 185–192; PAG New York.
66. Dudman, WF, and A Millerd (1975): Immunochemical behaviour of legumin and vicilin from *Vicia faba*: a survey of related proteins in the *Leguminosae* subfamily *Faboideae*. Biochem. System. Ecol. 3: 25–33.
67. Duk, M, and E Liskowska (1981): *Vicia graminea* anti-N lectin: Partial characterization of the purified lectin and its binding to erythrocytes. Eur. J. Biochem. 118: 131–136.
68. Duranti, M, and P Cerletti (1979): Amino acid composition of seed proteins of *Lupinus albus*. J. Agric. Food Chem. 27: 977–978.
69. Entlicher, G, and J Kocourek (1975): Studies on phytohemagglutinins. XXIV. Isoelectric point and hybridization of the pea (*Pisum sativum* L.) isophytohemagglutinins. Biochim. Biophys. Acta 393: 165–169.
70. Entlicher, G, JV Kostir, and J Kocourek (1970): Studies on phytohemagglutinins. III. Isolation and characterization of hemagglutinins from the pea (*Pisum sativum* L.). Biochim. Biophys. Acta 221: 272–281.
71. Evans, AM (1973): Genetic improvement of *Phaseolus vulgaris*. Nutritional Improvement of Food Legumes by Breeding (Milner, M ed.): 107–115; PAG New York.
72. Evans, CS, and EA Bell (1978): 'Uncommon' amino acids in the seeds of 64 species of *Caesalpinieae*. Phytochem. 17: 1127–1129.
73. Evans, IM, and D Boulter (1980): Crude protein and sulphur amino acid contents of some commercial varieties of peas and beans. J. Sci. Food Agric. 31: 238–242.
74. Felsted, RL, MJ Egorin, RD Leavitt, and NR Bachur (1977): Recombinations of subunits of *Phaseolus vulgaris* isolectins. J. Biol. Chem. 252: 2967–2971.
75. Felsted, RL, RD Leavitt, C Chen, NR Bachur, and RMK Dale (1981): Phytohemagglutinin isolectin subunit composition. Biochim. Biophys. Acta 668: 132–140.
76. Felsted, RL, J Li, G Pokrywka, MJ Egorin, J Spiegel, and RMK Dale (1981): Comparison of *Phaseolus vulgaris* cultivars on the basis of isolectin differences. Int. J. Biochem. 13: 549–557.
77. Flink, J, and I Christiansen (1973): The production of a protein isolate from *Vicia faba*. Lebensm.-Wiss.u.Technol. 6: 102–106.

78. Foriers, A, C Wuilmart, N Sharon, and AD Strosberg (1977): Extensive sequence homologies among lectins from leguminous plants. Biochem. Biophys. Res. Comm. 75: 980–986.
79. Fowden, L (1964): The chemistry and metabolism of recently isolated amino acids. Ann. Rev. Biochem. 173: 173–204.
80. Frankel, OH (1974): Genetic conservation: our evolutionary responsibility. Genetics 78: 53–65.
81. Frankel, TN, and ED Garber (1965): Esterases in extracts from germinating seeds of twelve pea varieties. Bot. Gaz. 126: 221–222.
82. Gatehouse JA, RRD Croy, and D Boulter (1980): Isoelectric-focusing properties and carbohydrate content of pea (*Pisum sativum*) legumin. Biochem. J. 185: 497–503.
83. Gates, P, and D Boulter (1979a): Nitrogen regime and isoenzyme changes in *Vicia faba*. Phytochem. 18: 1789–1791.
84. Gates, P, and D Boulter (1979b): The use of seed isoenzymes as an aid to the breeding of field beans (*Vicia faba* L.). New Phytol. 83: 783–791.
85. Gates, P, and D Boulter (1980): The use of pollen isoenzymes as an aid to the breeding of field beans (*Vicia faba* L.). New Phytol. 84: 501–504.
86. Gbikpi, PJ, and RK Crookston (1981): Effect of flowering date on accumulation of dry matter and protein in soybean seeds. Crop Sci. 21: 652–655.
87. Gilroy, J, DJ Wright, and D Boulter (1979): Homology of basic subunits of legumin from *Glycine max* and *Vicia faba*. Phytochem. 18: 315–316.
88. Gonzales, E, A Callejar, D Seidl, and WG Jaffé (1979): Subtilisin inhibitor activity in legume seeds. J. Agric. Food Chem. 27: 912–913.
89. Gottschalk, W (1982): The flowering behaviour of *Pisum* genotypes under phytotron and field conditions. Biol. Zbl. 101: 249–260.
90. Gottschalk, W, and HP Müller (1970): Monogenic alteration of seed protein content and protein pattern in X-ray-induced *Pisum* mutants. Improving Plant Protein by Nuclear Techniques: 201–215; IAEA Vienna.
91. Gottschalk, W, and HP Müller (1974): Quantitative and qualitative investigations on the seed proteins of mutants and recombinants of *Pisum sativum*. Theor. Appl. Genet. 45: 7–20.
92. Gottschalk, W, and HP Müller (1979): The reaction of an early-flowering *Pisum* recombinant to environment and genotypic background. Seed Protein Improvement in Cereals and Grain Legumes, I: 259–272; IAEA Vienna.
93. Gottschalk, W, and HP Müller (1982): Seed proteins of *Pisum* mutants and recombinants. Qual. Plant. in press.
94. Gottschalk, W, HP Müller, and G Wolff (1975a): Relations between protein production, protein quality and environmental factors in *Pisum* mutants. Breeding for Seed Protein Improvement Using Nuclear Techniques: 105–123; IAEA Vienna.
95. Gottschalk, W, HP Müller, and G Wolff (1975b): The genetic control of seed protein production and composition. Egypt. J. Genet. Cytol. 4: 453–468.
96. Gorman, MB, and YT Kiang (1977): Variety-specific electrophoretic variants of four soybean enzymes. Crop Sci. 17: 963–965.
97. Gorman, MB, and YT Kiang (1978): Models for the inheritance of several variant soybean electrophoretic zymograms. The J. Hered 69: 255–258.
98. Grant, DR, and JM Lawrence (1964): Effects of sodium dodecyl sulfate and other dissociating reagents on the globulins of peas. Arch. Biochem. Biophys. 108: 552–561.
99. Grant, DR, AK Sumner, and J Johnson (1976): An investigation of pea seed albumins. Can. Inst. Food Sci. Technol. 9: 84–91.
100. Gridley, HE, and AM Evans (1979): Prospects for combining high yield with increased protein production in *Phaseolus vulgaris* L. Seed Protein Improvement in Cereals and Legumes, II: 47–58: IAEA Vienna.
101. Guldager, P (1978): Immunoelectrophoretic analysis of seed proteins from *Pisum sativum* L. Theor. Appl. Genet. 53: 241–250.

102. Hall, TC, RC McLeester, and FA Bliss (1972): Electrophoretic analysis of protein changes during the development of the French bean fruit. Phytochem. 11: 647–649.
103. Hall, TC, RC McLeester, and FA Bliss (1977): Equal expression of the maternal and paternal alleles for the polypeptide subunits of the major storage protein of the bean *Phaseolus vulgaris* L. Plant Physiol. 59: 1122–1124.
104. Hall, TC, SM Sun, Y Ma, RC McLeester, JW Pyne, FA Bliss, and BU Buchbinder (1979): The major storage protein of French bean seeds: characterization in vivo and translation in vitro. In: Rubenstein, IR, RL Philips, ChE Green, and BG Gengenbach (eds.): The Plant Seed: Development, Preservation, and Germination: 3–26; Academic Press New York, London, Toronto, Sydney, San Francisco.
105. Hamblin, J, and SP Kent (1974): Possible role of phytohaemagglutinin in *Phaseolus vulgaris* L. Nature New Biol. 245: 28–30.
106. Hankins, CN, and LM Shannon (1978): The physical and enzymatic properties of a phytohemagglutinin from mung beans. J. Biol. Chem. 253: 7791–7797.
107. Hanson, WD, AH Probst, and BE Caldwell (1967): Evaluation of a population of soybean genotypes with implications for improving self-pollinated crops. Crop Sci. 7: 99–103.
108. Harlan, JR (1975): Our vanishing genetic resources. Science 188: 618–621.
109. Harlan, JR (1976): Genetic resources in wild relatives of crops. Crop Sci. 16: 329–333.
110. Hartwig, EE (1979): Breeding productive soybeans with a higher percentage of protein. Seed Protein Improvement in Cereals and Grain Legumes, II: 59–66; IAEA Vienna.
111. Hildebrand, DF, and T Hymowitz (1980a): The Sp1 locus in soybean codes for ß-amylase. Crop Sci. 20: 165–168.
112. Hildebrand, DF, and T Hymowitz (1980b): Inheritance of ß-amylase nulls in soybean seed. Crop Sci. 20: 727–730.
113. Hildebrand, DF, JH Orf, and T Hymowitz (1980): Inheritance of an acid phosphatase and its linkage with the Kunitz trypsin inhibitor in seed protein of soybeans. Crop. Sci. 20: 83–85.
114. Hill, JE, and RW Breidenbach (1974a): Proteins of soybean seeds. I. Isolation and characterization of the major components. Plant Physiol. 53: 742–746.
115. Hill, JE, and RW Breidenbach (1974b): Proteins of soybean seeds. II. Accumulation of the major protein components during seed development and maturation. Plant Physiol. 53: 747–751.
116. Hiraiwa, S, and S Tanaka (1978): Effects of successive irradiation and mass screening for seed size, density and protein content of soybean. Seed Protein Improvement by Nuclear Techniques: 265–274; IAEA Vienna.
117. Horstmann, C, A Rudolph, and P Schmidt (1978): Isolation, characterization, and subunit structure of a phytohemagglutinin from seed of *Vicia Faba* L. Biochem. Physiol. Pflanzen 173: 311–321.
118. Howard, IK, HJ Sage, and MD Stein (1971): Studies on a phytohemagglutinin from the lentil. J. Biol. Chem. 246: 1590–1595.
119. Hurich, J, H Parzysz, and J Przybylska (1977): Comparative study of seed proteins in the genus *Pisum*. II Amino acid composition of different protein fractions. Genet. Pol. 18: 241–252.
120. Hussein, HAS, and MMF Abdalla (1978): Protein and yield traits of field bean mutants induced with gamma rays, EMS and their combination. Seed Protein Improvement by Nuclear Techniques: 253–264; IAEA Vienna.
121. Hymowitz, T, and HH Hadley (1972): Inheritance of a trypsin inhibitor variant in seed protein of soybeans. Crop Sci. 12: 197–198.
122. Imam, MM (1979): Variability in protein content of locally cultivated *Phaseolus* and *Vigna* ssp. Seed Protein Improvement in Cereals and Grain Legumes, II: 119–126; IAEA Vienna.
123. Ishino, K, and ML Ortega (1975): Fractionation and characterization of major reserve proteins from seeds of *Phaseolus vulgaris*. J. Agric. Food Chem. 23: 529–533.
124. Jackson, P, D Boulter, and DA Thurman (1969): A comparison of some properties of vicilin and

legumin isolated from seeds of *Pisum sativum, Vicia faba* and *Cicer arietinum*. New Phytol. 68: 25–33.

125. Jaffé, WG (1973): Factors affecting the nutritional value of beans. Nutritional Improvements of Food Legumes by Breeding (Milner, M ed.): 43–48; PAG New York.
126. Jaffé, WG, A Levy, and DI Gonzalez (1974): Isolation and partial characterization of bean phytohemagglutinins. Phytochem. 13: 2685–2693.
127. Jakubek, M, and J Przybylska (1979): Comparative study of seed proteins in the genus *Pisum*. III. Electrophoretic patterns and amino acid composition of albumin fractions separated by gel filtration. Genet. Pol. 20: 369–380.
128. Janzen, DH, HB Juster, and IE Liener (1976): Insecticidal action of the phytohemagglutinin in black beans on a bruchid beetle. Science 192: 795–796.
129. Johnson, GB (1974): Enzyme polymorphism and metabolism. Science 184: 28–37.
130. Johnson, VA, and CL Lay (1974): Genetic improvement of plant protein. J. Agric. Food Chem. 22: 558–566.
131. Johnston, AWB, V Brewster, and DR Davies (1977): Seed proteins of peas in relation to nitrogen fixation. Ann. Bot. 41: 381–385.
132. Kaul, AK (1973): Mutation breeding and crop protein improvement. Nuclear Techniques for Seed Protein Improvement: 1–106; IAEA Vienna
133. Kelly, JD, and FA Bliss (1975a): Heritable estimates of percentage seed protein and available methionine and correlations with yield in dry beans. Crop Sci. 15: 753–757.
134. Kelly, JD, and FA Bliss (1975b): Quality factors affecting the nutritive value of bean seed protein. Crop Sci. 15: 757–760.
135. Kelly, JF (1973): Increasing protein quantity and quality. Nutritional Improvement of Food Legumes by Breeding (Milner, M. ed.): 179–184; PAG New York.
136. Khan, MRI, JA Gatehouse, and D Boulter (1980): The seed proteins of cowpeas (*Vigna unguiculata* L. Walp.). J. Exp. Bot. 31: 1599–1611.
137. Kiang, YT (1981): Inheritance and variation of amylase in cultivated and wild soybeans and their wild relatives. The J. Hered. 72: 382–386.
138. King, KW (1964): Development of all-plant food mixture using crops indigenous to Haiti: amino acid composition and protein quality. Econ. Bot. 18: 311–322.
139. Koshiyama, I, and D Fukushima (1979): Soybean globulins. In: Müntz, K (ed.): Seed Proteins of Dicotyledonous Plants: 21–43; Abh. Akad. Wiss. DDR, N4, Akad. Verlag Berlin.
140. Ladizinsky, G (1979): Species relationships in the genus *Lens* as indicated by seed-protein electrophoresis. Bot. Gaz. 140: 449–451.
141. Ladizinsky, G, and A Adler (1976): The origin of chickpea *Cicer arietinum* L. Euphytica 25: 211–217.
142. Ladizinsky, G and T Hymowitz (1979): Seed protein electrophoresis in taxonomic and evolutionary studies. Theor. Appl. Genet. 54: 145–151.
143. Lantz, EM, HW Gough, and AM Campbell (1958): Effect of variety, location, and years on the protein and amino acid content of dried beans. J. Agric. Food Chem. 6: 58–60.
144. Larsen, AL (1967): Electrophoretic differences in seed proteins among varieties of soybean (*Glycine max*. (L.) Merill). Crop Sci. 7: 311–313.
145. Larsen, AL, and WC Benson (1970): Variety-specific variants of oxidative enzymes from soybean seed. Crop Sci. 10: 493–495.
146. Larsen, AL, and BE Caldwell (1968): Inheritance of certain proteins in soybean. Crop Sci. 8: 474–476.
147. Leavitt, RD, RL Felsted, and NR Bachur (1977): Biological and biochemical properties of *Phaseolus vulgaris* isolectins. J. Biol. Chem. 252: 2961–2966.
148. Leleji, OI, MH Dickson, LV Crowder, and JB Bourke (1972): Inheritance of crude protein percentage and its correlation with seed yield in beans, *Phaseolus vulgaris* L. Crop Sci. 12: 168–171.

149. Leng, ER (1973): University of Illinois international soybean program. Nutritional Improvement of Food Legumes by Breeding (Milner, M ed.): 101–106; PAG New York.
150. Li, WH (1976): A mixed model of mutation for electrophoretic identity of proteins within and between populations. Genetics 83: 423–432.
151. Liener, I (1973): Antitryptic and other antinutritional factors in legumes. Nutritional Improvement of Food Legumes by Breeding (Milner, M ed.): 239–258; PAG New York.
152. Liener, I (1976): Phytohemagglutinins (Phytolectins). Ann. Rev. Plant Physiol. 27: 291–319.
153. Lis, H, and N Sharon (1973): The biochemistry of plant lectins (phytohemagglutinins). Ann. Rev. Biochem. 42: 541–574.
154. Luse, RA, and KO Rachie (1979): Seed protein improvement in tropical food legumes. Seed Protein Improvement in Cereals and Grain Legumes, II: 87–104; IAEA Vienna.
155. Ma, Y, and FA Bliss (1978): Seed proteins of common bean. Crop Sci. 18: 431–437.
156. Malhotra, VV, and KB Singh (1976): Genetic variability and path coefficient analysis for protein in green gram *Phaseolus aureus* Roxb. Egypt. J. Genet. Cytol. 5: 170–173.
157. Malley, A, L Baecker, B Mackler, and F Perlman (1975): The isolation of allergens from the green pea. J. Allerg. Clin. Immunol. 56: 282–290.
158. Manen, J-F, et M-N Miège (1977): Purification et caractérisation des lectines isolées dans les albumines et les globulines de *Phaseolus vulgaris*. Physiol. vég. 15: 163–173.
159. McLeester, RC, TC Hall, SM Sun and FA Bliss (1973): Comparison of globulin proteins from *Phaseolus vulgaris* with those from *Vicia faba*. Phytochem. 2: 85–93.
160. Meiners, JP, and SC Litzenberger (1973): Breeding for nutritional improvement. Nutritional Improvement of Food Legumes by Breeding (Milner, M ed.): 131–141; PAG New York.
161. Mies, DW, and T Hymowitz (1973): Comparative electrophoretic studies of trypsin inhibitors in seed of the genus *Glycine*. Bot. Gaz. 134: 121–125.
162. Millerd, A (1975): Biochemistry of legume seed proteins. Ann. Rev. Plant Physiol. 26: 53–72.
163. Millerd, A, M Simon, and H Stern (1971): Legumin synthesis in developing cotyledons of *Vicia faba* L. Plant Physiol. 48: 419–425.
164. Millerd, A, JA Thomson, and HE Schroeder (1978): Cotyledonary storage proteins in *Pisum sativum*. III. Patterns of accumulation during development. Aust. J. Plant Physiol. 5: 519–534.
165. Müller, HP (1978a): Gene mapping on chromosomes and some aspects of gene regulation in eukaryotic cells. Nucleus 21: 135–142.
166. Müller, HP (1978b): The seed proteins in gene-ecological investigations. Legume Res. 2: 29–40.
167. Müller, HP (1978c): Die genetische Steuerung der Zusammensetzung von Samenproteinen. Z. Pflanzenkrankh. Pflanzenschutz 85: 210–217.
168. Müller, HP (1979): The genetic control of seed protein polymorphism in *Pisum*. Proc. 1st Medit. Conf. Genet.: 747–764.
169. Müller, HP (1980): Biochemische Darstellung und gelelektrophoretische Charakterisierung der Samenproteine von Leguminosen. Göttinger Pflanzenzüchter-Seminar 4: 105–116.
170. Müller, HP (1982): Genotype-specific zymograms after direct isoelectric focusing of cotyledonous tissue. 4th International Congress on Isozymes, Austin, Texas.
171. Müller, HP, and W Gottschalk (1973): Quantitative and qualitative situation of seed proteins in mutants and recombinants of *Pisum sativum*. Nuclear Techniques for Seed Protein Improvement: 235–253; IAEA Vienna.
172. Müller, HP, and W Gottschalk (1978): Gene-ecological investigations on the protein production of different *Pisum* genotypes. Seed Protein Improvement by Nuclear Techniques: 301–314; IAEA Vienna.
173. Müller, HP, and S. Werner (1979): Seed protein characteristics of *Pisum* varieties, mutants and recombinants. In: Müntz, K (ed.): Seed Proteins of Dicotyledonous Plants: 189–209; Abh. Akad. Wiss. DDR, N4, Akad. Verlag Berlin.
174. Müntz, K, H Bäumlein, R Bassüner, R Manteuffel, M Püchel, P Schmidt, und W Wobus (1981):

Regulation von Biosynthese und Akkumulation der Reserveproteine während der Entwicklung pflanzlicher Samen. Biochem. Physiol. Pflanzen 176: 401–422.

175. Müntz, K, C Horstmann, und G Scholz (1972): Proteine und Proteinbiosynthese in Samen von *Vicia faba* L. Die Kulturpflanze 20: 277–326.

176. Murray, DR (1979): A storage role for albumins in pea cotyledons. Plant, Cell and Environment 2: 221–226.

177. Mutschler, MA, and FA Bliss (1981): Inheritance of bean seed globulin content and its relationship to protein content and quality. Crop Sci. 21: 289–294.

178. Mutschler, MA, FA Bliss, and TC Hall (1980): Variation in the accumulation of seed storage protein among genotypes of *Phaseolus vulgaris* (L.). Plant Physiol. 65: 627–630.

179. Nag, S, G Talukder, and A Sharma (1981): Plant lectins and their blastogenic and mitogenic activities. Nucleus 24: 16–38.

180. Nagl, K (1978): Breeding value of radio-induced mutants of *Vicia faba var. minor*. Seed Protein Improvement by Nuclear Techniques: 243–252; IAEA Vienna.

181. Orf, JH, and T Hymowitz (1977): Inheritance of a second trypsin inhibitor variant in seed protein of soybeans. Crop Sci. 17: 811–813.

182. Orf, JH, T Hymowitz, SP Pull, and SG Pueppke (1978): Inheritance of a soybean seed lectin. Crop Sci. 18: 899–900.

183. Otoul, E (1969): Répartition des principaux acides aminés dans les différentes parties de la graine d'un cultivar de *Phaseolus vulgaris* L. Bull. Rech. agron. Gembloux 4: 287–301.

184. Pandey, MP, M Frauen, and C Paul (1979): Selection for methionine by GLC after CNBr treatment in a germplasm collection and mutagen-treated population of *Vicia faba* L. Seed Protein Improvement in Cereals and Grain Legumes, II: 37–46; IAEA Vienna.

185. Pandey, S, and ET Gritton (1975): Inheritance of protein and other agronomic traits in a diallel cross of pea. J. Amer. Soc. Hort. Sci. 100: 87–90.

186. Pandey, S, and ET Gritton (1976): Observed and predicted response to selection for protein and yield in peas. Crop Sci. 16: 289–292.

187. Panton, CA, LB Coke, and RE Pierre (1973): Seed protein improvement in certain legumes through induced mutation. Nuclear Techniques for Seed Protein Improvement: 269–271; IAEA Vienna.

188. Pate, JS, and AM Flinn (1973): Carbon and nitrogen transfer from vegetative organs to ripening seeds of field pea (*Pisum arvense* L.). J. Exp. Bot. 24: 1090–1099.

189. Patel, KM, and JA Johnson (1974): Horsebean as protein supplement in breadmaking. I. Isolation of horsebean protein and its amino acid composition. Cereal Chem. 51: 693–701.

190. Planqué, K, and JW Kijive (1977): Binding of pea lectins to a glycan type polysaccharide in the cell walls of *Rhizobium leguminosarum*. FEBS Lett. 73: 64–66.

191. Polhill, RM, and PH Raven (eds.) (1981): Advances in Legume Systematics (I, II). Botanic Gardens, Kew; Ministry of Agriculture, Fisheries and Food, London.

192. Przybylska, J, S Blixt, J Hurich and Z Zimniak-Przybylska (1977): Comparative study of seed proteins in the genus *Pisum*. I. Electrophoretic patterns of different protein fractions. Genet. Pol. 18: 27–38.

193. Pueppke, SG (1979): Distribution of lectins in the Jumbo Virginia and Spanish varieties of the peanut, *Arachis hypogaea* L. Plant Physiol. 64: 575–580.

194. Pull, SP, SG Pueppke, T Hymowitz, and JH Orf (1978): Soybean lines lacking the 120000 Dalton seed lectin. Science 200: 1277–1279.

195. Pusztai, A, and WB Watt (1970): Glycoprotein II. The isolation and characterization of a major antigenic and non-hemagglutinating glycoprotein from *Phaseolus vulgaris*. Biochim. Biophys. Acta 207: 413–431.

196. Pusztai, A, and WB Watt (1974): Isolectins of *Phaseolus vulgaris*. A comprehensive study of fractionation. Biochim. Biophys. Acta 365: 57–71.

197. Pusztai, A, EMW Clarke, and TP King (1979a): The nutritional toxicity of *Phaseolus vulgaris* lectins. Proc. Nutr. Soc. 38: 115–120.
198. Pusztai, A, EMW Clarke, TP King, and JC Stewart (1979b): Nutritional evaluation of kidney beans (*Phaseolus vulgaris*): Chemical composition, lectin content and nutritional value of selected cultivars. J. Sci. Food Agric. 30: 843–848.
199. Pusztai, A, G Grant, and JC Stewart (1981): A new type of *Phaseolus vulgaris* (cv. Pinto III) seed lectin: Isolation and characterization. Biochim. Biophys. Acta 671: 146–154.
200. Raacke, ID (1957a): Protein synthesis in ripening peas. 1. Analysis of whole seeds. Biochem. J. 66: 101–110.
201. Raacke, ID (1957b): Protein synthesis in ripening peas. 2. Development of embryos and seed coats. Biochem. J. 66: 110–116.
202. Rachie, KO (1973): Improvement of food legumes in tropical Africa. Nutritional Improvement of Food Legumes by Breeding (Milner, M ed.): 83–92; PAG New York.
203. Randall, PJ, JA Thomson, and HE Schroeder (1979): Cotyledonary storage proteins in Pisum sativum. IV. Effects of sulfur, phosphorus, potassium and magnesium deficiencies. Aust. J. Plant Physiol. 6: 11–24.
204. Reddy, LJ, JM Green, U Singh, SS Bisen, and J Jambunathan (1979): Seed protein studies on *Cajanus cajan*, *Atylosis* spp. and some hybrid derivates. Seed Protein Improvement in Cereals and Grain Legumes, II: 105–117; IAEA Vienna.
205. Richardson, M (1977): The proteinase inhibitors of plants and microorganisms. Phytochem. 16: 159–169.
206. Romero, J, SM Sun, RC McLeester, FA Bliss, and TC Hall (1975): Heritable variation in a polypeptide subunit of the major storage protein of the bean, *Phaseolus vulgaris* L. Plant Physiol. 56: 776–779.
207. Roy, DN (1980): Trypsin inhibitor from *Lathyrus sativus* seeds: Final purification, separation of protein components, properties, and characterization. J. Agric. Food Chem. 28: 48–54.
208. Royes, WV (1973): Amino acid profiles of *Cajanus cajan* protein. Nutritional Improvement of Food Legumes by Breeding (Milner, M ed.): 193–196; PAG New York.
209. Rüdiger, H (1977): Purification and properties of bloodgroup-specific lectins from *Vicia cracca*. Eur. J. Biochem. 72: 317–322.
210 Rüdiger, H (1978): Lektine, pflanzliche zuckerbindende Proteine. Naturwiss. 65: 239–244.
211. Rutger, JN (1968): Variation in protein content and its relation to the other characters in beans (*Phaseolus vulgaris* L.). Agron. Abstracts, Amer. Soc. Agron.: 20.
212. Sahai, S, and RS Rana (1977): Seed protein homology and elucidation of species relationships in *Phaseolus* and *Vigna* species. New Phytol. 79: 527–534.
213. Sandhu, SS, WF Keim, HF Hodges, and WE Nyquist (1974): Inheritance of protein and sulfur content in seeds of chickpeas. Crop Sci. 14: 649–654.
214. Scandalios, JG (1974): Isozymes in development and differentiation. Ann. Rev. Plant Physiol. 25: 225–258.
215. Scandalios, JG, and LG Espiritu (1969): Mutant aminopeptidases of *Pisum sativum*. I. Developmental genetics and chemical characteristics. Molec. Gen. Genet. 105: 101–112.
216. Scharpé, A, and R van Parijs (1973): The formation of polyploid cells in ripening cotyledons of *Pisum sativum* L. in relation to ribosome and protein synthesis. J. Exp. Bot. 24: 216–222.
217. Schroeder, HE (1982): Quantitative studies on the cotyledonary proteins in the genus *Pisum*. J. Sci. Food Agric. in press.
218. Seidl, DS, H Abreu, and WG Jaffé (1978): Purification of a subtilisin inhibitor from black bean seeds. FEBS Lett. 92: 245–250.
219. Seidl, DS, WG Jaffé, E Gonzalez, and A Callejas (1978): Microelectrophoretic method for the detection of proteinase inhibitors. Analyt. Biochem. 88: 417–424.
220. Sgarbieri, VC, and MAM Galeazzi (1978): Some physico-chemical and nutritional properties of a sweet lupin (*Lupinus albus var. multolupa*) protein. J. Agric. Food Chem. 26: 1438–1442.

221. Shaik, MAQ, AK Kaul, MM Mia, MH Choudhury, and AD Bhuiya (1978): Screening of natural variants and induced mutants of some legumes for protein content and yielding potential. Seed Protein Improvement by Nuclear Techniques: 223–233; IAEA Vienna.
222. Sharma, RP, KS Nandpuri, and JC Kumar (1974): Mode of inheritance of ascorbic acid and protein content in pea (*Pisum sativum* L.). Veg. Sci. 1: 18–21.
223. Sharon, N, and H Lis (1972): Lectins: cell-agglutinating and sugar-specific proteins. Science 177: 949–964.
224. Siegel, BZ, and AW Galston (1967): The isoperoxidases of *Pisum sativum*. Plant Physiol. 42: 221–226.
225. Sikka, KC, AK Gupta, R Singh, and DP Gupta (1978): Comparative nutritive value, amino acid content, chemical composition, and digestability in vitro of vegetable and grain-type soybeans. J. Agric. Food Chem. 26: 312–316.
226. Simon, EW, and RM Raja Harun (1972): Leakage during seed imbibition. J. Exp. Bot. 23: 1076–1085.
227. Singh, OP, RB Singh, and F Singh (1980): Combining ability of yield and some quality traits in pea (*Pisum sativum* L.). Z. Pflanzenzüchtg. 84: 133–138.
228. Singh, OP, RB Singh, and F Singh (1981): Genetics of yield, seed weight and quality traits in pea (*Pisum sativum* L.). Genet. Agr. 35: 115–120.
229. Singh, L, CM Wilson, and HH Hadley (1969): Genetic differences in soybean trypsin inhibitors separated by disc electrophoresis. Crop Sci. 9: 489–491.
230. Sjödin, J (1974): Induzierte Variabilität von Qualitätseigenschaften der Ackerbohne. Göttinger Pflanzenzüchter-Seminar 2: 101–105.
231. Slinkard, AE (1972): Breeding and protein studies. Crop Development Centre, University of Saskatchewan S7N OWO, Canada.
232. Smith, MH (1966): The amino acid composition of proteins. J. Theor. Biol. 13: 261–282.
233. Sulser, H und R Stute (1974a): γ-Hydroxyarginin und γ-Hydroxyornithin, zwei ungewöhnliche Aminosäuren in den Samen der Linse (*Lens culinaris* Med.). I. Nachweis und quantitative Bestimmung von γ-Hydroxyarginin und γ-Hydroxyornithin in Linsen. Lebensmitt.-Wiss. u. Technol. 7: 322–326.
234. Sulser, H, und F Sager (1974b): γ-Hydroxyarginin und γ-Hydroxyornithin, zwei ungewöhnliche Aminosäuren in den Samen der Linse (*Lens culinaris* Med.). II. Zur Synthese und Struktur von γ-Hydroxyornithin und γ-Hydroxyarginin. Lebensmitt.-Wiss. u. Technol. 7: 327–329.
235. Sulser, H, and F Sager (1976): Identification of uncommon amino acids in the lentil seed (*Lens culinaris* Med.). Experientia 32: 422–423.
236. Sulser, H, M Beyeler, und F Sager (1975): γ-Hydroxyarginin und γ-Hydroxyornithin, zwei ungewöhnliche Aminosäuren in den Samen der Linse (*Lens culinaris* Med.). III. Isolierung von γ-Hydroxyarginin mittels Ionen-austauschchromatographie. Lebensm.-Wiss. u. Technol. 8: 161–162.
237. Sun, SM, MA Mutschler, FA Bliss, and TC Hall (1978): Protein synthesis and accumulation in bean cotyledons during growth. Plant Physiol. 61: 918–923.
238. Swaminathan, MS, and HK Jain (1973): Food legumes in Indian agriculture. Nutritional Improvement of Food Legumes by Breeding (Milner, M ed.): 69–82; PAG New York.
238a. Swiecicki, WK, Z Kaczmarek, and M Surma (1981): Inheritance and heritability of protein content in seeds of selected crosses of pea (*Pisum sativum* L.). Genet. Pol. 22: 189–195.
239. Tandon, OB, R Bressani, NS Scrimshaw, and F Le Beau (1957): Nutrients in Central American beans. J. Agric. Food Chem. 5: 137–142.
240. Thomson, JA, and H Doll (1979): Genetics and evolution of seed storage proteins. Seed Protein Improvement in Cereals and Grain Legumes, I: 109–124; IAEA Vienna.
241. Thomson, JA, and HE Schroeder (1978): Cotyledonary storage proteins in *Pisum sativum*. II. Hereditary variation in compounds of the legumin and vicilin fraction. Aust. J. Plant Physiol. 5: 281–294.

242. Thomson, JA, A Millerd, and HE Schroeder (1979): Genotype-dependent patterns of accumulation of seed storage proteins in *Pisum*. Seed Protein Improvement in Cereals and Grain Legumes, I: 231–240; IAEA Vienna.
243. Thomson, JA, HE Schroeder, and WF Dudman (1978): Cotyledonary storage proteins in *Pisum sativum*. I. Molecular herogeneity. Aust. J. Plant Phys. 5: 263–279.
244. Toms, GC, and A Western (1971): Phytohaemagglutinins. In: Harborne, JB, D Boulter, and BL Turner (eds.): Chemotaxonomy of the Leguminosae: 367–462; Academic Press, London and New York.
245. Turner, RH, and IE Liener (1975): The effect of the selective removal of hemagglutinins on the nutritive value of soybeans. J. Agric. Food Chem. 23: 484–487.
246. Waines, JG (1975): The biosystematics and domestication of peas (*Pisum sativum* L.). Bull. Torrey Bot. Club 102: 385–395.
247. Walbot, V, M Clutter, and IM Sussex (1972): Reproductive development and embryogeny in *Phaseolus*. Phytomorphology 22: 59–68.
248. Wall, JR, and SW Wall (1975): Isozyme polymorphisms in the study of evolution in *Phaseolus vulgaris-Phaseolus coccineus* complex of Mexico. In: Markert, CL (ed.): Isozymes, IV, Genetics and Evolution: 287–305; Academic Press New York, San Francisco, London.
249. Warsy, AS, G Norton, and M Stein (1974): Protease inhibitors from broad bean isolation and purification. Phytochem. 13: 2481–2486.
250. Weber, E, R Manteuffel, M Jakubek, and D Neumann (1981): Comparative studies on protein bodies and storage proteins of *Pisum sativum* L. and *Vicia faba* L. Biochem. Physiol. Pflanzen 176: 342–356.
251. Weeden, NF, and LD Gottlieb (1979): Distinguishing allozymes and isozymes of phosphoglucoisomerases by electrophoretic comparisons of pollen and somatic tissues. Biochem. Genet. 17: 287–296.
252. Weimberg, R (1968): An electrophoretic analysis of the isozymes of malate dehydrogenase in several different plants. Plant Physiol. 43: 622–628.
253. Wolpert, JS, and P Albersheim (1976): Host-symbiont interactions. I. The lectins of legumes interact with the O-antigen-containing lipopolysaccharides of their symbiontic *Rhizobia*. Biochem. Biophys. Res. Comm. 70: 729–737.
254. Wolff, G (1980): Investigations on the relations within the family *Papilionaceae* on the basis of electrophoretic banding patterns. Theor. Appl. Genet. 57: 225–232.
255. Wood, DR, EA Nowick, HJ Fabian, and PE McClean (1979): Genetic variability and heritability of available methionine in the Colorado dry bean breeding programme. Seed Protein Improvement in Cereals and Grain Legumes, II: 69–85; IAEA Vienna.
256. Woolfe, JA, and J Hamblin (1974): Within and between genotypes variation in crude protein content of *Phaseolus vulgaris* L. Euphytica 23: 121–128.
257. Wright, DJ, and D Boulter (1972): The characterization of vicilin during seed development in *Vicia faba*. Planta 105: 60–65.
258. Yoon, S, and BP Klein (1979): Some properties of pea lipoxygenase isoenzymes. J. Agric. Food Chem. 27: 955–962.
259. Youle, RJ, and AHC Huang (1978): Albumin storage proteins in the protein bodies of castor bean. Plant Physiol. 61: 13–16.
260. Youle, RJ, and AHC Huang (1979): Albumin storage protein and allergens in cottonseeds. J. Agric. Food Chem. 27: 500–503.

11. The Storage Proteins of *Phaseolus vulgaris* L., *Vicia faba* L. and *Pisum sativum* L.

DUNCAN R. ERSLAND, JOHN W. S. BROWN, ROD CASEY*, and TIMOTHY C. HALL.

INTRODUCTION

In recent years seed storage proteins have been the subject of extensive investigation due, firstly, to their economic importance as major protein sources for animals, and secondly, to their biochemical importance as model systems in the study of gene isolation, characterization and expression by molecular techniques. Seed storage proteins have evolved to provide a source of nutrition for germinating seeds and consequently make up a large percentage of the proteins in the mature seed. Such proteins are classified as storage proteins on the basis of several criteria, 1) they accumulate at a specific time in seed development, 2) they are relatively high in their nitrogen contents, and 3) they are sequestered in specialized inclusions (protein bodies) in cotyledon cells. Reviews dealing with the packaging of legume storage proteins into protein bodies have appeared elsewhere [8, 68, 77].

In legumes, the major seed storage proteins are globulins. In their classic studies, Osborne and Danielsson (reviewed in Derbyshire *et. al.* [38]) separated these globulins into two fractions, legumin and vicilin, on the basis of solubility. Sedimentation analyses of these two fractions under defined conditions showed legumins from several species to have sedimentation coefficients ranging from 10.1S to 14.0S (approximately 330,000 daltons molecular weight) and vicilin from 7.1S to 8.7S (approximately 180,000 daltons molecular weight) [32]. These studies also demonstrated the widespread occurrence of legumin and vicilin in leguminous plant seeds. The results of these early studies have been greatly extended by the purification and detailed characterization of the storage proteins and in light of this new information the similarities and differences between the seed storage proteins of different species can be re-evaluated.

In particular, we wish to take the opportunity in this review to compare the storage proteins of three legume species, *Vicia faba* L., *Pisum sativum* L., and *Phaseolus vulgaris* L. Those of *V. faba* and *P. sativum* are similar and the literature on these species complement each other well. Much of the early work on the characterization of the protein classes, and the formulation of the structural model for legumin was accomplished by studying *V. faba* proteins, while most of the more recent approaches have concentrated on *P. sativum* proteins. A number of references to the storage proteins of *Glycine max* (L.) Merr. will be made due to the similarity of the soybean proteins to those of the pea and broad bean. The differences between the storage globulins of these species and those of the common bean, *P. vulgaris* will be examined.

THE STORAGE PROTEINS OF VICIA FABA

The storage globulins of *V. faba* fall into two classes: legumin and vicilin. Together these proteins make up approximately 20% of the mature seed dry weight [110]. Another seed protein fraction, albumin, probably contains the metabolic proteins and is not thought to represent a storage protein component [76, 109] although this view is not universally held [78].

Legumin

Legumin, the principal seed storage protein of *Vicia faba*, can be isolated from low salt extracts of immature or dry seeds. The methods for purification have been reviewed extensively by Derbyshire *et al.* [38]. These authors reported that rapid initial purification of legumin was possible by precipitation at the isoelectric point, approximately pH 4.7, leaving vicilin in suspension. Further purification has been achieved by sucrose gradient centrifugation [70, 107], DEAE-Sephadex chromatography [105], or hydroxylapatite chromatography [43]. Millerd *et al.* [70] used immunodiffusion to show that pure preparations of legumin, containing no vicilin, could be obtained. Amino acid analysis of legumin showed it to contain high proportions of arginine, glutamic acid, and aspartic acid together with low levels of cysteine and methionine [53]. *Vicia* legumin does not appear to be glycosylated [27].

The structure of *V. faba* legumin has been thoroughly investigated [27, 37, 64, 104, 105, 107, 113]. Characterization of urea-dissociated legumin using anion-exchange chromatography identified two fractions, β or basic and α or acidic proteins [113]. SDS-polyacrylamide gels (SDS-PAGE; [58]) separated the β fraction into 3 bands (23,800, 20,900 and 20,100 daltons), while urea gel electrophoresis resolved the α fraction into 2 components (37,000 daltons each) [113]. When legumin was analyzed using SDS-PAGE in the absence of β-mercaptoethanol or urea, larger proteins with molecular weights of 49,500 and 50,500 daltons were observed [113]. These proteins represented undissociated complexes of one α and one β polypeptide joined by a disulfide bond.

More recent analyses of *V. Faba* legumin have revealed greater heterogeneity at both the polypeptide and holoprotein level. SDS-PAGE and isoelectric focusing techniques have revealed the presence of extensive α and β polypeptide heterogeneity in dissociated legumin [64, 105, 107, 108]. The studies of Matta *et al.* [64] described the α polypeptides as ranging from 23,000 to 58,000 daltons molecular weight while the β polypeptides ranged from 21,000 to 23,000 daltons. The α polypeptides also exhibited greater charge heterogeneity. In the absence of β-mercaptoethanol a number of undissociated proteins are observed. The major class of these proteins had a molecular weight of 53,000 daltons [64]. Further analysis of $\alpha\beta$ complexes has indicated nonrandom association of specific α polypeptides with specific β polypeptides such that 7–11 complexes in the range of 37,000 to 79,000 daltons were distinguished [64, 105, 107].

Heterogeneity in legumin was also indicated by N-terminal amino acid analysis of α polypeptides although only one β polypeptide N-terminal amino acid, glycine, has been identified to date [45, 113]. Microheterogeneity suggesting the presence of 3 different β polypeptides has been detected in the sequences of the N-terminal 29 amino acids by Gilroy *et al.* [45].

Wright and Boulter [113] have proposed that legumin has the subunit structure $\alpha_6\beta_6$. These authors calculated a molecular weight of approximately 343,000 daltons for the $\alpha_6\beta_6$ species which agrees well with the molecular weight determined for legumin by Danielsson [32]. The $\alpha_6\beta_6$ model accomodates the heterogeneity observed in legumin subunit and holoprotein molecular weights [64, 105, 107]. Acidic and basic polypeptides are synthesized together from a single mRNA as a precursor polypeptide which is cleaved to yield the disulphide-bonded α–β complex [64]. This complex forms the subunit of the legumin holoprotein with six such subunits being associated into the 11S protein. Approximately 80% of the *V. faba* legumin is formed from the major 53,000 dalton class of $\alpha\beta$ complex; small proportions of the minor complexes are also present [64]. A minor legumin holoprotein was found to be composed solely of one $\alpha\beta$ complex which was larger than the major class and which was not immunoprecipitable by *P. sativum* legumin antibody [64]. Thus, several legumin holoprotein size classes are possible [64, 105]. Matta *et al.* [64] have suggested that, based on the considerable charge and molecular weight heterogeneity in the α and β polypeptides, the numbers of legumin types described to date are probably underestimates. The compositional heterogeneity of legumin also varies among *V. faba* cultivars [43, 106, 107].

Vicilin

Vicilin, the minor globulin storage protein fraction of *V. faba* [109, 112] has not been thoroughly studied in *V. faba* due to difficulties involved in its purification and to its compositional heterogeneity. Vicilin has been isolated from globulin preparations after isoelectric precipitation of legumin, but contamination by residual legumin presented a problem [38]. Legumin-free vicilin fractions have been prepared by centrifugation of globulins in sucrose-NaCl density gradients as shown by immunodiffusion assays [70].

Several N-terminal amino acids have been identified in vicilin, the major ones being serine, leucine, and threonine [1], indicating substantial heterogeneity. The amino acid composition is similar to that of legumin [38]. Dissociation of 7S complexes of vicilin using SDS-PAGE reveals the presence of several protein bands ranging in size from 43,400 to 55,000 molecular weight [1, 38], with the major class being approximately 50,000 daltons [108]. Analysis of tryptic peptide maps indicated that some subunits are related [1], hence some components of the vicilin fraction of *V. faba* may be encoded as a multigene family. Vicilin preparations contain covalently bound carbohydrate moieties [1, 109]. Two types of 7S complex within the vicilin fraction have been identified on the basis of subunit composition [38]. Some of the proteins of the *V. faba* vicilin fraction appear to be capable of association into multimers at low ionic strength, as do some of the 7S storage proteins of *G. max* and *P. sativum* [89, 99]. This property reflects homologies with β-conglycinin of *G. max*, although *V. faba* vicilin and β-conglycinin do not cross-react immunologically [41].

Appearance of Storage Proteins During Seed Development

After anthesis, embryogenesis and seed formation in legumes occur in three physiologically distinct stages: a) cell division, b) cell expansion, c) maturation and desiccation [81]. Cell division ceases in *V. faba* when the young cotyledons are about 10 mm long, and is followed by a burst of protein synthesis during the cell expansion stage [70, 77]. Vicilin appears first, in the cell division stage, [46, 69, 77, 112] and some legumin may also be present at this time [70]. However, rapid accumulation of legumin occurs at the onset of cell expansion [70, 77]. Synthesis of both storage proteins is confined to cells of the cotyledons [46, 70].

Processing, Maturation, and Regulation of Storage Proteins

At the time that cell division in the developing seed ceases (10 mm length), nuclear DNA in cotyledon cells undergoes several rounds of endoreduplication concomitant with poly(A) RNA accumulation. The storage protein messenger RNA reaches a maxiumum level of synthesis just prior to the period of maximum storage protein synthesis [69, 81, 84]. Although it is not known what triggers the DNA endoreduplication events and the increase in endoplasmic reticulum in *V. faba* cotyledons, evidence presented by Püchel *et al.* [84] suggested that the appearance of storage proteins is regulated at the level of mRNA transcription. When *V. faba* cotyledons that were actively synthesizing poly(A) mRNA and storage proteins *in vitro* were incubated in the presence of RNA polymerase II-inhibiting concentrations of α-amanitin, they ceased synthesis of poly(A) RNA, vicilin, and legumin. If storage protein synthesis were regulated post-transcriptionally, storage protein synthesis would be expected to continue until stored mRNA precursors were exhausted in the cotyledons.

In the case of legumin, one membrane-bound 19S poly(A) mRNA encodes both the α and β subunits [64, 77, 108]. Synthesis and maturation of this protein is thus similar to legumin of *P. sativum* and glycinin of *G. max* [31, 77, 103, 108]. The major component of vicilin is initially synthesized as a 52,000 dalton precursor including a methionine-containing signal sequence [108]. The data of Weber *et al.* [108] suggest that the signal sequence is rapidly removed as vicilin polypeptide is transferred into the lumen of the rough endoplasmic reticulum.

THE STORAGE PROTEINS OF PISUM SATIVUM

The storage proteins of pea seeds are globulins and fall into three main classes: legumin, vicilin and convicilin. Murray [78] has pointed out that albumins may contribute significantly to the total protein of pea seeds and has suggested that they might therefore be storage proteins. However, they are not considered in this review because they have yet to be purified and clearly defined biochemically and are inherently complex [54]. All three globulins have been purified to a reasonable degree of homogeneity and the structure of each defined at the level of native protein and subunit molecular weights, amino-terminal groups [16, 28], amino acid compositions [24, 28, 38] and, in some cases, partial protein sequences [20, 21, 31].

Legumin

Legumin is a hexameric protein consisting of six morphological subunits of molecular weight about 60.000, arranged to give a trigonal bipyramid [19] of molecular weight 350,000–400,000 [4, 9, 16, 27, 32, 38, 47, 55, 56, 101, 113]. Each monomeric subunit consists of an acidic (α) and basic (β) subunit, of molecular weights about 40,000 and 20,000 respectively, linked by disulphide bonds; the acidic and basic subunits show no immunological cross-reactivity [19] and acidic-basic pairing is non-random [39, 65] as in *Vicia faba* legumin [64, 105] and *Glycine max* glycinin [93]. Although the majority of legumin species conform to the above pattern, there exist minor legumin species which differ slightly in whole protein and subunit molecular weight [65], possibly giving rise to the minor legumin subunits previously observed by others [17, 101]. Present evidence suggests that legumin does not contain covalently-linked carbohydrate [16, 42].

Both the acidic and basic subunits exhibit apparent size and charge heterogeneity [16, 27, 42, 57, 101]. Amino-terminal sequence analysis of the mixture of β subunits of legumin from a single cultivar [20] indicates the existence of a small gene family for the legumin basic subunits [18]. Some β subunit heterogeneity may also result from post-translational modification [92] and/or from storage artefacts in samples purified from dry, mature seed [63]. The α subunits also show heterogeneity on SDS, isofocusing, or two-dimensional gels [17, 42, 57, 65, 101]. In addition to the presence of minor (α^m) species, multiple forms of the major (α^M) subunits are also usually observed, although this is a function of genotype [17]; some genotypes have only one major acidic subunit [21]. Aminoterminal sequence analysis of a *Pisum* legumin α subunit has shown striking homology to the acidic subunits of glycinin [21, 74, 75] and N-terminal group analysis of legumins from a wide range of *Pisum* genotypes has shown conservation of the characteristic leucine (major) and threonine (minor) end groups [17]. Genetical studies of α subunits suggest that the α loci are closely linked [16] and would be consistent with the existence of a small gene family for *Pisum* α subunits.

It is now clear (see below) that legumin subunits are synthesised as combined $\alpha\beta$ precursor molecules of molecular weight about 60,000 in which an acidic and basic subunit are covalently linked by a bond other than a disulphide bridge [28, 91]. Thus, the specificity of $\alpha\beta$ pairing [39, 65] is explained by the fact that acidic and basic subunits are synthesised as pre-linked pairs which remain associated through a disulphide bridge after the bond has been 'nicked' [28]; the precursors are probably the products of a small gene family, each member of the family coding for both an α and a β sequence.

Although the above is an accurate description of the majority of the legumin molecules in any given pea genotype, it is nonetheless an oversimplification. Legumin in both *Pisum* and *Vicia* is heterogeneous at the native protein level, containing molecules both larger and smaller than average; these larger and smaller species are minor in amount and have a different subunit composition to the majority of the legumin [64, 65, 105]. Thus in addition to subunit heterogeneity there is also whole protein heterogeneity resulting from different subunit combinations. Through the use of 'multi-dimensional' gel electrophoresis, Matta *et al.* [65] concluded that there may be as many as 22 separate α subunits and 11 separate β subunits in a legumin preparation from a single *Pisum* genotype.

Vicilin

Vicilin prepared from *Pisum* by classical solubility-based procedures is apparently a mixture of several protein species differing in subunit composition, electrophoretic mobility and serological behaviour; there is clear evidence for vicilin heterogeneity at the subunit level [44, 72, 73, 101, 102]. The major subunits in a given preparation of vicilin have molecular weights of approximately 50,000 daltons; polypeptides of about 33,000 and 17,000 daltons are of intermediate abundance and those of 14,000 and 12,000 daltons are present in minor amounts [44, 101]. The 14,000 and 12,000 molecular weight subunits are structurally related [34] and virtually all the covalently linked carbohydrate to be found in vicilin resides on one of these small subunits [33, 44]. The most recent assessment of vicilin structure suggests it to be a multimer, possibly a trimer, of subunits with molecular weights of about 50,000 daltons and consequently a holoprotein molecular weight of about

150,000 daltons. The subunits of mulecular weight of less than 50,000 daltons arise from post-synthetic proteolysis of some of the approximately 50,000 molecular weight subunits (which have already assembled into vicilin molecules); the cleaved polypeptides remain associated in the native molecule [44]. The 50,000 dalton subunits display charge heterogeneity; the sequence heterogeneity implied by this observation may account for both the variety of the proteolysis products and the heterogeneity of molecular properties associated with vicilin, because different combinations of the initial polypeptides would be expected to result in vicilin molecules with different properties [44].

At present little is known concerning the amino acid sequence of proteins in the vicilin fraction, other than the N-terminal group analysis of Jackson *et al.* [53]. Therefore, formal proof of the structural relationships between the various vicilin subunits is still awaited.

Convicilin

Convicilin is a 7S protein of molecular weight 220,000–290,000 which is composed of three or four subunits of molecular weights of approximately 70,000 [22, 30]. Although convicilin appears to be immunologically very similar to vicilin [30], it contains no vicilin polypeptides and can be readily separated from vicilin under non-denaturing conditions. Unlike vicilin it contains some methionine and cysteine but no covalently-linked carbohydrate [22, 30]. Isoelectric focusing and N-terminal group analysis both suggest that convicilin has limited subunit sequence heterogeneity [30]. Convicilin has also been purified from *Vicia faba minor*; it appears to have a lower subunit molecular weight than that from *Pisum sativum* [30], although it should be remembered that the apparent subunit molecular weights of convicilin from *Pisum* show genotypic variability [22, 83, 100].

Variability in storage protein structure

The relative proportions of legumin, vicilin and convicilin are genotype-dependent and any one of the three can be the major species in a given genotype [23]; the relative proportions of the proteins are also affected by the environment in which the plant is grown [86, 90]. All three major storage proteins of *Pisum* show subunit heterogeneity on SDS and two-dimensional gels, but the nature and extent of this heterogeneity is genotype-dependent [17, 22, 27, 36, 82, 83, 100]; genotypic variability in *Vicia faba* legumin has also been demonstrated [107]. The description of the seed proteins as outlined above is, therefore, only a guideline to structure because the fine details will depend on the genotype being examined.

Appearance of storage proteins during seed development

Embryo development in *Pisum*, as in *Vicia* and *Phaseolus*, consists initially of a stage of intensive cell division, following which the cotyledonary cells expand and finally the seed dries out [50, 71]. It is around the transition from division to expansion of the bulk of the cell population that the major onset of storage protein synthesis occurs [71], although low levels of legumin synthesis can be demonstrated at very early developmental stages [40], long before endosperm absorption occurs [50].

All the storage proteins discussed above are synthesised on membrane-bound polysomes of the rough endoplasmic reticulum (RER) [28, 29, 39, 52, 90], from whence they are transported to the protein bodies, their site of deposition in the mature seed [25, 26, 101]. The major onset of synthesis of vicilin precedes that of legumin (and convicilin) in all genotypes so far examined [28, 30, 47, 73]; (see also Figure 1). This is not merely a reflection of the relative proportions of vicilin and

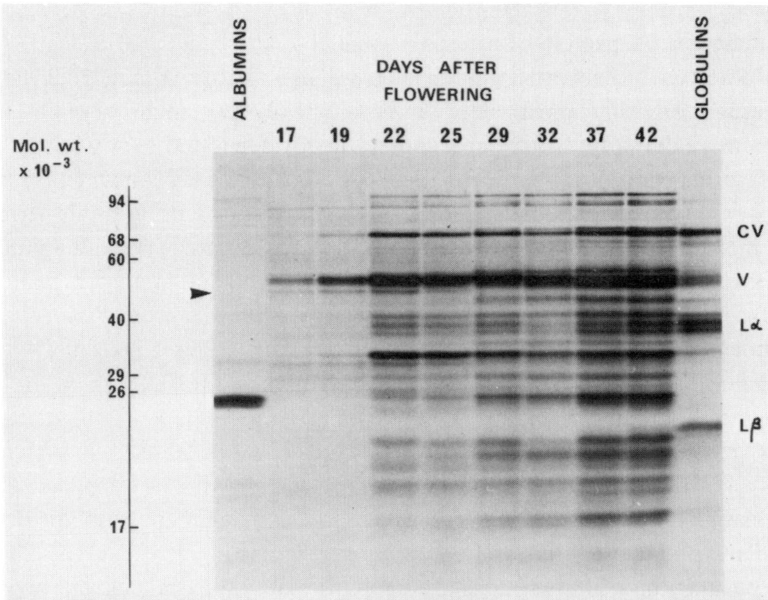

Figure 1. Appearance of pea seed storage proteins during seed development. Peas (cv. Dark Skinned Perfection) were grown under controlled conditions and cotyledons harvested at various intervals after flowering, frozen in liquid N_2, freeze-dried and ground to a fine powder. Known weights of this meal were extracted with boiling SDS/β-mercaptothanol and the polypeptides separated on a 15% polyacrylamide gel containing SDS. Note the 47,000 molecular weight vicilin subunit (arrowed) which is found only at early developmental stages [28] and the difference in mobility of the legumin β subunits from developing seeds compared with the globulin fraction from mature seed [63, 92].
cv = convicilin 70,000 molecular weight subunits;
v = vicilin 50,000 molecular weight subunits;
Lα and Lβ = legumin α and β subunits.

legumin synthesised by these genotypes, because that used by Croy *et al.* [30] is a high-legumin type [43].

Processing, maturation and regulation of storage proteins

Legumin, vicilin and convicilin are produced from polyadenylated mRNAs which are to be found in the membrane-bound polysomes of the RER [28, 29, 51]. *In vitro* and *in vivo* biosynthetic labelling experiments [28, 29, 44, 52, 91] have resulted in a clear picture of the stages of processing through which each protein passes on its journey from the RER to the protein body. The simplest case is convicilin, the 70,000 molecular weight subunits of which are synthesised in slightly larger form than the final product, having a leader sequence which can be removed in a cell-free system by the addition of microsomal membranes [52].

Legumin is also probably synthesised with a leader sequence which is removed in the ER but the pattern is more complex in that the initial product is a pre-pro-protein; after removal of the leader sequence, the 60,000 molecular weight legumin ρα precursor is subsequently cleaved after several hours [28, 91] to give the final ρ and α subunits. This cleavage, or 'nicking', possibly at or near an Arg-Arg recognition sequence [31], takes place after the 60,000 molecular weight subunits have begun to associate into a 9S intermediate halfmolecule [91]. The ρα nature of the 60,000 molecular weight precursor is inferred from its molecular weight, its reactivity against anti-legumin and from gel mapping experiments [28, 91]; formal proof that the precursor contains both ρ and α sequences can be obtained by tryptic peptide mapping of labelled precursor and isolated ρ and α subunits ([39]; and see Figure 2).

Vicilin is synthesised as subunits of molecular weight around 50,000 and 47,000; the latter is transient in nature ([29, 44]; and see Figure 1) and is possibly 'nicked' after assembly into the native conformation to give the range of smaller subunits found in vicilin from mature seeds. Both the 50,000 and 47,000 molecular weight subunits are synthesised with leader sequences [44, 52] which are removed in the RER; subsequent 'nicking' occurs several hours later, possibly in or at the protein body.

At present, we have little understanding of what controls the onset, or rate, of storage protein synthesis in developing peas. Neither is it known how the relative proportions of the various storage proteins in the mature seed are arrived at in a given genotype.

There is no selective amplification of legumin genes [31] during the endoreduplication of cotyledonary cell nuclear DNA which occurs during pea seed development [35, 71]. Measurements of gene numbers [31] suggest that there are not large numbers of genes coding for the major storage proteins, neither are there multiple copies of individual genes. The genes for legumin are, and those for vicilin may be, located on linkage group 7 in the region of the r_a locus ([36]; D. Boulter, personal

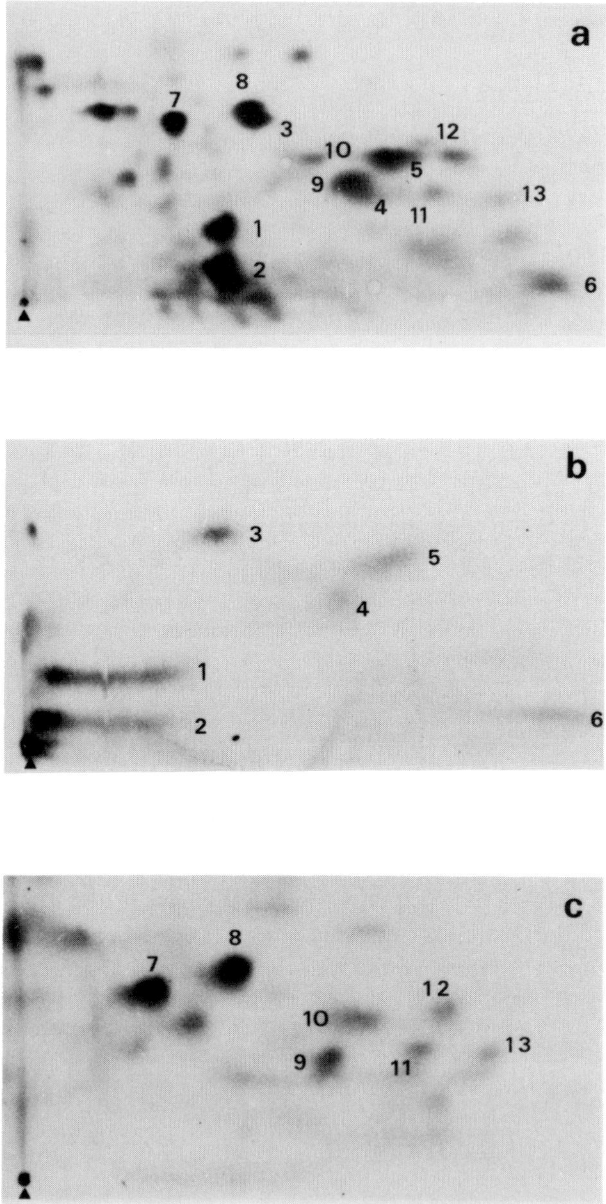

Figure 2. The legumin 60,000 molecular weight precursor contains both α and β sequences. Legumin precursor was labelled in developing embryos (cv. Birte) by standing the embryos on ^{14}C amino acids for 1.5 h [39, 92]; the α and β subunits were labelled by culturing embryos in ^{14}C amino acids for 1 week [39]. Total protein was extracted and the α and β subunits and precursor isolated by immunoaffinity

communication). It is clearly going to be of some considerable interest to understand the molecular basis of genotypic variability in the relative proportions of the storage proteins [23] and to examine the reasons for the effect of the r_a locus on legumin synthesis [36].

THE STORAGE PROTEINS OF PHASEOLUS VULGARIS

The storage proteins of *P. vulgaris*, like those of *V. faba* and *P. sativum* are globulins, but there is only one major class, phaseolin. A second globulin fraction, globulin-2 (G2), making up only 5–12% of the total seed protein [60] is composed mainly of lectin proteins [13, 14, 15].

Phaseolin

Phaseolin, the major storage protein fraction, comprises some 36–46% of the total seed protein [60]. In the past, phaseolin has been termed G1 protein [67, 96], glycoprotein II [85], and vicilin [5]. It contains 37.6% acidic residues and amides, and about 1% sulphur-containing residues [96]. Phaseolin is glycosylated [49, 85, 95], containing about 3–5% sugars [95]. Phaseolin will undergo pH-dependent reversible and irreversible association and dissociation between peptide (3S), protomer (7S), and tetramer (18S) configurations. Polypeptides (molecular weights ranging from 45,000 to 51,000 daltons) at high pH, associated at neutral pH to protomeric forms (approximately 163,000 daltons) which formed tetramers (approximately 600,000 daltons) at pH 4 [3, 96]. The protomeric form contains three polypeptides [3], but it is not yet known whether there is random or non-random association of polypeptides into the protomeric form or in what structure the protomers are associated to form the tetrameric configuration.

The polypeptides of phaseolin exhibit both molecular weight (45,000–51,000 daltons) and charge (pH 5.6-pH 5.8) heterogeneity [7, 12, 49, 88]. Two-dimensional polyacrylamide gel electrophoresis has resolved only three different electrophoretic profiles for common bean cultivars studied to date [11, 12]. These electrophoretic types were named after the cultivars 'Tendergreen' (T), 'Sanilac' (S), and 'Contender' (C). The T and S patterns have no polypeptides in common by virtue of molecular weight and isoelectric point, and consist of five and eight individual

---

chromatography [16, 39] followed by SDS-PAGE. The precursor and subunits were located by autoradiography, excised from the gels, digested with trypsin and 'fingerprinted' on thin-layer cellulose plates [111]. Peptides were located by fluorography [87].
(a) 60,000 molecular weight precursor
(b) 40,000 molecular weight (α) subunits and
(c) 20,000 molecular weight (β) subunits.
▲ denotes the point of sample application.

polypeptides respectively. The C phaseolin pattern represents a rare recombinant type between the T and S types. It contains the five polypeptides of the T pattern, two polypeptides of the S pattern, and one unique polypeptide [12].

The narrow ranges of molecular weight and isoelectric point of the phaseolin polypeptides suggest that they are all similar proteins. This is supported by the homologies demonstrated for the polypeptides of 'Tendergreen' by peptide mapping [61], hybrid-select translations [98], and sequence analysis of phaseolin complementary DNA clones (J.L. Slightom, personal communication).

Genetic analyses of the inheritance of the phaseolin patterns from 292 F_2 seed of the three crosses between Tendergreen, Sanilac, and Contender types showed that the genes controlling these polypeptides were linked, co-dominant, allelic, and were inherited in a block as a single Mendelian gene [10]. Thus, the phaseolin genes represent a small, multi-gene family.

Appearance of Storage Proteins During Seed Development

The accumulation of phaseolin and the G2/albumin lectin polypeptides has been followed using electrophoretic and immunological techniques. Phaseolin was detected in the seeds of 'Tendergreen' fourteen days after flowering (seeds 9 mm long) with the major accumulation occurring between sixteen and twenty-eight days after flowering (seeds 12–19 mm long) [97]. This exponential accumulation of phaseolin coincides with the time of rapid proliferation of rough endoplasmic reticulum in the cotyledon cells [6, 80]. After synthesis on endoplasmic reticulum-bound polysomes, phaseolin is firstly sequestered into the endoplasmic reticulum and then into the protein bodies [2, 6]. Accumulation of the G2/albumin polypeptides followed closely that of phaseolin [97]. The rate and amount of phaseolin accumulation varies among different cultivars [79].

Processing, Maturation, and Regulation of Storage Proteins

When mRNA from the cultivar 'Tendergreen' is translated *in vitro* by the wheat germ cell-free protein synthesising system, polypeptides, immunoprecipitable with phaseolin antibody, are produced [48]. One-dimensional SDS-PAGE showed only two polypeptide bands [48]. These two bands were clearly shown to be glycosylated to give the authentic T phaseolin pattern upon injection of the mRNA into *Xenopus laevis* L. oocytes [66]. Hybrid selection of 'Tendergreen' 16S mRNA with a cloned phaseolin cDNA isolated mRNAs encoding both bands typically seen by *in vitro* translation, thus showing homology among the phaseolin mRNAs [98]. The presence or absence of a signal polypeptide has not yet been determined. DNA-DNA hybridization analysis has suggested a phaseolin gene number estimate of 5–10 copies (Ersland D.R., Chee, P., Hoffman, L.H., Slightom, J.L., and Hall, T.C. unpublished).

Little is known about the regulation of phaseolin synthesis. However, Sullivan [94] developed 'single-gene deviate' lines carrying genes showing a major enhancement effect on phaseolin expression. Heritable suppression of phaseolin synthesis and/or accumulation has been transferred from a wild *P. vulgaris* accession (J. Romero and F.A. Bliss, personal communication). The effect of this gene would be analagous to the *opaque* and *floury* loci of maize and sorghum. The discovery of such genes will provide excellent systems for studying gene expression in legumes.

DISCUSSION

It is clear that legumin from *P. sativum* is similar to that of *V. faba* and also to glycinin, the 11S storage protein from *G. max*. Jackson *et al.* [53] first showed similarities between legumin subunits from *V. faba* and *P. sativum* by urea gel electrophoresis. These authors also compared amino acid compositions and tryptic digest peptide maps of legumin from *V. faba* and *P. sativum* which again showed similarity between these proteins. Immunological identity between these proteins was demonstrated by Dudman and Millerd [41]. This identity has recently been shown to be only partial, *Pisum* having more antigenic determinants than *Vicia* [27, 109].

Structural similarity [six pairs of disulphide-bonded α (acidic) and ß (basic)] subunits has been found for legumin from *P. sativum* and *V. faba* [27, 64, 65] and for glycinin from *G. max* [20, 21, 45, 74, 75]; for a review, see Larkins [59]. *V. faba* legumin, *P. sativum* legumin and *G. max* glycinin are all synthesised as 60,000 molecular weight precursor subunits [28, 77, 91, 103] which are proteolytically cleaved to give the S-S- bonded αß complex found in the native protein. The isolation of non-random αß associations in glycinin and legumin is consistent with this model [64, 93].

Vicia legumin and *Glycine* glycinin are sufficiently similar to be able to form hybrid molecules [105]. Furthermore, Casey *et al.* [20], have shown that there is extensive homology among the ß subunit N-terminal amino acid sequences of *V. faba*, *P. sativum* and *Glycine max*. It is, then, surprising that no protein showing immunological cross-reactivity with *Vicia* storage proteins can be detected in *Glycine* [41]. However, as pointed out by Derbyshire *et al.* [38] homologous storage proteins may not cross react immunologically because single amino acid substitutions can have drastic serological effects (see [62]). The extensive similarities between legumin of *P. sativum* and *V. faba*, and glycinin of *G. max* in sedimentation values, molecular weight, protein structure, amino acid composition, and N-terminal amino acid and amino acid sequence of acidic and basic polypeptides suggest a common genetic ancestry for these storage proteins.

Isolation of a minor 11S legumin-like component from *P. vulgaris* has been reported [37]. It had a molecular weight of 330,000–350,000 daltons and N-terminal amino acids were detected as Leu, Thr, Gly, and Met (cf. *Vicia*/*Pisum* legumin with

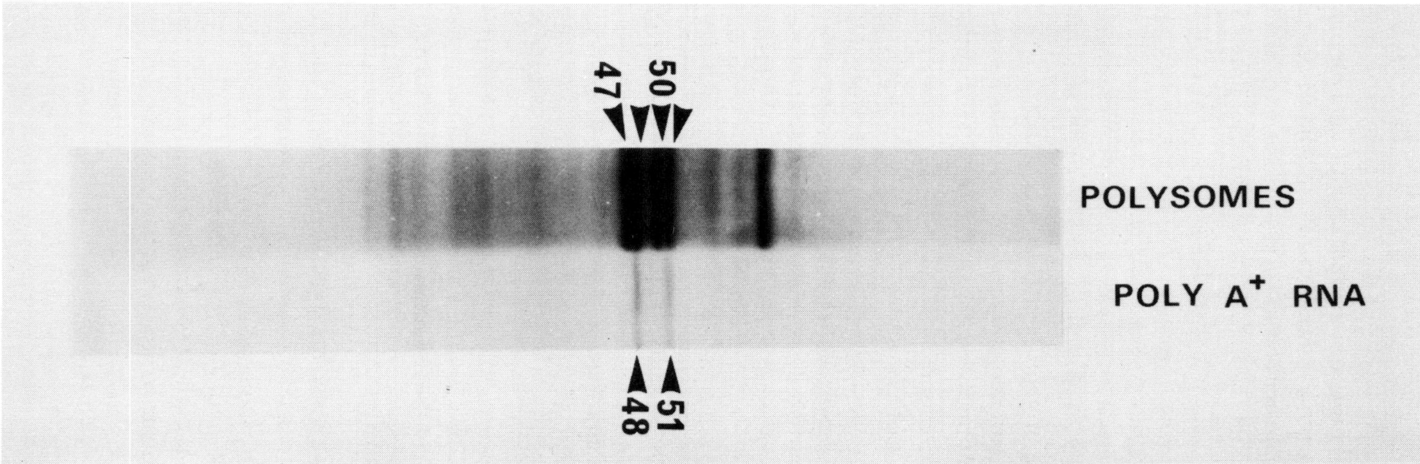

Figure 3. Vicilin polypeptides are synthesized with a leader sequence. Polysomes or poly A^+ RNA from seeds of cv. Dark Skinned Perfection at an early stage of development were translated in a reticulocyte lysate cell-free translation system using [^{3}H] leucine as label. Translation products were separated on a 15% SDS gel and located by fluorography (Bonner and Laskey, 1974). The products from poly (A^+) RNA have not been processed, whereas those from polysomes (which were membrane-bound in the seed) show both the unprocessed (48,000 and 51,000 mol. wt.) and processed forms of the 47,000 and 50,000 molecular weight subunits.

Leu, Thr, Gly). However, no immunological identity was shown between *P. vulgaris* and *V. faba* proteins [41]. As yet, no acidic or basic subunits of the legumin-like protein have been detected with two-dimensional electrophoresis of *P. vulgaris* total protein extracts [12]. Legumin has, however, been observed in lines of *P. coccineus* (R. Bollini, personal communication).

Tryptic peptide fingerprints of *V. faba* and *P. sativum* vicilins were similar [53]; and these proteins also showed immunological identity [41]. Partial immunological identity has been reported by Weber *et al.* [109]. It is likely, in view of the similarities in subunit structure and biosynthetic patterns, that vicilin from *Pisum* has a homologous counterpart in *Vicia faba* [44, 112] and it is known that convicilin exists in *Vicia faba* [30]. It has been suggested that *Pisum* vicilin and *Phaseolus* phaseolin (G1 protein; glycoprotein II) might be structurally related [44], but this suggestion is made only on the basis of similarities in whole protein and subunit molecular weights. Immunological studies [41, 109] do not support an antigenic relationship between either *V. faba* and *P. vulgaris* storage proteins, or *V. faba* and *G. max* storage proteins. No immunological reactivity was observed between phaseolin antibodies and *G. max* storage proteins (J.W.S. Brown, unpublished data).

Phaseolin exhibits a pH-dependent, reversible association between its 7S protomeric (3 polypeptides) and 18S tetrameric (12 polypeptides) forms [96]. The 7S vicilins of *V. faba* and *P. sativum*, and the 7S storage proteins of *G. max* (3 polypeptides) are capable of undergoing association to dimeric forms (6 polypeptides) at low ionic strengths [56, 89, 99]. Some components have been observed to associate to a tetrameric (12 polypeptide) form like phaseolin, and phaseolin has been reported to occur in a dimeric (10S) form [3, 85]. This association-dissociation behaviour may suggest homology among proteins of the 7S fractions of these different species.

Recently, extensive homology between β-conglycinin and phaseolin has been observed (M.A. Schuler, J.D. Doyle, and R.N. Beachy, personal communication). Firstly, the hybridization of a cloned α-α′ conglycinin DNA sequence to *P. vulgaris* DNA showed a similar pattern of hybridization as was obtained with a phaseolin complementary DNA sequence probe; the conglycinin probe also hybridized to *Vicia faba* genomic DNA. Secondly, comparison of the DNA sequences of the cloned conglycinin and phaseolin sequences revealed clear homology. Thus, although some of the chemical, physical and immunological properties of *P. vulgaris* phaseolin and *G. max* conglycinin differ, they appear to be homologous proteins probably with a common genetic ancestry. It also seems apparent that proteins in the 7S fractions of these species from the subfamily *Phaseoleae* bear the same homology to the proteins of the 7S vicilin fractions of *P. sativum* and *V. faba* of the subfamily *Faboideae*.

ACKNOWLEDGEMENTS

The authors wish to acknowledge Dr. C.A.M. Domoney for Figure 2. We are grateful for Grant RG 155–80 from N.A.T.O. which made the interactions leading to the development of this review possible. Aspects of this work were also facilitated by a grant from the Herman Frasch Foundation.

REFERENCES

1. Bailey, CJ, and D Boulter (1972): The structure of vicilin of *Vicia faba*. Phytochem. 11: 59–64.
2. Baumgartner, B, KT Tokuyasu, and MJ Chrispeels (1980): Immunocytochemical localization of reserve protein in the endoplasmic reticulum of developing bean (*Phaseolus vulgaris*) cotyledons. Planta 150: 419–425.
3. Blagrove, RJ, PM Colman, GG Lilley, A Van Donkelaar, and JN Vargnese (1981): Physiochemical and structural studies of phaseolin from French bean seed. Austr. Biochem. Soc. Abstr.
4. Blagrove, RJ, GG Lilley, and R Davey (1980): Molecular weight of legumin from *Pisum sativum*. Aust. J. Plant Physiol. 7: 221–225.
5. Bollini, R, and MJ Chrispeels (1978): Characterization and subcellular localization of vicilin and phytohaemagglutinin, the two major reserve proteins of *Phaseolus vulgaris* L. Planta 142: 291–298.
6. Bollini, R, and MJ Chrispeels (1979): The rough endoplasmic reticulum is the site of reserve protein synthesis in developing *Phaseolus vulgaris* cotyledons. Planta 146: 487–501.
7. Bollini, R, and A Vitale (1981): Genetic variability in charge microheterogeneity and polypeptide composition of phaseolin, the major storage protein of *Phaseolus vulgaris*; and peptide maps of its three major subunits. Physiol. Plantarum 52: 96–100.
8. Boulter, D (1979): Structure and biosynthesis of legume storage proteins. Seed Protein Improvement in Cereals and Grain Legumes I: 125–134; IAEA Vienna.
9. Brand, BP, DAI Goring, and P Johnson (1955): The attempted preparation of monodisperse seed globulins. Trans. Farad. Soc. 51: 872–876.
10. Brown, JWS, FA Bliss, and TC Hall (1981): Linkage relationships between genes controlling seed proteins in French bean. Theor. Appl. Genet. 60: 251–259.
11. Brown, JWS, FA Bliss, and TC Hall (1982): Genetic divergence among commercial classes of beans in relation to phaseolin type. Hort. Sci.; in press.
12. Brown, JWS, Y Ma, FA Bliss, and TC Hall (1981): Genetic variation in the subunits of globulin-1 storage protein of French bean. Theor. Appl. Genet. 59: 83–88.
13. Brown, JWS, TC Osborn, FA Bliss, and TC Hall (1981): Genetic variation in the subunits of globulin-2 and albumin seed proteins of French bean. Theor. Appl. Genet. 60: 245–250.
14. Brown, JWS, TC Osborn, FA Bliss, and TC Hall (1982a): Bean Lectins I.: Relationships between agglutinating activity and electrophoretic variation in the lectin-containing G2/albumin seed proteins of French Bean (*Phaseolus vulgaris* L.). Theor. Appl. Genet.; in press.
15. Brown, JWS, TC Osborn, FA Bliss, and TC Hall (1982b): Bean Lectins II.: Relationship between qualitative lectin variation in *Phaseolus vulgaris* L. and previous observations on purified bean lectins. Theor. Appl. Genet.; in press.
16. Casey, R. (1979a): Immunoaffinity chromatography as a means of purifying legumin from *Pisum* (pea) seeds. Biochem. J. 177: 509–520.
17. Casey, R (1979b): Genetic variability in the structure of the α-subunits of legumin from *Pisum* – a two-dimensional gel electrophoresis study. Heredity 43: 265–272.
18. Casey, R (1982): The genetics of pea storage proteins. Qualitas Plantarum; in press.

62. Margoliash, E, A Nisonoff, and M Reichlin (1970): Immunological activity of cytochrome c. J. Biol. Chem. 245: 931–946.
63. Matta, NK, and JA Gatehouse (1981): Modification of the basic subunits of pea legumin on storage. Phytochemistry 20: 2621–2624.
64. Matta, NK, JA Gatehouse, and D Boulter (1981a): The structure of legumin of *Vicia faba* L. – a reappraisal. J. Exp. Bot. 32: 183–197.
65. Matta, NK, JA Gatehouse, and D Boulter (1981b): Subunit heterogeneity and structure of legumin of *Pisum sativum* L. – a multi-dimensional gel electrophoresis study. J. Exp. Bot. 32: 1295–1307.
66. Matthews, JA, JWS Brown, and TC Hall (1981): Bean seed protein (phaseolin) mRNA is translated to yield glycosylated polypeptides by *Xenopus* oocytes. Nature 294: 175–176.
67. McLeester, RC, TC Hall, SM Sun, and FA Bliss (1973): Comparison of globulin proteins from *Phaseolus vulgaris* with those from *Vicia faba*. Phytochemistry 12: 85–93.
68. Miflin, BJ, and PR Shewry (1981): Seed storage proteins: Genetics, synthesis, accumulation, and protein quality. In: Bewley, J (ed.): The Physiology and Biochemistry of Plant Productivity: 195–248; Martinus Nijhoff, Den Haag.
69. Millerd, A (1975): Biochemistry of legume seed proteins. Ann. Rev. Plant Physiol. 26: 53–72.
70. Millerd, A, M Simon and H Stern (1971): Legumin synthesis in developing cotyledons of *Vicia faba* L. Plant Physiol. 48: 419–425.
71. Millerd, A, and D Spencer (1974): Changes in RNA-synthesizing activity and template activity in nuclei from cotyledons of developing pea seeds. Aust. J. Plant Physiol. 1: 331–341.
72. Millerd, A, JA Thomson, and PJ Randall (1979): Heterogeneity of sulphur content in the storage proteins of pea cotyledons. Planta 146: 463–466.
73. Millerd, A, JA Thomson, and HE Schroeder (1978): Cotyledonary storage proteins in *Pisum sativum*. III. Patterns of accumulation during development. Aust. J. Plant. Physiol. 5: 519–534.
74. Moreira, MA, MA Hermodson, BA Larkins, and CN Nielsen (1979): Partial characterization of the acidic and basic polypeptides of glycinin. J. Biol. Chem. 254: 9921–9926.
75. Moreira, MA, MA Hermodson, BA Larkins, and NC Nielsen (1981): Comparison of the primary structure of the acidic polypeptides of glycinin. Arch. Biochem. Biophys. 210: 633–642.
76. Morris, GFI, DA Thurman, and D Boulter (1970): The extraction and chemical composition of aleurone grains (protein bodies) isolated from seeds of *Vicia faba*. Phytochemistry 9: 1707–1714.
77. Müntz, K, H Baumlein, R Bassüner, R Manteuffel, M Püchel, P Schmidt, and U Wobus (1981): The regulation of biosynthesis and accumulation of storage proteins during plant seed development. Biochem. Physiol. Pflanzen 176: 401–422.
78. Murray, DR (1979): A storage role for albumins in pea cotyledons. Plant, Cell and Environm. 2: 221–226.
79. Mutschler, MA, FA Bliss, and TC Hall (1980): Variation in the accumulation of seed storage protein among genotypes of *Phaseolus vulgaris* L. Plant Physiol. 65: 627–630.
80. Öpik, H (1968): Development of cotyledon cell structure in ripening *Phaseolus vulgaris* seeds. J. Exp. Bot. 19: 64–76.
81. Payne, ES, A Brownrigg, A Yarwood, and D Boulter (1971): Changing protein synthetic machinery during development of seeds of *Vicia faba*. Phytochemistry 10: 2299–2303.
82. Przybylska, J, J Hurich, and S Blixt (1981): Variation of basic legumin components in the genus *Pisum* revealed by isoelectrofocusing. Pisum Newsletter 13: 44–45.
83. Przybylska, J, J Hurich, and Z Zimniak-Przybylska (1979): Comparative study of seed proteins in the genus *Pisum*. IV. Electrophoretic patterns of legumin and vicilin components. Genetica Polonica 20: 517–528.
84. Püchel, M, K Müntz, B Parthier, O Aurich, R Bassüner, R Manteuffel, and P Schmidt (1979): RNA metabolism and membrane-bound polysomes in relation to globulin biosynthesis in cotyledons of developing field beans (*Vicia faba* L.). Eur. J. Biochem. 96: 321–329.
85. Pusztai, A, and WB Watt (1970): Glycoprotein II. The isolation and characterization of a major

antigenic and non-haemagglutinating glycoprotein from *Phaseolus vulgaris*. Biochem. Biophys. Acta. 207: 413–431.

86. Randall, PJ, JA Thomson, and HE Schroeder (1979): Cotyledonary storage proteins in *Pisum sativum*. IV. Effects of sulfur, potassium and magnesium deficiencies. Aust. J. Plant Physiol. 6: 11–24.
87. Randerath, K (1970): An evaluation of film detection methods for weak β-emitters, particularly tritium. Analyt. Biochem. 34: 188–205.
88. Romero, J, MS Sun, RC McLeester, FA Bliss, and TC Hall (1975): Heritable variation in a polypeptide subunit of the major storage protein of the bean (*Phaseolus vulgaris* L.). Plant Physiol. 56: 776–779.
89. Schlesier, B, R Manteuffel, and G Scholz (1978): Studies on seed glubulins from legumes. VI. Association of vicilin from *Vicia faba* L. Biochem. Physiol. Pflanzen 172: 285–290.
90. Spencer, D, and TJV Higgins (1979): Molecular aspects of seed protein biosynthesis. Curr. Adv. Plant Sci. 34: 1–15.
91. Spencer, D, and TJV Higgins (1980): The biosynthesis of legumin in maturing pea seeds. Biochem. Internat. 1: 501–509.
92. Spencer, D, TJV Higgins, SC Button, and RA Davey (1980): Pulselabelling studies on protein synthesis in developing pea seeds and evidence of a precursor form of legumin small subunit. Plant Physiol. 66: 510–515.
93. Staswick, PE, MA Hermodson, and NC Nielsen (1981): Identification of the acidic and basic subunit complexes of glycinin. J. Biol. Chem. 256: 8752–8755.
94. Sullivan, JG (1981): Recurrent selection for increased seed yield and percentage seed protein in the common bean (*Phaseolus vulgaris* L.) using a selection index; and isolation and analysis of major genes controlling phaseolin. Thesis, University of Wisconsin, Madison.
95. Sun, SM, and TC Hall (1975): Solubility characteristics of globulins from *Phaseolus vulgaris* in regard to their isolation and characterization. J. Agric. Fd. Chem. 23: 184–189.
96. Sun, SM, RC McLeester, FA Bliss, and TC Hall (1974): Reversible and irreversible dissociation of globulins from *Phaseolus vulgaris* seed. J. Biol. Chem. 249: 2118–2120.
97. Sun, SM, MA Mutschler, FA Bliss, and TC Hall (1978): Protein synthesis and accumulation in bean cotyledons during growth. Plant Physiol. 61: 918–929.
98. Sun, SM, JL Slightom, and TC Hall (1981): Intervening sequences in a plant gene – comparison of the partial sequence of cDNA and genomic DNA of French bean phaseolin. Nature 289: 37–41.
99. Thanh, VH, and K Shibasaki (1978): Major protein of soybean seeds. Subunit structure of β-conglycinin. J. Agr. Food Chem. 26: 692–695.
100. Thomson, JA, and HE Schroeder (1978): Cotyledonary storage proteins in *Pisum sativum*. II. Hereditary variation in components of the legumin and vicilin fractions. Aust. J. Plant Physiol. 5: 281–294.
101. Thomson, JA, HE Schroeder, and WF Dudman (1978): Cotyledonary storage proteins in *Pisum sativum*. I. Molecular heterogeneity. Aust. J. Plant Physiol. 5; 263–249.
102. Thomson, JA, HE Schroeder, and AM Tassie (1980): Cotyledonary storage proteins in *Pisum sativum*. V. Further studies on molecular heterogeneity in the vicilin series of holoproteins. Aust. J. Plant Physiol. 7: 271–282.
103. Tumer, NE, VH Thanh, and NC Nielsen (1981): Purification and characterization of mRNA from soybean seeds. Identification of glycinin and β-conglycinin precursors. J. Biol. Chem. 256: 8756–8760.
104. Utsumi, S, H Inaba, and T Mori (1980): Formation of pseudo and hybrid – 11S globulins from subunits of soybean and broad bean 11S globulins. Agric. Biol. Chem. 44; 1891–1896.
105. Utsumi, S, and T Mori (1980): Heterogeneity of broad bean legumin. Biochim. Biophys. Acta. 621: 179–189.
106. Utsumi, S, and T Mori (1981): Subunit composition of molecular species of legumin from the broad bean. Agric. Biol. Chem. 45: 2273–2276.

107. Utsumi, S, Z Yokoyama, and T Mori (1980): Comparative studies of subunit compositions of legumins from various cultivars of *Vicia faba* L. seeds. Agric. Biol. Chem. 44: 595–601.
108. Weber, E, J Ingversen, R Manteuffel, and M Püchel (1981): Transfer of *in vitro* synthesized *Vicia faba* globulins and barley prolamins across the endoplasmic reticulum of *Vicia faba*. Carlsberg Res. Commun. 46: 383–393.
109. Weber, E, R Manteuffel, M Jakubek, and D Neumann (1981): Comparative studies on protein bodies and storage proteins of *Pisum sativum* L. and *Vicia faba* L. Biochem. Physiol. Pflanzen 176: 342–356.
110. Weber, E, R Manteuffel, and D Neumann (1978): Isolation and characterization of protein bodies of *Vicia faba* seeds. Biochem. Physiol. Pflanzen 172: 597–614.
111. Whittaker, RG, and BA Moss (1981): Comparative peptide mapping at the nanomole level. Analyt. Biochem. 110: 56–60.
112. Wright, DJ, and D Boulter (1972): The characterization of vicilin during seed development in *Vicia faba* (L.) Planta 105: 60–65.
113. Wright, DJ, and D Boulter (1974): Purification and subunit structure of legumin of *Vicia faba* L. (broad bean) Biochem. J. 141: 413–418.

12. Seed Protein Production of *Pisum* Mutants and Recombinants

W. GOTTSCHALK

Abstract The seed protein production of 23 radiation-induced *Pisum* mutants and 89 recombinants was investigated. The traits which are commonly responsible for this character, are highly influenced by environmental factors. Therefore, reliable results are only obtained if the genotypes are studied over many generations. Some high-yielding genotypes with stem fasciation or stem bifurcation produce considerably more protein than the mother variety in spite of having a reduced grain size. No significant correlation between seed size and protein content of the seed meal exists in the material studied. Between number of seeds per plant and protein content of the seed meal, however, a slight but statistically significant negative correlation was found. Only one of the mutant genes studied was found to influence the protein content of the seed flour positively. It is not yet clear whether the abnormally low mean values of some genotypes for this trait are genetically conditioned. Our findings demonstrate that the protein production of a variety can easily be improved by improving its seed production directly by means of mutant genes or indirectly by producing high-yielding recombinants homozygous for several mutant genes. It is, however, very difficult to reach this aim by means of specific mutant genes causing an increase of the protein content of the seed meal.

INTRODUCTION

The seed protein production of a plant is a complex trait dependent on seed production, seed size and protein content of the seed meal. Each of these quantitative characters is highly influenced by environmental factors; moreover, there are certain interactions between them. As a consequence of this difficult situation, an extraordinarily broad variability of the mean values for the character 'seed protein production per plant' is obtained, if we evaluate the material in subsequent years or at different locations in the same year. These difficulties appear already within the same genotype. The various characters mentioned above are not only influenced by the environment, but they are furthermore controlled by specific genes of the genome which can be changed in mutation experiments. In some cases, it is already difficult to discern mutant genes or alleles controlling the seed size. It is much more difficult to analyze differences in the protein content of different genotypes reliably. Because of manifold interactions between distinct environmental factors and distinct genes, which commonly influence the seed proteins, it is very difficult to select mutant genes which are involved in the control of seed protein production.

A voluminous collection of radiation-induced *Pisum* mutants and recombinants is available at our institute containing more than 200 fertile mutants and about 450 recombinants of known genetic constitution. A small proportion of these genotypes is of interest for mutation breeding because of having characters which could be utilized agronomically. This holds true for earliness and some other traits, especially for increased yield. More than a hundred genotypes of this group were studied with regard to their protein production. Moreover, the relations between seed size, seed production and protein content of the seed meal of the material were investigated.

MATERIAL AND METHODS

Twenty three radiation-induced mutants, 89 recombinants and the commercial variety 'Dippes Gelbe Viktoria' (DGV) of *Pisum sativum* were evaluated with regard to the following characters:

plant height,
flowering time,
number of seeds per plant,
seed size (thousand grain weight),
seed weight per plant,
protein content of the seed meal,
protein production per plant.

All the mutants derive from the variety 'Dippes Gelbe Viktoria' which was used as initial line for our X-ray and neutron treatments. The derivation of the recombinants is not uniform; on the contrary, they can be subdivided into three different groups as follows:

Group I was selected after having crossed either two different mutants with each other or fasciated mutants of complicated genotypic constitution with the mother variety (R-numbers).

Group II was selected after having crossed a mutant with a recombinant (RM-numbers).

Group III was selected after having crossed two different recombinants with each other (RR-numbers).

The recombinants selected in F_2 or F_3 families were developed into pure lines. Most of the genotypes evaluated (mutants, recombinants and mother variety) were grown in the form of 3 or 4 replications over several years with 30 to 50 plants per replication. The mean values for the various criteria mentioned above were related to the corresponding control values of the mother variety grown in the same year at the same location. From some recombinants, which have been selected only a few years ago, only limited amounts of seed were available. Therefore, they could not yet be tested in the form of yielding analyses. This holds true for some genotypes having a low seed production.

The amount of the seed proteins was determined by means of the Kjeldahl method using 30 milled seeds per genotype per sample. The seed flour was dried for 3 days at 105°C. One gram of this material was used for analysis. The protein content was calculated by multiplying the nitrogen value obtained by 6.25.

I thank my technicians, Mrs. E. Knob and Miss S. Faxel, for carrying out the protein analyses.

RESULTS

The Protein Production per Plant

The protein production of a plant is a very variable criterion depending not only on its genotypic constitution but also on different criteria which are highly influenced by environmental factors. This holds true for the number of seeds per plant, the seed size and the protein content of the seed flour. In some mutants, the environmentally conditioned variability of these traits is so great that the protein production per plant can be judged reliably only if mean values of a large number of generations are available. Some fasciated mutants of our *Pisum* collection are impressive examples for demonstrating these difficulties. We have analyzed the protein production of 20 mutants, 83 recombinants and the mother variety 'Dippes Gelbe Viktoria' used as initial line for our radiation genetic experiments. According to the organization of the shoot system, the material tested can be subdivided into the following three groups:

genotypes with apical stem fasciation;
genotypes with dichotomous stem bifurcation;
genotypes with normal stem structure.

In the following, these three groups are separately discussed.

The protein production of fasciated genotypes. The strongly fasciated mutant 489C belongs to the highest-yielding genotypes of our collection. The plants are homozygous for at least 16 genes which have mutated during irradiation in the embryonic growing point of that seed which gave rise to the mutant. Its genome contains the following genes [1, 7, 9, 10]:

3, possibly 4 genes for different kinds of stem fasciation,
1 gene for apical stem bifurcation,
6 genes for different stem and internode lengths,
3 genes for different flowering and ripening times,
1 gene for a slightly reduced chlorophyll content,
1 gene for a reduced grain size (gene *sg*),
1 gene for the control of the photoperiodic reaction (gene *fis*).
The plants do not flower under short-day conditions.

Only 5 genes of this group are discernible in the mutant; the other ones appear only in segregating families after 489C has been crossed with non-fasciated genotypes. Thus, the mutant has a very complicated genotypic constitution not only because of the large number of mutant genes present but mostly because these genes belong to a system of epi- and hypostatic units.

The plants of mutant 251A are likewise fasciated, but their uppermost internodes are not shortened. Therefore, the flowers and pods are distributed over a longer part of the stem. This anomaly is called 'linear stem fasciation' in contrast to 'strong fasciation' of 489C. Mutant 251A is likewise homozygous for a large group of genes, most of them being identical with the mutant genes of 489C [9].

The agronomic value of fasciated pea lines lies in their high seed production due to the strongly increased number of pods per plant as a consequence of the apical stem fasciation. Moreover, the flowering period is considerably condensed because all the flowers are clustered in the apical part of the stem due to the strongly shortened internodes of this region. This is an advantage with regard to pest control. The clustering of leaves at the top of the plants leads furthermore to an accumulation of tendrils resulting in good standing ability when grown in the field [24, 25]. The two German fodder pea varieties 'Ornamenta' and 'Rosakrone' are fasciated and have been developed by means of a spontaneously arisen mutant [26, 27]. The disadvantages of our experimentally produced fasciated mutants are their tallness, their lateness and the reduced grain size.

Some yielding criteria of the fasciated mutants 489C and 251A are presented in figures 1 and 2. In mutant 489C, the mean values for the character 'number of seeds per plant' varied between 118 and 212% of the control values of the mother variety considering 15 generations grown in Bonn. The seeds, however, are so small that the mean values for the trait 'seed weight per plant' vary only around the corresponding means of the initial line. Details for the M_{17} to M_{19} generations are given in figure 1. With regard to the protein content of the seed flour, the mutant does not deviate significantly from the mother variety. From the data just mentioned, the mean values for the trait 'seed protein production per plant' can be calculated. They varied between 83 and 108% of the control values of the mother variety in the 3 generations just mentioned. Thus, the high seed production of the mutant did not result in an improved protein production due to the strongly reduced grain size as a consequence of the presence of gene *sg* in the genome of mutant 489C.

We have studied the seed proteins of this mutant over a period of 12 generations; the mean values for the protein production per plant are given in the righthand part of figure 1. In the first three generations studied (M_8 to M_{10}), these values were considerably higher than the control values. It was therefore our conviction that the improved protein production is a permanent character of this genotype due both to a strongly increased seed production and to an increased protein content of the seed meal [11, 20]; the M_8 values are given in the lefthand part of figure 1. This favourable situation, however, could not be confirmed in later generations, especially not with regard to the increased protein content of the seed meal [16]. On the contrary, the mutant was found to have an immensely broad environmentally conditioned variability in its protein production ranging between 65 and 169% of the corresponding mean values of the mother variety. This inconstency was only discerned because the material was evaluated in so many generations.

In the second fasciated genotype tested over many generations – mutant 251A – the situation is essentially more favourable (figure 2). With regard to the number of seeds per plant and the protein content of the seed meal, there are no significant

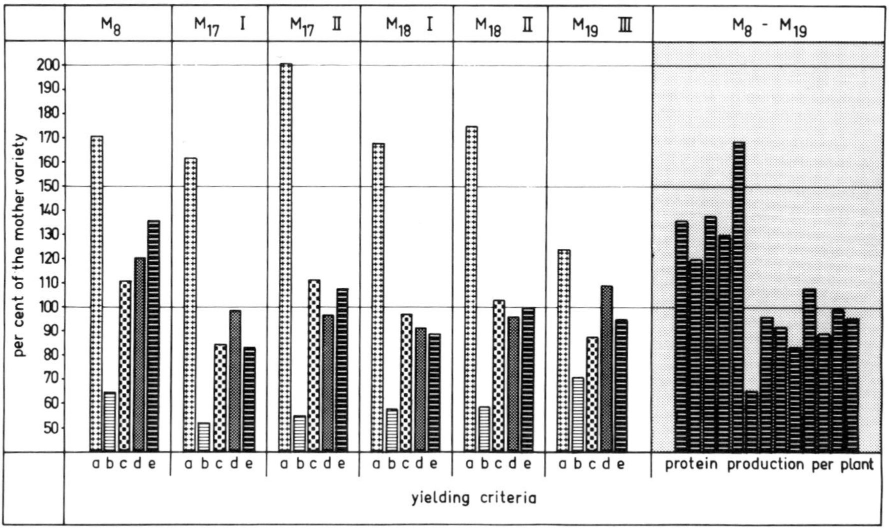

Figure 1. The dependence of the seed production of the strongly fasciated mutant 489C on different yielding criteria in subsequent generations and at different locations. The following criteria were evaluated:

a: number of seeds per plant
b: seed size
c: seed weight per plant
d: protein content of the seed meal
e: seed protein production per plant.

Righthand part of the figure: The protein production per plant of the mutant in 12 generations. All the mean values are related to the corresponding values of the mother variety = 100%.

differences between 251A and 489C, but the seeds of 251A are bigger. Therefore, not only better mean values for the trait 'seed weight per plant', but also for the character 'protein production per plant' are reached (figure 2; righthand part). The mean values for the M_7 to M_{16} generations lay 23 to 67% over the corresponding means of the mother variety demonstrating a high amount of constancy of this agronomically valuable character in successive generations.

Unexpectedly, the values obtained for mutant 251A in M_{17}/1980 lay 35% below the control values. This low productivity was due both to a relatively low number of seeds per plant as compared to the other generations evaluated and to abnormally small seeds. As these two negative criteria were not compensated by an increased protein content of the seed meal, the protein production per plant was very low. This is undoubtedly a reaction of the mutant to the extreme weather conditions in 1980. During the flowering and ripening period, it was abnormally cool and wet in West Germany and the seed size of many genotypes of our collection was smaller than in climatically more normal years.The comparison of figures 1 and 2 shows, that the two fasciated mutants 251A (M_{17}/1980) and 489C (M_{19}/1980) reacted differently to the abnormal weather conditions. The number of seeds per plant was relatively low

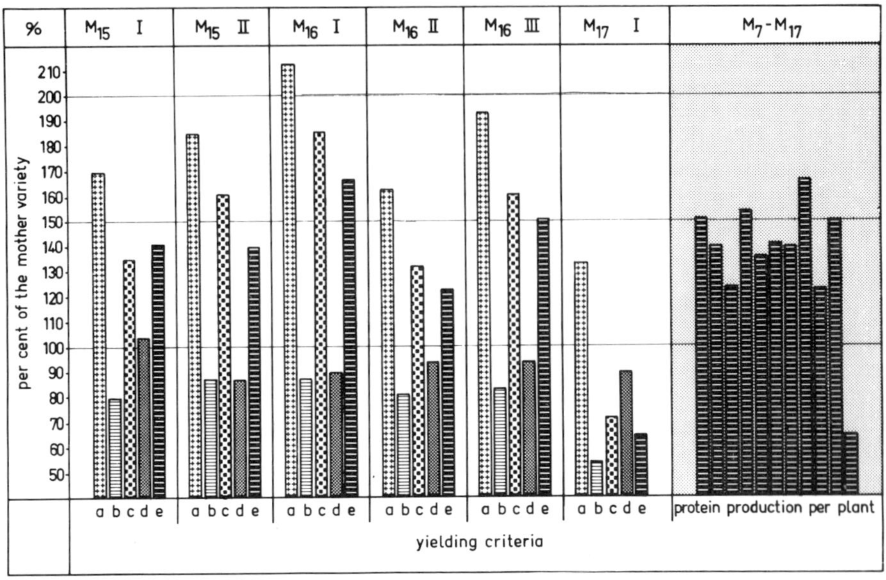

Figure 2. The dependence of the seed production of the linearly fasciated mutant 251A on different yielding criteria in subsequent generations and at different locations. Righthand part of the figure: The protein production per plant of the mutant in 11 generations. Arrangement of the details as in figure 1. All the mean values are related to the corresponding means of the mother variety = 100%.

in both the genotypes. The seeds of 251A, however, were abnormally small, those of 489C were abnormally large. Therefore, the protein production of 489C was higher than that of 251A. Differences in the seed proteins under the influence of different ecological conditions were also observed for other genotypes of our collection which were comparatively grown at different German localities or in West Germany and India, respectively [12, 14, 21].

The complicated genotypic constitution of the fasciated mutants 489C and 251A has already been mentioned. A similar mutation genetic situation is realized in some other fasciated mutants of our collection as well as in the fodder pea variety 'Ornamenta' developed by SCHEIBE [26, 27] by means of a spontaneously arisen fasciated mutant [7, 9]. If one of these fasciated genotypes is used for crosses, highly heterozygous hybrids arise because of the large number of gene pairs involved. They show complicated segregations in their progenies. Many different recombinant types appear in the F_2 and F_3 families in which many characters are specifically combined with each other. Hundreds of them have been selected during the past years. They are either already available in the form of pure lines or they are still segregating for distinct traits because of the high degree of heterozygosity of the F_1 plants. Eighty nine of them have been analyzed with regard to their protein production; the findings obtained are given in figures 3 to 7.

Of particular interest are those recombinants which are homozygous for one or

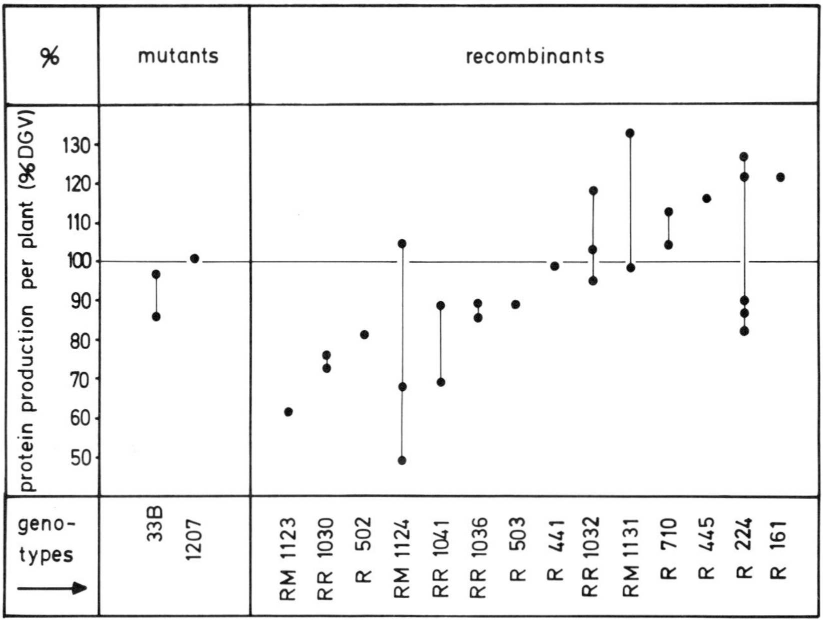

Figure 3. The protein production per plant of two pea mutants and 14 recombinants with strong or linear stem fasciation. The plant height of these genotypes is similar to that of the mother variety or shorter. Each dot gives the mean value for one generation or for one of the replications grown. All the means are related to the control values of the mother variety = 100%.

several of the 3 or 4 *fasciata* genes present in the genomes of the fasciated mutants. In *Pisum*, stem fasciation often results in high seed production because of the strongly increased number of flowers and pods per plant. In many recombinants of our collection, however, stem fasciation is combined with characters which reduce the seed production. This holds true for instance for strongly shortened internodes or for extreme lateness, the respective genes deriving from the parental fasciated mutants. A similar negative effect is often observed if *fasciata* genes are combined with gene *efr* for earliness deriving from mutant 46 of our collection. Genotypes of this group are not considered in the present paper. On the contrary, we have preferably considered recombinant types with favourable yielding properties.

The protein production of 52 different fasciated recombinant lines is graphically presented in figures 3 and 4. Most of our fasciated mutants are taller than the mother variety. In some recombinants the genes for long or very long internodes have been replaced by genes for shorter internodes. The protein production of 14 genotypes of this group is given in figure 3. Only a few of them are superior to the mother variety. Of particular interest is recombinant RM 1131 selected after having crossed a fasciated recombinant with a non-fasciated mutant having extremely shortened internodes, its plant height ranging between 10 and 15 centimeters. The seeds of RM 1131 are very small but the number of seeds per plant is very high. In the two

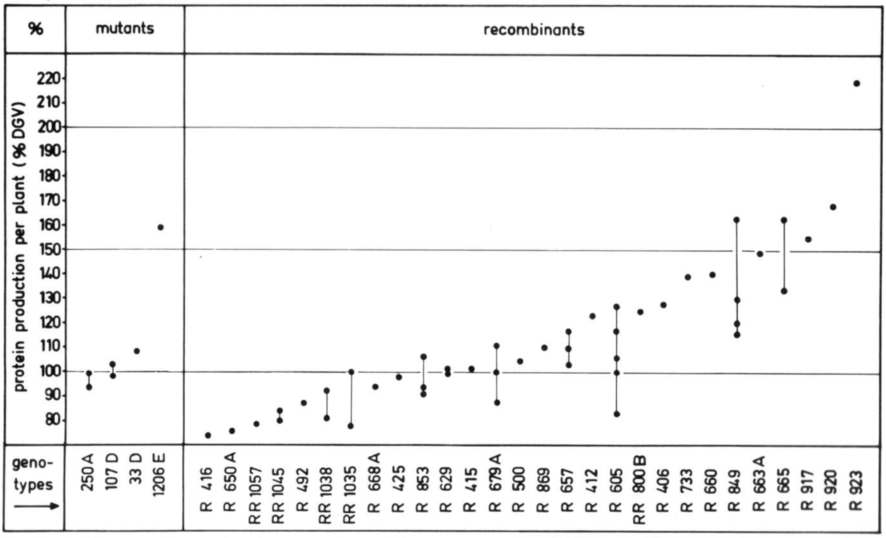

Figure 4. The protein production per plant of 4 pea mutants and 28 recombinants with strong or linear stem fasciation. The plants of these genotypes are longer than those of the mother variety. (Arrangement as in figure 3).

generations tested, the protein content of the seed meal was higher than in the mother variety and the parental mutant 489C. Material from more generations has to be analyzed until definite conclusions can be drawn. This holds also true for recombinant R 710 derived from crosses between mutant 489C and the *cochleata* mutant 5137 of Blixt's collection in Landskrona. The plants of this recombinant are somewhat shorter than those of mutant 489C. Their seeds are small having normal protein content, but the number of seeds per plant is very high resulting in a good protein production per plant. An additional advantage is a certain degree of earliness not only with regard to flowering but also to harvesting. The data obtained for the fasciated recombinant R 224 clearly demonstrate the difficulties of this kind of investigations. Yielding criteria of five generations were evaluated. In three of them, the protein production ranged between 80 and 90% of the control values of the mother variety, whereas considerably higher values were obtained in the remaining two generations. Thus, five generations are not enough for being able to judge the capacity of this recombinant reliably.

Long-stemmed pea mutants and recombinants show in general better yielding properties than most of the short-stemmed ones. This holds also true for long-stemmed fasciated recombinants, the protein production of which is given in figure 4. Even if we consider that many recombinant types which have been developed into pure lines only one or two years ago, could only be analyzed in one generation, there is no doubt that some of them surpass the mother variety considerably in protein production. Recombinant R 657 derives from the cross between mutant 489C and a short-stemmed *afila* mutant. Both these parents are late flowering and ripening

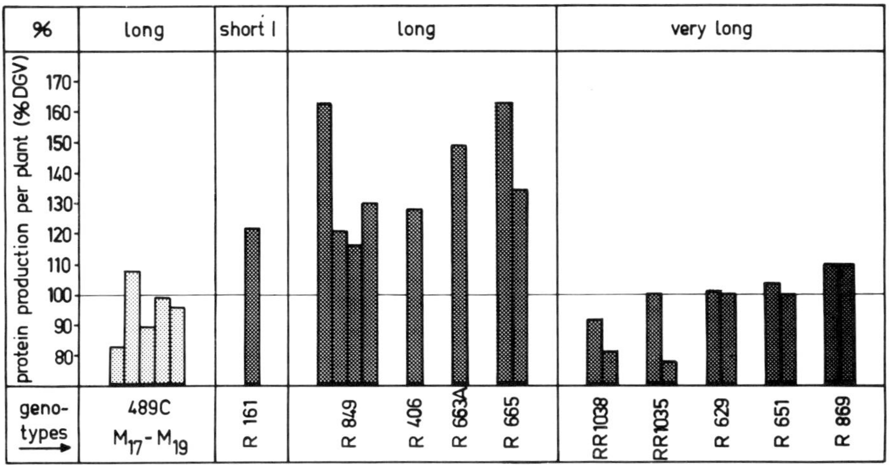

Figure 5. The protein production of fasciated recombinants as compared to that of the fasciated mutant 489C. Each column represents the mean value for one generation or one location, respectively, related to the control value of the mother variety = 100%.

whereas the linearly fasciated plants of R 657 are considerably earlier. In addition, their protein production was 3 to 17% higher than that of the mother variety considering three generations.

Of particular interest are the recombinants R 917, R 920 and R 923 because of their extraordinarily high protein production. R 917 and R 920 derive from crosses between the fasciated mutant 250A which is genetically closely related to 489C, with the mutant 46 homozygous for gene *efr* causing earliness. This gene is present in the genome of R 917, but it is not present in R 920. Nevertheless, the plants are earlier than those of the fasciated parent 250A. The early flowering recombinant R 923 was selected in an F_2 family of the cross between the fasciated fodder pea variety 'Ornamenta' and our early flowering mutant 46. The seed size of the three recombinants just mentioned is strongly reduced due to the presence of gene *sg* derived from the fasciated parents. The protein content of the seed meal is normal; the high protein production is exclusively due to the strongly increased number of seeds per plant. So far, only values of one generation are available. Analyses of the next generations will show whether these favourable properties are constant characters of these genotypes.

The aim of crosses consists normally in combining useful traits from different parents in the genome of the recombinant wanted. This has been done for instance in the last mentioned recombinants with regard to the characters stem fasciation and earliness deriving from different cross partners. If a mutant organism is homozygous for a large number of mutant genes, favourable recombinants can arise even when the second partner has a strongly negative selection value. An interesting example is recombinant R 605 derived from the cross of mutants 489C × 94A. Mutant 94A has greenish flowers showing a characteristic anomaly

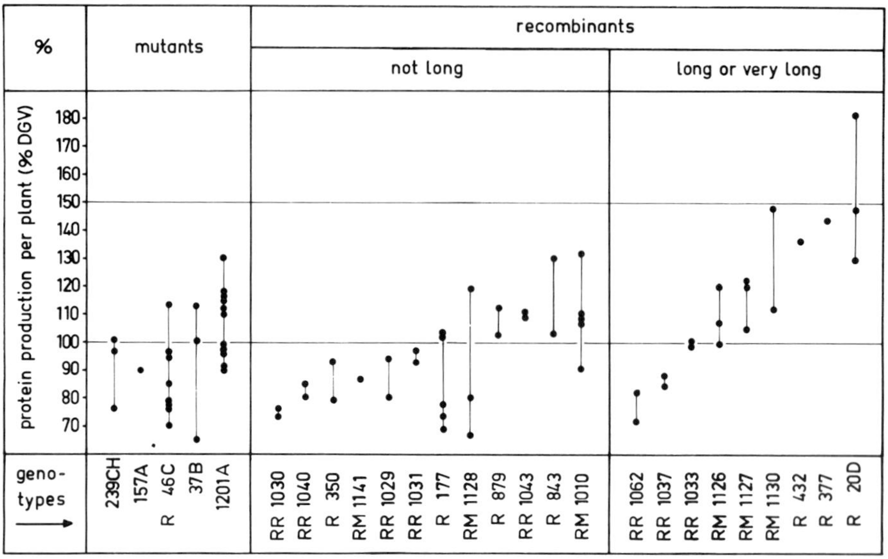

Figure 6. The protein production per plant of 4 mutants and 22 recombinants with dichotomous stem bifurcation. (Arrangement as in figure 3).

with regard to their sex organs [6]. The seed production of these plants is very low; they are without any agronomic value. The plants of recombinant R 605 are long, linearly fasciated and late. With regard to their protein production, they were found to be better than 489C, in three out of five generations tested even better than the mother variety.

The limiting factor in the protein production of many high-yielding fasciated genotypes is their reduced grain size. This trait is not part of the pleiotropic pattern of one of the mutant genes present in the fasciated mutants. On the contrary, in 489C it is due to the action of a specific mutant gene designated *sg*. A similar but considerably weaker gene action becomes discernible in the fasciated mutant 251A. According to the seed size of hybrids between 489C and 251A, both genotypes seem to be homozygous for the same gene *sg*. In most generations tested, mutant 251A had somewhat bigger seeds than 489C. A characteristic feature of 251A, however, is a very strong variability of the seed size, which is much stronger than in the other small-grained fasciated mutants of our collection. It is certain that the reduced seed size of mutants 489C and 251A is not caused by different genes. It is most probably due to the same allele *sg*, possibly to a different allele of a multiple series. By crossing these fasciated mutants with non-fasciated genotypes, fasciated recombinants with seeds of normal size were selected. At least some of them should have an improved protein production provided that the protein content of their seed meal is not reduced.

Because of the large number of mutant genes involved in the crosses just men-

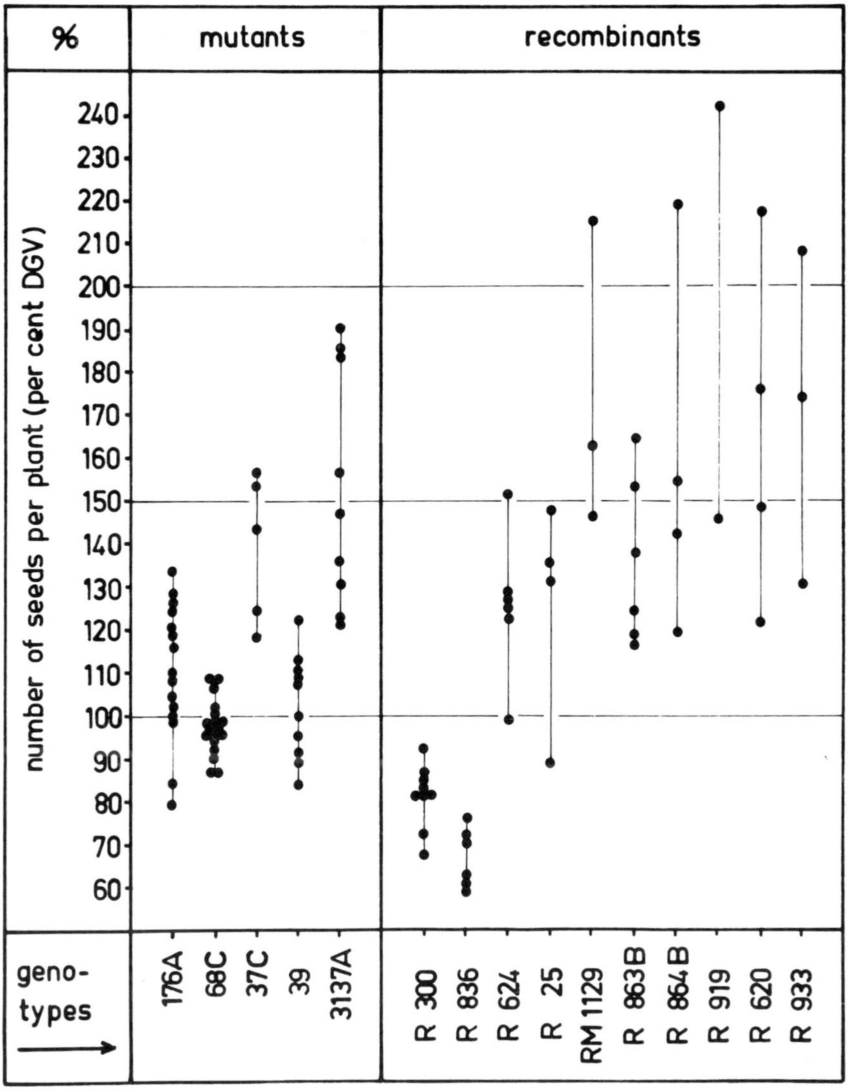

Figure 7. The seed production (number of seeds per plant) of some of the genotypes considered in figure 8. Each dot represents the mean value for one generation as related to the control value of the mother variety = 100%.

tioned, only a small number of pure recombinant lines is already available homozygous for *fasciata* genes and gene *Sg* for normal seed size. The genotypes of this group, derived from crosses between mutant 489C and non-fasciated genotypes of our collection, are considered in figure 5. They are compared to 489C with regard to the protein production per plant. Five of these recombinants have very long interodes; the plants are considerably higher than those of mutant 489C. In spite of having seeds of normal size, their protein production is not better than that of 489C. This is partly due to a reduced protein content of the seed meal (R 629), but mainly to the reduced number of seeds per plant. A second group of recombinants, however, shows a better situation. The stem length of these genotypes is similar to that of 489C but the protein production is considerably higher. The plants of recombinant R 849 are linearly fasciated and early flowering due to the presence of gene *efr* deriving from mutant 46. They were grown in several replications in 1978 and 1979 at two different locations in each year. Their protein production per plant was 16 to 63% higher than that of the mother variety 'Dippes Gelbe Viktoria' and 23 to 93% higher than that of the parental mutant 489C. This is an improvement with regard to two agronomically important traits:

The gene for lateness of mutant 489C has been replaced by gene *efr* for earliness. The plants begin flowering 15–16 days prior to those of 489C.

Gene *sg* for reduced grain size has been replaced by its allele for normal seed size. The improved protein production is due to both the increased seed size and the increased number of seeds per plant.

Recombinant R 665 derives from the cross of mutants 489C × 189. The plants have the following characters:

long internodes (from 489C),
stem linearly fasciated and bifurcated (from 489C),
greenish flower colour (gene *gf* from 189),
late flowering and ripening (probably gene *lat* from 189).

Unfortunately, gene *ipc* of mutant 189, causing an increased protein content of the seed flour, is not present in this recombinant. Nevertheless, the protein production per plant was 34 and 63% higher than that of the mother variety considering two subsequent generations. From the other two recombinants (R 161, R 406), only values of one generation are available so far, not yet allowing to draw reliable conclusions.

The protein production of bifurcated genotypes. In our radiation genetic experiments, four mutants arose showing a stem bifurcation in the upper part of the shoot thus having a dichotomous stem structure instead of the monopodium characteristic for the species *Pisum sativum*. Mutants 1201A and 239CH of this group were found to be identical both being homozygous for gene *bif-1*. Mutant 157A contains gene *bif-2* polymeric to *bif-1*. A peculiarity of these two genes is their unstable penetrance which is highly influenced by both environmental factors and other mutant genes. They can either suppress the action of the *bif*-genes or stabilize their penetrance [8]. The gene present in mutant 37B shows full penetrance; it was found to be allelic to *bif-1*. The plants of recombinant R 46C contain gene *efr* for earliness derived from mutant 46A and gene *bif-1* from mutant 1201A. The recombinant arose probably in combination with a spontaneous hybridization between the two mutants mentioned. The advantage of the bifurcated mutants consists in the increased number of pods per plant as a consequence of the stem bifurcation resulting in the increase of the seed production. Because of the unstable penetrance of the *bif* genes, the breeding value of these genotypes is reduced. In spite of this disadvantage, they can be regarded to be useful mutants. The seed production of mutant 1201A, for instance, was in most of the 23 generations tested somewhat better than that of the mother variety.

The protein production of the 4 bifurcated mutants just mentioned and of 22 bifurcated recombinants is given in figure 6. Mutant 1201A shows a favourable protein yield considering 11 generations tested. In 5 out of these 11 generations, the protein production per plant varied between 90 and 99%, in the remaining 6 generations between 110 and 130% of the control values of the mother variety. Seed size and protein content of the seed flour are normal. Thus, the protein production of this genotype is equivalent to its seed production per plant. A similar situation is realized in the early flowering recombinant R 46C. Its seed production is lower than that of the initial line resulting in a lower protein production.*

This big group of bifurcated recombinants has been divided into two sub-groups:

in genotypes the plant height of which is similar to that of the initial line or shorter,

and in genotypes with long or very long internodes being considerably higher than the mother variety.

Most of the bifurcated recombinants have been selected after having crossed mutant 1201A or recombinant R 46C with the fasciated mutant 489C containing gene *sg* for reduced grain size. Many of these genotypes are homozygous for *sg*. Thus, they have two peculiarities influencing their agronomic usefulness negatively:

the reduced grain size giving relatively low mean values for the seed weight per plant;

the unstable penetrance of gene *bif-1* resulting in a reduction of the number of bifurcated plants in the lines, which leads likewise to a reduction of the seed production.

The low protein production of many recombinants considered in figure 6 is caused by these two negative effects. But there are some interesting exceptions which should be discussed in detail.

The plants of recombinant R 177 (from 489C × 1201A) are homozygous for genes *bif-1* and *sg*, the latter one stabilizing the penetrance of *bif-1*. Thus, all the plants of R 177 show the stem bifurcation. They have favourable mean values for the number of seeds per plant, but the seed size is reduced. Therefore, the values for the protein production per plant are considerably lower than the corresponding values of mutant 1201A. Recombinant R 177 was crossed with mutant 1000 with big

*R 46C has been incorporated into the group 'mutants' in Figure 6 because its origin is not completely clear. It has not been produced by means of specific hybridizations like the other recombinants considered in the graph.

seeds giving rise to recombinant RM 1010. The seed size of these plants is nearly normal. With regard to the penetrance of gene *bif-1*, there is no difference between recombinants R 177 and RM 1010, i.e. all the plants are bifurcated. The protein production of RM 1010, however, is essentially better than that of R 177 reaching the values of the parental mutant 1201A.

The genetic constitution of recombinant R879 is not yet completely clear. It was selected in the F_2 of a testcross between mutant 1201A (*bif-1 bif-1*) and the early flowering recombinant R 46C (*efr efr / bif-1 bif-1*). The plants are early and bifurcated. Morphologically, they agree with R 46C, but the penetrance of *bif-1* is regularly higher in this combination than in R 46C. Recombinant R 879 contains obviously a modifier gene which has no morphologically discernible effect, but which influences the penetrance of *bif-1* positively. Consequently, the protein production of these plants is better than that of R 46C.

Some long-stemmed bifurcated recombinants have a clearly better protein production than the mother variety. This holds true for the obviously identical genotypes RM 1126, RM 1127 and RM 1130. Their seeds are small due to the presence of gene *sg*, but this disadvantage is compensated by the high seed production per plant. A similarly favourable protein production was observed for recombinants R 432 and R 377, the latter one containing gene *efr* for earliness.

Recombinant R 20D is of particular interest in this concern. It was selected after having crossed the fasciated mutant 489C with the early flowering recombinant R 46C. The plants show the following characters:

very long internodes (from 489C),
slightly reduced chlorophyll content (from 489C),
stem repeatedly bifurcated (gene *bif-1* from R 46C),
early flowering (gene *efr* from R 46C),
normal seed size (from R 46C).

Thus, R 20D is homozygous for 4 mutant genes. There is full penetrance of *bif-1* in this combination. Moreover, the point of the dichotomous bifurcation is very low at the stem resulting in two or more corresponding pod-bearing branches. This unusual shoot structure is the result of positive interactions between *bif-1* and the gene for long internodes which are not observed in other long-stemmed recombinants of this group. As in the parental recombinant R 46C, the first flowers are already formed at the 4th to 6th foliage leaf in R 20D, but these flowers do generally not produce any pods. In spite of being genetically early, the plants have a very long flowering period and are very late with regard to seed ripening. Their seed production is very high. As the seed size is nearly normal, excellent mean values for the protein production per plant were obtained. Investigations on the protein quality of *Pisum* mutants and recombinants have shown that an increased protein production is in most cases due to the increase of the globulins whereas the amount of the more valuable albumins remains more or less stable. This holds, however, not true for recombinant R 20D. The seed proteins of this genotype have the same proportion of albumins as those of the mother variety [30]. Thus, the plants of R 20D are capable to produce high amounts of seed proteins of good nutritional quality.

The protein production of genotypes with normal stem structure. So far, preferably genotypes with stem fasciation and bifurcation with high seed production have been discussed. Most of the mutants selected in our mutagenic treatments show a negative selection value, their seed production being considerably worse than that of the initial line. As the protein production depends highly on the seed yield, very unfavourable values were obtained in this material. Therefore, only a small group of genotypes with normal shoot structure has been chosen for this paper which are of agronomic interest because of distinct peculiarities. Their seed and protein production is graphically illustrated in figures 7 and 8.

The seed production of mutant 176A, homozygous for gene *dim-1*, was found to be equivalent or somewhat better than that of the mother variety in most generations tested so far (figure 7). The seed size, however, is strongly reduced; therefore, the values for the trait 'seed weight per plant' are considerably lower than the control values. The protein content of the seed meal is normal, but the protein production per plant is low due to the small grains. The mean values ranged between 49 and 91% of the control values considering 4 generations (figure 8).

Mutant 68C is homozygous for gene *ion* causing an increase of the number of ovules per carpel by about 30% resulting in an increased number of seeds per pod.

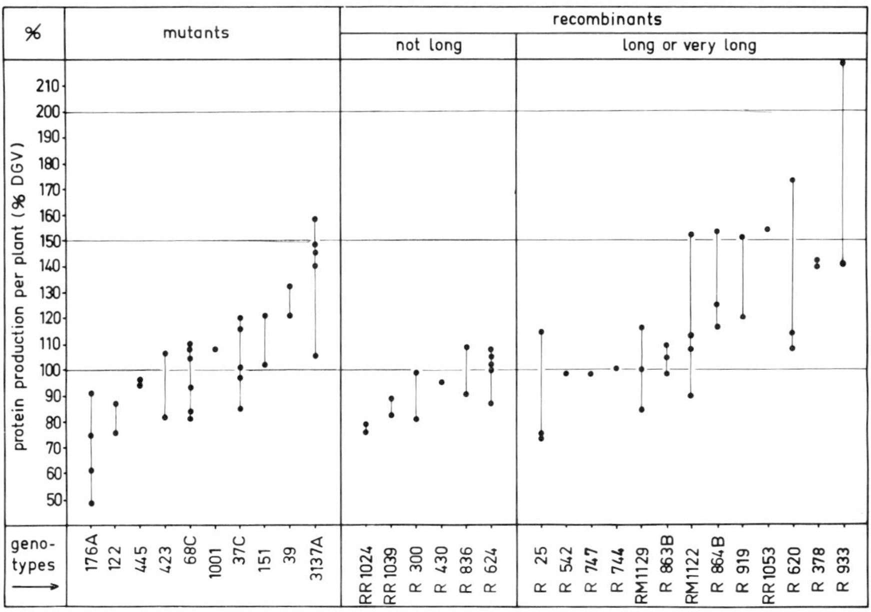

Figure 8. The protein production per plant of 10 pea mutants and 19 recombinants with normal shoot structure as related to the control values of the mother variety = 100%. (Arrangement as in figure 3).

The number of pods per plant, however, is regularly reduced due to the pleiotropic action of gene *ion*. As in many other cases, the pleiotropic pattern of the mutant gene consists of a positive and a negative component the latter one hindering the agronomic utilization of the mutant. Figure 7 shows that the seed production of mutant 68C was about equivalent to that of its mother variety in 18 generations tested so far. This holds in principle also true with regard to the protein production.

Mutant 37C is a typical micromutant. The plants are morphologically so similar to the control plants, that it would not be possible to identify them reliably in segregating families. They begin flowering somewhat later and their seed production was regularly higher in all generations tested so far. The protein content of their seed flour is normal but the seed size is reduced. In spite of this disadvantage, favourable values for the protein production were obtained in most generations due to the high number of seeds per plant.

The plants of the micromutant 39 show a slight deviation in the leaflet shape. Seed size and protein content are normal; the mean values for the seed yield per plant vary around the control values of the intial line. The favourable values for the protein production will certainly not be confirmed if a larger number of generations is evaluated. The high-yielding plants of the neutron-induced mutant 3137A have several disadvantageous characters which hinder their agronomic utilization: they are very tall and late; moreover, they have smaller seeds. Their seed production, however, is so high that excellent values for the protein production per plant were obtained exceeding the control values of the mother variety considerably. We have tried to remove at least one of the negative traits by crossing mutant 3137A with mutant 1001 having somewhat bigger seeds than the initial line. In the F_2 generation, recombinant R 933 was selected and developed into a pure line. The protein production of this recombinant seems to be clearly better than that of the parental mutant 3137A (righthand part of figure 8), but material from more generations has to be evaluated before reliable conclusions can be drawn.

This holds true for most of the recombinants considered in figures 7 and 8. They have been selected only a few years ago and only limited amounts of seeds were available for carrying out yielding analyses. The mean values for the number of seeds per plant, the seed size, the protein content of the seed meal and the protein production per plant varied considerably in subsequent generations. The righthand parts of figures 7 and 8 demonstrate conspicuously that even mean values of 3 or 4 generations, obtained from yielding analyses with 4 to 6 replications, do not yet suffice for judging the productivity of a strain reliably. This may at least partly be due to specific reactions of mutants or recombinants to climatic differences in successive growth seasons different from the reaction of the mother variety.

Recombinant R 300 was selected in the F_2 following crosses between mutant 68C and recombinant R 46C. The plants are homozygous for three mutant genes which are theoretically responsible for favourable traits, but which nevertheless do not result in an agronomically useful strain:

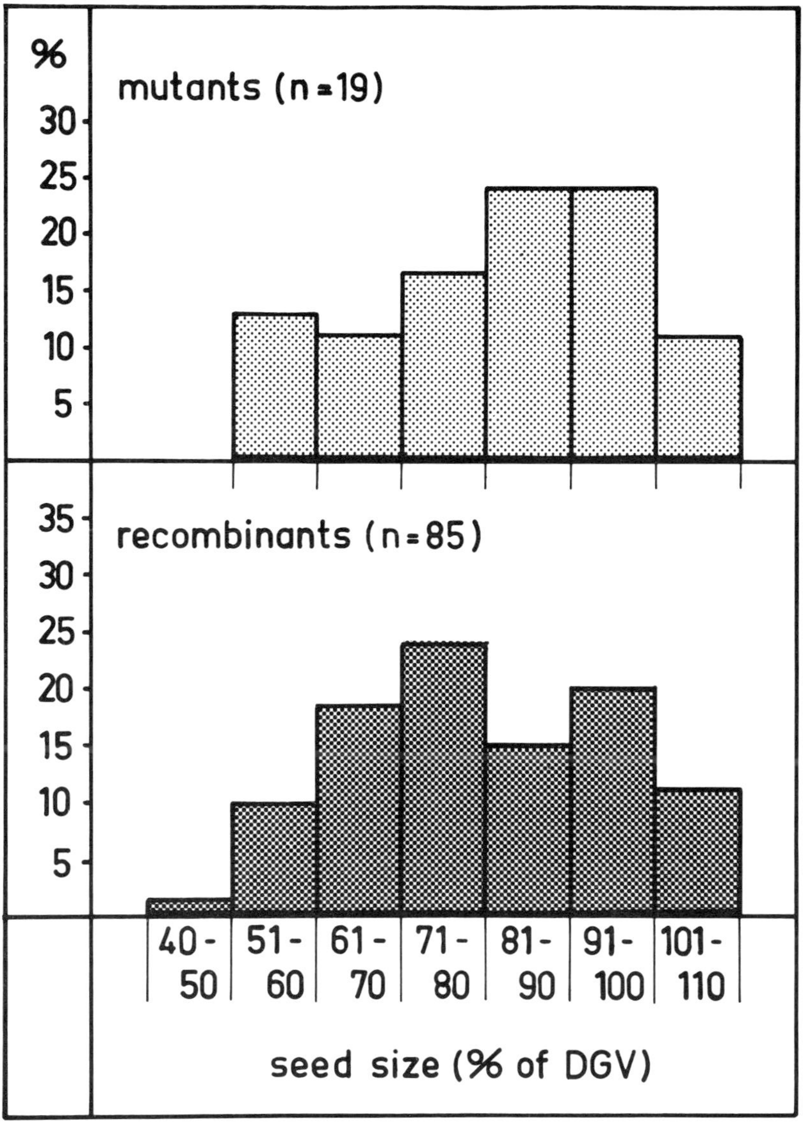

Figure 10. The seed size of the genotypes studied as related to the control values of the mother variety.
Top: 54 mean values of 19 different mutants.
Bottom: 172 mean values of 85 different recombinants.

essentially higher. This is mainly due to the low values obtained in 1980 as a consequence of the abnormal weather conditions. Many genotypes of our collection reacted to the low temperatures and the high amounts of rainfall during the ripening period by forming abnormally small seeds. The distribution of the mean values for 104 different genotypes is given in figure 10. They represent the same material which

Table 1. The seed size of 21 pea genotypes in successive generations and the protein content of the seed meal.

genotypes	thousand kernel weight (g)			protein content of the seed meal (%)		
	1978	1979	1980	1978	1979	1980
mother variety						
Dippes Gelbe Viktoria	284.35	304.22	256.81	24.38	24.15	24.75
Mutants						
37B	245.07	318.82	183.45	24.83	21.55	26.28
251A	237.19	254.03	138.88	23.26	22.30	22.24
1201A	256.00	269.40	204.47	25.30	24.65	25.24
Recombinants from crosses between different mutants						
R 177	167.75	-	142.47	25.66	-	23.34
R 406	-	325.67	267.95	-	26.65	22.99
R 416	-	313.55	209.47	-	25.18	25.82
R 441	-	256.75	195.43	-	24.96	24.02
R 624	226.88	232.13	173.95	24.87	24.50	24.26
R 853	211.48	-	172.81	24.15	-	20.38
R 863B	232.30	238.80	185.05	21.26	23.11	21.74
R 864B	251.44	-	177.51	23.22	-	24.34
Recombinants from crosses between mutants and recombinants						
RM 1126	208.28	226.10	172.73	22.12	24.61	22.54
RM 1127	214.98	224.73	180.84	24.34	23.41	21.74
RM 1128	186.64	192.40	157.20	25.78	26.32	24.10
RM 1129	190.35	178.14	149.23	22.95	20.98	23.01
Recombinants from crosses between different recombinants						
RR 1032	212.86	-	176.64	26.59	-	25.54
RR 1043	227.03	-	188.30	23.56	-	26.36
RR 1045	225.30	-	170.70	24.03	-	23.25
RR 1053	-	314.93	238.44	-	27.64	25.04
RR 1062	-	277.91	208.31	-	25.47	23.86

is considered in figure 9. The graph demonstrates a shift towards a reduced grain size as compared to the mother variety for most of the genotypes studied. Some detailed examples are given in Table 1. The table shows that the strong differences in the seed size in successive generations are not combined with corresponding alterations of the protein content, i.e. that there are no correlations between these two characters in the material studied. This seems to be generally valid for the species *Pisum sativum*. Already a few years ago, 148 mutants of our collection were investigated in this respect [15]. The situation of another 19 mutants and 73 recombinants is graphically presented in figure 11, most of them studied in several generations. In exceptional cases, there may be a negative correlation between seed size and protein content, but also the opposite correlation was observed in a few genotypes. If we consider the whole material evaluated, no significant correlation between the two criteria was found. It would thus not be possible in *Pisum* to select protein-rich

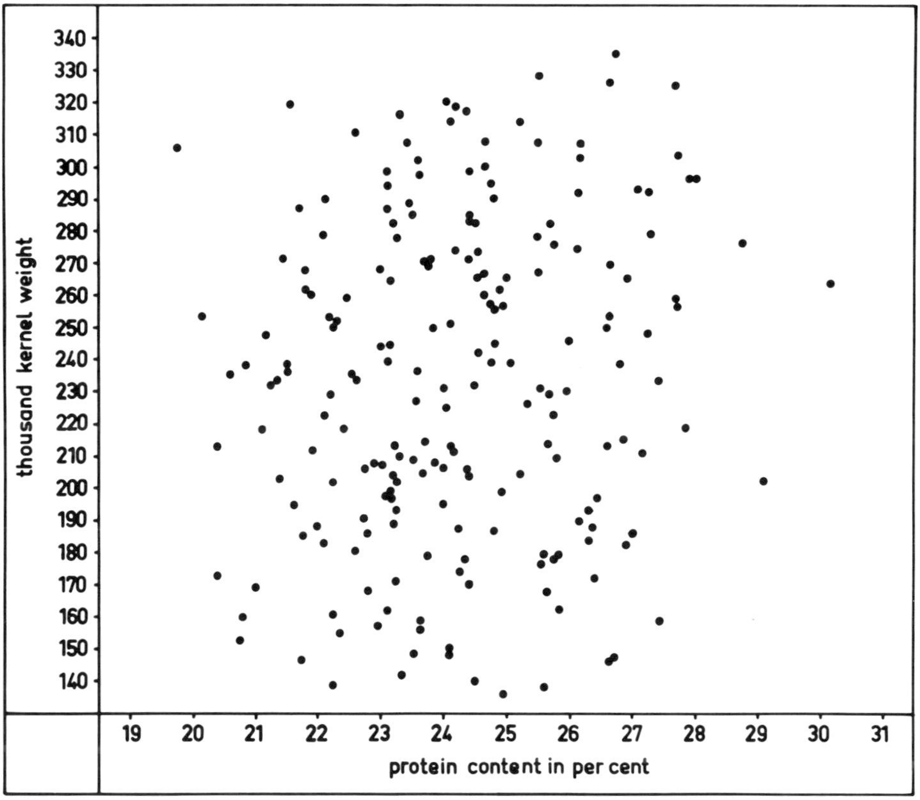

Figure 11. The relations between seed size and protein content of the seed meal in 19 *Pisum* mutants, 73 recombinants and the mother variety 'Dippes Gelbe Viktoria'.

genotypes by preselecting small-grained mutants and recombinants. Similar findings are available in barley [4]. In some other crops, however, a negative correlation between seed size and protein content of the seed flour was observed. This holds true for different wheat species [3, 5], for rice [19, 29] and for rye [22].

Relations between Seed Production and Protein Content of the Seed Meal

With regard to the relations between seed production and protein content, the findings obtained in different crops are not uniform. In *Vicia faba* and *Vigna radiata*, the grain protein content does not show a significant correlation to the seed yield [17, 23]. In *Hordeum vulgare* and *Oryza sativa*, however, a negative correlation between the two criteria was observed [19, 28]. *Pisum sativum* seems to belong to the second group. Many pea varieties and strains show a negative correlation between seed production and protein content of the seed flour [2, 18]. The pea plant seems to be only capable to produce a certain maximum amount of seed proteins. If the number of seeds per plant is small, their protein content can be relatively high whereas it is lower in high-yielding genotypes. Such a correlation was also found in some mutants and recombinants of our collection [30]. Certain findings, however, show that these relations are not so simple. If they would be exclusively due to the physiological capacity of the pea plant in general, this negative correlation should already become visible within the same genotype. This is, however, not the case. We have investigated material of 12 generations of the variety 'Dippes Gelbe Viktoria', the mother variety of our mutants, in this respect, but we did not find a significant negative correlation between the two criteria (see figure 10 of GOTTSCHALK and WOLFF's paper in the present book). The same situation was found in the fasciated mutant 489C studied over a period of 12 generations in this respect. It is one of the highest yielding genotypes of our collection, but the mean values for the trait 'protein content of the seed meal' do not differ significantly from the corresponding means of the initial line.

If we, however, consider the values of a great number of different genotypes, a slight but statistically significant negative correlation between number of seeds per plant and protein content of the seed flour becomes discernible (figure 12). As this correlation does not appear within the same genotype, it cannot only have physiological causes. On the contrary, it seems to be to some extent genetically con-

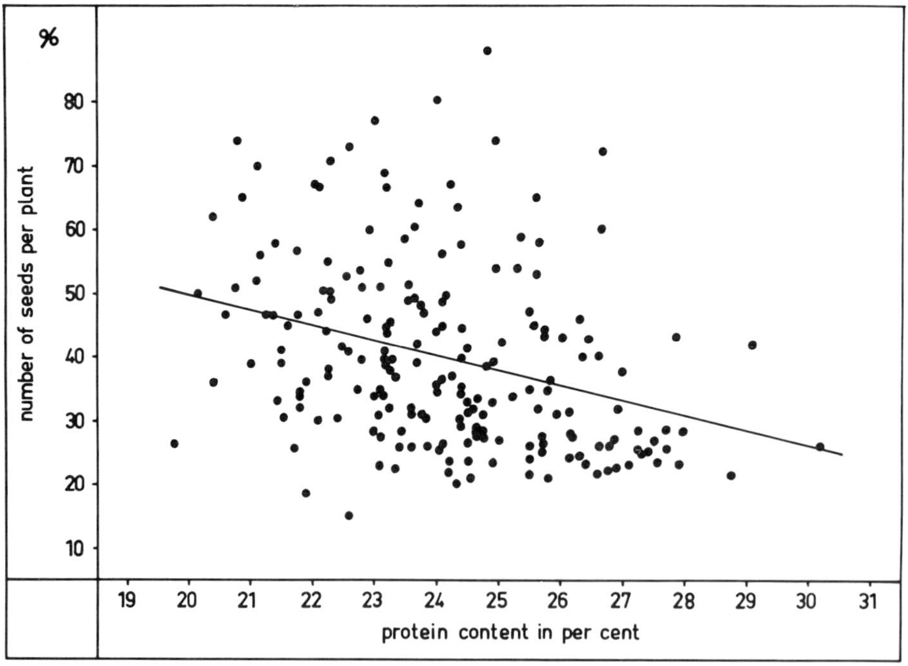

Figure 12. The relations between seed production and protein content of the seed meal in the mutant material considered in figure 11.

ditioned. Because of manifold interactions between different environmental factors and genes controlling or influencing the seed protein production, it is very difficult to discern these genes in mutation treatments and to select mutants or recombinants homozygous for them. Gene *ipc* of the *Pisum* genome seems to belong to this group; it increases the protein content of the seed meal by about 20% [13]. Mutant 189 of our collection carrying this gene, however, is very unreliable in its seed production in successive generations. Thus, the high protein content of its seed flour does not result in an increased protein production per plant because of the low seed set of the mutant in most generations tested so far. This holds also true for the small number of recombinants homozygous for *ipc* available so far. But this does not mean that it is generally not possible to combine *ipc* with distinct genes for high seed production. This programme is just at work. High mean values for the protein content of the seed meal were also found in some other mutants of our collection, most of them showing a very poor seed production [15]. It is not yet clear whether this is merely the negative correlation between the two criteria just mentioned or whether the increased protein content is genetically conditioned at least in some of these genotypes.

CONCLUDING REMARKS

Our investigations have shown that it is theoretically possible to improve the protein production per plant by means of two different methods:

by improving the seed production per plant;

by improving the protein content of the seed meal.

The first method is much better suited for reaching this aim than the second one. In mutation treatments, the first method can be used in two different ways as follows:

Mutants with higher yield can be selected.

Recombinants can be produced outyielding not only the mother variety but also their parental mutants with regard to the seed production.

Both these ways have been used in our treatments. An important factor in this concern is the seed size. A genetically reduced seed size considerably reduces the seed weight per plant of a genotype. In some crops such as rice [19, 29], wheat [3, 5] and rye [22] this negative effect is compensated to some extent by an increased protein content of the small grains as compared to genotypes with bigger seeds. This is, however, not the case in peas. Unfortunately, the highest yielding mutants of our collection – a group of fasciated genotypes – contain a gene for reduced kernel size. In these particular cases, this unfavourable situation is fully compensated by a very high seed production due to a considerably increased number of pods per plant as a result of the apical stem fasciation. Moreover, it is possible to replace the gene for small grains by the dominant allele for normal seed size or even by a mutant gene causing an increased seed size. In this way, genotypes with high protein production were obtained.

The fasciated *Pisum* mutants just mentioned are not only of interest because of their productivity. Their most important genetic peculiarity consists in their complicated genotypic constitution, each of them containing at least 16 different mutant genes. Thus, they are ideal partners for hybridizations. If they are crossed with non-fasciated mutants having characters of agronomic interest, hybrids arise heterozygous for a large number of different gene pairs. They give rise to a great variety of recombinant types in which agronomically useful characters are combined in different ways. Many of our recombinant lines have been produced in this way, some of them showing a protein production essentially higher than that of the mother variety.

A very small number of mutants of our collection showed regularly an increased protein content of the seed meal in all the generations tested so far. Their common feature is the low seed production. Therefore, a negative correlation between these two traits cannot be excluded in these genotypes. Intensive investigations of a mutant of this group and of its recombinants have shown that the strongly increased protein content is obviously genetically conditioned in that particular case (see the following paper by GOTTSCHALK and WOLFF). Theoretically, it should be possible to combine this favourable 'protein gene' with genes for normal or even increased seed production by means of recombination. In this way, high-yielding genotypes with improved protein content could be produced. So far, we did not yet succeed in selecting such a genotype, but a large number of recombinants of different genotypic constitution is available, arisen after having crossed the protein-rich mutant with high-yielding genotypes of our collection. They will be studied with regard to their protein production.

It is probable that some of the mutant genes present in our collection cause an insignificant increase of the protein content of the seed flour. It is very difficult to discern these 'protein genes' because of the overlapping effect of the respective genes and distinct environmental factors with regard to the protein production during seed development. Moreover, we have to clarify whether the abnormally low protein content of some mutants and recombinants is genetically conditioned.

REFERENCES

1. Bandel, G and W Gottschalk (1978): Recombinants from crosses between fasciated and non-fasciated pea mutants. II. Late flowering recombinants. Z. Pflanzenzüchtung 81: 60–76.
2. Blixt, S (1979): Natural and induced variability for seed protein in temperate legumes. Seed Protein Improvement by Nuclear Techniques I: 3–21; IAEA Vienna.
3. Dhaliwal, HS (1977): Genetic variability and improvement of seed proteins in wheat. Theor. Appl. Genet. 51: 71–79.
4. Doll, H (1972): Variation in protein quantity and quality induced in barley by EMS treatment. Induced Mutations and Plant Improvement: 331–342; IAEA Vienna.

5. Dumanović, J, L Ehrenberg and M Denić (1970): Induced variation of protein content and composition in hexaploid wheat. Improving Plant Protein by Nuclear Techniques: 107–120; IAEA Vienna.
6. Gottschalk, W (1975): Investigations on the co-operation of mutated genes. I. Polymeric genes. Egypt. J. Genet. Cytol. 4: 336–344.
7. Gottschalk, W (1977): Fasciated peas – Unusual mutants for breeding and research. J. Nuclear Agric. Biol. 6: 27–33.
8. Gottschalk, W (1978): The dependence of the penetrance of mutant genes on environment and genotypic background. Genetica 49: 21–29.
9. Gottschalk, W (1981a): Genetic constitution of seven fasciated pea mutants. A mutator gene in *Pisum?* Pulse Crops Newsletter 1: 54–55.
10. Gottschalk, W (1981b): Induced mutations in gene-ecological studies. Induced Mutations – a Tool in Plant Breeding: 411–436; IAEA Vienna.
11. Gottschalk, W and HP Müller (1974): Quantitative and qualitative investigations on the seed proteins of mutants and recombinants of *Pisum sativum*. Theor. Appl. Genet. 45: 7–20.
12. Gottschalk, W and HP Müller (1979): The reaction of an early-flowering *Pisum* recombinant to environment and genotypic background. Seed Protein Improvement in Cereals and Grain Legumes I. 259–272; IAEA Vienna.
13. Gottschalk, W and G Wolff (1982): The behaviour of a protein-rich *Pisum* mutant in crossing experiments. In: Gottschalk, W, and HP Müller (eds.): Seed Proteins – Biochemistry, Genetics, Nutritional Value; Martinus Nijhoff, Dr. W Junk Publ., Den Haag; in press.
14. Gottschalk, W, HP Müller, and G Wolff (1975a): Relations between protein production, protein quality and environmental factors in *Pisum* mutants. Breeding for Seed Protein Improvement Using Nuclear Techniques: 105–123; IAEA Vienna.
15. Gottschalk, W, HP Müller, and G. Wolff (1975b): The genetic control of seed protein production and composition. Egypt. J. Genet. Cytol. 4: 453–468.
16. Gottschalk, W, HP Müller, and G Wolff (1976): Further investigations on the genetic control of seed protein production in *Pisum* mutants. Evaluation of Seed Protein Alterations by Mutation Breeding: 157–177; IAEA Vienna.
17. Griffiths, DW, and DA Lawes (1978): Variation in the crude protein content of field beans (*Vicia faba* L.) in relation to the possible improvement of the protein content of the crop. Euphytica 27: 487–495.
18. Jermyn, WA, and AE Slinkard (1977): Variability of percent protein and its relationship to seed yield and seed shape in peas. Legume Research 1: 33–37.
19. Kaul, MLH (1980): Seed protein variability in rice. Z. Pflanzenzüchtung 84: 302–313.
20. Müller, HP, and W Gottschalk (1973): Quantitative and qualitative situation of seed proteins in mutants and recombinants of *Pisum sativum*. Nuclear Techniques for Seed Protein Improvement: 235–253; IAEA Vienna.
21. Müller, HP and W Gottschalk (1978): Gene ecological investigations on the protein production of different *Pisum* genotypes. Seed Protein Improvement by Nuclear Techniques: 301–314; IAEA Vienna.
22. Ruebenbauer, T and S Kaleta (1980): Inheritance of protein content of kernels of inbred lines of rye. Genet. Polonica 21: 1–15.
23. Sandhu, TS, BS Bhullar, HS Cheema and AS Gill (1979): Variability and inter-relationships among grain protein, yield and yield components in mungbean. Indian J. Genet. Plant Breed 39: 480–484.
24. Scheibe, A (1954a): Der *fasciata*-Typus bei *Pisum*, seine pflanzenbauliche und züchterische Bedeutung. Z. Pflanzenzüchtung 33: 31–58.
25. Scheibe, A (1954b): Die phänophasisch bedingte Typenresistenz der Erbsensorten gegen der Erbsenwickler. Phytopathol. Z. 21: 433–448.
26. Scheibe, A (1965): Die neue Mähdrusch-Futtererbse 'Ornamenta'. Saatgutwirtschaft 17: 116–117.

27. Scheibe, A (1968): Der *fasciata*-Erbsentypus im Rahmen der Sortenanerkennung. Saatgutwirtschaft 20: 126–128.
28. Scholz, F (1972): Induced high protein mutants of barley. Problems in breeding for protein content. Proc. Symp. Breeding and Productivity of Barley; Kroměřiž: 255–265.
29. Tanaka, S and Y Takagi (1970): Protein content of rice mutants. Improving Plant Protein by Nuclear Techniques: 55–62; IAEA Vienna.
30. Wenzel, U (1981): Untersuchungen über die Samenproteine sowie deren Fraktionen bei verbänderten Genotypen von *Pisum sativum*. Thesis Bonn; 91 pp.

13. The Behaviour of a Protein-Rich *Pisum* Mutant in Crossing Experiments

W. GOTTSCHALK, G. WOLFF

Abstract A low-yielding *Pisum* mutant with increased protein content of the seed meal was crossed with a high-yielding fasciated mutant with normal protein content. Because of the large number of mutant genes involved (about 20), highly heterozygous F_1 hybrids arose giving rise to complicated segregations in their progenies. Twenty four recombinant lines were studied with regard to their protein production. A mutant gene, designated as *ipc*, was found to be responsible for the increased protein content. Moreover, genes for green flower colour, lateness and tallness are present in the mutant.

In five recombinants, the protein production per plant exceeded that of the high-yielding parental mutant and the mother variety considerably. In four of them, this was due to the strongly increased seed production; gene *ipc* is obviously not present in their genomes. The fifth one contains *ipc* contributing to the favourable protein yield. In three other recombinants, *ipc* is present, but it does hardly improve their protein production because of the extremely low seed yield.

INTRODUCTION

During the past ten years, the attempt was made to improve the seed proteins of many crops quantitatively and qualitatively by means of induced mutations. The intensity of this research becomes visible from the findings presented in 7 protein symposia organized by the International Atomic Energy Agency [19–25]. In spite of the tremendously high number of mutants analyzed so far, only a few genes were isolated which influence the seed proteins directly. The classical examples are the *opaque-2* and *floury-2* genes of maize [30, 33]. The *lyse-1* gene (formerly hily) of an Ethiopian barley line is in its action obviously parallel to the two maize genes just mentioned [31, 32]. Gene *lys-2* of the barley mutant 1508, isolated after ethyleneimine treatment [7], and gene *hl* of grain sorghum [1] belong to the same category. The outstanding character of the respective genotypes is the qualitative alteration of their seed proteins in that way that the lysine content is increased. Mutants of this group are only known so far in cereals but not in legumes. In addition, a relatively large number of mutants with quantitative alterations of their seed proteins is available. Examples for an increased protein content of the seed flour are known in rice [46], barley [8, 37], wheat [3], soybean [2] and *Phaseolus vulgaris* [5] amongst others. Many details on these findings are given in the symposium volumes of the International Atomic Energy Agency initially mentioned. In all these cases, the genetic basis of the quantitative alterations of the seed proteins has not yet been analyzed in an exact way. The results and difficulties in studying these problems are discussed in some contributions of the present book.

Some mutants and recombinants of our *Pisum* collection were found to have an increased protein content in their seed meal, but they have likewise not yet been analyzed genetically in this respect [10, 15–17, 35, 47]. Studies of this problem were begun by crossing a protein-rich mutant with a high-yielding genotype and by analyzing a large number of different recombinants. The findings obtained can be interpreted in that way, that a specific gene is responsible for the increased protein content. Its behaviour in the parental mutant and in some recombinants is discussed in the present paper.

MATERIAL AND METHODS

The X-ray induced mutant 189 of the variety 'Dippes Gelbe Viktoria' (DGV) of *Pisum sativum*, containing genes for green flower colour, lateness and high protein content of the seed meal, was crossed with the fasciated mutant 489C homozygous for at least 16 mutant genes. In the F_2 to F_4 generations, more than 50 recombinant types were selected and developed into pure lines. In 1980, twenty four of them were grown together with their parental mutants and the mother variety DGV in the form of several replications with 25 plants per replication. In this material, the various traits of the two mutants are combined in different ways. The following characters are involved:

- 2 flower colours
- 4 different kinds of stem fasciation
- dichotomous stem bifurcation
- 6 different plant heights
- 3 flowering and ripening times
- 2 seed sizes
- strongly diverging seed productions.

The following criteria were evaluated in detail:

- plant height
- number of seeds per plant
- thousand kernel weight
- seed weight per plant
- protein content of the seed meal
- protein production per plant.

The determination of the seed protein content was carried out by means of the Kjeldahl method. Thirty fully ripe seeds of each genotype were milled. One gram of this material was used for analysis after the seed meal had been dried for 3 days at 105°C. The protein content was calculated by multiplying the nitrogen value by 6.25. For the determination of the amino acid composition, 500 mg of the seed meal were hydrolysed in 6n HCl under nitrogen [29]. The amino acid composition of the seed proteins and the amounts of the single amino acids were determined by means of an amino acid analyzer of the firm Optica following the method of SPACKMAN *et al.* [43].

RESULTS

From the small group of *Pisum* genotypes with altered amounts of seed proteins, available in our collection, mutant 189 is under study since several years. The plants showed a clearly increased protein content of the seed flour in all generations tested so far. It was therefore of interest to clarify the genetic constitution of this mutant by means of crosses and to study the behaviour of the 'protein gene' in different gene combinations.

The Characteristics of Mutant 189

Mutant 189 of our collection belongs to the group of flower mutants. The plants have flowers with greenish corollas due to a low amount of chlorophyll in the petals. Moreover, the shape of the corolla deviates slightly from normal pea blossoms showing the tendency of being 'open'. In these flowers, the keel does not enclose all the stamens present thus favouring natural cross-fertilization of the normally cleistogamous pea plant. The plants are tall, vigorous and otherwise do not differ phenotypically from the parental variety. They enter the flowering period later than the initial line, their seeds ripening with considerable delay. Many of the flowers formed do not show further development; they dry and are dropped. The recessive gene responsible for these deviations is designated as *gf* (= green flowers).

A characteristic peculiarity of the mutant is the extraordinarily broad variation of its seed production. This is valid for both different generations as well as different replications of the same generation grown at the same location. Details concerning 17 generations studied are given in figure 1. Most of the mean values obtained are far below the control values of the mother variety due to a strong reduction of the number of seeds per pod whereas the number of pods per plant is often increased. In M_{17}/1976, the mean value for the character 'number of seeds per plant' was only 0.16 (= 1.1% of the corresponding value of the initial line). In 1977, more than one hundred plants of this genotype were grown. They were healthy and vigorous but they did not produce any seeds. The opposite situation was observed in M_{19}/1978 when the mutant surpassed its parental variety by about 50% with regard to its productivity. This behaviour shows that the mutant reacts very susceptible to the environment in a way not yet known in detail.

The relations between different yielding criteria of the mutant are graphically presented in figure 4 concerning data of three or four replications of the M_{21}/1980. In summer 1980, the seed size of mutant 189 was slightly reduced in comparison to that of the initial line whereas a strong reduction of the number of seeds per plant was observed. These two negative factors resulted in very low mean values for the trait 'seed weight per plant' ranging between 35 and 53% of the control values.

A special characteristic of the mutant is the increased protein content of its seed meal. Values of the mutant and the initial line are given in figure 2 for 8 generations.

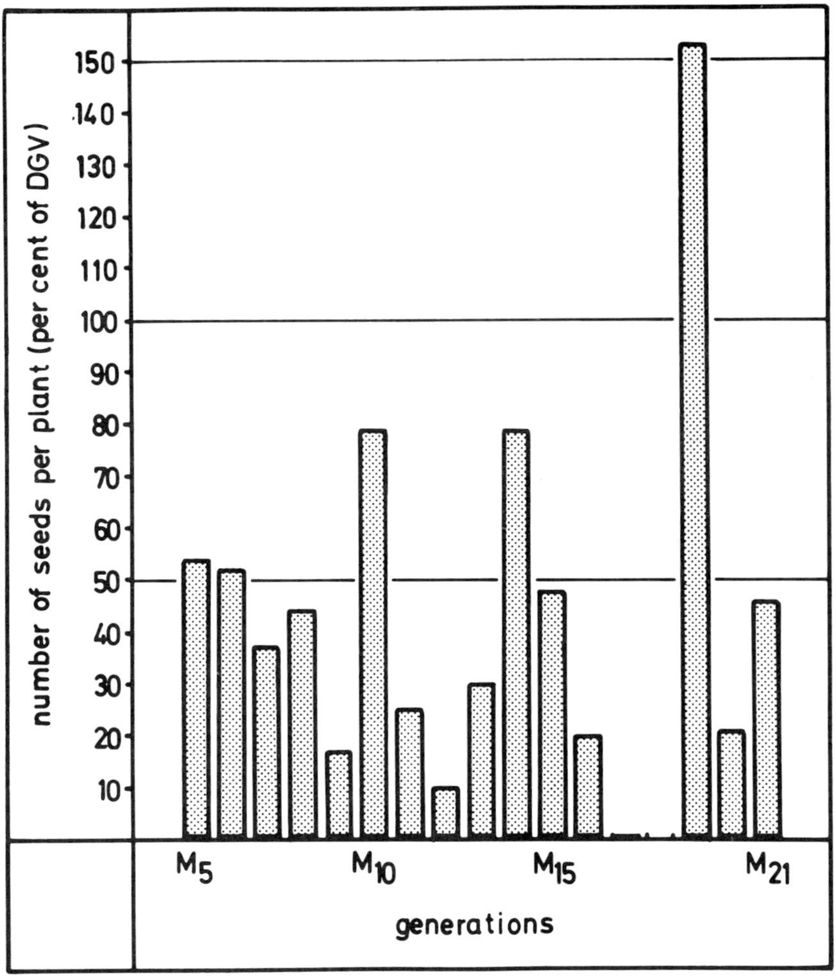

Figure 1. The seed production of mutant 189 in 17 generations as related to the corresponding values of the initial line = 100%.

In the mother variety, the mean values ranged between 22.6 and 25.5%. The corresponding values of the mutant ranged between 23.8 and 30.3%. The differences between the two genotypes become especially clear by comparing the mean values for the total material analyzed (24.05 and 28.58, respectively). Thus, the protein content of the seeds of the mutant is almost 20% higher than that of the initial line. The value of mutant 189 for 1971 was exceptionally low for reasons not known to us (23.80). The corresponding value of the mother variety was even lower (23.10). Thus, the difference between the two genotypes becomes also discernible in this generation.

The seed proteins of the two genotypes were analyzed with regard to their amino

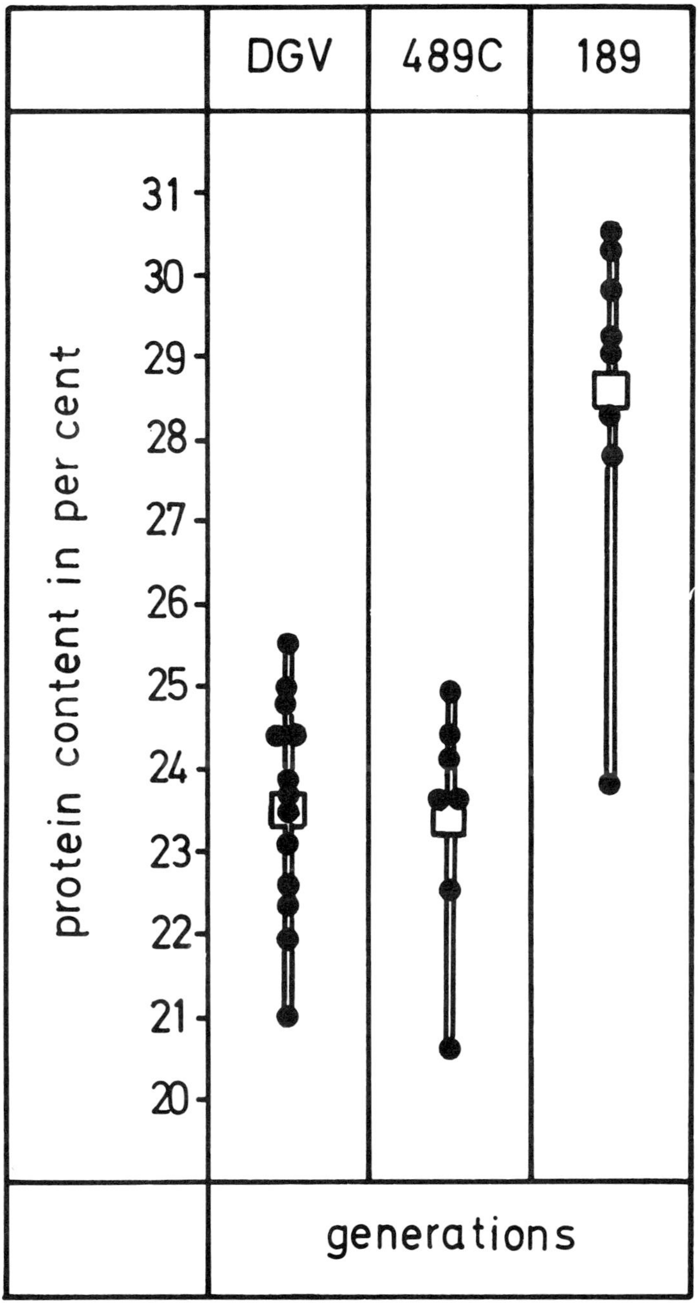

Figure 2. Comparison of the protein content of the seed meal of mutants 189 and 489C with the corresponding values of their mother variety 'Dippes Gelbe Viktoria' (DGV). Each dot gives the mean value for one generation. The squares represent the means for the total material analyzed.

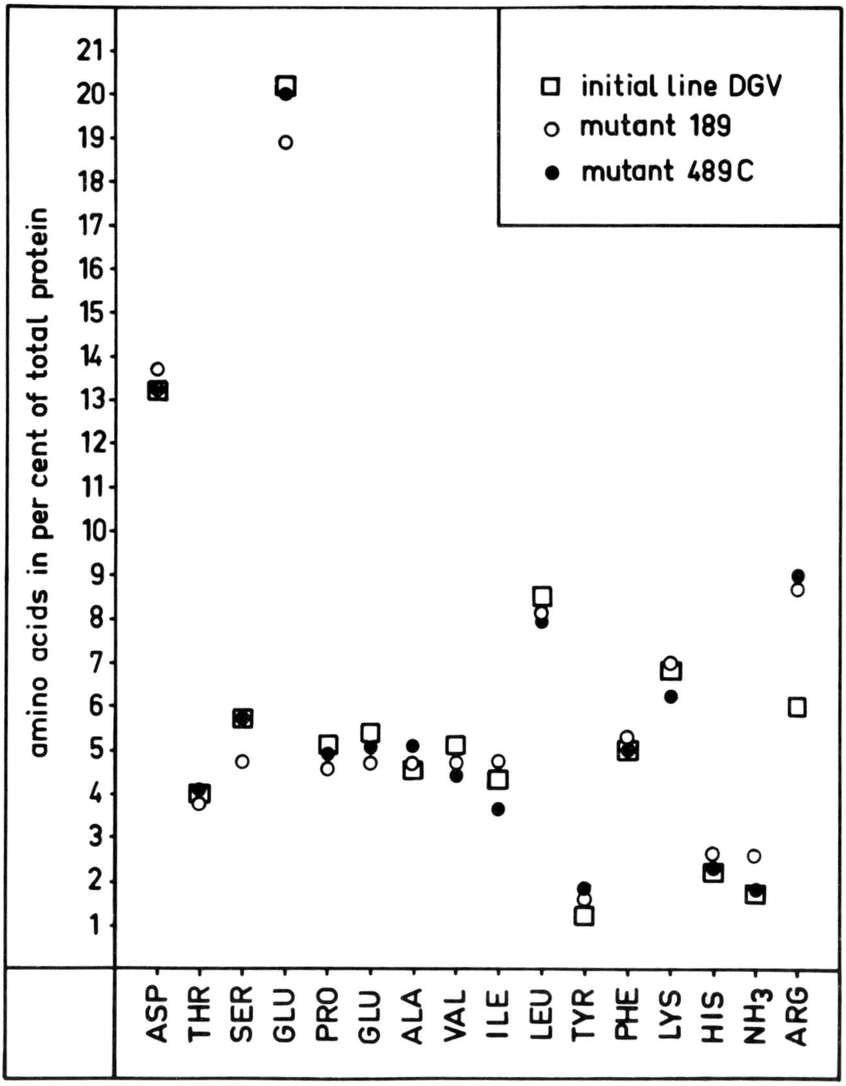

Figure 3. The amino acid composition of the seed proteins of mutants 189, 489C and their mother variety 'Dippes Gelbe Viktoria'.

acid composition in order to see whether the quantitative alterations are combined with qualitative changes. This is not the case. Figure 3 shows that the proportion of the single amino acids in the seed proteins is principally equal in mutant 189 and its mother variety.

From the values for the seed weight per plant and the protein content of the seed meal, the mean values for the character 'protein production per plant' can be calculated. This is shown in figure 4 for three replications of the M_{21}/1980. In spite

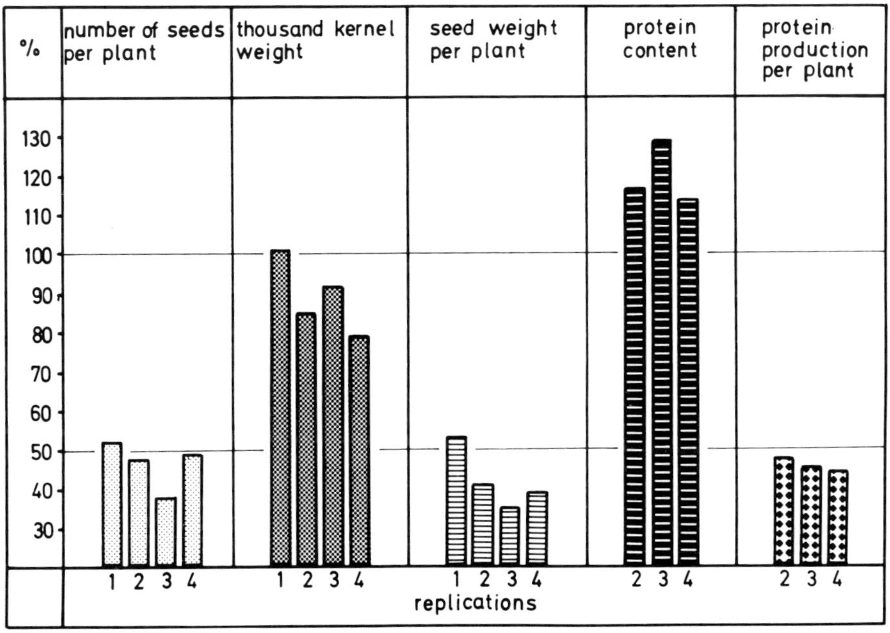

Figure 4. Different yielding criteria of mutant 189 as related to the control values of the initial line = 100%. Each column represents the mean value for one replication grown in M_{21}/1980.

of the high protein content of the seed flour, the protein production per plant is extraordinarily low as a consequence of the strongly reduced seed production of the mutant. Thus, the favourable protein situation of mutant 189 does hardly influence its negative breeding value. Therefore, it was tried to transfer the increased protein content into genotypes with higher seed production.

If we assume that only gene *gf* is present in mutant 189, causing all the characters described, the increased protein content would be part of the pleiotropic pattern of *gf*. But it is also conceivable that the high protein content is caused by a second gene acting independently from *gf*. Finally, it is possible that the increased protein content is a consequence of the reduced seed production of the mutant thus being not directly conditioned genetically. It is very difficult to decide which of these three possibilities is realized in the mutant. This problem will be discussed in connection with the recombinants obtained in crossing experiments.

The Characteristics of Mutant 489C

The plants of mutant 489C, the second partner of our crosses, are strongly fasciated in the upper part of the stem. Their apical internodes are strongly shortened and all the flowers and pods are accumulated in the top region of the plants. They are taller and later than the initial line. Their seed size is reduced due to the action of gene *sg* of the genome. As a consequence of the stem fasciation, the number of flowers and pods per plant is considerably increased resulting in high mean values for the trait 'number of seeds per plant'. They ranged between 125 and 213% of the control values of the mother variety considering 14 generations. This productivity is negatively influenced by the reduced seed size. Nevertheless, the mean values for the character 'seed weight per plant' were higher than the control values in most of the generations tested so far.

The genetic peculiarity of the mutant consists in being homozygous for at least 16 different mutant genes which have mutated in the embryonic growing point giving rise to mutant 489C [11, 13]. Most of them are not discernible in the mutant because of being hypostatic. This holds true for three further genes for different kinds of stem fasciation, for one gene causing stem bifurcation, for 6 genes influencing internode length, furthermore for genes controlling the flowering and ripening behaviour of the plants, their chlorophyll content and their response to the photoperiod. The hypostatic genes become only discernible in crossing experiments when the respective epistatic genes have been eliminated.

With regard to the seed proteins, mutant 489C did not differ significantly from the initial line neither quantitatively nor qualitatively in most of the generations tested. In 1980, the mutant showed a somewhat diverging behaviour obviously due to the abnormal climatic conditions in West Germany. It was grown in the form of 4 replications; the mean values for different yielding criteria are given in figures 5–8. The values for the number of seeds per plant and the protein content of the seed meal were higher, the thousand kernel weight was lower than the corresponding control values. This resulted in a protein production similar to that of the mother variety.

Because of the complicated genotypic constitution of mutant 489C, highly heterozygous hybrids arise when the mutant is crossed with other genotypes. In their offspring, complicated segregations occur and a large number of different recombinant types can be selected. That is the reason why we have used this genotype for crosses with mutant 189.

Recombinants with Green Flowers

In the F_2 to F_4 families following crosses between mutants 489C × 189, fifty six recombinant types have been selected and developed into pure lines. From 23 of them, sufficient seed material was available for growing them in the form of several replications in 1980. The mean values for some yielding criteria are given in figures 5–8. Most of the remaining recombinants have such a low seed production that they could not be included into this yielding trial. This holds also true for some recombinants which were only selected in the F_4 generation and which have not yet been propagated sufficiently.

A characterization of the recombinants considered in the present paper is given in table 1. The character 'stem fasciation' is not subdivided in the table into the various

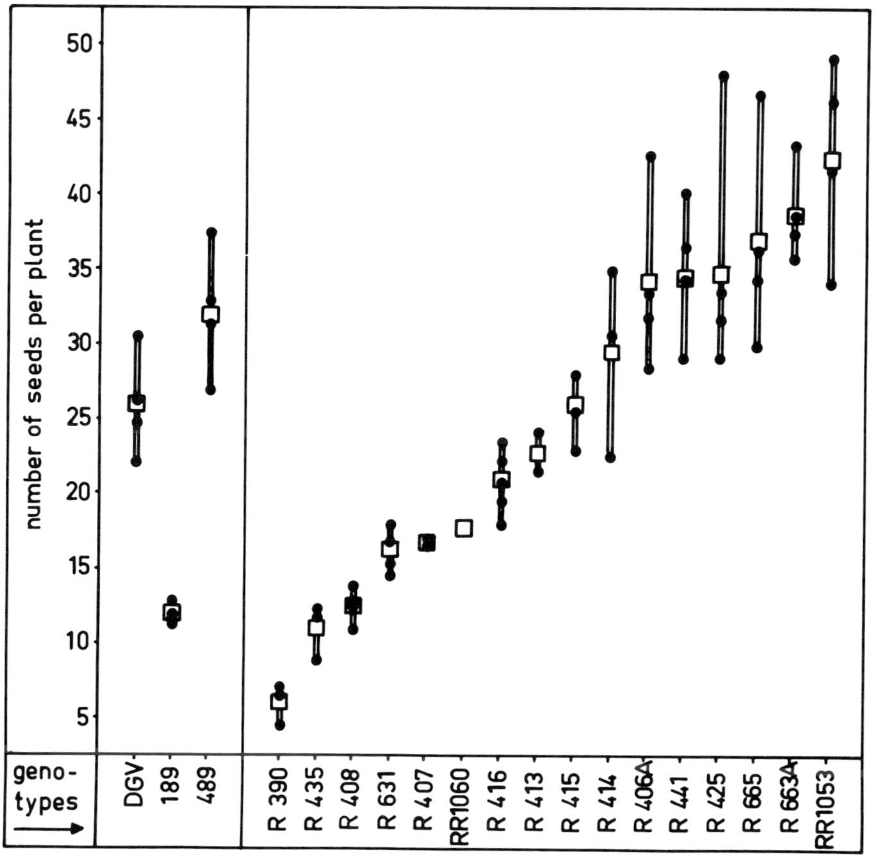

Figure 5. The number of seeds per plant of the variety 'Dippes Gelbe Viktoria', mutants 189 and 489C and of 16 recombinants homozygous for gene *gf* of 189 and for different genes derived from 489C. Each dot represents the mean value for one replication grown in 1980; the squares give the total mean of the genotypes.

types of fasciation controlled by different genes of this polygenic group. The genotypic constitution of most of the fasciated recombinants is not yet completely clear with regard to the number and the combination of the *fasciata* genes involved. The first group of the recombinants studied is homozygous for gene *gf* of mutant 189 and for groups of different genes derived from mutant 489C. While the flowers of these genotypes are greenish, those of the second group are white not possessing gene *gf*. Most of the recombinants contain at least one gene for stem fasciation. With regard to the seed production, a subdivision into three groups has been made. One of them is similar to the mother variety, the other ones are worse or better, respectively. Out of the six genes for different stem and internode lengths of mutant 489C, five are present in the material studied. Details concerning the distribution of the various characters of the different recombinant types can be seen from the table.

Table 1. Characteristics of the initial line, mutants 189, 489C and of 23 recombinant types selected after having crossed the two mutants with each other.

genotypes	green flowers	stem fascia-tion	number of seeds per plant			seed size		plant height					begin of flowering		
	(gene gf)		<20	21-30	>30	normal	small (gene sg)	short I 70-90cm	= DGV 100cm	long I 120cm	long II 140-160 cm	long III 180-230 cm	earlier than 189	= 189	later than 189
AF	-	-	-	+	-	+	-	-	+	-	-	-	+	-	-
189	+	-	+	-	-	+	-	-	-	+	-	-	-	+	-
489C	-	+	-	-	+	-	+	-	-	+	-	-	-	+	-
R 663A	+	+	-	-	+	+	-	-	-	-	-	+	+	-	-
R 665	+	+	-	-	+	+	-	-	-	-	-	+	-	+	-
R 406A	+	+	-	-	+	+	-	-	-	-	-	+	+	-	-
R 425	+	+	-	-	+	-	+	-	-	-	-	+	-	+	-
R 441	+	+	-	-	+	-	+	-	-	-	-	+	-	+	-
R 415	+	+	-	+	-	+	-	-	-	-	-	+	-	+	-
R 416	+	+	-	+	-	-	+	-	-	-	+	-	+	-	-
R 413	+	+	-	+	-	-	+	-	-	-	+	-	-	-	+
R 414	+	+	-	+	-	-	+	-	-	-	+	-	-	+	-
R 390	+	+	+	-	-	+	-	+	-	-	-	-	-	-	+
R 407	+	+	+	-	-	+	-	+	-	-	-	-	-	-	+
R 408	+	+	+	-	-	-	+	+	-	-	-	-	-	-	+
R 631	+	+	+	-	-	-	+	-	-	-	+	-	+	-	-
RR 1053	+	-	-	-	+	+	-	-	-	-	-	+	-	-	+
R 435	+	-	+	-	-	+	-	-	+	-	-	-	-	+	-
RR 1060	+	-	+	-	-	+	-	-	+	-	-	-	-	+	-
R 544	-	+	-	-	+	-	+	-	+	-	-	-			
R 492	-	+	-	-	+	-	+	-	-	-	+	-			
R 445	-	+	-	-	+	-	+	-	-	-	+	-	-	+	-
R 401	-	+	+	-	-	-	+	+	-	-	-	-	-	+	-
R 747	-	-	-	-	+	-	+	-	-	-	-	+	-	-	+
R 410	-	-	-	+	-	-	+	+	-	-	-	-	-	+	-
R 726	-	-	+	-	-	-	+	-	+	-	-	-			

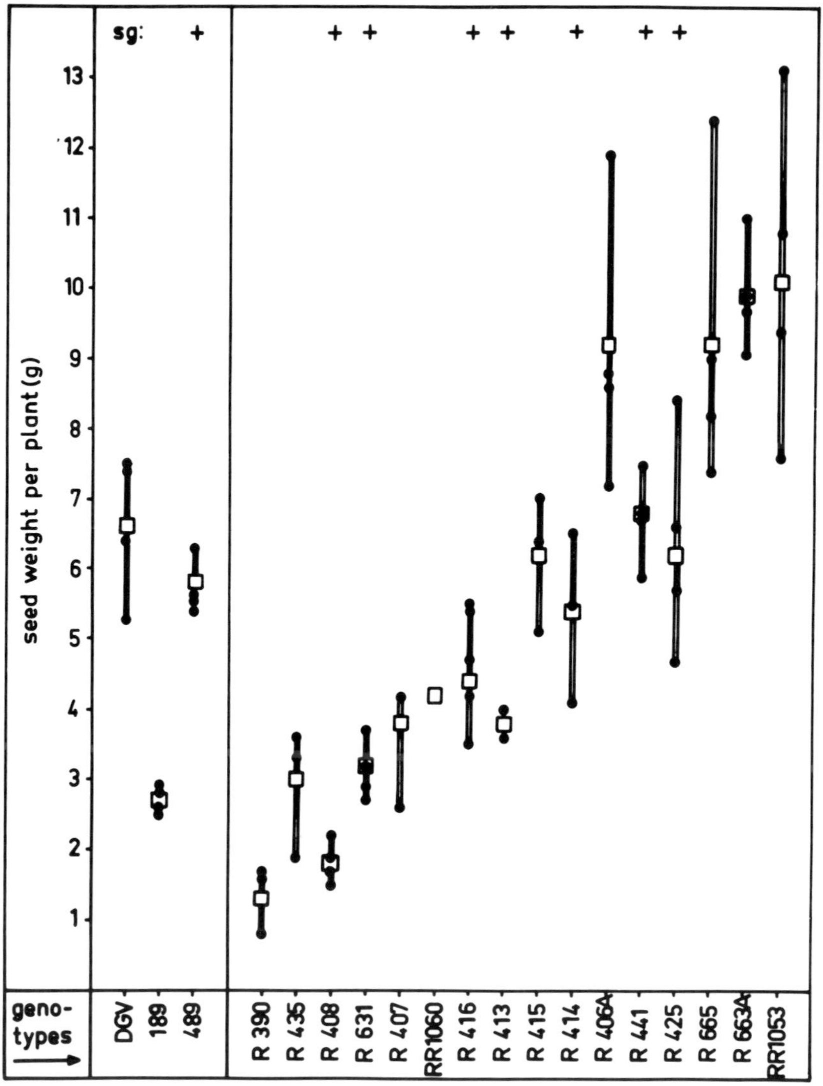

Figure 6. Mean values for the trait 'seed weight per plant' of the recombinants studied, their parental mutants and of the mother variety. Values obtained in 1980; arrangement as in figure 5.

The Number of Seeds per Plant. The seed production of the recombinants homozygous for gene *gf*, of the two parental mutants and of the mother variety 'Dippes Gelbe Viktoria' is comparatively presented in figure 5. The mean values refer to the yielding trials carried out in 1980. Mutant 189 produced only small amounts of seed the total mean reaching less than 50% of the control value of DGV whereas the fasciated mutant 489C produced about 25% more seeds than the control. The recombinants studied show a broad variation of their mean values

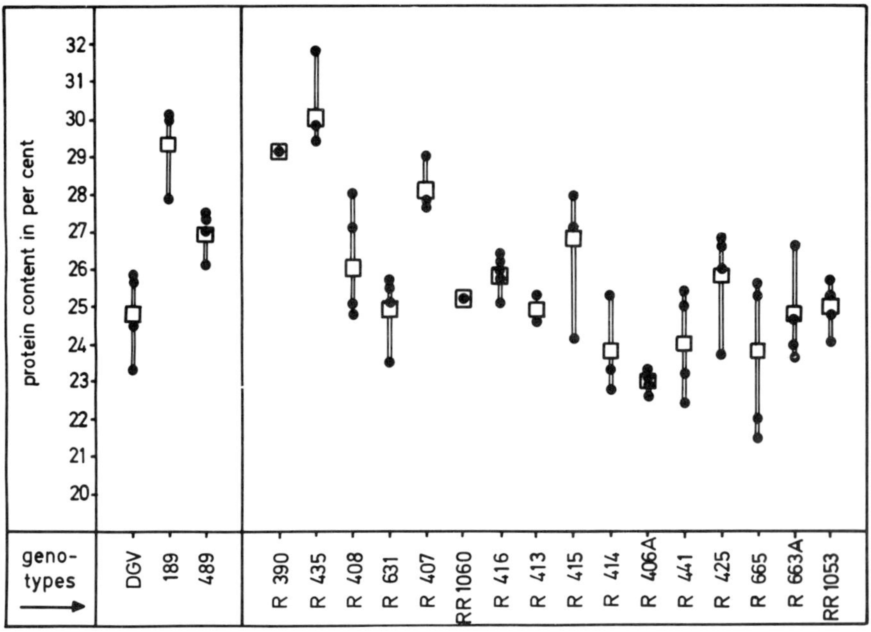

Figure 7. The protein content of the seed flour of the genotypes studied. Values obtained in 1980; arrangement as in figure 5.

ranging between 5 and 40 seeds per plant, i.e. between 23 and 164% of the corresponding values of the mother variety. Thirteen out of the 16 recombinants considered in the graph produced clearly more seeds than mutant 189. Five recombinants showed a similar yielding capacity as the second cross parent, mutant 489C. Two recombinants were even better whereas the values of the remaining ones lay between the control values of the two parental genotypes. Thus, the presence of gene *gf* in the genome does not necessarily cause low productivity. On the contrary, the combination of *gf* with specific mutant genes of 489C leads to genotypes which even surpass the high-yielding mutant 489C with regard to their productivity (R 663A, RR 1053). The high seed production of 489C is due to the gene for strong stem fasciation. The findings just mentioned, however, show that the improvement is not in all the cases caused by genes for strong or linear stem fasciation. The plants of the high-yielding recombinant R 663A contain a gene for the weakest degree of fasciation known in *Pisum*, those of RR 1053 are non-fasciated. Both these recombinants are very long. Their tallness combined with a high degree of vitality is obviously responsible for their productivity.

A similar situation was found with regard to the flowering behaviour. In 1980, the plants of mutants 189 and 489C began flowering about 10 days later than those of the mother variety. The lateness, however, is not in all the cases combined with the presence of gene *gf*. Three recombinants, homozygous for *gf*, flowered a few days earlier than mutant 189 (R 406A, R 631, R 663A). The recombinant R 416 was even similar to the mother variety with regard to its flowering and ripening behaviour thus being about 10 days earlier than mutant 189.

The Seed Weight per Plant. So far, only the number of seeds per plant was considered. If we furthermore consider the seed size, the seed weight per plant can be calculated. The fasciated mutant 489C, used for the crosses with 189, contains gene *sg* for reduced seed size. In 1980, the thousand kernel weight of 489C was 71 and 80% of the corresponding values of 'Dippes Gelbe Viktoria' or mutant 189, respectively. Out of the sixteen recombinant types homozygous for gene *gf*, seven were furthermore homozygous for gene *sg* (for details, see table 1). This gene reduces the breeding value of the respective genotypes. Its negative influence on the productivity becomes discernible by comparing figures 5 and 6. In both these figures, the genotypes are arranged in the same order and the increasing values for their productivity are visible in both the graphs for most of the recombinants studied. This agreement, however, holds not true for those genotypes which contain gene *sg*. Mutant 489C has higher values for the trait 'number of seeds per plant' than its mother variety 'Dippes Gelbe Viktoria' (figure 5), but its mean values for the character 'seed weight per plant' are lower due to the negative influence of *sg* (figure 6). A similar situation was found in most of the recombinants containing gene *sg* (see for instance R 408, R 413, R 441, R 425). In those cases, in which an increased seed number is combined with normal or even increased seed size, very high mean values for the seed weight per plant were reached. Recombinants R 406A, R 665, R 663A and RR 1053 were far more productive than their parental mutants and the mother variety (table 2). The improvement of these genotypes in relation to the high-yielding mutant 489C ranged between 60 and 70%, in relation to the low-yielding mutant 189 even between 240 and 270%.

The Protein Content of the Seed Meal and the Protein Production per Plant. It was already mentioned that the protein content of the seed meal of mutant 189 is regularly higher than that of the initial line. In 1980, the increase was 18% while it was about 9% for mutant 489C. The mean values for the single replications are given in figure 7. An important question with regard to the problems investigated consists in studying whether the increased protein content appears in all those genotypes which are homozygous for gene *gf*. This is not the case (figure 7). On the contrary, most of them do not differ significantly from the values obtained for 'Dippes Gelbe Viktoria' and mutant 489C. Out of the 16 recombinants analyzed, there are only three the protein content of which is as high as that of mutant 189 (R 390, R 435, R 407). Thus, the increased protein content of mutant 189 is not necessarily connected with the presence of gene *gf*.

From the mean values for the seed weight per plant and the protein content of the seed meal, the protein production per plant can be calculated (figure 8). Unfor-

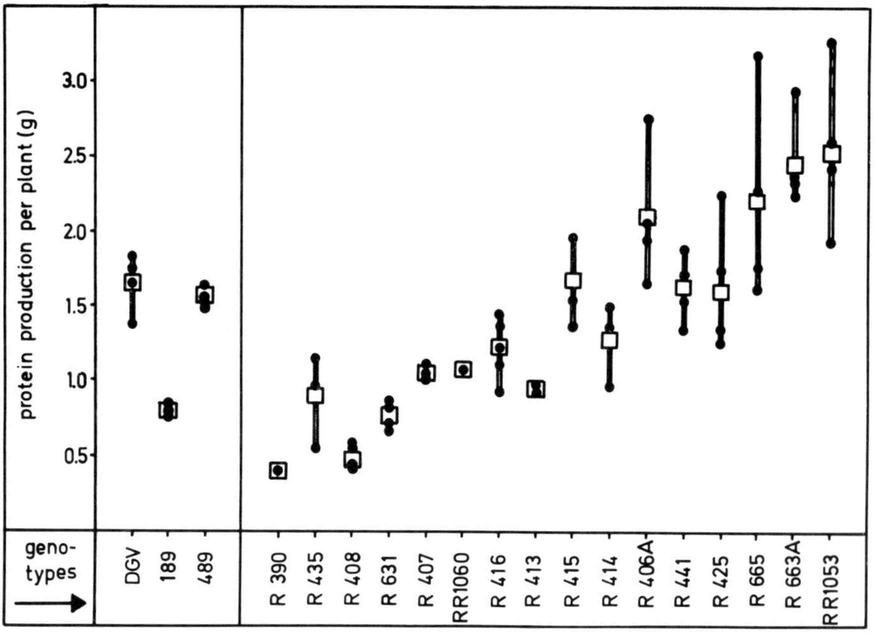

Figure 8. The protein production per plant of the genotypes studied. Values obtained in 1980; arrangement as in figure 5.

tunately, the three recombinants just mentioned are very bad in this respect. The plants of R 390 produced even less protein than those of the parental mutant 189. The mean values of recombinants R 435 and R 407 are similar to those of 189. All the three recombinants have normal seed size not containing gene *sg*; their low protein production is exclusively due to their low seed production.

It can, however, not be concluded from these findings, that the crossing experiments have only led to negative results in this concern. On the contrary, the recombinants R 406A, R 665, R 663A and RR 1053 lay about 30 to 60% over the

Table 2. Seed weight and protein production per plant of four recombinant types as related to the control values of the parental mutants and the mother variety 'Dippes Gelbe Viktoria' (DGV).

recombinant	seed production per plant as related to:			protein production per plant as related to:		
	189	489C	DGV	189	489C	DGV
R 406A	337.4%	157.1%	138.0%	264.9%	134.2%	128.2%
R 665	339.0%	157.8%	138.7%	275.9%	139.8%	133.5%
R 663A	363.8%	169.3%	148.8%	308.2%	156.2%	149.1%
RR 1053	372.3%	173.3%	152.2%	318.4%	161.3%	154.1%

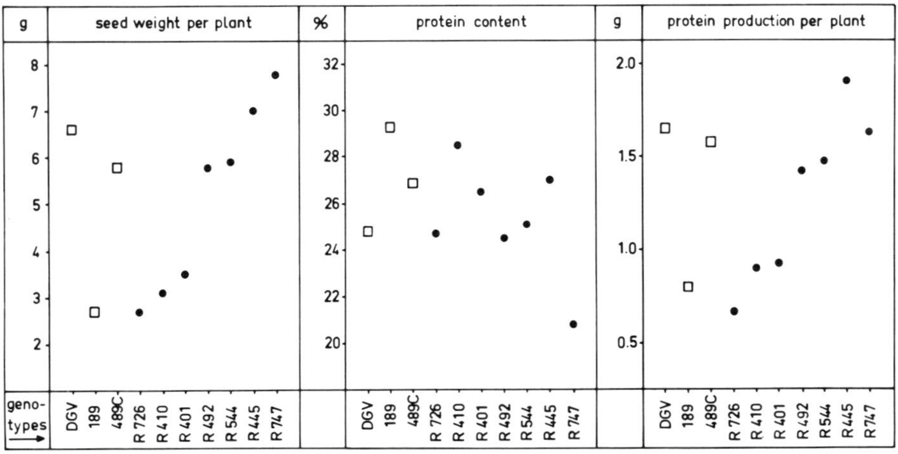

Figure 9. Seed weight per plant, protein content and protein production per plant of several white flowering recombinants, their parental mutants and the mother variety.

corresponding values of mutant 489C and 'Dippes Gelbe Viktoria'. If we relate them to mutant 189, an increase of 160 to 220% has been reached (table 2). This improvement, however, is not due to an increased protein content but exclusively to the increased number of seeds per plant. The high protein production of these genotypes as such cannot be utilized agronomically because they are not suited for field cultivation due to their strongly increased plant height. All the four genotypes contain gene *long III* for very long internodes derived as a hypostatic gene from the fasciated mutant 489C. Three of the four recombinants are fasciated. Two of them are considerably earlier than mutant 189 (R 406A, R 663A), whereas recombinant RR 1053 belongs to the latest genotypes of the whole group.

Recombinants with White Flowers

In the F_2 to F_4 generations, following crosses between mutants 489C × 189, not only green- but also some white-flowering recombinants were selected and developed into pure lines. From these genotypes, only small seed samples were available; therefore, they could not be included into our yield trials. As they are of interest for the interpretation of the findings obtained, they are briefly mentioned. They do not contain gene *gf*, but all of them are homozygous for gene *sg*. Four of them are fasciated, the other ones are non-fasciated. With regard to the plant height, they can be subdivided into four groups (for details, see table 1). Some of their yielding characters are given in figure 9. In spite of their small seeds, recombinants R 445 and R 747 have a high seed weight per plant due to the increased number of seeds per plant surpassing the control values of the parental mutants as follows:

R 445: 20.7% more than mutant 489C
159.4% more than mutant 189

R 747: 33.4% more than mutant 489C
186.7% more than mutant 189.

Of particular interest in this concern is the protein content of the seed flour. In 5 out of the 7 recombinants studied, the values lay within the range of the control values of mutant 489C and mother variety 'Dippes Gelbe Viktoria'. The protein content of the high-yielding recombinant R 747 was extraordinarily low, namely only 20.75%, i.e. 77 and 71% of the corresponding values of mutants 489C or 189, respectively. Recombinant R 410, on the other hand, had a value very close to the value of mutant 189:

protein content: 28.45 = 97.1% of mutant 189
= 105.8% of mutant 489C

With regard to the protein production per plant, the high-yielding recombinant R 747 reached only values similar to those of mutant 489C due to the very low protein content of the seed meal. The plants of the recombinant R 445, however, exceeded 489C by 21, 'Dippes Gelbe Viktoria' by 16%. This is due both to a high seed production per plant and to a relatively high protein content of the seed flour. Recombinants R 410 and R 445 will be analyzed more in detail when sufficient seed material is available.

DISCUSSION

The most interesting character of mutant 189 consists in the strongly increased protein content of its seed meal. Furthermore, some other deviating traits such as green flowers, lateness, tallness and low yield are present. It should therefore be discussed whether these characters are controlled by a single pleiotropic gene or whether several mutant genes are involved. But before discussing this question, it should be clarified whether the altered protein content of the mutant is at all genetically conditioned.

The protein production of mutant 189 and of its recombinants

For some crops, a negative correlation between seed production and protein content of the seed meal is known. Findings are available in bread wheat [3, 6], barley [37–41] and rice [45]. It is possible that such a correlation exists also in the pea. This dependence could be explained in that way, that the pea plant provides a distinct maximum amount of precursors which are used for protein production in the seeds. A large number of seeds per plant would thus automatically imply low protein content of the seed flour whereas a reduced seed production could be combined with high protein content.

We have studied this question using seed material from several generations

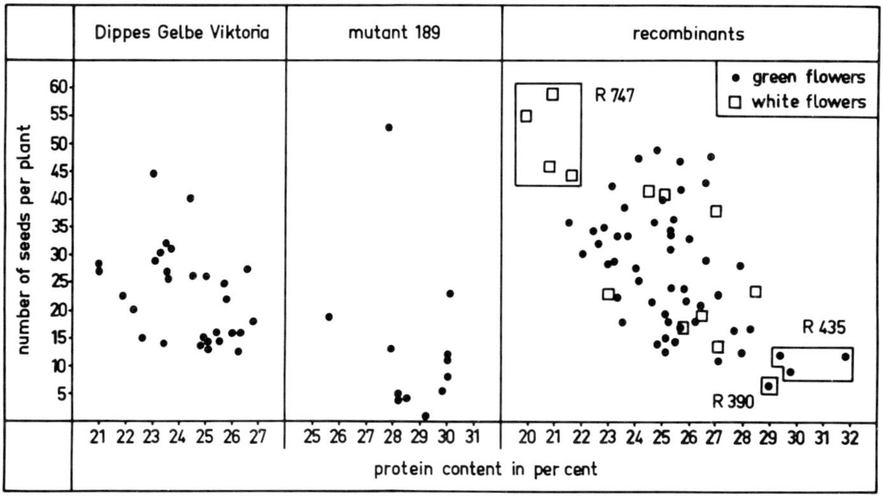

Figure 10. The relations between seed production and protein content of the seed meal in the mother variety 'Dippes Gelbe Viktoria', mutant 189 and in 24 recombinants obtained from crosses between mutants 489C × 189. The dots represent mean values either for one generation or for several replications within the same generation, respectively.

and/or several replications (figure 10). In the lefthand part of this figure, the relations between seed production and protein content in the mother variety 'Dippes Gelbe Viktoria' are presented. The distribution of the values gives no indication for any correlation between the two traits. This holds also true with regard to the values of mutant 189 (figure 10, middle part). Thus, the high protein content of this genotype is obviously not due to its low yield. This becomes particularly clear from the values obtained in 1978. In that year, the seed production of the mutant was extraordinarily high (153% of the control values of 'Dippes Gelbe Viktoria'). The protein content of these seeds, however, was not lower than that of those plants which had a poor yield. A similar situation was found in a replication grown in 1980. The seed production of the respective plants was a little higher than that of the control plants (i.e. they had an abnormally high yield as related to the average values of the mutant; see figure 1), but the protein content was one of the highest values ever found in 189. These findings could not be expected, if there were a negative correlation between seed production and the protein content of the seed meal in this genotype.

So far, the correlation within the same genotype was discussed. Similar investigations were carried out in 24 different recombinant types obtained after having crossed mutant 189 with 489C (righthand part of figure 10). The distribution of the dots seems to point to a negative correlation between the two criteria. However, it should be considered that the extreme values of this part of the graph belong to several replications of only three genotypes (R 390, R 435, R 747). If we do

genotype	pattern of characters						
	flower colour	protein content	seed production	seed size	plant height	flowering time	shoot structure
189	▼	●	■	▲	LI	L	NF
489C	▽	○	□	△	LI	L	F
R 441	▼	○	□	△	LIII	L	F
R 445	▽	●	□	△	LII	L	F
R 401	▽	○	■	△	SI	L	F
R 408	▼	○	■	△	SI	VL	F
R 631	▼	○	■	△	LII	E	F
RR1053	▼	○	□	▲	LIII	VL	NF
R 665	▼	○	□	▲	LIII	L	F
R 663A	▼	○	□	▲	LIII	E	F
RR1060	▼	○	■	▲	DGV	L	NF
R 435	▼	●	■	▲	DGV	L	NF
R 407	▼	●	■	▲	SI	VL	F

Figure 11. The combination of different characters of mutants 189 and 489C in different recombinant types.

LI, II, III:	different groups of long internodes
SI:	short internodes
F:	stem fasciated
NF:	non-fasciated
L:	late-flowering
VL:	very late flowering
E:	early flowering
DGV:	Dippes Gelbe Viktoria

not consider these recombinants, there is a random distribution of all the other dots which does not allow to interpret our findings in the sense of a negative correlation between seed yield and protein content. The three recombinants just mentioned show a specific situation with regard to the relations between the two criteria, which will be studied more in detail in further generations. From the findings available, it can be concluded that the increased protein content of mutant 189 is genetically conditioned.

The genotypic constitution of mutant 189

In agreement with many other mutants, it was initially presumed that all the anomalies of mutant 189 are caused by the pleiotropic action of a single recessive gene designated as *gf* [14]. This is obviously not the case because various traits could easily be separated from each other (figure 11). After having crossed mutant 189 with 489C, a large number of different recombinants were selected in which distinct characters of both the mutants are combined in different ways. In recombinant R 441, for instance, only the green flower colour derives from 189 whereas all the other characters are from 489C. A similar situation is valid with regard to the trait 'increased protein content' in recombinant R 445. Recombinant R 407 has the green flowers and the increased protein content of 189, but the flowering behaviour is controlled by a gene for extreme lateness deriving from 489C. Examples for further different combinations of characters from both the parental mutants are given in figure 11. It is evident that the three traits

greenish flower colour,
increased protein content of the seed meal,
lateness

of mutant 189 behave independently from each other in crosses. Thus, they cannot be caused by a single pleiotropic gene. Mutant 189 is obviously homozygous for three genes which are provisionally designated as follows:

gf (greenish flower colour),
ipc (increased protein content),
lat (lateness).

Possibly, a fourth gene causing a slight increase of internode length is involved. These genes have commonly mutated during the X-irradiation in that embryonic growing point which gave rise to the mutant. They have not yet been localized, but they cannot be closely linked because they could be separated easily from each other by crossing.

Figure 11 shows that the character 'number of seeds per plant' behaves in principle similar to the three or four other traits just mentioned. Mutant 189 had a very low seed production in most of the generations studied so far, but some of the recombinants containing genes of 189 show a very high seed yield. We have therefore to ask whether a specific gene of its genome is responsible for the reduced seed production. Investigations on the selection value of mutants genes in barley [18] and peas [9] have shown that more than 95% of all the mutants tested are not competitive to their mother varieties because of their low seed production. Similar results were obtained in many other species. As this effect appears regularly in each generation, it is certainly genetically conditioned. This should, however, not be understood in that sense that specific mutant genes are responsible for this negative reaction. On the contrary, this is obviously an unspecific action of most of the mutant genes which appears beside their specific action. This behaviour can be interpreted by assuming that the presence of a mutant gene as such disturbs the genic harmony of a given genome thus resulting in the reduced productivity observed in many mutants. If we make use of this hypothesis for interpreting the genetic constitution of mutant 189, it is not necessary to make an additional mutant gene responsible for the low seed production.

The mode of action of gene *ipc* with regard to the physiological causes of the increased protein production is not yet known. In distinct wheat and rice varieties, differences in the protein content seem to be due to differences in the translocation of nitrogen compounds from leaves to developing grains [27, 34]. It could be possible that these physiological differences are genetically conditioned. In a specific rice mutant, nitrogen compounds are deposited into the grains during a longer period than in the mother variety resulting in a higher protein content [44]. Similar relations could exist in mutant 189 of our *Pisum* collection, but investigations of this kind have not yet been made.

Recombination as an improving process

Investigations in barley [36, 37, 40, 42], rice [26, 28], wheat [4] and peas [12, 14] have shown that the selection value of mutants can be improved by recombination. Similar results were obtained in the present crossing experiments with regard to the seed yield (figure 6, tables 1, 2). Of particular interest in connection with the problems studied is the question whether the protein production could be increased as well. Theoretically, this is possible in two ways:

The seed production is increased whereas the protein content of the seed meal remains unchanged.

The protein content is increased whereas the seed production is either unchanged or improved.

Figure 8 shows that some of our recombinants have a higher protein production per plant as related to the control values of the two parental mutants and the mother variety. However, these genotypes do not have an increased protein content of the seed flour. Thus, gene *ipc* is obviously not present in their genomes (figure 7). Their increased protein production is due exclusively to the increased seed production (figures 5, 6). Those recombinants, on the other hand, which are homozygous for gene *ipc* (R 390, R 435, R 407), have a very low protein production as a consequence of a strongly reduced seed production. The three recombinants just mentioned contain not only gene *ipc* but also gene *gf* for green flowers. However, some white-flowering recombinants are available in our collection, in which the presence of *ipc* is not combined with low productivity. This holds true to some extent for recombinant R 410. Its mean value for the trait 'number of seeds per plant' lay between the control values of the two parental mutants. However, the size of these seeds was even more reduced than that of 489C resulting in a very low mean value for the seed weight per plant. The increased protein content of the seed flour is not able to compensate this negative feature. There was an improvement by 13% as related to mutant 189, but the protein production of R 410 was only 57% of that of mutant 489C. A considerably more favourable situation was realized in recombinant R 445. In 1980, the plants of this genotype produced about as many seeds as 489C. In spite of being homozygous for gene *sg* causing small seeds, its protein production per plant was about 20% higher than that of 489C due to the presence of gene *ipc*.

Our interpretations are based on the assumption that gene *ipc* is only present in those genotypes which show the strongly increased protein content. It can, however, not be excluded, that it is also present in some other recombinants in which it is unable to express its action. This could be due to negative interactions between *ipc* and specific other genes of the respective genotypes.

Because of the complicated genetic constitution of the mutants 189 and 489C, it is conceivable that further recombinant types exist in which gene *ipc* is combined with genes which do not influence the seed production negatively. They have not yet been selected in the segregating families. We shall, however, try to select as many new recombinant types as possible and to test them with regard to their protein production.

REFERENCES

1. Axtell, JD, and W Lafayette (1976): Naturally occurring and induced genotypes of high lysine sorghum. Evaluation of Seed Protein Alterations by Mutation Breeding: 45–53; IAEA Vienna.
2. Bazavluk, JM, and VB Enken (1976): Minor mutations as a method of increasing the content and improving the quality of protein in soya beans. Genetika (USSR) 12: 46–54.
3. Bhatia, CR, RM Desai, and KN Susulan (1978): Attempts to combine high lysine and increased grain protein in wheat. Seed Protein Improvement by Nuclear Techniques: 51–57; IAEA Vienna.
4. Borojević, K, and S Borojević (1972): Mutation breeding in wheat. Induced Mutations and Plant Improvement: 237–251; IAEA Vienna.
5. Crocomo, OJ, AT Neto, A Ando, S Blixt, and D Boulter (1978): Breeding for improved protein content and quality in the bean (*Phaseolus vulgaris*). II. Seed Protein Improvement by Nuclear Techniques: 207–222; IAEA Vienna.
6. Denić, M (1978): Some characteristics of proteins in mutant lines of hexaploid wheat. Seed Protein Improvement by Nuclear Techniqes: 365–381; IAEA Vienna.
7. Doll, H (1975): Genetic studies of high lysine barley mutants. Barley Genetics III: 542–546.
8. Favret, EA, R Solari, L Manghers, and A Anila (1969): Genetic control of the qualitative and

quantitative production of endosperm proteins in wheat and barley. New Approaches to Breeding for Improved Plant Protein: 87–107; IAEA Vienna.

9. Gottschalk, W (1971): Die Bedeutung der Genmutationen für die Evolution der Pflanzen. Fischer Verlag, Stuttgart: 296pp.
10. Gottschalk, W (1975): The influence of mutated genes on quantity and quality of seed proteins. Indian Agric. 19: 205–223.
11. Gottschalk, W (1977): Fasciated peas – Unusual mutants for breeding and research. J. Nuclear Agric. Biol. 6: 27–33.
12. Gottschalk, W (1979): Differential behaviour of a mutant gene in *Pisum* recombinants. Genetika (Beograd) 11: 15–28.
13. Gottschalk, W (1981): Induced mutations in gene-ecological studies. Induced Mutations – a Tool in Plant Breeding: 411–436; IAEA Vienna.
14. Gottschalk, W (1981): The improvement of radiation-induced *Pisum* mutants through recombination. (in press).
15. Gottschalk, W, HP Müller, and G Wolff (1975): The genetic control of seed protein production and composition. Egypt. J. Genet. Cytol. 4: 453–468.
16. Gottschalk W, HP Müller, and G Wolff (1975): Relations between protein production, protein quality and environmental factors in *Pisum* mutants. Breeding for Seed Protein Improvement using Nuclear Techniques: 105–123; IAEA Vienna.
17. Gottschalk, W, HP Müller, and G Wolff (1976): Further investigations on the genetic control of seed protein production in *Pisum* mutants. Evaluation of Seed Protein Alterations by Mutation Breeding: 157–177; IAEA Vienna.
18. Gustafsson, Å (1954): Mutations, viability, and population structure. Acta Agric. Scand. 4: 601–632.
19. International Atomic Energy Agency (1969): New Approaches to Breeding for Improved Plant Protein; 193pp.
20. International Atomic Energy Agency (1970): Improving Plant Protein by Nuclear Techniques; 458pp.
21. International Atomic Energy Agency (1973): Nuclear Techniques for Seed Protein Improvement; 422pp.
22. International Atomic Energy Agency (1975): Breeding for Seed Protein Improvement using Nuclear Techniques; 229pp.
23. International Atomic Energy Agency (1976): Evaluation of Seed Protein Alterations by Mutation Breeding, 216pp.
24. International Atomic Energy Agency (1978): Seed Protein Improvement by Nuclear Techniques; 582pp.
25. International Atomic Energy Agency (1979): Seed Protein Improvement in Cereals and Grain Legumes; I and II; 421 and 472pp.
26. Ismachin, M, and K Mikaelsen (1976): Early maturing mutants for rice breeding and their use in cross-breeding programmes. Induced Mutations in Cross-Breeding: 119–121; IAEA Vienna.
27. Johnson, VA, PJ Mattern, DA Whited, and JW Schmidt (1969): Breeding for high protein content and quality in wheat. New Approaches to Breeding for Improved Plant Protein: 29–40; IAEA Vienna.
28. Kawai, T (1968): Genetic studies on short-grain mutants in rice. Mutations in Plant Breeding II: 161–182; IAEA Vienna.
29. Lein, KA, K Brunkhorst, and WJ Schön (1973): Assessment of different techniques for estimation of total nitrogen and lysine in cereals. Nuclear Techniques for Seed Protein Improvement: 363–369; IAEA Vienna.
30. Mertz, ET, LS Bates, and OE Nelson (1964): Mutant gene that changes protein composition and increases the lysine content of maize endosperm. Science 145: 279–286.

31. Munck, L (1970): Increasing the nutritional value in cereal protein. Basic research on the hily character. Improving Plant Protein by Nuclear Techniques: 319–330; IAEA Vienna.
32. Munck, L, KE Karlsson, A Hagberg, and BO Eggum (1970): Gene for improved nutritional value in barley seed proteins. Science 168: 985–987.
33. Nelson, OE, E Mertz, L Bates (1965): Second mutant gene affecting the amino acid pattern of maize endosperm proteins. Science 150: 1469–1470.
34. Perez, CM, GB Cagampang, BV Esmama, RK Monserrate and BO Juliano (1973): Protein metabolism in leaves and developing grains of rices, differing in grain protein content. Plant Physiol. 51: 537–542.
35. Quednau, HD, and G Wolff (1978): Investigations of the seed protein content of several pea genotypes grown in two different years. Theor. Appl. Genet. 53: 181–190.
36. Scholz, F (1967): Utility of induced mutants of barley in hybridization. Erwin-Baur-Gedächtnisvorlesungen 4: 161–168.
37. Scholz, F (1971): Utilization of induced mutations in barley. Barley Genetics II: 94–105.
38. Scholz, F (1972): Induced high-protein mutants of barley – Problems in breeding for protein content. Proc. Symp. Breed. Product. Barley, Kroměřiž: 255–265.
39. Scholz, F (1975): Problems of breeding for high protein yield in barley. Barley Genetics III: 548–555.
40. Scholz, F (1976): Experience and opinions on using induced mutants in cross-breeding. Induced Mutations in Cross-Breeding: 5–19; IAEA Vienna.
41. Scholz, F (1976): Zur Frage der Kombination von hohem Eiweißgehalt mit hohem Kornertrag bei Gerste. Tag. Ber. Akad. Landwirtsch. Wiss. Berlin 143: 173–189.
42. Sigurbjörnsson, B (1976): The improvement of barley through induced mutation. Barley Genetics III: 84–95.
43. Spackman, DH, WH Stein, and S Moore (1958): Automatic recording apparatus for use in the chromatography of amino acids. Anal. Chem. 30: 1190–1206.
44. Tanaka, S (1978): Difference in nitrogen absorption between a radiation-induced high-protein rice mutant and its original variety. Seed Protein Improvement by Nuclear Techniques: 199–201; IAEA Vienna.
45. Tanaka, S, and S Hiraiwa (1978): Induction of high-protein mutants in rice. Seed Protein Improvement by Nuclear Techniques: 191–198; IAEA Vienna.
46. Tanaka, S, and S Tamura (1968): A short report on γ-ray induced rice mutants having high protein content. Japan. Agric. Res. Quart. 3(3): 1–4.
47. Wolff, G (1975): Quantitative Untersuchungen über den Proteingehalt in Samen von *Pisum sativum*. Z. Pflanzenzüchtung 75: 43–54.

14. Storage Proteins of Soybean

IKUNORI KOSHIYAMA

INTRODUCTION

Soybean, a native of China but now the most production in the United States, is well-known and much utilized throughout the world as an oil seed. Interest in soybeans for food uses has stimulated research and development in many laboratories. In particular, the protein constituent has been very important to our Japanese as the traditional soy foods such as soy sauce (shoyu), miso, tofu and so on. In contrast with the active application of soybean proteins for food use, the fundamental researches on soybean proteins (soybean globulins), owes much to works done during the past about twenty years. In this paper, I shall summarize the results of the biochemical investigations on soybean storage proteins, the composition, isolation and characterization of the protein constituents, their molecular structures and related problems.

THE COMPOSITION OF SOYBEAN STORAGE PROTEINS

The storage proteins of legume seeds are generally represented with the two major fractions, vicilin and legumin. The proteins have sedimentation constants of approximately 7S and 11S. Although a definition of storage protein is strictly very difficult, a recognition of storage protein for the proteins is that they are located in cotyledonous subcellular particles called 'protein bodies' which are filled with protein in addition to no biological activities.

The protein bodies of soybean seeds have been briefly isolated by sucrose density gradient centrifugation [86] and also by differential centrifugation in cottonseed oil-carbon tetrachloride mixture [75], almost simultaneously. Preparatively large particles of the protein bodies have been effectively obtained by formation of cotyledonous protoplast with cellulase, pectinase and so on prior to density gradient centrifugation [18].

(A)

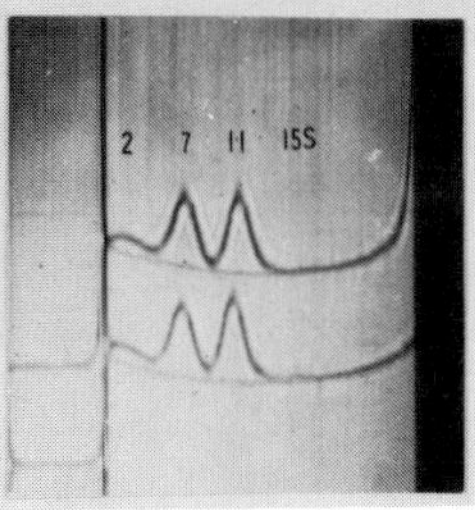

(B)

	2S	7S	11S	15S
Globulins (A)	9.4%	43.0%	43.6%	4.0%
Protein bodies (B)	8.3	42.0	44.1	4.8

Figure 1. Sedimentation patterns of soybean globulins and proteins of soybean protein bodies at 0.5 ionic strength [5].

On the other hand, soybean globulins have been generally recognized as storage proteins for a long time since Osborne and Campbell [72] separated them as the 'classical glycinin' in 1898. Soybean globulins are insoluble in the region of their isoelectric points about pH 4.5 and occupied nearly 90% of the extractable proteins with water or various aqueous solvents.

Comparison of soybean globulins with the proteins found in the protein bodies revealed no significant differences as measured by gel filtration, ultracentrifugation and isoelectric focusing [50]. Figure 1 shows the ultracentrifugal patterns of soybean globulins and the proteins in the protein bodies. Both are composed of two major ultracentrifugal fractions, 7S and 11S, in addition to the minor fractions, 2S and 15S, at 0.5 ionic strength. But a portion of the 7S fraction aggregates to form a 9S peak at 0.1 ionic strength. Accordingly, there are two kinds of ultracentrifugally different 7S components as a basis of the association reaction with the change of ionic strength.

In 1968, immunological techniques were applied to the characterization of soybean storage protein components by Catsimpoolas et al. [7, 12, 13]. They found that the protein bodies contained six different antigenic components by disc immunoelectrophoresis and four by immunoelectrophoresis. The same results were also obtained from soybean globulins [9]. Ultracentrifugal and immunological compositions of soybean globulins are shown in Table 1. They are comparatively simple and the proportions of each component determined by both methods are parallel.

However, it has been recently shown that the ultracentrifugally and immunologically simple two major components of soybean storage proteins, the 7S (β-ConGl) and the 11S globulin (Gl), are not always composed of a single protein component by separating techniques based on electric property of protein. Although ultracentrifuge still remains a powerful analytical tool, the powers of resolution are inferior to those of electrophoresis.

Soybean proteins are referred to most commonly by sedimentation constants. But Catsimpoolas and Ekenstam [9] proposed the new names to them. Glycinin (Gl) is identical with the 11S protein, β- (β-ConGl) and γ-conglycinin (γ-ConGl) with the 7S proteins and α-conglycinin (α-ConGl) is a 2S protein.

Table 1. Content of immunological and ultracentrifugal components in soybean globulins

	Immunological content [56]	Ultracentrifugal content [47]
2S (α-ConGl)	13.8%	15.0%
7S	30.9	34.0
(γ-ConGl)	3.0	3.1
(β-ConGl)	27.9	30.9
11S	40.0	41.9
15S	–	9.1

At first, the preliminary demonstration of the heterogeneity for the 7S globulin fraction was done [23, 27]. Afterwards, the globulin capable of association (conversion into a 9S form) at 0.1 ionic strength separated at least three bands on polyacrylamide discontinuous gels and the three bands were fractionated by sucrose density gradient sedimentation at 0.5 ionic strength [29]. Whereas the 7S globulin is unstable against the change of ionic strength, it is a surprising observation that the 9S fraction still consisted of three gel electrophoretic compounds when the centrifugation was conducted at 0.1 ionic strength [29]. Recently, the existence of the three 7S globulin isomers capable of the association reaction with the change of ionic strength were also confirmed [30]. In particular, the three bands estimated to be the 7S globulin by disc electrophoresis are distinguishable in the fresh soybean globulin preparation [59]. These distinguishable three bands change into diffuse broad bands in disc electrophoresis when the preparation is stored in the cold even at 0.5 ionic strength. Consequently, the proteins are unstable.

Moreover, the complication is that the six isomers of β-ConGl (the 7S globulin) (B_1 to B_6-conglycinins) were isolated by using a DEAE-Sephadex A-50 column [78, 79, 81]. In brief, it is possible that the 7S globulin (β-ConGl) which is only an ultracentrifugally and immunologically single component is apparently electrophoretically further separable.

As suggested previously, there is also a 7S protein that does not have an ability of the association reaction at 0.1 ionic strength in the ultracentrifugal 7S fraction at 0.5 ionic strength. The 7S protein is, as mentioned afterwards, γ-ConGl [56]. Gamma-conglycinin is a minor protein component of the 7S globulins because it was estimated to represent about 3% of the total soybean globulin as shown in Table 1. The sum of β- and γ-ConGl accounts for about 95% of the total 7S globulins. Therefore, some minor 7S globulins still appear to be unaccounted for.

More recently, the heterogeneity of the 11S globulin (Gl) was also found [88]. The molecular weights of the four molecular species of the 11S globulin were estimated to be 340000, 345000, 360000 and 375000 by gradient gel electrophoresis. The similar heterogeneity was also observed in the half-molecule of the protein.

The next complication is that polymorphism of the 11S globulin between soybean cultivars based on the difference of number and species of subunits which make up the 11S quaternary structure has been demonstrated [24, 25, 26, 39, 66].

The ultracentrifugal 2S fraction which has been recognized to be storage protein as the 2S globulin is not also composed of only one protein component. Up to the present, two ultracentrifugally pure proteins, 2.3S and 2.8S globulins [90, 92] and an antigenic protein, α-conglycinin [9] have been isolated from the 2S fraction. However, recently the existence of two antigenic components has been confirmed [61]. Further, several bands caused from the protein components were identified by disc electrophoresis using 15% polyacrylamide gel in the 2S fraction although the number of bands is different between soybean cultivars. On the contrary to the recognition that the 2S globulins are storage proteins, the two antigenically different proteins and most of the band components had some inhibitory acitivities against trypsin and/or α-chymotrypsin [61]. One of the two antigenic proteins in the 2S globulins immunologically coincided well with the commercial soybean Kunitz trypsin inhibitor and the other was a new protease inhibitor that had an ability of preferable inhibition against α-chymotrypsin [60]. Accordingly, the 2S globulins are not correctly storage proteins.

The 15S fraction is not yet well understood except a polymer of the 11S protein [28].

ISOLATION AND CHARACTERIZATION OF SOYBEAN STORAGE PROTEIN COMPONENTS

Isolation of the Protein Components

The innovation of ultracentrifugation, electrophoresis, ion exchange chromatography, gel filtration and more recently serological methods as analytical tools for high molecular weight substances led to a closer investigation of the seed globulins. It is not an exception as to the investigations of soybean storage proteins. At the initial stage of the investigation, great efforts have been made for ultracentrifugally pure isolation of the four sedimenting components of soybean globulins, 2S, 7S, 11S and 15S, which were shown by Naismith [67].

The 11S globulin. Among the ultracentrifugal components, the initial effort of the isolation was done for the 11S component. The partial purification and characterization of the 11S component were first studied by Wolf and coworkers [3, 94–99], and the first ultracentrifugal pure isolation was accomplished in 1965 [62].

The 11S protein is a cryoprotein. Enriched 11S fraction, therefore, can be easily obtained from an aqueous extract of defatted soybean meal by cooling to 4°C as the cold insoluble fraction (CIF) [94, 99]. The CIF also increases by adding calcium ion, because calcium ion tends to precipitate preferentially the 11S fraction [93]. Subsequently, the ultracentrifugally pure 11S globulin was isolated in many laboratories [1, 14, 17, 33, 51].

The most abundant form of the protein is the dimer which exhibits a sedimentation constant of 11S [8] and the protein is unstable to the dissociation and association reaction with the change of pH and ionic strength during the preparation. Accordingly, a mild procedure is very important to the protein preparation. Considering this point, gel filtration and above all affinity chromatography with concanavalin A (Con A)-Sepharose by which the difficult elimination of the contaminant of 7S protein is easily performed can be said to be a very desirable procedure for the purification of the 11S globulin.

For the large-scale preparation of the 11S globulin, a simple and conventional method based on differential solubility of 7S and 11S proteins in calcium chloride solutions (5 ~ 12.5 mM) is useful [76]. At the higher level of calcium chloride, 7S proteins are preferentially recovered. Moreover, another large-scale method was developed by selective precipitation at pH 6.6 from a dilute tris(hydroxymethyl)aminomethane (Tris) buffer (0.03M, pH 8.0, 0.07 ionic strength) extract of defatted soybean meal [80]. This method utilizes skilfully the solubility of the 11S globulin to ionic strength in the isoelectric pH range. At ionic strengths below 0.06M in Tris buffer, the 11S globulin shows minimum solubility around pH 6 where the 7S globulins are quite soluble.

As mentioned previously, heterogeneity and polymorphism of the 11S molecule have been recently found in succession. However, the isolation procedure established to the protein can also be usefully applied to the isolation of the molecular species.

The 7S globulins. About 80% of the total 7S fraction in the water extract from defatted soybean meal precipitates at pH 4.5 as the 7S globulins [47]. The 7S globulins are the mixture of protein components, β- and γ-ConGls, as mentioned afterwards. For the isolation of the ultracentrifugally pure 7S preparation, the 7S globulin (β-ConGl) capable of the association reaction with the change of ionic strength, greater effort than that of the 11S globulin was made in the initial study of the protein, because the 7S globulin is more unstable than the 11S globulin with the change of ionic strength.

In 1965, an ultracentrifugally 90% pure 7S protein was first isolated [74]. Immediately thereafter, the 7S protein was isolated in ultracentrifugally pure state [41]. The successful purification was performed by gel filtration only using a long column with Sephadex G-100 and G-200 after elimination of the contaminated 11S protein adding calcium ion in the cold in order to promote the cryoprecipitation of the 11S protein. Subsequently, the 7S globulin was also purified by an another method [52]. This method utilizes the following phenomenon: the 7S globulin is completely soluble in 0.01N HCl containing sodium chloride below the concentration of 0.8M, whereas the 11S globulin is insoluble above 0.5M of sodium chloride [53].

In conclusion, the 7S protein of Roberts and Briggs [74] and Koshiyama [41, 52] is identical. However, Catsimpoolas and Ekenstam [9, 12] suggested that their β-ConGl was the protein of the former and their γ-ConGl the latter. Later the suggestion was corrected by serological methods and the comparison of characteristics between both proteins [55].

The purification of γ-ConGl was done by using of affinity chromatography with monospecific antibody against β-ConGl [56]. This type of affinity chromatography is very useful in order to isolate storage protein components even though the production of antibody is time consuming.

The 2S globulins. The 2S globulin fraction is able to be easily prepared by gel filtration with a Sephadex G-100 column. The protein components were briefly isolated on a DEAE-Sephadex A-50 column by a linear gradient system [61].

The Characteristics of Soybean Globulin Components.

Physico-chemical properties of soybean globulin components are summarized in Table 2. The amino acid compositions are compared in Table 3. In the amino acid compositions, the most significant differences between the two major globulins, the 7S and the 11S globulins, are the 5- to 6-fold higher contents of tryptophan, methionine and half cystine in the latter.

One of the distinct different characteristics between the 7S and the 11S globulins is that the former are glycoproteins [42]. A large glycopeptide of which the carbohydrate moiety corresponded with that in the original 7S protein, β-ConGl [45]. The molar ratio of mannose and N-acetyl-glucosamine was found to be about 3 to 1. All the β-ConGls isolated by Thanh and Shibasaki were also found to contain mannose and glucosamine with a ratio of about 3 to 1, respectively [81].

Afterwards the properties of the 7S globulin as a glycoprotein were investigated in detail by Yamauchi and coworkers [101, 103, 104]. They separated three L-β-aspartamido-carbohydrates (Asn-carbohydrates) (designated as II_a, II_b and II_c) on a column of Dowex 50W-X2 with 1mM sodium-acetate buffer (pH 2.6) from the adequate pronase digestion of the 7S globulin (β-ConGl) [101]. Their components were Asn-$(GlcNAc)_2(Man)_9$ for II_a, Asn-$(GlcNAc)_2(Man)_8$ for II_b and Asn-$(GlcNAc)_2(Man)_7$ for II_c. Furthermore, 1-L-β-aspartamido-2-acetamido-1, 2-dideoxy-β-glucose, which corresponds to the protein-carbohydrate linkage, was separated by partial hydrolysis of the Asn-carbohydrates. Glycopeptides from the pronase digest were further also separated into five fractions [103]. Four of them were tripeptide-carbohydrates. Two kinds of tripeptides were Asn-Gly-Thr and Asn-Ala-Thr with asparagine as the N-terminal amino acid residue to which the carbohydrate moiety was attached.

Subsequently, the sequences of the carbohydrate moiety of the three Asn-carbohydrates were determined by using the methods of methylation analysis, partial acetolysis and Smith degradation [104]. In each Asn-carbohydrate methyla-

Table 2. Physico-chemical properties of soybean globulin components

	2S[90, 92]		7S		11S	
	2.3S	2.8S	β-ConG1[43]	γ-ConG1[11]	Ref[1]	Ref [51]
1. General						
Nitrogem content, %					16.3	17.12[96]
Extinction coefficient, $E^{1\%}_{280nm,1cm}$		9.06	5.47 4.16[83]		8.1	8.04
Partial specific volume, V (ml/g)			0.0725		0.73	0.708
Sedimentation constant, $s^{O}_{20,w}$(S)	2.28	2.80	7.92 7.20[83]		12.35	12.05
Diffusion constant, $D^{O}_{20,w}$($\times 10^{-7} cm^2/sec$)			3.95 4.52[83]		3.44	3.26
Stokes radius (Å)			59		58.5	59
Isoelectric point, pI (pH)		4.4	4.90[44]			4.64
Frictional ratio, f/f0			1.37			1.40
Intrinsic viscosity (dl/g)			0.0638			0.0485
2. Molecular weight						
Gel filtration			210000 175000[83]		302000	297000
Sedimentation equilibrium	18200	32600	180000	104000	317000	312000
Sedimentation-diffusion			181000 150000[83]	102000	322000	313100
3. Size						
Electron microscopy					100x100x70Å* 110x110x80Å**	
X-ray scattering					110x110x75Å	

* As observed, ** Allowing for hydrophobic region.

Table 3. Amino acid composition of soybean globulin components

Amino acid	2S	7S								11S	
		β-ConG1							γ-ConG1[102]	G1	
	2.82[92]	Ref[43]	B1[83]	B2[83]	B3[83]	B4[83]	B5[83]	B6[83]		Ref[1]	Ref[6]
Aspartic acid (D)	13.5	14.1	11.9	13.6	11.1	11.3	10.9	11.0	10.0	13.1	13.9
Threonine (T)	3.4	2.8	2.2	2.2	2.1	2.1	2.0	1.9	4.2	3.4	4.1
Serine (S)	4.1	6.8	5.6	5.6	5.4	5.5	5.4	5.3	6.5	4.2	6.5
Glutamic acid (E)	9.4	20.5	24.5	24.5	26.3	26.6	28.6	28.2	17.4	18.0	25.1
Proline (P)	4.5	4.3	4.2	4.1	4.7	4.9	5.0	5.3	5.9	5.4	6.9
Glycine (G)	3.9	2.9	2.7	2.6	2.7	2.5	2.6	2.5	6.1	4.0	5.0
Alanine (A)	2.5	3.7	3.5	3.5	3.2	3.2	3.0	3.1	4.7	3.6	4.0
Valine (V)	6.8	5.1	3.7	3.5	3.7	3.5	3.2	3.3	6.4	5.1	4.9
Isoleucine (I)	8.1	6.4	4.5	4.5	4.6	4.7	4.3	4.6	4.4	4.7	4.9
Leucine (L)	6.4	10.3	8.8	9.1	8.3	8.4	7.7	8.1	7.6	7.2	8.1
Tyrosine (Y)	2.2	3.6	3.4	3.2	3.3	3.2	3.0	3.1	2.1	4.1	4.5
Phenylalanine (F)	6.5	7.4	6.8	6.7	6.3	6.2	5.9	5.9	5.5	5.7	5.5
Histidine (H)	0.6	1.7	2.6	1.7	2.1	1.5	2.0	1.2	2.8	2.2	2.6
Lysine (K)	5.6	7.0	6.8	6.6	7.3	7.0	7.3	6.9	6.8	4.9	5.7
Arginine (R)	6.5	8.8	8.8	8.8	9.0	9.4	9.2	9.8	6.3	7.8	8.9
1/2Cystine (C)	1.9	0.3	–	–	–	–	–	–	1.1	1.4	1.7
Methionine (M)	0.6	0.3	–	–	–	–	–	–	1.4	1.8	1.3
Amide ammonia	–	1.7	–	–	–	–	–	–	–	2.9	1.6
Tryptophan (W)	2.3	0.3	–	–	–	–	–	–	0.7	1.6	1.5
Carbohydrate	0.4	4.9	4.0	3.8	4.3	4.6	5.1	5.4	5.5[56]	–	–

(g/100g protein)

tion analysis gave 3,6-di-0-methyl derivative from the N-acetyl-glucosamine residue, and 2,3,4,6-tetra-0-methyl, 3,4,6-tri-0-methyl and 2,4-di-0-methyl derivatives from the mannose residues. The first Smith degradation of all the Asn-carbohydrates gave Asn-$(GlcNAc)_2(Man)_2$ and the second degradation Asn-$(GlcNAc)_2$. Partial acetolysis of the Asn-carbohydrates followed by deacetylation yielded mannobiose and mannotriose from II_a, and mannose and mannobiose from II_b and II_c. From these results, they proposed the following two core structures of II_a, II_b and II_c.

$M^1 - ({}^2M_1)_x$ (A)
$M^1 - ({}^2M^1)_y - {}^3M_1^6$
$M^1 - ({}^2M^1)_z - {}^3M^{61} - {}^4GlcNAc^1 - {}^4GlcNAc - Asn$
II_a, $x = 1, y = 1$ and $z = 2$, or $x = 2, y = 0$ and $z = 2$
II_b, $x = y = 0, z = 3$
II_c, $x = y = 0, z = 2$

$M_1 \quad M_1$ (B)
$(M_1^2)_y(M_1^2)_x$
$M^1 - ({}^2M^1)_z - {}^3M^6 - {}^3M^{61} - {}^4GlcNAc^1 - {}^1GlcNAc - Asn$
II_a, $x = 1, y = 2$ and $z = 1$, or $x = 2, y = 1$ and $z = 1$
II_b, $x = 0, y = 1$ and $z = 2$, or $x = 1, y = 0$ and $z = 2$
II_c, $x = 0, y = 1$ and $z = 1$, or $x = 1, y = 0$ and $z = 1$

Gamma-conglycinin contains also about 5.5% carbohydrate [56], but the 11S globulin is not a glycoprotein [57].

For N-terminal amino acids of the 11S protein two different results were reported. They were $(glycine)_8$, $(phenylalanine)_2$ and (leucine or isoleucine)$_2$ [14], and $(glycine)_6$, $(phenylalanine)_2$, $(leucine)_2$ and $(isoleucine)_2$ [1]. The complication is that arginine in addition to phenylalanine, leucine and isoleucine was isolated as the NH_2 termini of acidic subunits while only glycine as those of basic subunits in the 11S globulin using the cultivar CX635-1-1-1 by structural analyses [63]. Still more kinds of N-terminal amino acids may be found by the discovery of heterogeneity and polymorphism that is originated from cultivar differences in the 11S globulin [24, 25, 26, 39, 66, 88].

On the other hand, nine N-terminal residues composed of eight different amino acids, aspartic acid, alanine, glycine, tyrosine, glutamic acid, valine, leucine (isoleucine) and two moles of serine per mole of the protein, were first found in the 7S globulin [43]. Afterwards, the N-terminal analysis for all the β-ConGls, B_1 to B_6 conglycinins, generally gave the same qualitative results as those of the previous report, but a great difference exists with regard to the qualitative data [81]. Further investigations will be needed.

N-Terminal amino acid of γ-ConGl is reported to be only isoleucine [102], and that of 2S globulin components is all aspartic acid up to now [60, 92].

Among soybean globulin components, the association-dissociation properties of the 7S globulin (β-ConGl) with the change of ionic strength are characteristic. On the basis of the properties, vicilin-type proteins (generally 7S globulins) from legume seeds can be divided into three types [16]. One type dimerizes to a 9S–12S form at 0.1 ionic strength and neutral pH, the second retains a 7S form at low ionic strength and the third is insensitive to changes of ionic strength but associates an 18S form (probably a tetramer of 7S) at pH values near its isoelectric point. The 7S globulin (β-ConGl) belongs to the first type and γ-ConGl to the second type.

The characteristic conformational changes of the 7S globulin as a function of pH and ionic strength are shown in Figure 2 [46]. Between pH 2 and 10, the protein undergoes a distinctive reversible reaction with changes of ionic strength. The 7S globulin has a molecular weight of about 180000 at 0.5 ionic strength [43], but it sediments at a rate of about 9S at 0.1 ionic strength. 9S form has a molecular weight of 370000 [46]. Thus, 9S form is a dimer of the 7S globulin.

However, recently Thanh and Shibasaki [85] showed that the protomer (7S) predominates at ionic strength greater than 0.5 and the dimer $s^0_{20,w}$ of a 9S form at 0.1 ionic strength is 10.7 [85] at ionic strength less than 0.2. In the ionic strength region of 0.2 ∼ 0.5, varying amounts of the protomer and the dimer accounted for the gradual changes in sedimentation coefficient which approached 7S or 10S as equilibrium was shifted to either side.

The monomer-dimer reaction of the 7S globulin was also confirmed by Iibuchi and Imahori [31]. But the protomer of which $s^{0.16\,0\%}_{20,w}$ was estimated to be 5.6S was obtained at ionic strength greater than 0.8 and the dimer at ionic strength less than 0.2. In the ionic strength region between 0.2 and 0.8, the gradual change in sedimentation of the protein was found.

MOLECULAR STRUCTURE

Higher Structure

Conformational studies about secondary and tertiary structures of the 7S (β-ConGl) and the 11S globulins (Gl) have been made in detail [19, 20, 32, 51, 53, 54].

$$(2 \sim 3S + 5 \sim 6S) \underset{(3)}{\overset{(1)}{\rightleftharpoons}} 7S \underset{(4)}{\overset{(2)}{\rightleftharpoons}} 9S \text{ (Dimer of 7S)}$$

Fig. 2. Schematic pattern of association and dissocation reaction of the 7S globulin (β-ConG1)[46].

(1) acidic; ionic strength (I) ⩾0.1, alkaline; I ⩾0.5
(2) alkaline; I ⩽0.2[31, 85]
(3) acidic; I <0.1
(4) alkaline; I ⩾0.5
Acidic; 2 ⩽pH <isoelectric point (pH 4.90)
Alkaline; isoelectric point <pH <10

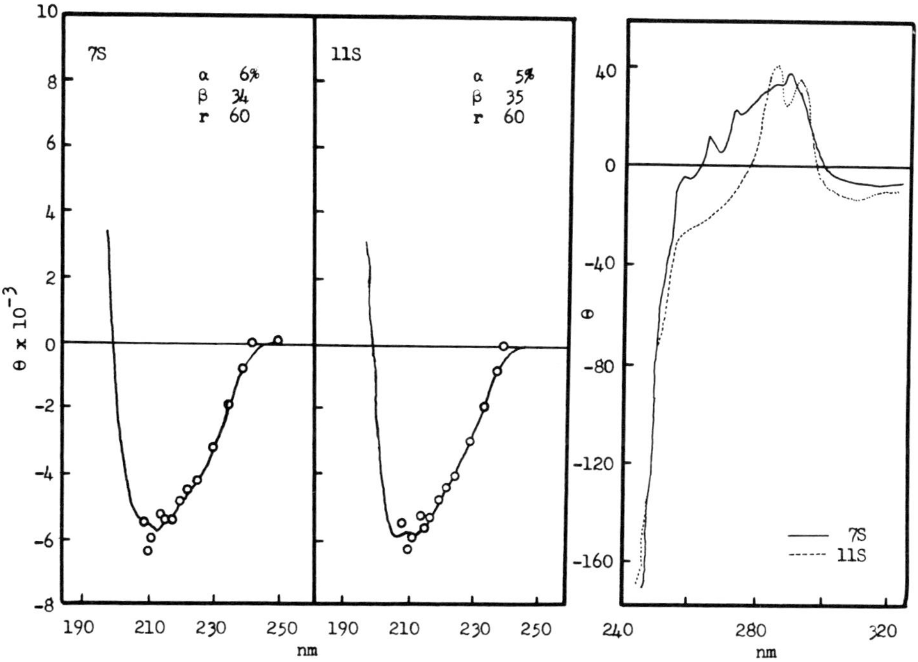

Figure 3. The circular dichroism spectra of the 7S and the 11S globulins in the buffer of 0.5 ionic strength. [54]

The open circles are molecular ellipticities calculated from Moffitt's parameters of a_0 and b_0 for poly-L-lysine reference spectra in water using the above-illustrated parameters for the both proteins.

Figure 3 shows circular dichroism (CD) spectra of the two globulins [54]. They are fairly similar. Their experimental CD patterns between 210 and 240 nm agreed well with those calculated from optical rotatory dispersion (ORD) parameters, a_0 and b_0, according to the method of Greenfield and Fasman [22]. The secondary structures of the two globulins appeared analogous and the contents of α-helix, β-structure and random coil in the proteins were estimated to be about 5, 35 and 60%, respectively. Further, the two proteins clearly have the amide V band at 698 cm^{-1}, characteristic of β-form by infrared (IR) absorption measurements [15, 19, 20]. Consequently, it can be concluded that the two globulin molecules in aqueous solution are compactly folded ones with hardly any α-helix, and they seem to be of an intramolecular cross-β-structure which is also stabilized by hydrophobic bonds in addition to hydrogen bonds. However, CD spectra between 250 and 320 nm differed in the proteins as shown in Figure 3. These CD bands may originate from tyrosyl and tryptophyl residues. Probably, the state of aromatic amino acid side chain residues may be different between the molecules.

Approximately 14% of the total ionizable groups in the 11S globulin, tyrosine residue, appear to be buried by hydrogen titration [6]. The presence of buried ionic groups in the protein, tyrosine and tryptophan residues, was demonstrated by ultraviolet difference spectra [15].

On the other hand, all the tyrosine residues in the 7S globulin ionized with a pK value (about pH 11). Accordingly, they seem to exist in the same state buried in the interior of the molecule. And disulfide bonds were not involved in binding between subunits [49].

In particular, marked dissimilarities between the 7S and the 11S globulin were observed by ultraviolet difference spectra, ultracentrifugation and ORD in acid solution at 0.1 ionic strength [53].

Subunit Structure

Among soybean globulin components, the 7S [10, 48, 70] and the 11S globulin [10, 70, 73, 91] had been shown to have quaternary structures, subunit structures. In 1970, Catsimpoolas clearly showed that the 11S protein (Gl), β- and γ-ConGl had quaternary structures by disc electrophoresis of their proteins after dissociation in the solvent system of phenol-acetic acid-0.2M 2-mercaptoethanol (2-ME) (2:1:1 w/v/v) with 5M urea [5].

The 11S globulin. Above all, the subunit structure of the 11S globulin has been well investigated. After valuable pioneering studies on the subunits of the 11S globulin made by Okubo and coworkers [10, 68, 69, 71], Catsimpoolas and coworkers [4, 11] isolated three acidic and three basic subunits from the purified 11S protein by isoelectric focusing in urea-2-ME, and determined their isoelectric points, molecular weights and amino acid compositions. On the basis of these results and electron microscopy experiments the dimer which is the most stable form of the 11S globulin was proposed to be composed of two annular-hexagonal structures each composed of six subunits. The two annular-hexagonal structures are packed the one on top of the other. An attractive hypothesis would be that the acidic and basic subunits are alternating in the annular structure thus contributing to the stability of the molecule in terms of ionic interactions. This speculation was confirmed by Badley *et al.* using X-ray scattering and electron microscopy [1]. As shown in Figure 4, the dimer of the 11S molecule is constructed by six acidic and basic subunits. The basic subunits have molecular weight of 19600, but the acidic subunits are almost twice as large at 34800. Analyses also revealed three different kinds of acidic and basic subunits. These twelve subunits are packed in two identical hexagons, placed one on the other, yielding a hollow oblate cylinder estimated to be 110 × 10 × 75 Å. Hydration was 0.36 g/g. Staining characteristics observed by electron microscopy suggested that the region between the two hexagonal rings are hydrophobic and consequently, the interaction is speculated to involve electrostatic and/or hydrogen bonding. Interactions between subunits within a hexagonal ring are proposed to be alternating acidic (A) and basic (B) subunits. Alternation of

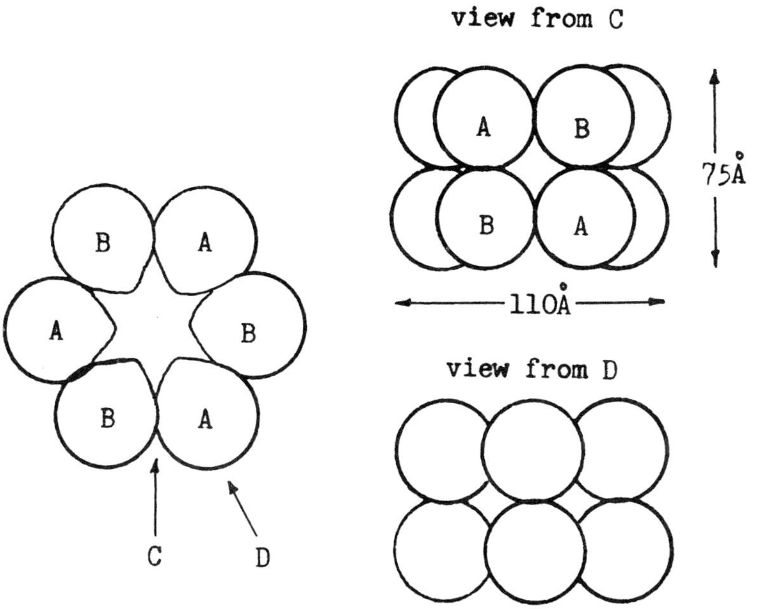

Figure 4. Schematic representation of the 11S molecule [1].

acidic and basic subunits between the two layers of adjacent hexagonal ring structure is also considered likely.

On the other hand, the 11S molecule was estimated to be an oblate ellipsoid with a rigid and nearly anhydrous state in solution according to the procedure of Shimha and Perrin by measuring the partial specific volume, diffusion constant, molecular weight and volume fraction intrinsic viscosity [58]. Subsequently, the same result was estimated according to the treatment of Scheraga and Mandelkern [77] by calculating the shape factor β, and hydrodynamically effective volume. The length of the major axis of the molecule is about 180 Å and that of the minor axis about 20 Å. Although the hydration and dimensions of the protein are cross between the previous results [1, 58], its shape may be concluded to be oblate.

However, recently further complexity of the subunit structure was found by Kitamura and Shibasaki [34, 37]. They found four kinds of acidic and four kinds of basic subunits designated as A_1 to A_4 for acidic subunits and B_1 to B_4 for basic subunits from the variety of Raiden. These subunits were present in the molar ratio of 1:1:2:2 in order of the designated number, respectively. Intermediary subunits of the 11S globulin were also isolated in the absence of reducing reagents in 6M urea. Disulfide bonds appeared to precipitate in binding the acidic and basic subunits in the molar ratio of 1:1 with the following combinations: A_1 and A_2 combined with B_3, A_3 with B_1 and B_2, and A_4 with B_4. From these results, it was proposed that the 11S protein molecule consisted of two similar, but not identical monomers and had the same following dissociation scheme as that proposed for the legumin of Vicia faba (broad bean) [100].

$$11S \xrightarrow{\text{urea or SDS}} \overset{\text{Intermediary subunit}}{(A_1B_1 + A_2B_3 + A_3B_1 + A_3B_2 + 2A_4B_4)}$$

$$\xrightarrow{\text{urea or SDS}} (A_1 + A_2 + 2A_3 + 2A_4 + B_1 + B_2 + 2B_3 + 2B_4)$$

The molecular weights estimated by sodium dodecyl sulfate (SDS)-gel electrophoresis were 22500 for the basic subunits, 37000 for A_1 to A_3 and 45000 for A_4 acidic subunit. The molecular weight of the 11S globulin calculated from those of the subunits was 362000. The value is in a relatively good agreement with that of Wolf and Briggs [96], but about 10% larger than the other previous results indicated [1, 51, 58]. This discrepancy may be elucidated from the heterogeneity of the protein as reported by Utsumi *et al.* [88] as well as that shown in broad bean [89].

Further, similarities and dissimilarities of amino acid sequences among the acidic subunits were analysed by quantitative immunological precipitin with antibodies to the subunits, gel filtration of the cyanogen bromide (CNBr)-fragments and ion exchange chromatography of the tryptic peptides [35, 36]. Quantitative immunological precipitin revealed that the A_1 to A_3 subunits were strongly immunologically related, but that the A_4 subunit was shown to be serologically different from the other subunits. The fact was also supported by the analyses of tryptic peptides and CNBr-fragments of these acidic subunits. The occurrence of A_4 and B_4 subunits indicates the presence of genetic variants of the protein.

Recently, six acidic and four basic subunits of the 11S globulin from a genetically well defined soybean cultivar, CX635-1-1-1, were isolated [63]. The amino acid compositions of these subunits are shown in Table 4. Their observation seems to have significance in efforts to improve soybean quality genetically with respect to methionine. As shown in Figure 5, the four residue sequences around the only cysteine residue in the NH_2 termini of A_1 to A_3, F_2-(1) and F_2-(2) are substantially identical. In particular, polypeptides A_1 and A_2 were clearly homologous to one another except the first seven residues. But the A_4 subunit was the least similar of the six acidic polypeptides. The basic subunits were more homologous to one another than the acidic ones. The NH_2-terminal sequences of these basic subunits nearly coincided with the results given to the basic subunit mixtures by Gilroy *et al.* [21].

More recently, the subunit compositions of the 11S globulin from various cultivars have been investigated in detail by polyacrylamide gel electrofocusing in the presence of urea and 2-ME [66]. To our surprise, many other acidic and basic subunits were found. From the analyses, the subunit compositions could be classified into five groups according to the differing molecular charges of the subunits: group I contained 7 acidics and 8 basics; group II, 7 acidics and 7 basics; group III, 6 acidics and 7 basics; group IV, 6 acidics and 5 basics; and group V, 6 acidics and 3 basics.

As a very interesting study on the subunit structure of the 11S globulin, the reconstruction of the 11S subunit structure was done [38]. The yield of the renaturated 11S globulin was 70% from the denaturated state in the absence of 2-ME

Table 4. Amino acid composition of the glycinin subunits[63]

	A_1	F2-(1)	A_2	A_3	F2-(2)	A_4	B_1	B_2	B_3	B_4
B*	36.8	34.8	42.1	45.5	9.0	50.8	25.5	24.3	19.2	20.7
T	12.0	12.2	12.3	15.5	3.9	11.8	8.1	9.1	6.2	5.4
S	18.3	19.0	16.4	27.1	7.7	23.5	13.5	12.4	12.1	12.4
Z**	85.3	85.3	86.4	91.6	14.9	92.6	22.5	22.7	24.8	21.0
P	24.0	25.0	21.3	33.9	8.1	27.3	10.5	10.8	10.2	9.1
G	31.0	27.3	29.9	29.5	7.9	22.4	11.1	10.4	13.4	16.1
A	14.4	15.9	18.1	10.9	5.0	6.2	15.6	14.3	12.4	11.2
V	11.9	12.4	15.3	17.4	3.8	12.1	11.4	10.8	17.0	19.2
M	3.6	4.1	5.8	2.4	1.1	1.4	2.3	2.7	0	1.3
I	17.6	16.6	15.3	12.2	5.1	10.4	9.2	9.8	7.0	7.3
L	20.1	18.7	20.0	21.8	10.7	14.0	17.9	17.4	18.1	18.1
Y	7.3	8.7	6.6	5.6	2.0	4.4	2.8	2.5	5.8	8.4
F	12.2	17.9	12.3	12.0	0.9	7.7	8.6	9.1	6.0	5.7
H	6.0	4.8	2.6	14.1	2.8	9.5	2.1	2.7	4.8	4.2
K	21.2	15.9	14.9	14.8	3.9	18.8	5.9	5.9	7.0	6.5
R	18.1	21.2	22.7	22.2	3.1	28.4	8.9	9.9	10.9	12.5
C	4.5	N.D.	4.3	3.6	N.D.	0.7	1.7	1.5	0.2	1.5

*Asparagine, ** Glutamine. (The number of residues per subunit)

```
Acidic
subunits
                         5        10        15        20        25
A4     NH2– R R G S R S Z K Z Z L Z D(S)H Z K I(R)H F B E G D G

                    5        10        15        20        25
A1       NH2– F S S R E Z P Z Z B E C Z I Z KL B A L K P D B( )I

                    5        10
F2(1)    NH2– F S F R E ZP Z Z B E(C)Z I Z

                    5        10        15        20
A2            NH2– L R E Z A Z Z B E C Z I Z K L B A L K P D B(R)I

                    5           10        15        20
A3     NH2– I T S S K F---------- B E C Z L B B L B A L E P D H R V E( )E G

                    5          10         15
F2 (2) NH2– I S S S K L---------- B E C Z L B B L B A L
Basic
subunits
                    5        10        15
B1     NH2– G I D E T I C T M R L R Z B I G Z

B2     NH2– G I D E T I C T M R L R H B I G Z

B3     NH2– G V E E B I C T L K L H E B I A R

B4     NH2– G V E E B I C T M K L H E B I A R
```

Figure 5. The NH_2-terminal amino acid sequences of the 10 glycinin subunits aligned for maximal homology [63].

whereas only 20% in the reductively denaturated state. This result suggests that the formation of a disulfide bridge between the acidic and basic subunits is important in the process of the formation of the 11S globulin molecule.

Further recently, reconstitution of the intermediary subunits formed between the acidic and the basic subunits involving a disulfide bridge was investigated in detail [65]. The combination between the acidic and the basic subunits was not defined, but the formation of the intermediary subunits was widely observed. Surprisingly, the intermediary complexes were observed when native acidic and basic subunits of soybean 11S globulin and sesame 13S globulin. Further, pseudo- and hybrid-11S globulins were also reconstituted from native acidic and basic subunits of soybean and broad bean 11S globulin [87].

The 7S globulins. With respect to the subunit structure of β-ConGl, Thanh and Shibasaki [81–83] reported that six β-conglycinin isomers (B_1 to B_6-conglycinin) were composed of three major (designated as α, α′ and β) and one minor subunit (designated as γ). The molecular weights and N-terminal amino acid for the α, α′ and β subunits were 57000, 57000 and 42000, and valine, valine and leucine, respectively [83]. The amino acid compositions of the three subunits are compared in Table 5. The subunit structures of the six β-ConGls are proposed as follows: B_1, $\alpha'\beta_2$; B_2, $\alpha\beta_2$; B_3, $\alpha\alpha'\beta$; B_4, $\alpha_2\beta$; B_5, $\alpha_2\alpha'$; B_6, α_3. The schematic representation of the subunit structure of β-conglycinin isomers is illustrated in Figure 6. Based on these subunit compositions, the six isomers can be classified into three groups: group A (B_1 and B_2) which has two β subunits per molecule (7S); group B (B_3 and B_4), one β subunit; and group C (B_5 and B_6), no β subunit. The proposed groups were supported by the result of N-terminal amino acid composition, group A; Val(1)Leu(2), group B; Val(2)Leu(1) and group C; Val(3) since α and α′ subunits had valine and β subunit leucine as N-terminal amino acid. Accordingly, the molecular weights from group A to C are calculated to be 141000, 156000 and 171000 from the subunit sizes, respectively. The order of the molecular weights may account for the order of elution, group C, B and A, from the gel filtration column of Sepharose 6B [79]. The elution order of B_1 to B_6 conglycinin with a DEAE-Sephadex A-50 column [81] agreed with a gradually increasing negative charge in the order of B_1 to B_6. This is nothing but an increase (from B_1 to B_6) in the distribution of α and α′ subunits to the proposed subunit structures. The isoelectric points of the subunits are in the order of $\alpha < \alpha' < \beta$ [100].

On the other hand, Iibuchi and Imahori have found that the isolated three 7S globulin isomers having an ability of the dimerization reaction at 0.1 ionic strength all consist of the combination of two kinds of subunits, α and β [30]. The subunit construction of the three isomers was estimated to be α_3, $\alpha_2\beta$ and $\alpha\beta_2$, respectively. The molecular weight was given to be 68000 and 52000 for the α and β subunit, respectively. These values are slightly different from the previous results [82]. Since the N-terminal amino acids of the α_3, $\alpha_2\beta$ and $\alpha\beta_2$ isomers were determined to be valine, valine and leucine (isoleucine), respectively, the N-terminal amino acid of the α subunit must be valine and that of the β subunit leucine (isoleucine). Therefore, the α and β subunits in this report correspond to α and α′, and β subunits of Thanh and Shibasaki [82].

Table 5. Amino acid composition of the isolated subunits[82].

	α	α'	β
D	55.27	51.70	45.72
T	10.29	10.90	9.44
S	31.02	30.55	25.39
E	96.30	100.39	58.92
P	33.23	30.55	17.93
G	20.77	23.41	15.53
A	21.48	20.02	18.77
C	trace	0	0
V	20.77	22.65	19.02
M	1.97	2.26	0.39
I	27.03	22.94	21.87
L	40.80	35.01	36.70
Y	10.86	10.58	9.08
F	24.49	23.83	22.51
K	29.23	34.36	18.81
H	5.50	17.15	7.11
R	40.89	33.65	24.97

The number of residues per subunit

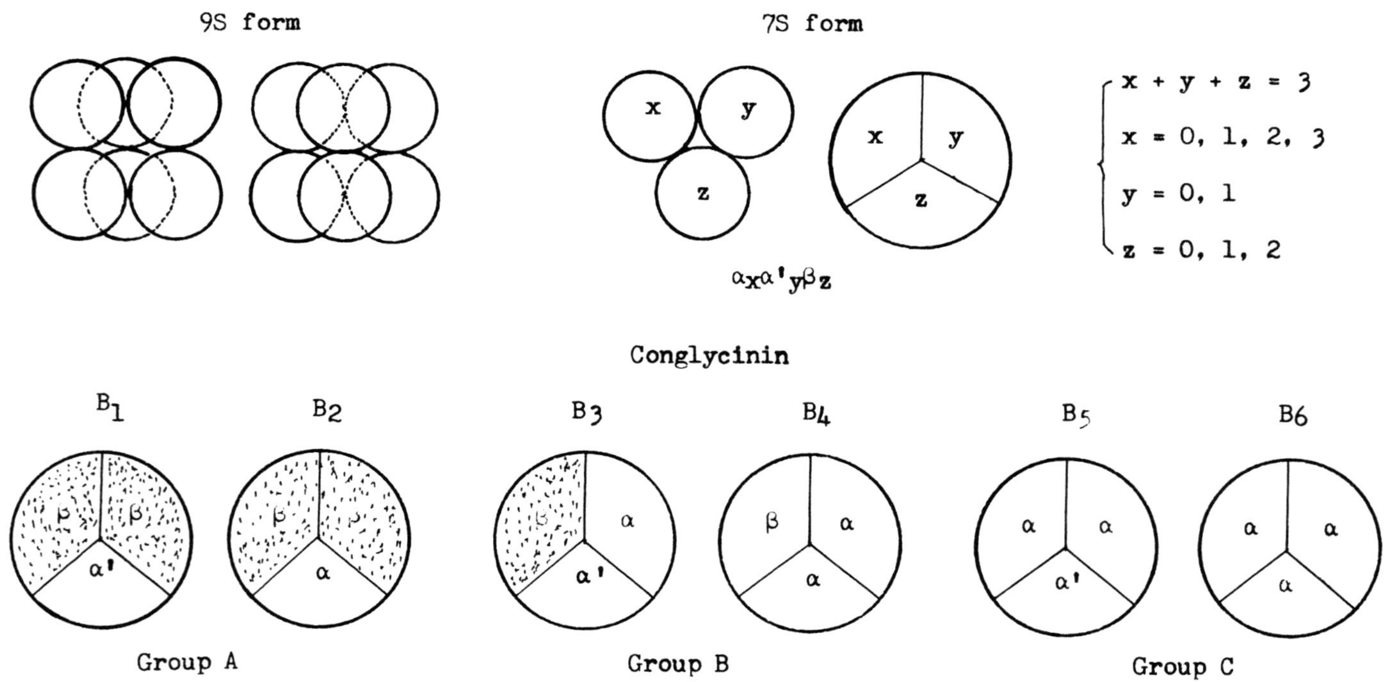

Figure 6. Schematic representation of the subunit structure of β-conglycinin and its six isomers [83].

Most protein oligomers generally have an even number of subunits. So, the subunit structures of β-ConGls, presenting three subunits, particularly nonidentical subunit, per molecule seem to be uncommon. Certainly, of the 300 proteins of which the subunit compositions have been established [40] only ten have three subunits per oligomer and only four of them (troponin, $\alpha_2\beta$; bovine procarboxypeptidase A, $\alpha\beta\gamma$; molybdoferredoxin, $\alpha_2\beta$; and cysteine synthetase, $\alpha_2\beta$) are composed of nonidentical subunits. But the proposed dimer form (9S-form) seems to be common since it possesses a hexameric structure, an even number of subunits.

If the general structure of β-conglycinin isomers was represented as $\alpha_x\alpha'_y\beta_z$ where x, y and z were the number of α, α′ and β subunits, respectively, in the molecules and the three kinds of subunits were able to associate at random to form trimers, they could constitute ten molecular species. Since six molecular species have been isolated already, the other four molecular species corresponding to the values y = 2 and 3, and z = 3, $\alpha'_2\beta$, $\alpha'_2\alpha$, α'_3 and β_3, would be possible. Thanh and Shibasaki [84] skilfully discussed the possibility of these remained molecular species by reconstruction of the multiple molecular species from the three subunits. In fact, the six molecular species already isolated could be reconstituted by mixing the three subunits in urea solution and subsequent dialysis of the solution against phosphate buffer. However, most α′ subunits recombined to build a 7S aggregate which had no ability to dimerize at 0.1 ionic strength. The β subunit associated to form a 16S aggregate (probably $(\beta_3)_4$) at 0.05 ionic strength, pH 8.4. Therefore, the homotrimer β_3 and α'_3 which belong to β-ConGl that have the 7S$\rightleftharpoons$9S dimerization characteristic seem to be unlikely in vivo. Although the possibility that the remained two molecular species containing α′ dimer ($\alpha'_2\beta$ and $\alpha'_2\alpha$) may exist in soybean, cannot be ruled out, they might, like the homotrimer of α'_3, lack the dimerization ability. If so, the two possible molecular species might be excluded from β-conglycinin preparation.

From the consideration of the above smart clear results, the molecular and subunit structures of the 7S globulin (β-ConGl) seem to become completely clear. Nevertheless, there are still some obscure points to be settled. For example, the charge heterogeneity of β subunit [11, 27], the role of γ subunit [11, 27] to the subunit structures, the complicated interaction between the 7S globulin isomers during preparation [56] remain to be solved.

As to the subunit structure of γ-ConGl, it was found to consist of three identical subunits with the molecular weight of about 50000 [43].

As the most novel studies on soybean storage proteins, the in vitro synthesis of a complete form of glycinin (the 11S globulin) in a wheat germ cell-free system has been demonstrated by using the polyribosomes isolated from the developing soybean seeds [64]. More recently, the in vitro synthesis of the α and α′ subunit of the 7S globulin was also demonstrated with wheat germ extract using messenger RNAs from polysomes extracted from immature soybean seeds [2].

Consequently, it may fairly be said that the whole aspect of the very complicated soybean storage proteins has been almost clarified by very active researches in many laboratories for the past about twenty years, particularly in the 7S (β-ConGl) and the 11S globulins. Probably, on the basis of the results in the structural investigations the rapid progress of genetic analyses in the storage proteins that are not always well investigated must be expected in the near future.

REFERENCES

1. Badley, RA, D Atkinson, H Hauser, D Oldani, JP Green, and JM Stubbs (1975): The structure, physical and chemical properties of the soy bean protein glycinin. Biochim. Biophys. Acta 412: 214–228.
2. Beachy, RN (1980): In vitro synthesis of the α and α' subunits of the 7S storage proteins (conglycinin) of soybean seeds. Plant Physiol. 65: 990–994.
3. Briggs, DR, and WJ Wolf (1957): Studies on the cold-insoluble fraction of the water-extractable soybean proteins. I. Polymerization of the 11S component through reactions of sulfhydryl groups to form disulfide bonds. Arch. Biochem. Biophys. 72: 127–144.
4. Catsimpoolas, N (1969): Isolation of glycinin subunits by isoelectric focusing in urea-mercaptoethanol. FEBS Letters 4: 259–261.
5. Catsimpoolas, N (1970): A note on dissimilar subunits present in dissociated soybean globulins. Cereal Chem. 47: 70–71.
6. Catsimpoolas, N, T Berg, and EW Meyer (1971): Hydrogen in titration of ionizable side chains in native and denatured glycinin. Int. J. Protein Res. 3: 63–71.
7. Catsimpoolas, N, TG Campbell, and EW Meyer (1968): Immunochemical study of changes in reserve proteins of germinating soybean seeds. Plant Physiol. 43: 799–805.
8. Catsimpoolas, N, TG Campbell, and EW Meyer (1969): Association-dissociation phenomena in glycinin. Arch. Biochem. Biophys. 131: 577–586.
9. Catsimpoolas, N, and C Ekenstam (1969): Isolation of alpha, beta and gamma conglycinins. Arch. Biochem. Biophys. 129: 490–497.
10. Catsimpoolas, N, C Ekenstam, DA Rogers, and EW Meyer (1968): Protein subunit and dormant and germinating soybean seeds. Biochim. Biophys. Acta 168: 122–131.
11. Catsimpoolas, N, JA Kenny, EW Meyer, and BF Szuhaj (1971): Molecular weight and amino acid composition of glycinin subunits. J. Sci. Food Agric. 22: 448–450.
12. Catsimpoolas, N, N Leuthner, and EW Meyer (1968): Studies on the characterization of soybean proteins by immunoelectrophoresis. Arch. Biochem. Biophys. 127: 338–345.
13. Catsimpoolas, N, and EW Meyer (1968): Immunological study of soybean proteins. J. Agr. Food Chem. 16: 128–131.
14. Catsimpoolas, N, DA Rogers, SJ Circle, and EW Meyer, (1967): Purification and structural studies of the 11S component of soybean proteins. Cereal Chem. 44: 631–637.
15. Catsimpoolas, N, J Wang, and T Berg (1971): Spectroscopic studies on the conformation of native and denatured glycinin. Int. J. Protein Res. 3: 277–284.
16. Derbyshire, E, DJ Wright, and D Boulter (1976): Legumin and vicilin, storage proteins of legume seeds. Phytochem. 15: 3–24.
17. Eldridge, AC, and WJ Wolf (1967): Purification of the 11S component of soybean protein. Cereal Chem. 44: 645–652.
18. Fukazawa, C (1976): Preparation of plant protein bodies. Protein Nucleic Acid Enzyme (Suppl.), February: 134–145; Kyoritsu Shuppan Co.; Tokyo.
19. Fukushima, D (1967): Optical rotatory dispersion (far ultraviolet region), infrared absorption and deutration studies of soybean proteins (7S and 11S). Agr. Biol. Chem. 31: 130–132.

20. Fukushima, D (1968): Internal structure of 7S and 11S globulin molecules in soybean proteins. Cereal Chem. 45: 203–224.
21. Gilroy, J, DJ Wright, and D Boulter (1979): Homology of basic subunits of legumin from *Glycine max* and *Vicia faba*. Phytochem. 18: 315–316.
22. Greenfield, N, and GD Fasman (1969): Computed circular dichroism spectra for the evaluation of protein conformation. Biochemistry 8: 4108–4116.
23. Harada, K (1968): Varietal difference in globulin composition of soybean cotyledon. II. The ratio of 7S–11S fraction. Japan J. Breed. 18 (Suppl. 2): 195–196.
24. Harada, K (1970): Varietal difference in globulin composition of soybean cotyledon. III. Genetic polymorphism of 11S fraction. Japan J. Breed. 20 (Suppl. 2): 166–167.
25. Harada, K (1971): Varietal difference in globulin composition of soybean cotyledon. IV. Isolation of the three 11S fractions. Japan J. Breed. 21 (Suppl. 2): 54–55.
26. Harada, K (1972): Varietal difference in globulin composition of soybean cotyledon. V. Subunits of four 11S fractions. Japan J. Breed. 22 (Suppl. 2): 108–109.
27. Harada, K (1973): Varietal difference in globulin composition of soybean cotyledon of 7S fraction. Japan J. Breed. 23 (Suppl. 1): 138–139.
28. Hashizume, K, K Kakiuchi, E Koyama, and T Watanabe (1971): Denaturation of soybean protein by freezing. Part I. Agr. Biol. Chem. 35: 449–459.
29. Hill, JE, and RW Breidenbach (1974): Proteins of soybean seeds. I. Isolation and characterization of the major components. Plant Physiol. 53: 742–746.
30. Iibuchi, C, and K Imahori (1978): Heterogeneity and its relation to the subunit structure of the soybean 7S globulin. Agr. Biol. Chem. 42: 31–36.
31. Iibuchi, C, and K Imahori (1978): Interconversion between monomer and dimer of the 7S globulin of soybean seed. Agr. Biol. Chem. 42: 25–30.
32. Jacks, TJ, RH Barker, and DE Weigang Jr. (1973): Conformations of oilseed storage proteins (globulins) determined by circular dichroism. Int. J. Peptide Protein Res. 5: 289–291.
33. Kitamura, K, K Okubo, and T Shibasaki (1974): The purification of soybean 11S globulin with Con A-Sepharose 4B and Sepharose 6B. Agr. Biol. Chem. 38: 1083–1085.
34. Kitamura, K, and K Shibasaki (1975): Isolation and some physico-chemical properties of the acidic subunits of soybean 11S globulin. Agr. Biol. Chem. 39: 945–951.
35. Kitamura, K, and K Shibasaki (1975): Homology between the acidic subunits of soybean 11S globulin. Agr. Biol. Chem. 39: 1509–1510.
36. Kitamura, K, and K Shibasaki (1977): Homology among the acidic subunits of soybean 11S globulin. Agr. Biol. Chem. 41: 351–357.
37. Kitamura, K, T Takagi, and K Shibasaki (1976): Subunit structure of soybean 11S globulin. Agr. Biol. Chem. 40: 1837–1844.
38. Kitamura, K, T Takagi, K Shibasaki (1977): Renaturation of soybean 11S globulin. Agr. Biol. Chem. 41: 833–840.
39. Kitamura, K, Y Toyokawa, and K Harada (1980): Polymorphism of glycinin in soybean seeds. Phytochem. 19: 1841–1843.
40. Klotz, IM, DW Darnall, and NR Langerman (1975). In: Neurath, H, and RL Hill (eds.): The Proteins 1: 293–411; Academic Press, New York.
41. Koshiyama, I (1965): Purification of the 7S component of soybean proteins. Agr. Biol. Chem. 29: 885–887.
42. Koshiyama, I (1966): Carbohydrate component in 7S protein of soybean casein fraction. Agr. Biol. Chem. 30: 646–650.
43. Koshiyama, I (1968): Chemical and physical properties of a 7S protein in soybean globulins. Cereal Chem. 45: 394–404.
44. Koshiyama, I (1968): Chromatographic and sedimentation behavior of a purified 7S protein in soybean globulins. Cereal Chem. 45: 405–412.

45. Koshiyama, I (1968): Isolation of a glycopeptide from a 7S protein in soybean globulins. Arch. Biochem. Biophys. 130: 370–373.
46. Koshiyama, I (1968): Factors influencing conformation changes in a 7S protein of soybean globulins by ultracentrifugal investigations. Agr. Biol. Chem. 32: 879–887.
47. Koshiyama, I (1969): Distribution of the 7S proteins in soybean globulins by gel filtration with Sephadex G-200. Agr. Biol. Chem. 33: 281–284.
48. Koshiyama, I (1970): Dissociation into subunit of a 7S protein in soybean globulins with urea and sodium dodecyl sulfate. Agr. Biol. Chem. 34: 1815–1820.
49. Koshiyama, I (1971): Some aspects of subunit structure of a 7S protein in soybean globulins. Agr. Biol. Chem. 35: 385–392.
50. Kohiyama, I (1972): A comparison of soybean globulins and the protein bodies in the protein composition. Agr. Biol. Chem. 36: 62–67.
51. Koshiyama, I (1972): Purification and physico-chemical properties of the 11S globulin in soybean seeds. Int. J. Peptide Protein Res. 4: 167–176.
52. Koshiyama, I (1972): A newer method for isolation of the 7S globulin in soybean seeds. Agr. Biol. Chem. 36: 2255–2257.
53. Koshiyama, I (1972): Comparison of acid-induced conformation changes between 7S and 11S globulin in soybean seeds. J. Sci. Food Agric. 23; 853–859.
54. Koshiyama, I, and D Fukushima (1973): Comparison of conformation of 7S and 11S soybean globulins by optical rotatory dispersion and circular dichroism studies. Cereal Chem. 50: 114–121.
55. Koshiyama, I, and D Fukushima (1976): Identification of the 7S globulin with ß-conglycinin in soybean seeds. Phytochem. 15: 157–159.
56. Koshiyama, I, and D Fukushima (1976): Purification and some properties of γ-conglycinin in soybean seeds. Phytochem. 15: 161–164.
57. Koshiyama, I, and D Fukushima (1976): A note on carbohydrates in the 11S globulin of soybean seeds. Cereal Chem. 53: 768–769.
58. Koshiyama, I, and D Fukushima (1976): Physico-chemical studies on the 11S globulin in soybean seeds: Size and shape determination of the molecule. Int. J. Peptide Protein Res. 8: 283–289.
59. Koshiyama, I, and D Fukushima (1978): Soybean globulins. Proc. Symp. Seed Proteins of Dicotyledonous Plants: N4, 21–43.
60. Koshiyama, I, M Kikuchi, and D Fukushima: The 2S globulins of soybean seeds. II. Physico-chemical and biological properties of the protease inhibitors in the 2S globulins. J. Agric. Food Chem. in press.
61. Koshiyama, I, M Kikuchi, K Harada, and D Fukushima: The 2S globulins of soybean seeds. I. Isolation and characterization of the protein components. J. Agric. Food Chem.: in press.
62. Mitsuda, H, T Kusano, and K Hasegawa (1965): Purification of the 11S component of soybean proteins. Agr. Biol. Chem. 29: 7–12.
63. Moreia, MA, MA Hermodson, BA Larkins, and NC Nielsen (1979): Partial characterization of the acidic and basic polypeptides of glycinin. J. Biol. Chem. 254: 9921–9926.
64. Mori, T, S Takagi, and S Utsumi (1979): Synthesis of glycinin in a wheat germ cell free system. Biochem. Biophys. Res. Comm. 87: 43–49.
65. Mori, T, S Utsumi, and H Inaba (1979): Interaction involving disulfide bridges between subunits of soybean seed globulin and between subunits of soybean and sesame seed globulins. Agr. Biol. Chem. 43: 2317–2322.
66. Mori, T, S Utsumi, H Inaba, K Kitamura, and K Harada: Differences in subunit composition of glycinin among cultivars. J. Agric. Food Chem.: in press.
67. Naismith, WEF (1955): Ultracentrifuge studies on soya bean protein. Biochim. Biophys. Acta 16: 203–210.
68. Okubo, K, M Asano, Y Kimura, and K Shibasaki (1969): On basic subunits dissociated from C (11S) component of soybean proteins with urea. Agr. Biol. Chem. 33: 436–465.

69. Okubo, K, G Sagara, and K Shibasaki (1969): The relationship between the ultracentrifugal fraction and the gel electrophoretical bands of soybean proteins. Tohoku J. Agric. Res. 20: 222–230.
70. Okubo, K, and K Shibasaki (1966): Starch gel electrophoresis of soybean proteins in high concentration of urea. Tohoku J. Agric. Res. 16: 317–329.
71. Okubo, K, and K Shibasaki (1967): Fractionation of main components and their subunits of soybean proteins. Agr. Biol. Chem. 31: 1276–1282.
72. Osborne, TB, and GF Campbell (1898): Proteids of the soy bean (Glycine hispida). J. Am. Chem. Soc. 20: 419–428.
73. Puski, G, and P Melnychyn (1968): Starch-gel electrophoresis of soybean globulins. Cereal Chem. 45: 192–201.
74. Roberts, RC, and DR Briggs (1965): Isolation and characterization of the 7S component of soybean globulins. Cereal Chem. 42: 71–85.
75. Saio, K, and T Watanabe (1966): Preliminary investigation on protein bodies of soybean seeds. Agr. Biol. Chem. 30: 1133–1138.
76. Saio, K, and T Watanabe (1973): Food use of soybean 7S and 11S proteins. Extraction and functional properties of their fractions. J. Food Sci. 38: 1139–1144.
77. Scheraga, HA, and L Mandelkern (1953): Consideration of the hydrodynamic properties of proteins. J. Am. Chem. Soc. 75: 179–184.
78. Thanh, VH, K Okubo, and K Shibasaki (1975): Isolation and characterization of the multiple 7S globulins of soybean seeds. Plant Physiol. 56: 19–22.
79. Thanh, VH, K Okubo, and K Shibasaki (1975): The heterogeneity of the 7S soybean protein by Sepharose gel chromatography and disc electrophoresis. Agr. Biol. Chem. 39: 1501–1503.
80. Thanh, VH, and K Shibasaki (1976): Major proteins of soybean seeds. A straight-forward fractionation and their characterization. J. Agric. Food Chem. 24: 1117–1121.
81. Thanh, VH, and K Shibasaki (1976): Heterogeneity of ß-conglycinin. Biochim. Biophys. Acta 439: 326–338.
82. Thanh, VH, and K Shibasaki (1977): Beta-conglycinin from soybean proteins. Isolation and immunological and physico-chemical properties of the monomeric forms. Biochim. Biophys. Acta 490: 370–384.
83. Thanh, VH, and K Shibasaki, (1978): Major proteins of soybean seeds. Subunit structure of ß-conglycinin. J. Agric. Food Chem. 26: 692–695.
84. Thanh, VH, and K Shibasaki (1978): Major proteins of soybean seeds. Reconstitution of ß-conglycinin from its subunits. J. Agric. Food Chem. 26: 695–698.
85. Thanh, VH, and K Shibasaki (1979): Major proteins of soybean seeds. Reversible and irreversible dissociation of ß-conglycinin. J. Agric. Food Chem. 27: 805–809.
86. Tombs, MP (1967): Protein bodies of the soybean. Plant Physiol. 42: 797–813.
87. Utsumi, S, H Inaba, and T Mori (1980): Formation of pseudo- and hybrid-11S globulins from subunits of soybean and broad bean 11S globulins. Agr. Biol. Chem. 44: 1891–1896.
88. Utsumi, S, H Inaba, and T Mori: Heterogeneity of soybean glycinin. Phytochem.: in press.
89. Utsumi, S, and T Mori (1980): Heterogeneity of broad bean legumin. Biochim. Biophys. Acta 621: 179–189.
90. Vaintraub, IA (1965): Isolation of the 2S component of soybean globulins. Biokhymiya 30: 628–633.
91. Vaintraub, IA (1967): The heterogeneity of the subunits of the 11S protein of soybean seeds. Mol. Biol. (USSR) 1: 671–676.
92. Vaintraub, IA, and AD Shutov (1969): Isolation and certain properties of the 2.8S protein of soybean seeds. Biokhymiya 34: 984–992.
93. Wolf, WJ (1978): In: Smith, AK, and SJ Circle (eds.): Soybeans; Chemistry and Technology. AVI Publishing Co., Westport CT; 93 pp.

94. Wolf, WJ, GE Babcock, and AK Smith (1962): Purification and stability studies of the 11S component of soybean proteins. Arch. Biochem. Biophys. 99: 265–274.
95. Wolf, WJ, and DR Briggs (1958): Studies on the cold-insoluble fraction of the water-extractable soybean proteins. II. Factors influencing conformation changes in the 11S component. Arch. Biochem. Biophys. 76: 377–393.
96. Wolf, WJ, and DR Briggs (1959): Purification and characterization of the 11S component of soybean proteins. Arch. Biochem. Biophys. 85: 186–199.
97. Wolf, WJ, JJ Rackis, AK Smith, HA Sasame, and GE Babcock (1958): Behavior of the 11S protein of soybeans in acid solutions. I. Effects of pH, ionic strength and time on ultracentrifugal and optical rotatory properties. J. Am. Chem. Soc. 80: 5730–5735.
98. Wolf, WJ, and DA Sly (1965): Chromatography of soybean proteins on hydroxylapatite. Arch. Biochem. Biophys. 110: 47–56.
99. Wolf, WJ, and DA Sly (1967): Cryoprecipitation of soybean 11S protein. Cereal Chem. 44: 653–668.
100. Wright, DR, and D Boulter (1974): Purification and subunit structure of legumin of Vicia faba L. (broad bean). Biochem. J. 141: 413–418.
101. Yamauchi, F, M Kawase, M Kanbe, and K Shibasaki (1975): Separation of the ß-aspartamido-carbohydrate fractions from soybean 7S protein: Protein-carbohydrate linkage. Agr. Biol. Chem. 39: 873–878.
102. Yamauchi, F, W Sato, Y Kamata, and K Shibasaki (1978): Isolation and subunit structure of γ-conglycinin in soybean globulin. Proc. Fifth Intern. Cong. Food Sci. Technol.: 179.
103. Yamauchi, F, VH Thanh, M Kawase, and K Shibasaki (1976): Separation of the glycopeptides from soybean 7S protein: Their amino acid sequences. Agr. Biol. Chem. 40: 691–696.
104. Yamauchi, F, and T Yamaguchi, (1979): Carbohydrate sequence of a soybean 7S protein. Agr. Biol. Chem. 43: 505–510.

15. Possibilities of Seed Protein Improvement in Tropical and Sub-Tropical Legumes

C. R. BHATIA

INTRODUCTION

Grain legumes or what are commonly known as pulses or beans are next to the cereals in importance as food plants for man. Their cultivation is very ancient and goes back to prehistoric times. Lentils and chickpeas, two important legumes have been found at sites dating back to 7000–6000 B.C. and 5450 B.C. respectively [101]. It is believed that the domestication of lentil and chickpea started either simultaneously with that of cereals like wheat and barley or soon afterwards [101]. In the New World also grain legumes are under cultivation for, at least, 4000 years [15].

As the name indicates, they belong to the family Leguminosae which approximately includes 650 genera and 18000 species [95]. At present, out of these, about 25 are cultivated in tropical and sub-tropical regions for their grain. Species that are extensively cultivated are listed in Table 1, which also gives the regions where these crops are mainly grown. Soybean, *Phaseolus* and peas, though extensively grown in tropical and sub-tropical regions are not considered in this chapter as they are dealt with elsewhere by other authors in this book. Other grain legume species cultivated in the tropics are listed in Table 2. Little information with respect to seed proteins is available for these species [1, 15, 70, 83, 159, 95]. Their shortcomings and breeding objectives are more or less similar to those of the major grain legume crops.

IMPORTANCE OF GRAIN LEGUMES IN TROPICAL AND SUB-TROPICAL AGRICULTURE

Grain legumes are extremely important in tropical and sub-tropical agriculture not only for the high protein grain they provide, but equally for their ability to convert elemental nitrogen into ammonia and other soluble forms of nitrogen that can be utilized by plants. In traditional farming systems still widely practised in the tropical countries, their cultivation helps in maintaining soil productivity in the absence of extensive use of chemical fertilizers. The conversion of atmospheric N_2 to

Table 1. Grain legume species widely cultivated in tropical and sub-tropical regions.

Botanical name	Common name	Areas where grown
Arachis hypogaea L.	Groundnut	Southern United States, India, West Africa in semi-arid regions.
Cajanus cajan (L.) Millsp.	Pigeonpea, Congobean, Red gram	Indian sub-continent, East and West Africa and Caribbean, mainly in semi-arid to humid regions at low to intermediate elevations.
Cicer arietinum L.	Chickpea, Bengal gram	Middle East, Indian sub-continent in semi-arid, cooler regions.
Glycine max Merr.	Soybean	Sub-humid to humid areas in South-East Asia, China and North and South America.
Lens culinaris Medik	Lentil	Near East, North Africa, North India, Central and South America.
Phaseolus vulgaris L.	Common bean, Haricot bean, Kidney bean	North, Central and South America, East Africa.
Pisum sativum L.	Pea	Areas of temperate climate in all tropical, sub-tropical regions.
Vigna mungo (L.) Wilczek	Black gram	South East Asia, Indian sub-continent in semi-arid to humid areas.
Vigna radiata (L.) Wilczek	Mungbean, Green gram	South East Asia, Indian sub-continent in semi-arid to humid areas.
Vigna unguiculata (L.) Walp	Cowpea	Asia, tropical Africa, West-Indies in sub-humid and humid lowlands.

ammonia is brought about by the *Rhizobium* species which live in symbiotic association with legume plants forming nodules on roots. These bacteria depend upon the host legume plant for carbon assimilates [68]. In turn they provide nitrogen to the plant which ultimately contributes towards the harvest of protein in the grain. Thus, unlike other plant protein sources, like cereals and non-leguminous oilseeds, which require nitrogenous inputs in the form of chemical fertilizers, grain legumes meet most of their nitrogen requirement by their own fixation. Some of the fodder

Table 2. Other grain legume species cultivated in tropical, sub-tropical regions.

Botanical name	Common name
Canavalia ensiformis (L.) DC.	Jack bean
Canavalia gladiata (Jacq.) DC.	Sowrd bean
Cyamopsis tetragonoloba (L.) Taub	Cluster bean
Lablab purpureus (L.) Sweet	Kidney bean Hyacinth bean
Lathyrus sativus L.	Grasspea
Lupinus spp	Lupines
Macrotyloma uniflorum (Lam.) Verdc.	Horse gram
Phaseolus acutifolius A. Gray	Tepary bean
Phaseolus lunatus L.	Lima bean
Psophocarpus tetragonolobus (L.) DC.	Wing bean
Vigna aconitifolia (Jacq.) Marechal.	Moth bean
Vigna angularis (Willd) Ohwi & Ohashi	Adzuki bean
Vigna umbellata (Thunb.) Ohwi & Ohashi	Rice bean
Voandzeia subterranea (L.) Thouars var. *subterranea*.	Bambara groundnut

legumes can add up to 500 kg N/ha/year to soil [27, 28]. However, in grain legumes where protein rich grains are harvested, nitrogen added to the soil could vary from 40–200 kg/ha/crop [31]. Due to the increased cost of nitrogenous fertilizers produced using non-renewable fossil fuels, currently there is worldwide interest in legume crops as a renewable source of nitrogen in farming systems [54, 68].

Grain legumes are grown either as a sole crop or in mixed cropping systems, mainly with the cereal crops, under rainfed conditions. When grown in the same season as the principal cereal crop, legumes are generally sown in poor soils. In irrigated areas, they form an important component of crop rotations. In many tropical areas, legume crops are grown along with tuber crops like cassava and sweet potato. The average yields of legumes are lower than those of cereal crops not only under traditional farming in tropical areas but also under advanced farming systems, and temperate climate [84]. Mean yields of grain legumes in tropical and sub-tropical regions vary from 500–2000 kg/ha though the highest yields recorded are much higher [52, 84]. Area, production, yield, protein production and protein yield of the major grain legume crops are given in Table 3.

Table 3. World area, production, yield, protein production yield of major grain legumes.

Crop	Area[1] x 10^6 ha	Production[1] x 10^6 metric tons	Yield[1] kg/ha	Protein[2] Production x 10^6 mt	Protein[2] Yield kg/ha
Soybean	56.7	94.2	1660	35.8	630
Dry beans	25.4	14.8	580	3.3	128
Groundnut[3] (in shell)	18.9	19.2	1016	3.4	182
Chickpea	10.4	7.4	714	1.5	142
Peas	10.5	12.2	1169	2.7	257
Lentil	1.8	1.1	583	0.3	141
Pigeonpea[4]		1.8	630	0.4	132
Cowpea[4]	3.1	1.1	390	0.3	91
Total pulses	72.5	51.8	715		

1 Area, production and yield from PAO Production Year Book 1979.

2 Protein production and yield is calculated based on protein percentage values from Siegel and Pawcett (78).

3 Shelling Percentage for groundnut is taken as 79%.

4 Pigeonpea and cowpea values are from Meiners and Litzenberger (53).

IMPORTANCE OF GRAIN LEGUMES AS FOOD

Grain legumes are nutritionally important as the main source of proteins in the diet of people who cannot afford to buy more expensive animal proteins. For millions of vegetarians in India who, for religious beliefs, do not eat meat products, legumes form the principal source of proteins. Legumes are also an important source of protein for poultry and other monogastric animals that provide meat for human consumption. Besides proteins, grain legumes also provide calories, vitamins and minerals important in human nutrition [15]. At present, in most tropical countries there is a shortage of legumes which is partly due to their low productivity.

End-use: Traditionally grain legumes are processed and consumed as human food in a variety of different ways [15, 62, 78]. Fresh or canned unripe pods or seeds are consumed as vegetable. Pounding, grinding and milling of dry grain is commonly practised to obtain split seeds or cotyledons with or without seed coat. The most common method of cooking is to boil grain, as such, or the dehusked cotyledons in water with salt to make soups or porridges. Whole, husked or unhusked seeds are roasted, parched or puffed. Flour is used in many different ways – baking, steaming or deep frying in oil. Sprouted seeds are widely used as vegetable in oriental foods. Modern processing methods have been developed using flour or isolated protein concentrates to make instant foods, simulated food products and meat analogs especially from soybean and groundnut meal after oil extraction [15, 63, 89]. Besides soybean and groundnut other legumes have a great potential to provide protein concentrates for processed foods, including infant foods that has not been exploited. Soybean and groundnut meal after oil extraction are used as poultry and animal feed. During the main meals, legumes are consumed along with the cereals and in some tropical areas with tubers like cassava or sweet potato. Snacks and other items for munching are often prepared entirely out of legume seeds. The consumption pattern of grain legumes is an important point to be considered in any breeding programme aiming at the improvement of their nutritional qualities [14, 19].

IMPORTANCE AS A SOURCE OF PROTEINS

Seed composition for principal grain legume crops is given in Table 4. In the grain, protein concentration is highest in the embryo, followed by cotyledons and least in the seed coats [14]. Because of their size, cotyledons account for the maximum amount of protein. Grain protein concentration varies with the cultivar and the same cultivar grown at different locations may show variation in protein percentage [10, 58, 72, 91]. Grain protein concentration, expressed as per cent, reflects the ratio of proteins to carbohydrates in most pulse crops, and the ratio of proteins, carbohydrates and lipids in oil crops like groundnut. A decrease in the proportion of carbohydrates or lipids or both would show increased grain protein percentage without any increase in the actual amount of protein. Amino acid composition of the proteins depends upon the species and some cultivaral variations have been noted [14, 42, 52, 100]. Essential amino acid content of tropical grain legumes is given in Table 5.

Nutritional parameters such as the biological value (BV) of proteins, protein efficiency ratio (PER) and net protein utilization (NPU) vary both between and within a species (Table 4). Some of the cultivaral variation in nutritional parameters seen in the published reports could be partly due to differences in the experimental procedures used in different laboratories. At the same time, large inter-cultivaral variations have been observed in experiments carried out in the same laboratory [5]. Addition of methionine in the diet improves PER [14, 62]. In pigeonpea there was no improvement in PER with addition of methionine alone, slight increase with addition of tryptophan but a marked increase was observed when both methionine and tryptophan were added [14].

Storage, processing and cooking considerably alter the nutritional qualities of legume seeds [14, 62]. As mentioned previously, legume seeds are processed and cooked in a variety of different ways. However, little information is available on the protein concentration, amino acid availability and nutritional qualities of the final products consumed.

As also stated earlier, legumes are mostly consumed with cereals and hence it is necessary to consider the nutritive value of mixed cereal-legume diets. The relative proportion of legumes in such combinations varies in different countries. In Latin America, it is estimated to be about 10 per cent on dry weight basis [14]. In Asian

Table 4. Proximate analyses[1], range of protein percentage and nutritional values of tropical grain legumes.

	Moisture %	Ash	Fibre	Fat	Carbo-hydrate	Protein[2] %	Digestibility[3] %	Nutritional values NPU[3]	PER[3]	BV[4] %
Black gram	9.7	4.8	3.8	1.0	57.3	23.4(21.2-31.1)				60-64
Chickpea	9.8	2.7	3.9	5.3	61.2	17.1(10.6-31.1)	86(77-88)	58(52-64)	1.7(1.1-2.2)	52-78
Cowpea	11.0	3.6	3.9	1.3	56.8	23.4(20.4-24.6)	79(73-86)	45(35-51)	–	45-72
Groundnut[5]	5.0	3.7	2.4	48.2	15.9	24.8(20.7-29.1)				
Mungbean (green gram)	9.7	4.0	3.3	1.2	58.2	23.6(19.5-28.5)	81	46	2.1	39-66
Pigeonpea	10.1	3.8	8.1	1.5	57.3	19.2(15.1-31.5)	78(73-85)	52	1.5(1.3-1.7)	46-74

Values in brackets indicate the range.

Source: 1. Siegel and Fawcett (78).
2. Range of protein values are from Swaminathan and Jain (91) for black gram; Jambunathan and Singh (43) for chickpea; Luse and Rachie (52) for cowpea; Samson (75) for mungbean; Jambunathan (42) for pigeonpea.
3. Luse and Rachie (52) except for mungbean where the source is Engel (30).
4. Bressani (14).
5. Groundnut values are recalculated from Peanut Culture & Use (66) considering 5% moisture in raw, dehusked seed.

Table 5. Essential amino acid content of tropical grain legume proteins (g/16 g N).

	Arginine	Histidine	Isoleucine	Leucine	Lysine	Methio-nine	Cystine	Phenyl-alanine	Threo-nine	Tyrosine	Trypto-phan	Valine
Black gram[1]	5.70	2.70	5.50	7.20	6.00	1.10	0.70	5.41	4.30	2.67	0.49	6.40
Chickpea[1]	8.50	1.90	9.98	8.48	7.82	1.23	–	7.86	3.58	–	0.40	3.86
Cowpea[1]	8.00	3.41	5.09	7.44	7.76	1.26	0.51	4.20	4.02	1.98	1.09	5.02
Groundnut[2]	11.58	3.17	3.40	6.53	3.87	0.51	0.33	6.38	2.67	4.09	–	3.88
Mungbean[1] (green gram)	6.30	2.70	6.30	7.69	6.99	0.99	0.61	5.90	3.50	3.86	0.40	6.40
Pigeonpea	6.15[3]	3.55[3]	6.22	8.67	8.73	1.78	–	4.13	3.54	2.00	0.42	4.51

1. Values calculated from Bressani and Elias (15).
2. Values from Basha and Cherry (6).
3. Arginine and histidine values from Bhagwat *et al* (7).

countries it is generally higher [62] for those who can afford to buy the pulses which are more expensive than the cereals. In cereal-legume mixed diet net protein retention (NPR) values increase as the amino acids in cereal and legume proteins are complementary [14, 15, 19, 38]. Cereal proteins, in general, are deficient in lysine while legume proteins are limiting in sulfur amino acids. Maximal increase in nutritive value is obtained in case of cereals which are poor in protein quality like maize and sorghum, followed by wheat, rice and oats. Further, increasing the proportion of legumes in the diet or feeding legumes with increased protein percentage or feeding legume with improved lysine content, enhanced protein utilization in mixed diets [14, 16]. The results from feeding experiments suggest that the limiting amino acid in cereal legume mixed diets depends upon their respective proportion [15, 19]. Legumes with higher protein concentration or higher levels of methionine and lysine improve the nutritional value of mixed diets [15]. It is implied that raising the level of other amino acids except the limiting one in the mixed diets is of no value [19]. Further, increasing the level of the limiting amino acid can improve the nutritive value of the mixture to a level at which the next essential amino acid becomes limiting.

Digestibility. It is generally recognized that grain legumes, with the exception of green gram, are difficult to digest. A greater proportion of N consumed is passed out as faecal N following consumption of legumes in comparison to animal proteins. This is due to poor digestibility of legume seed proteins. Cultivaral variation in digestibility has been observed. In pigeonpea and cowpea cultivars digestibility varied between 59–90 and 86–90 per cent respectively [14]. Grain size has a significant influence on protein digestibility. Small seeds have heavy cotyledons which are difficult to digest and hence, there is a natural consumer preference for larger seed size. Prolonged storage of seeds causes their discoloration and hardening. The latter enhances cooking time and reduces digestibility [18]. Digestibility is one of the most important factors influencing the availability of amino acids and hence the nutritional value of grain legumes [50].

Flatulence. Consumption of most legume seeds increases flatulence [17]. Low molecular weight carbohydrate fractions stachyose and raffinose are responsible for the production of gas by the anaerobic bacteria in the gastrointestinal tract [17, 40].

Toxic and anti-nutritional factors. Several types of toxic and anti-nutritional factors have been reported from legume seeds. They have been extensively reviewed [1, 45, 49, 51]. Some like protease inhibitors are heat labile and are destroyed during cooking. Heat stable factors like cyanogens and alkaloids are generally absent, or present in insignificant amount, in the tropical grain legumes considered here and do not appear to seriously impair their nutritional qualities. However, such factors are a major limitation in the utilization of some minor tropical legumes like *Lathyrus sativus*.

Storage proteins of legume seeds. Seed proteins whose only known function is to serve as a store of nitrogen are defined as storage proteins [25]. They constitute the bulk of proteins in seeds. In case of legumes, they are deposited in the cotyledonary cells in membrane bound protein bodies. On seed germination protein bodies gradually disappear, storage proteins are broken down and utilized by the growing seedling [98]. Storage proteins of legume crops like pea, soybean *Phaseolus* and groundnut have been extensively investigated. However, the same is not true for the other tropical legume species [25]. Even in well investigated crops, the knowledge regarding the storage proteins, their sub-units, synthesis and deposition in protein bodies is still inadequate for complete understanding and manipulation to achieve nutritional improvement in seed proteins [93].

A major portion of legume seed proteins are salt soluble and are classified as globulins. Albumins, prolamins and glutelins form only a small fraction of the legume seed proteins. Originally pea seed globulins were separated by Osborne into legumin and vicillin, based on their solubility in dilute salt solutions and heat coagulability [25]. All other legume seed proteins examined, so far, also fall in these two broad categories. Ultracentrifugation technique led to the identification of two major globulin fractions from 34 legume species having sedimentation coefficients ranging between 6.6 – 8.3s and 11 – 13s [23]. They are commonly referred as 7s and 12s fractions and correspond to vicillin and legumin respectively. Globulins sedimenting at 2s and 10s are also found in many legume species. The molecular weights of proteins sedimenting at 2, 7, 10 and 12s are 20, 180, 300 and 350 $\times$ 10^3 respectively [12]. Carbohydrate content of the 7s and 12s protein is in the range of 2–5 and 1 or less than 1 percent respectively [12].

With change in ionic strength and pH both 7s and 12s fractions show reversible association and disassociation of subunits. On reduction with 2-mercaptoethanol 12s proteins separate into larger, acidic and smaller basic subunits. After polyacrylamide gradient electrophoresis and isoelectrofocussing further heterogeneity has been observed [12]. These, N-terminal amino acid analyses of subunits and inheritance studies point to genetic heterogeneity and polymorphism of 12s proteins. Subunit structure of 7s proteins is still less known. Nevertheless, subunit heterogeneity has been observed [94]. 12s and 7s fractions differ in their amino acid composition, 12s fraction, in general, has more sulfur amino acids [25].

INCREASED LEGUME PROTEIN PRODUCTION

The overall situation, in most of the tropical and sub-tropical developing countries, demands increased availability of legume proteins at a cost commensurate with the earning capacity of the weaker sections of their population. It has been shown that cereal-legume mixed diets, if available in proper balance and adequate quantity, can satisfy protein and calorie needs of adult human beings [36, 90, 97].

Protein deficiency in the diet arises due to inadequate amount of either legumes in the diet or the total food consumed, which is often due to low income and purchasing capacity. Hence the first priority in breeding programmes should be to aim at increasing yield potential and stability of yield. In many tropical and sub-tropical areas grain legumes can be grown throughout the year, if adequate irrigation facilities are available. Therefore, crop productivity in relation to the crop duration is significant in such areas. In other words, grain as well as protein productivity should be considered on per day basis. Resistance to stress factors like diseases, pests and environmental stresses is an important component for the stability of yield and should have high priority. However, in yield oriented breeding strategy it is important to constantly monitor the protein and nutritional qualities of the advanced lines. The high yielding types should, at least, maintain the nutritional properties of the cultivars presently grown. It is also important that the new cultivars and associated agronomy should be such as to keep the production costs low.

Improving protein concentration, amino acid composition and nutritional qualities. Most tropical and sub-tropical legume crops cultivated, at present, have not been directly selected for their grain protein concentration or amino acid composition. However, selection against toxic and anti-nutritional substances, grain appearance, end-use properties, taste, yield and resistance to stress factors has been practised since their domestication. This has, no doubt, contributed to increased yield of grain and indirectly also of protein. Even in the present day breeding programmes, grain composition has received little attention in legumes except in soybean and groundnut. Thus, there is considerable scope for the improvement of protein quantity, quality and digestibility in most tropical grain legume crops [50, 52, 59, 85]. However, there are several constraints which are discussed further.

Breeding constraints. One of the major constraints in initiating breeding programmes to improve protein characteristics is the absence of clear guidelines from the nutritionists as to what improvements are needed in specific situations. Such guidelines can be given only after nutritional surveys for specific areas, considering the predominant cereal – legume or tuber – legume mixed diets consumed. This type of information is not available for most tropical and sub-tropical regions though some generalized recommendations given in Table 6 have recently been made [16].

Breeders have to select for a large number of characters and each additional parameter reduces the choice of parents to be used in crossing programmes to generate variability, as well as the number of selections that can be made in segregating generations. This reduces the probability of success and enhances the time and effort necessary to reach the desired goal [19, 65, 99]. Therefore breeders are also hesitant to devote their resources to such programmes. Uncertainty of the future consumer demand for such nutritionally superior cultivars also contributes

Table 6. Proposed nutrient levels in grain legumes.

When supplemented to	Protein %	Lysine	Tryptophan (g/16 g N)	Total sulphur amino acids
Cereals	28–30	6.3–6.4	1.10–1.20	2.3–2.4
Cassava	28–30	6.3–6.4	1.10–1.20	3.2–3.4

Based on Bressani (16).

to this reluctance [99]. Consumers should also be willing to pay a higher price for improved nutritional qualities, as they always pay for better size and appearance of grain, which, of course, can be seen unlike the improved nutritional qualities.

Genetic Constraints. There are few studies on the inheritance of protein concentration or their amino acid composition in tropical legumes. Where genetic investigations on grain protein percentage have been carried out, they indicate polygenic inheritance. Protein percentage, in general, is negatively correlated with yield and with the amount of sulfur containing amino acids expressed as g/100 g protein [2, 3, 11, 37, 47]. Where legume proteins have been examined at the protein sub-unit level the indications are for the multiplicity of loci coding for storage protein sub-units [94]. Hence, any single mutation is not likely to bring about drastic alterations in the amino acid composition. Mutations similar to the *opaque-2* in maize, and other cereals with significant alterations in amino acid composition and enhanced lysine content of the grain have not been found in any of the legumes. Possiblities of genetic alteration in globulins which are the main seed storage proteins are limited [93]. Globulin fractions rich in sulfur containing amino acids have been found and increasing this fraction by genetic or agronomic means is perhaps the best possible alternative. Further legumins, in general, are nutritionally superior to the vicilins because of their better sulfur amino acid content and hence increase in the amount of legumins would improve the quantity of limiting amino acids [94]. Increase in albumin fraction, though rich in limiting amino acids, is likely to disturb the plant metabolism and may not be practical.

Physiological and Bioenergetic Constraints. Grain protein percentage examined from physiological and biochemical view points is a complex character which is the end product of a chain of metabolic events. Thus several component characters contribute to the end parameter. The most important are the rate and duration of N accumulation in developing seeds. Amino acids or amides are the main source of N for developing seeds. Fruit and seed growth has been investigated extensively in peas, soybean and *Phaseolus* spp. [29, 60]. However, little information is available on physiology and biochemistry of developing seeds in tropical grain legumes [82]. Duration of grain filling and N accumulation varies with the species and also depends upon the temperature and water stress during the period of pod growth. In peas externally supplied N^{15} reached the seeds via leaf and pod cover [48]. The available evidences indicate a similar path for N in other legumes [60]. Because of asynchronous flowering and fruit development and the limitations imposed by crop duration, it is generally not possible to extend the period of grain filling. Hence to meet the N demand of the developing seeds, the other alternative is to increase the rate of N accumulation. As such, the nitrogen demand of grain legume seeds is higher than is evident from Figure 1, where mg N/g photosynthate is plotted against seed biomass productivity which is defined as g seed/g photosynthate [79]. Limitations of space do not permit to elaborate on the derivation of these values. It would suffice to say that they are based on extensive examination of biochemical pathways and energy requirements in heterotrophic plant systems, indicating that 1 g of glucose can be used to produce 0.83 g carbohydrate, or alternatively 0.40 g of protein (assuming nitrate to be the N source) or 0.33 g of lipids [67].

The grain legume species considered here fall in three groups, chickpea, pigeonpea, cowpea and mungbean have biomass productivity in the range of 0.64–0.66 and N requirement of 23–28 mg/g photosynthate. Groundnut because of its high lipid content has the lowest seed biomass productivity though its N requirement is not as high as those of the other grain legumes. In contrast, soybean has even higher N requirement. The horizontal dotted line in Figure 1 is drawn considering N uptake and biomass production at the rate of 5 kg/ha/day and 250 kg/ha/day respectively, which represent the highest observed values for any crop. Thus, at best , 20 mg N/g of photosynthate can be available to the developing seed from soil. Most grain legume species fall above this line and are not able to meet N demand of their seeds by uptake from soil or by fixation. Nitrogen requirement of developing seeds is met by mobilization from foliage which, in turn, induces leaf senescence and abscission. This phenomenon has been described as 'self destructive' and it is suggested that it is responsible for low seed yield of grain legume species [79]. 'Self destructive' mechanism was first described for soybean [80]. Observations in mungbean also point to a similar mechanism [7] and it is likely that 'self destruction' operates in other grain legumes having annual growth habit.

It has been shown in cereals that increase in grain protein enhances the N demand for seed biomass [8]. Nitrogen requirement for seed biomass necessary to bring about 1% increase in grain protein is illustrated in Table 7 taking chickpea as an example. For each percent point increase in protein, N requirement is enhanced by about 3.5%. Similar results are obtained for other legumes. As brought out in the previous para, grain yield in 'self destructive', annual grain legume species is already partly limited by high N demand of their developing seeds. Genetic increase in grain protein (N) percentage would further enhance their N requirement. Nevertheless, higher seed protein percent can be obtained by ensuring adequate overall N supply

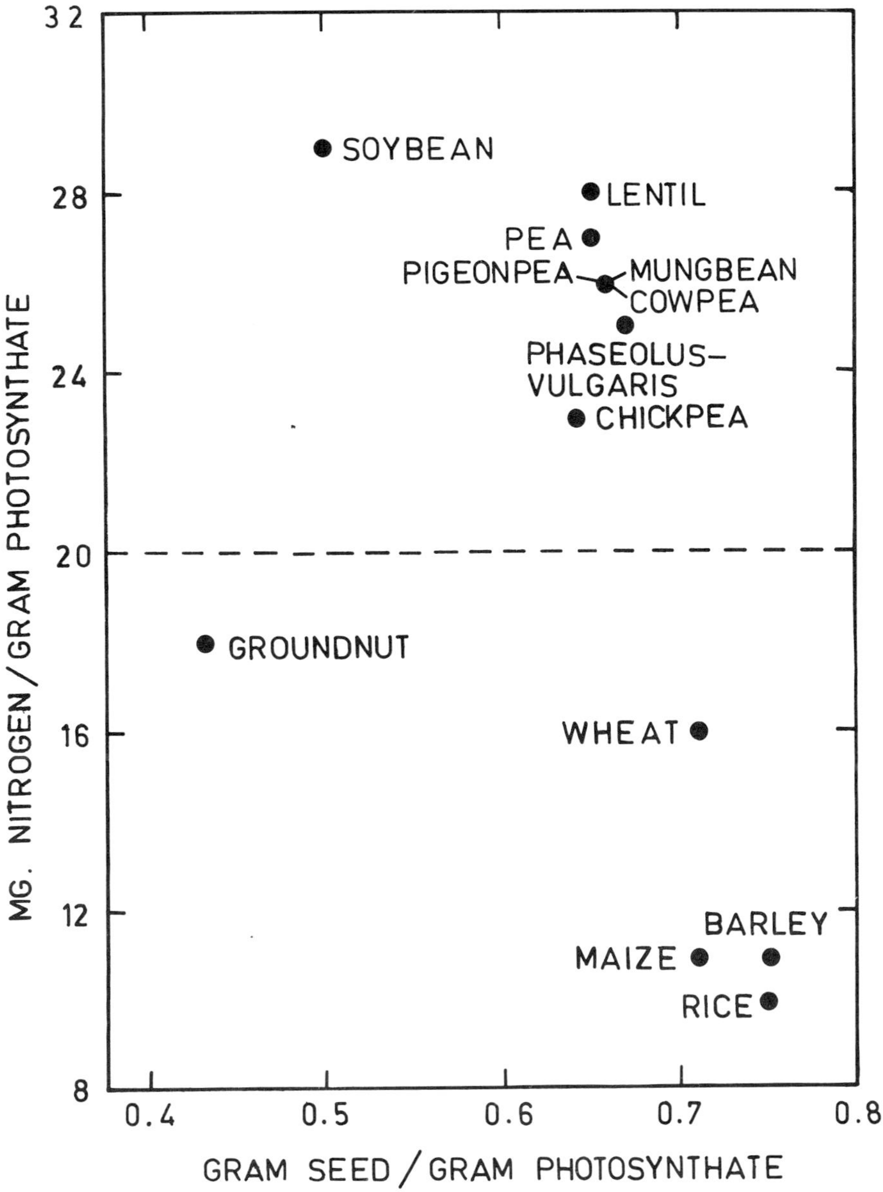

Figure 1. Nitrogen requirement (mg N/gram of photosynthate) of seeds is plotted against seed biomass productivity (gram of seed/gram of photosynthate) for important grain legumes. Cereals like wheat, maize, barley and rice are included for comparison. The dashed line represents nitrogen requirement when the nitrogen supply rate is 5 gram/ha/day and available photosynthate rate is 250 kg/ha/day.

Source: Sinclair and de Wit [79].

to the plant and its efficient translocation to the developing seeds. This can be achieved either by enhancing N fixation of the *Rhizobium*-legume interactions or by application of N fertilizers or more efficient mobilization from leaves in certain species. Both N fixation and mobilization consume energy [54]. It has been shown in soybean, cowpea and white clover that legume plants fixing their own N respire 11–13% more of their fixed carbon each day in comparison to similar plants lacking nodules and utilizing nitrate nitrogen [74]. This is mainly due to increased respiration of nodulated roots. Moreover, availability of adequate soil N decreases N_2 fixation in grain legumes [54, 68].

As illustrated in Table 7, 1% increase in grain protein at the expense of starch would enhance glucose requirement by 0.83%. This marginal increase in photosynthate requirement is not likely to cause significant decrease in yield, provided the plant's CO_2 fixation capacity is adequate. However, in legumes, developing fruits as well as the nodulated roots are two major sinks for carbon assimilates. Moreover, at least, in some short duration grain legumes, like mungbean, results indicate that yield is limited by low rates of plant dry matter increase [64].

It has been shown that the construction cost of methionine, the most limiting amino acid in grain legumes is 64% higher than that of glutamic acid which is the most predominant amino acid in legume proteins [55]. Hence, if methionine increase is brought about by a proportionate reduction of glutamic acid there will be some additional glucose requirement to meet the energetic cost.

The above discussion leads to the conclusion that improvement in protein quantity, protein quality and yield are feasible only by ensuring adequate and balanced supply of photosynthates and nitrogen. Since in the legume plants main source of N is from legume – *Rhizobium* symbiotic association which, in turn, is dependent on carbon assimilation, the latter becomes the key factor for yield and protein improvement. Improvement in digestibility of grain legumes can increase the overall availability of proteins [3, 50], and some of the constraints associated with increasing protein concentration may not apply, but then, at present very little is known about digestibility.

So far, the grain legume crops have been discussed as a group. In the following, available information on origin, main areas of cultivation, end uses and prospects for improving seed proteins are summarized for five major crops – groundnut, chickpea, pigeonpea, cowpea and mungbeans. These species widely vary in their distribution, growth habit and seed composition. Groundnut differs from others in having aerial flowers but the pod development is underground. Chickpea is adapted to lower temperature. Pigeonpea has a perennial growth habit and even as an annual crop, it produces a large phytomass. Cowpea and mungbeans have many features in common, taxonomically they are now included in the same genus *Vigna*.

Table 7. Nitrogen requirement and energetic cost of increasing grain protein concentration in chickpea.

Component	Amount[1] (g/100 g)	PV[2]	Equivalent[3] glucose required (g)	Nitrogen[4] requirement mg/g of photosynthate
Standard cultivar				
Carbohydrate	68	0.83	81.93	
Protein	23	0.40	57.50	
Lipid	5	0.33	15.15	
Minerals	4			
		Total	154.58	23.81
Cultivar with 1 percent more grain protein				
Carbohydrate	67	0.83	80.72	
Protein	24	0.40	60.00	
Lipid	5	0.33	15.15	
Minerals	4			
		Total	155.87	24.64
Percent increase			0.83	3.49

1. Standard chemical composition is from Spector (86).
2. Production value (PV) is calculated as (weight of the end product) / (weight or substrate required for carbon skeletons and energy production). PV's are taken from Penning de Vries *et al* (67).
3. Equivalent glucose requirement is calculated as (amount of component) / PV, where the amount is expressed as grams per 100 gram of grain biomass.
4. Nitrogen requirement is calculated by assuming that protein is 16 percent nitrogen by weight.

GROUNDNUT (*Arachis hypogaea* L.)

Among the grain legumes groundnut, known as peanut in America, is second to soybean in world production. It differs from the other grain legume crops considered here, in that seeds are rich in both oil and protein. It is cultivated in warm temperature and tropical semi-arid regions in Asia, Africa and America; India, China, United States, Nigeria and Senegal are the leading producers [66]. The genus *Arachis* is native of South America with 40–70 species. Brazil is considered as the centre of diversity for the genus *Arachis*, while Southern Bolivia is believed to be the centre of domestication for groundnuts [34]. At the time of the discovery of America, groundnut was widely grown there in the tropical and sub-tropical regions. Later it spread to Africa and Asia [66]. The earliest archaeological records date back to 2000–3000 B.C. in Peru [34].

Groundnut is one of the well investigated grain legumes next only to soybean.

Considerable breeding, agronomic and other investigations have been carried out in the U.S.A. in the past thirty years. In the fifties the average pod yield in the U.S.A. was around 800 kg/ha. Improved cultivars and agronomy have raised this to the current yield levels of over 2000 kg/ha [66]. In other countries where groundnut is extensively cultivated, yield levels are in the range of 1000 kg/ha, they are still lower around 800 kg/ha in India, which is the largest producer.

Unlike other pulses, groundnuts do not form a regular part of the diet, they are only added to the main cereal dishes or made into sauces. However, the main consumption throughout the world is in the form of roasted nuts. Peanut butter is manufactured in many countries which provides an excellent protein supplement in the diet. In India groundnut is an important source of oil and the crop is primarily considered as an oilseed crop. The meal after oil extraction is largely used as feed or added to soil as fertilizer. Its role as an important source of proteins for human nutrition has been recognized only recently. A large number of products have been developed at the Central Food Technological Research Institute, Mysore in India and in the U.S.A. based on groundnut meal or protein isolates. These include blended flours with cereals or cassava, multipurpose food supplement, weaning and infant foods and other high protein foods like biscuits and soft drinks [63, 89]. Groundnut has a great potential as a source of food proteins which has not been utilized, as yet, to the same extent as that of soybean [6]. Infection of nuts as well as the meal with *Aspergillus flavus* due to improper storage and consequent contamination with aflatoxin is a serious problem limiting the utilization of groundnut meal as food and feed [26].

Arachin and conarachin (also known as non arachin) are the two major storage proteins in groundnut cotyledons [87]. Their proportion depends upon the cultivar, the environment where it is grown and above all on the frationation procedures used [6]. Both proteins are composed of several sub-units of different molecular weights [76]. The two fractions also differ in their amino acid composition (Table 8). Conarachin fraction is relatively rich in lysine, methionine and cysteine in comparison to arachin which is deficient in sulfur amino acids, threonine being the most limiting. Chemical scores of the two fractions, obtained by comparing the essential amino acid content with the recommended FAO reference pattern were 31–38% and 68–82% respectively for arachin and conarachin [24].

Seasonal variation in 26 strains representing a large cross section of the elite germplasm for 8 years revealed considerable year to year variation in protein percentage [39]. Protein concentration was generally higher in years of low rainfall. Rise or fall in protein was inversely associated with increase or decrease in oil. Similarly high protein strains were low in oil and vice versa but at the same time there were some exceptions [39]. In another study, protein percentage in the wild species of *Arachis* and cultivated strains of *Arachis hypogaea* were in the range of 17.1–30.8% and 20.7 to 29.1% respectively [20]. A significant negative correlation is obtained between protein and oil percentage. In six cultivars and their F_2 seed

Table 8. Amino acid composition (g/16 g N) of arachin and conarachin fractions of groundnut.

Amino acid	Arachin	Conarachin
Lysine	2.49	4.13
Methionine	0.59	1.89
Cystine	0.88	1.85
Leucine	7.05	6.39
Isoleucine	3.78	3.33
Valine	4.51	3.86
Threonine	2.56	2.04
Tyrosine	5.13	3.04
Phenylalanine	6.20	4.28
Aspartic acid	12.63	11.29
Serine	5.10	4.86
Glutamic acid	22.23	21.61
Proline	5.56	4.50
Glycine	4.53	6.96
Alanine	4.45	3.36
Histidine	2.43	2.38
Arginine	12.69	12.34
NH_3	2.16	2.39

Mean of eight cultivars. Based on the values given by Dawson and McIntosh (24).

population, protein as well as oil were quantitatively inherited. Correlation coefficients between protein and oil were negative and highly significant in some crosses [92]. Amino acid composition was also quantitatively inherited. Application of sulfur dust as fungicide increased yield and protein concentration in crop grown in sulfur deficient soil [46].

Breeding for improved protein quantity or quality has not received serious attention in breeding programmes. Significant improvement in protein concentration is likely to be at the expense of reduction in oil. Simultaneous increase in both protein and oil has a high energetic cost [56]. Altering the amino acid composition seems feasible by increasing the proportion of conarachin proteins.

CHICKPEA (*Cicer arietinum* L.)

Chickpea also called Bengal gram is an important legume crop grown during the cooler months in semi-arid, sub-tropical regions in Asia and Africa. India and Pakistan are the major producers. Ethiopia and Morocco in Africa, Spain, Portugal and Italy in Europe and Mexico in North America are the other producers.

Chickpea originated in the Caucasian region and from there spread both towards east to India and to the west, along the Mediterranean [71]. The earliest archaeological records date back to 5450 B.C. [101]. Maximum diversity is found in western Asia and the Caucasus region.

Two types of chickpea are cultivated. The eastern ecotype, largely grown in the Indian sub-continent is characterized by thin stem, smaller leaflets, flowers and seeds which are often wrinkled. The Mediterranean ecotype has tall plants, large leaflets and flowers and produces large, light coloured seeds. This ecotype has been greatly modified by human selection. The types also differ in their chemical composition (Table 9). The mediterranean ecotype accounts for 10–15% of the world production.

Chickpea is generally grown under limited inputs. Highest average yields, in the range of 1000–1300 kg/ha, are obtained in Turkey, Greece and Bulgaria. In other parts, where the crop is extensively grown, average yields are below 1000 kg/ha.

Chickpea though considered as poor man's food is one of the most useful plants from which a wide range of dishes are prepared. Green, unripe seeds, dry seeds, sprouted seeds, and cotyledons are consumed. Flour obtained from whole seeds or cotyledons after removal of seed coat is either used as such or mixed with barley or wheat flour to make unleavened bread. Grain is widely used as feed for horses and cattle. It is a good source for making inexpensive protein rich multipurpose or infant foods for the poor.

Table 9. Composition of mediterranean and eastern ecotypes of chickpea.

	Whole seed		Cotyledons	
	Mediterranean	Eastern	Mediterranean	Eastern
Protein %	24.0	22.4	25.0	26.8
Starch %	48.6	43.7	55.6	54.4
Sugar %	6.1	5.4	5.4	5.2
Fibre %	3.2	9.2	1.2	1.1
Fat %	4.7	4.1	5.3	4.8
Ash %	3.2	3.3	3.1	2.9
100 seed weight (g)	22.7	17.6		
Seed coat (%)	7.1	16.0		
Seed coat N (%)	0.95	0.59		

Based on the data of Jambunathan and Singh (43). Values are mean of 7 mediterranean and 8 eastern ecotype cultivars grown at Hissar.

Table 10. Chickpea grain quality (cotydedons).[1]

Constituent	Number of samples	Range	Mean
Protein (%)	17679	10.6 –31.1[2]	20.5
S–amino acids[3] (Met + Cys) (% protein)	52	2.07– 2.95	2.31
Methionine (% protein)[4]	80	1.10– 1.63	1.27
Tryptophan (% protein)[5]	10	0.59– 0.95	0.74
Protein digestibility (%) (*in vitro*)	15	47.4 –64.1	52.1
Starch (%)	32	51.1 –58.1	55.6
Soluble sugars (%)	32	4.1 – 6.0	5.1
Wether extract (%)	32	3.5 – 6.8	5.5
Crude fibre (%)	32	0.7 – 1.3	1.1
Ash (%)	32	2.1 – 3.7	2.9

1 Based on the data of Jambunathan and Singh (43).
2 Mean of 4352 samples.
3 Ion exchange chromatography after performic acid oxidation.
4 Estimated by microbilogical method.
5 Estimated by ion exchange chromatography.

Besides the national breeding efforts, intensive crop improvement programmes are in progress at the International Crop Research Institute for Semi-Arid Tropics (ICRISAT), Hyderabad, India and at the International Centre for Agricultural Research in Dry Areas (ICARDA), Syria. Literature on chickpea published between 1930–1974, including protein quality aspects is collated in Chickpea Bibliography [81]. In the past, little efforts have been made to investigate and improve protein quality and quantity. Germplasm of over 11,000 accessions is maintained at ICRISAT. Part of this has been screened for seed protein content (Table 10). Grain protein varied from 10–31%. In an independent screening of 1320 germplasm lines methionine (mg/g sample) varied from 1–3.55 [77].

Sulfur amino acids (methionine and cystine) and tryptophan are the limiting amino acids. In a sampling of 30 cultivars protein (%) was negatively correlated with methionine and cystine (% of protein). Total sulfur in seed (g/100 g meal) was positively correlated with cystine + methionine [41]. Heritability for grain protein and methionine was estimated to be 48 and 23 percent respectively in a 5×5 diallel [77].

PIGEONPEA (*Cajanus cajan* Millsp.)

Pigeonpea also known as red gram or congo bean, is extensively grown in the Indian sub-continent, Africa, Caribbean islands and parts of Latin America. As a forage crop, it is grown in South-Eastern parts of the U.S.A., Hawaii, Australia and Italy. It is a hardy crop, well suited to the semi-arid regions. India accounts for nearly 90% of the world production.

Though it is grown as an annual crop, basically the plant has a perennial growth habit and grows as tall shrubs up to 5 m in height. Duration of the crop varies from 120 to over 300 days, though plants can survive up to 10–12 years under favourable environment. It is grown as a single crop, but more often in multiple cropping systems, with low inputs.

Cajanus is a monotypic genus which is believed to have originated from wild species of *Atylosia* in India [73] though pigeonpea seeds dating back to 2000 B.C. have been found in Egyptian tombs [70]. Some of the *Atylosia* species can be readily crossed with *Cajanus* cultivars [72].

Split cotyledons, after removal of the seed coat are mainly used to make soup like preparation which is eaten with cereals. Green pods are used as vegetable. Canned, unripe seeds are used on a limited scale. Seeds are generally free from metabolic inhibitors and flatulence producing sugars.

Little work has been done on the improvement of protein content or quality in this crop. Most intensive breeding programmes and germplasm screening for protein percentage are underway at ICRISAT. Protein percentage in dehusked seeds of 21403 cotyledon samples from germplasm collection ranges between 15.1 – 31.5% [42]. Variation in the same cultivar, grown at different locations and seasons ranged from 18.7 to 28.8 [72]. Other grain quality characters are given in Table 11. Protein percentage was higher in wild *Atylosia* species and ranged between 28.7 – 30.2%. Protein content in *Cajanus* × *Atylosia sericea* and *Cajanus* × *A. scarabaeoides* F_6 lines ranged between 20.5 – 33.7% [72].

Sulfur amino acids (methionine, cystine) and tryptophan are the limiting amino acids. In a sampling of 25 cultivars cystine and methionine (g/100 g protein) were negatively correlated with protein percent though the correlation was not significant. These amino acids, expressed as percentage of protein, decline with maturation [41]. Methionine and cystine content (g/16 g N) was higher in *Atylosia serecia* and *A. scarabaeoides*, and was intermediate in their hybrids with *Cajanus* [72]. These values were, however, in the same range as observed in *Cajanus* cultivars.

Seed proteins have been partially characterized [33], about 78% of seed proteins were salt soluble out of which 61% were globulins. These were further separated into α, β and γ fractions [33]. Globulins on SDS gels also separated into 9 subunits, while the α, β and γ fractions separated into 7, 3 and 3 subunits respectively. Many subunits were common. The amino acid compostition of the three fractions was

Table 11. Pigeonpea grain (cotyledons) quality.[1]

Constituent	Number of samples analyzed	Range	Mean
Protein %	21403	15.1 –31.5	23.2
S–amino acids[2] (Met + Cys) (% of protein)	58	0.96– 2.55	1.96
Tryptophan[3] (% of protein)	10	0.47– 0.63	0.53
Protein digestibility (%) *in vitro*	25	42.6 –70.0	58.1
Starch %	10	57.7 –63.2	60.7
Soluble sugars %	10	4.7 – 5.7	5.2
Ether extract %	10	1.2 – 2.2	1.6
Crude fibre %	10	1.0 – 1.2	1.1
Ash %	10	3.7 – 4.3	3.9

1 Based on the data of Jambunathan (42).

2 Estimated by ion exchange chromatography after performic acid of oxidation.

3 Ion exchange chromatography.

different (Table 12). The α fraction was deficient while the γ fraction was considerably rich in sulfur amino acids, and contained about four times the total sulfur amino acids compared to the α fraction. Identification of a globulin fraction rich in sulfur amino acids suggests possibilities for increasing methionine and cystine.

COWPEAS (*Vigna unguiculata* L. Walp)

Cowpeas are grown either as a pulse or as a vegetable crop throughout the tropical and sub-tropical regions of Africa, Asia and to a lesser extent in the New World. It is the most important grain legume of West Africa. In advanced tropical farming systems, it is often grown as a forage crop. Intensive research for the improvement of this crop is being carried out at the International Institute of Tropical Agriculture (IITA) at Ibadan, Nigeria. The major emphasis is to improve yield and stability of yield through control of diseases and pests [32]. A large germplasm collection is also being maintained at IITA.

It is believed that cowpea was domesticated in Ethiopia from where it spread to West Africa, India, South East Asia, and Far East [88]. It was introduced in the tropical regions of the New World by the Spanish. At present, maximum diversity is found in West Africa and India.

Table 12. Amino acid composition of the globulin fractions of *Cajanus Cajan* (g/16 g N).

Amino acid	Total globulin	fraction	fraction	fraction
Lysine	4.40	4.73	4.40	3.17
Histidine	2.76	2.68	2.00	1.33
Ammonia	6.13	7.24	4.48	7.57
Arginine	4.40	4.45	3.79	2.59
Methionine sulfoxide	0.13	TRACE	0.13	0.22
Aspartic acid	6.12	8.05	6.05	4.67
Threonine	2.67	2.99	2.47	2.26
Serine	3.43	4.02	3.71	2.85
Glutamic acid	11.15	14.58	9.86	5.03
Proline	3.32	3.60	3.20	2.70
Glycine	2.52	2.28	2.28	2.24
Alanine	2.99	3.05	2.86	2.28
Half cystine	0.45	–	0.32	1.72
Valine	2.93	3.06	2.79	3.25
Methionine	0.81	0.77	0.68	0.85
Isoleucine	2.57	3.11	2.75	2.14
Leucine	5.27	6.95	6.15	3.65
Tyrosine	2.37	2.42	1.84	2.21
Phenylalanine	4.91	7.15	5.49	2.50
Methionine sulfone	0.18	TRACE	TRACE	0.33

All values are ± 2%, except methionine ± 5%, 1/2 cystine, methionine sulphoxide and sulphones ± 10%.

Source: Gopala Krishna, Mitra and Bhatia (33).

Green pods, immature, dry and germinated seeds are consumed. Shelled green peas are canned in the U.S.A. Maximum consumption, however, is in the form of dry seeds, which, like other pulses, are eaten in a variety of different ways after steaming or frying.

Average yields are in the range of 400 kg/ha though the highest reported yield is 3500 kg/ha [52]. In the germplasm collection of over 5000 accessions protein in seed varied from 20.4–34.6 [52]. Cowpea proteins are rich in lysine (Table 13). Methionine, cystine and tryptophan are considered to be the limiting amino acids in that order [13]. In a sampling of 1200 cowpea introductions highest methionine values recorded were 3.26 mg/g of seed meal against 2.75 of the check variety which was in the normal range. There were other lines with 20–30% less methionine than the standard check [35]. Out of 2300 entries tested in the germplasm of the IITA,

0.13% and 0.0006% lines were higher than the average in protein and sulfur amino acids on mg/g flour basis [52]. In another study heritability of protein and sulfur amino acids was 0.54 and 0.46 [10].

Agronomic practices have been tried to increase protein or the limiting amino acids. Rhizobial inoculation of soil as well as application of nitrogenous fertilizers increased seed protein content [44]. Sulfur application to the crop did not enhance yield, methionine or cystine content of the grain, though total sulfur content and S-methyl cystine increased [57].

Storage proteins have been partially characterized [18]. 7s globulin which is heterogenous is the major fraction. 11s globulin is typical of the legumin proteins of the other legumes. Amino acid composition of three major sub-units was deter-

Table 13. Amino acid composition of cowpea var. 'prima'.

				Globulin subunits		
	Meal (g/100 g meal)	Albumin (g/100 g protein)	Globulin (g/100 g protein)	Mol. wt. 56000 (g/100 g protein)	Mol. wt. 54000 (g/100 g protein)	Mol. wt. 52000 (g/100 g protein)
Aspartic acid	9.65	12.26	11.29	11.5	8.42	12.00
Threonine	3.47	4.62	2.99	2.48	1.65	2.62
Serine	4.89	4.56	4.26	5.70	3.44	4.98
Glutamic acid	15.60	12.00	15.75	13.6	10.74	13.22
Proline	3.28	3.34	3.78	3.14	3.08	3.76
Glycine	4.01	3.76	2.72	2.96	1.97	2.21
Alanine	3.81	4.28	2.83	3.14	1.93	2.99
Cysteine[1]	1.01	2.69	0.55	0.6	0.37	0.23
Valine	5.09	3.75	4.96	3.40	2.52	4.30
Methionine	1.40	1.06	0.98	Trace	0.72	0.81
Isoleucine	4.46	2.43	3.86	3.06	2.13	4.15
Leucine	7.47	2.96	6.36	6.51	4.54	9.42
Tyrosine	2.86	2.65	2.23	2.37	1.68	3.12
Phenylalanine	5.05	2.78	6.66	4.64	2.53	7.05
Histidine	3.06	2.64	3.27	2.12	2.72	2.67
Lysine	6.27	5.55	6.75	4.03	4.25	5.56
Arginine	6.36	4.32	6.00	4.51	5.53	9.42
Tryptophan	0.85	n.d.	n.d.	n.d.	n.d.	n.d.

1 Determined as cysteic acid after performic oxidation.

n.d. = not determined.

Source: Carasco et al. (18).

mined [18], concentrations of sulphur containing amino acids were different in the three sub-units (Table 13). Studies on developing seeds indicated that individual globulin sub-units are probably under separate genetic control [18].

Little serious efforts have been made, besides screening of germplasm, for the improvement of protein concentration or limiting amino acids. An intensive programme was initiated at IITA but protein improvement has a low priority in their cowpea breeding programmes, at present [32].

MUNGBEAN

Mungbean is an important short duration (60–120 days) crop, widely grown in tropical and sub-tropical parts of Asia. They are also cultivated to a limited extent in Central America and in Australia. India is the largest producer followed by Thailand which is the biggest exporter not only to other Asian countries but also to Europe and North America. Two species now classified as *Vigna radiata* (L.) Wilczek (green gram or golden gram) and *Vigna mungo* (L.) Wilczek (black gram) are commonly referred to as mungbean in literature [70]. They have also been classified as variety *aureus* and *mungo* respectively of a single species *Vigna radiata* [96]. Previously these species were included in the genus *Phaseolus* and known as *Phaseolus aureus* Roxb (green gram) and *P. mungo* L. (black gram).

Both species have wide adaptability and are grown as pure crops or as inter crops in multiple cropping systems. Very early maturing types (60–80 days crop) are known in green gram which is often grown as a catch crop, in between the two main crops, when water is available for irrigation. As the common names indicate, most *V. radiata* (L.) Wilczek cultivars produce seeds with green, bright green or yellow testa while *V. mungo* (L.) Wilczek seeds are mostly black, though types with dull green seed coats are known. The two species can be differentiated on the basis of pod and seed characteristics. Pods are reflexed or pendant with short hairs in green gram whereas they are erect or sub-erect with long hairs in black gram [85]. The two species can be crossed [4, 22] with each other and also with *Vigna umbellata* (Thnub) Ohwi and Ohashi and *V. angularis* (Willd), Ohwi and Ohashi respectively known as rice bean and adzuki bean [4]. Mungbeans are believed to have originated in India from common ancestral species *V. radiata* var. *sublobata* and or *V. trinervis* [70]. Mungbeans are a highly nutritious source of protein and calories in the Asian diet [30]; they are consumed in a variety of different ways. Besides, sprouted mungbeans are widely used as a fresh vegetable in Chinese and Japanese food, also in Europe and America. Protein isolates from mungbean have been used for making noodles and other textured preparations [9, 21].

Green gram is easily digestible and is considered an excellent food for infants, in absence of milk and for the covalescent. In contrast black gram, though generally higher in grain protein percentage is not so easy to digest. Protein content in 1658 accessions of mungbean screened at the Asian Vegetable Research and Development Centre ranged between 19.5 and 28.5% [75], which was similar to the variation reported earlier in 313 cultivars [110] (Table 14). A wide range of variation for grain protein percent was observed in the induced mutants obtained in improved cultivars of both green and black gram [7]. However, the mutants showing increased protein percentage were lower in yield.

It is believed that protein content of the present cultivars is good enough for human diet with 75% rice and 25% mungbean. Even in this proportion, supplementation with lysine, methionine and threonine increased PER to 2.59 in comparison to 1.67 for rice + mungbean alone [75]. Lysine and methionine values as percent of protein are given in Table 14. None of the 313 strains screened had favourable combination of protein, lysine and methionine [100]. In general, methionine content is higher in black gram cultivars than in green gram and was intermediate in 172 lines of *aureus* × *mungo* crosses [75].

Seed proteins from black gram have been partially characterized. Globulins which formed 81% of the solubilized proteins were devoid of sulfur containing amino acids [61]. These amino acids and threonine were also deficient in total proteins. Globulins contain at least 2 basic and 7 acidic subunits with molecular weights ranging from $2.1 \times 10^4 - 2.0 \times 10^5$.

Oligosaccharides, especially raffinose and stachyose which are considered as primary flatulence producing factors were estimated in 32 strains (Table 14). Their amount was positively correlated (r = 0.72 and 0.82) with total sugar in seeds [40].

Improvement in protein content and quality is one of the objectives in the breeding programmes at the Asian Vegetable Research and Development Centre

Table 14. Protein, lysine, methionine, total and individual sugar content in mungbean seeds.

	Range
% Protein	19.1 –28.3[1]
Lysine (as % protein)	6.3 – 7.9[1]
Methionine (as % protein)	0.55– 1.78[1]
Total sugar	2.69– 5.88[2]
Mono saccharides	0.38– 1.00[2]
Sucrose	1.06– 2.19[2]
Raffinose	0.38– 0.69[2]
Stachyose	0.50– 1.50[2]

1 n = 313
2 n = 32

Based on Yohe and Poehlman (100), and Hymowitz, Collins and Poehlman (40).

and in many national breeding programmes. However, as yet improved protein cultivars are not available. Literature published until 1972 is included in the Mungbean Bibliography [69].

CONCLUSIONS

The above survey shows that, at present, breeding for seed protein improvement in five major tropical grain legumes – groundnut, chickpea, pigeonpea, cowpea and mungbean is at very early stages. Even the available germplasm has not been fully screened for protein and limiting amino acid content. Sporadic attempts have been made to improve seed protein characteristics by inter-species or intergeneric hybridization and inducing mutations in the elite cultivars. The priorities in most breeding programmes and rightly so, are to improve yield and stability of yield. Higher yield of grain legumes can directly contribute to improved nutrition. However, even in yield oriented breeding strategy, it is essential to maintain protein quantity, quality, and digestibility at the present level. At the same time, in long range programmes, it is desirable to bring about genetic improvement in the percentage of (a) proteins, (b) sulfur containing amino acids (methionine and cystine), (c) lysine, (d) tryptophan and (e) overall digestibility, without any appreciable loss in yield potential. With a better understanding of the various constraints, in the future, it should be possible to accomplish this. Nevertheless, taking lesson from the current status of breeding protein improvement in cereals where many efforts have been made, well characterized high protein and high lysine stocks are available and grain protein percentage can be enhaced by application of additional nitrogenous fertilizers. The progress towards the commercial cultivation of improved protein cultivars in tropical legumes, which lack most of the above mentioned advantages of the cereals, is likely to be at a gradual pace.

REFERENCES

1. Abrol, YP, and SR Chatterjee (1980): Nutritional quality of grain legumes. Plant Biochem. J., S.M. Sircar Memorial Vol.: 125–149.
2. Adams, MW (1973): On the quest for quality in the field bean. In: Milner M (ed.): Nutritional Improvement of Food Legumes: 143–149; UN New York.
3. Adams, MW (1976): Legume breeding – an assessment of problems and potentials for improvement in the edible bean. In: Wilcke, HL (ed.): Improving the Nutrient Quality of Cereals II: 204–211; USAID, Washington.
4. Ahn, CS, and RW Hartman (1978): Interspecific hybridization among four species on the genus *Vigna* savi. Proc. First Intern. Mungbean Symp.: 240–246; AVRDC, Shanhua, Taiwan.
5. Bajaj, S (1973): Biological value of legume proteins as influenced by genetic variation. In: Milner, M (ed.): Nutritional Improvement of Food Legumes by Breeding: 223–232; UN New York.
6. Basha, SMM, and JP Cherry (1976): Composition, solubility, and gel electrophoretic properties of

proteins isolated from florunner (*Arachis hypogaea* L.) peanut seeds. J. Agric. & Fd. Chem. 24: 359–365.

7. Bhagwat, SG, CR Bhatia, T Gopalakrishna, DC Joshua, RK Mitra, P Narahari, SE Pawar, and RG Thakare, (1979): Increasing protein production in cereals and grain legumes. Seed Protein Improvement in Cereals and Grain Legumes II: 225–236; IAEA Vienna.
8. Bhatia, CR, and R Rabson (1976): Bioenergetic considerations in cereal breeding for protein improvement. Science 194: 1418–1421.
9. Bhumiratana, A (1978): Mungbean and its utilization in Thailand. Proc. First Intern. Mungbean Symp.: 46–48; AVRDC, Shanhua, Taiwan.
10. Bliss, FA, LN Barker, JD Franckowiak, and TC Hall (1973): Genetic and environmental variation of seed yield, yield components and seed protein quantity and quality of cowpea. Crop Sci. 13: 656–662.
11. Blixt, S (1979): Natural and induced variability for seed protein in temperate legumes. Seed Protein Improvement in Cereals and Grain Legumes II: 3–21; IAEA Vienna.
12. Boulter, D (1979): Structure and biosynthesis of legume storage proteins. Seed Protein Improvement in Cereals and Grain Legumes I: 125–136; IAEA Vienna.
13. Boulter, D, IM Evans, A Thompson, and A Yarwood (1973): The amino acid composition of *Vigna unguiculata* (Cowpea) meal in relation to nutrition. In: Milner, M (ed.): Nutritional Improvement of Food Legumes by Breeding: 205–216; UN New York.
14. Bressani, R (1973): Legumes in human diets and how they might be improved. In: Milner, M (ed.): Improvement of Food Legumes by Breeding: 15–42; UN New York.
15. Bressani, R, and LG Elias (1974): Legume foods. In: Altschul, AM (ed.): New Protein Foods: 230–297; Acad. Press, New York.
16. Bressani, R, and LG Elias (1979): The world protein and nutritional situation. Seed Protein Improvement in Cereals and Grain Legumes I: 3–23; IAEA Vienna.
17. Calloway, DH (1975): Gas forming property of food legumes. In: Milner, M (ed.): Nutritional Improvement of Food Legumes by Breeding: 263–270; UN New York.
18. Carasco, JF, R Croy, E Derbyshire, and D Boulter (1978): The isolation and characterization of the major polypeptides of the seed globulin of cowpea (*Vigna unguiculata* L. Walp) and their sequential synthesis in developing seeds. J. Exper. Bot. 29: 309–323.
19. Carpenter, KJ (1970): Nutritional considerations in attempts to change the chemical composition of crops. Proc. Nutr. Soc. 29: 3–12.
20. Cherry, JP (1977): Potential sources of peanut seed proteins and oil in genus *Arachis*. J. Agric. Food Chem. 25: 186–193.
21. Coffman, CW, and VV Gracia (1978): Isolation and functional characterization of a protein isolate from mungbean flour. Proc. First Intern. Mungbean Symp.: 69–73; AVRDC, Shanhua, Taiwan.
22. Dana, S (1966): Cross between *Phaseolus aureus* Roxb and *P. mungo* L. Genetica 37: 259–274.
23. Danielsson, CE (1949): Seed globulins of the Gramineae and Leguminosae. Biochem. J. 44: 387–400.
24. Dawson, R, and AD McIntosh (1973): Varietal and environmental differences in the proteins of the groundnut (*Arachis hypogaea*). J. Sci. Fd. Agric. 24: 597–609.
25. Derbyshire, E, DJ Wright, and D Boulter (1976): Legumin and vicilin, storage proteins of legume seeds. Phytochemistry 15: 3–24.
26. Diener, UL (1973): Deterioration of peanut quality caused by fungi. In: Wilson, CT (ed.): Peanuts – Culture and Use: 523–557; Am. Peanut Res. Assoc., Stillwater, Oklahoma.
27. Dobereiner, J (1978): Potential for nitrogen fixation in tropical legumes and grasses. In: Dobereiner, J, RH Burris, and A Hollaender (eds): Limitations and Potentials for Biological Nitrogen Fixation in the Tropics: 13–24; Plenum Press, New York.
28. Dobereiner, J, RH Burris, and A Hollaender (1978): Limitations and Potentials for Biological Nitrogen Fixation in the Tropics; Plenum Press, New York; 398 pp.

29. Dure, LS (1975): Seed formation. Ann. Rev. Plant Physiol. 26: 259–278.
30. Engel, RW (1978): The importance of legumes as a protein source in Asian diets. Proc. First Intern. Mungbean Symp.: 35–39; AVRDC, Shanhua, Taiwan.
31. Franco, AA (1978): Contribution of the legume – *Rhizobium* symbiosis to the ecosystem and food production. In: Dobereiner, J, RH Burris, and A Hollaender (eds.): Limitations and Potentials for Biological Nitrogen Fixation in the Tropics: 65–74; Plenum Press, New York.
32. Golasworthy, PR. Assistant Director, Grain Legume Programme, IITA, Ibadan, Nigeria. Personal communication.
33. Gopala Krishna, T, RK Mitra and CR Bhatia (1977): Seed globulins of *Cajanus cajan*. Qual. Plant. – Pl. Fds. Hum. Nutr. 27: 313–317.
34. Gregory, WC, and MP Gregory (1976): Groundnut. In: Simmonds, NW (ed.): Evolution of Crop Plants: 151–154; Longmans, London.
35. Hannah, LC, J Ferrero, and J Dessauer (1976): High methionine lines of cowpea. Tropical Grain Legume Bull. 4: 9.
36. Harper, AE (1976): Widespread malnutrition – A protein problem, a food problem or a population problem. In: Wilcke, HL (ed.): Improving the Nutritional Quality of Cereals II: 255–264; USAID, Washington.
37. Hartwig, EE (1979): Breeding productive soybeans with a higher percentage of protein. Seed Protein Improvement in Cereals and Grain Legumes II: 59–66; IAEA Vienna.
38. Hellendoorn, EW (1979): Beneficial physiological activity of leguminous seeds. Qual. Plant. – Pl. Fds. Hum. Nutr. 29: 227–244.
39. Holley, KT, and RO Hammons (1968): Strain and seasonal effects on peanut characteristics. Res. Bull. 32: 27; University of Georgia.
40. Hymowitz, T, FI Collins, and JM Poehlman (1974): Relationship between the content of oil, protein and sugar in mungbean seed. Trop. Agric. (Trinidad) 52: 47–51.
41. ICRISAT (International Crops Research Institute for the Semi-Arid Tropics) 1980: Annual Report 1978–79; Patancheru, A.P. 502 324; INDIA.
42. Jambunathan, R, ICRISAT, Hyderabad, India (Personal communication).
43. Jambunathan, R, and U Singh (1979): Chemical composition of desi and kabuli chickpea (*Cicer arietinum* L.) cultivars. Proc. Intern. Workshop on Chickpea Improvement, ICRISAT, Hyderabad (in press).
44. Kang, BT, and RL Fox. Influence of soil fertility on the protein and sulphur content of grain legumes. IITA, Ibadan–Mimeographed Report.
45. Kaul, AK (1973): Mutation breeding and crop protein improvement. Nuclear Techniques for Seed Protein Improvement: 1–106; IAEA Vienna.
46. Laurence, RCN, RW Gibbons, and CT Young (1976): Changes in yield, protein, oil and maturity of groundnut cultivars with the application of sulphur fertilizers and fungicides. J. Agric. Sci., Camb. 86: 245–250.
47. Leleji, BI, MH Dickson, LV Crowder, and JB Bourke (1972): Inheritance of crude protein percentage and its correlation with seed yield in beans *Phaseolus vulgaris*. Crop Sci. 12: 168–171.
48. Lewis, OAM, and JS Pate (1973): The significance of transpirationally derived nitrogen in protein synthesis in fruiting plants of pea (*Pisum sativum* L.). J. Exper. Bot. 24: 596–606.
49. Liener, IE (1973): Antitryptic and other antinutritional factors in legumes. In: Milner, M (ed.): Nutritional Improvement of Food Legumes: 239–258; UN New York.
50. Liener, IE (1973): Summary and research needed. In: Milner, M (ed.): Nutritional Improvement of Food Legumes: 307–309; UN New York.
51. Liener, IE (1969): Toxic Constituents in Plant Foodstuffs. Academic Press, New York; 500 pp.
52. Luse, RA, and KO Rachie (1979): Seed protein improvement in tropical food legumes. Seed Protein Improvement in Cereals and Grain Legumes II: 87–104; IAEA Vienna.
53. Meiners, JP, and SC Litzenberger (1973): Breeding for nutritional improvement. In: Milner, M (ed.): Nutritional Improvement of Legumes by Breeding: 131–141; UN New York.

54. Minchin, FR, RJ Summerfield, P Hadley, EH Roberts, and S Rawsthorne (1981): Carbon and nitrogen nutrition of nodulated roots of grain legumes. Plant, Cell and Environment 4: 5–26.
55. Mitra, RK, CR Bhatia, and R Rabson (1979): Bioenergetic cost of altering the amino acid composition of cereal grains. Cereal Chem. 56: 249–252.
56. Mitra, R, and CR Bhatia (1979): Bioenergetic considerations in the improvement of oil content and quality in oilseed crops. Theor. Appl. Genet. 54: 41–47.
57. Nangju, D (1976): Effect of fertilizer management on seed sulfur content of cowpea (*Vigna unguiculata* L. Walp). Tropical Grain Legume Bull. 4: 6–8.
58. New Vistas in Pulse Production. Indian Agric. Res. Inst. (1971): 111.
59. Nutritional Improvement of Food Legumes by Breeding (1973). Milner, M (ed.). Protein Advisory Group of the United Nations System, New York; 389 pp.
60. Oliker, M, A Poljakoff-Mayber, and AM Mayer (1978): Changes in weight, nitrogen accumulation, respiration and photosynthesis during growth and development of seeds and pods of *Phaseolus vulgaris*. Amer. J. Bot. 65: 366–371.
61. Padhye, VW, and DK Salunkhe (1979): Biochemical studies on black gram (*Phaseolus mungo* L.) seed II. Amino acid composition and subunit constitution of fraction of proteins. J. Food Sci. 44: 606–610, 614.
62. Parpia, HAB (1973): Utilization problems in food legumes. In: Milner, M (ed.): Nutritional Improvement of Legumes by Breeding: 281–295; UN New York.
63. Parpia, HAB, M Swaminathan, and DS Bhatia (1965): Peanut protein foods for protein-poor countries. J. Food Sci. Tech. 2: 17–33.
64. Pawar, SE, and CR Bhatia (1980): The basis for grain yield differences in mungbean cultivars and identification of yield limiting factors. Theor. Appl. Genet. 57: 171–175.
65. Payne, PR (1976): Nutritional criteria for breeding and selection of crops: with special reference to protein quality. Plant Foods for Man 2: 95–112.
66. Peanuts – Culture and Use (1973). Wilson, CT (ed.): Am. Peanut Res. Educ. Assoc., Oklahoma State Univ., Stillwater, Oklahoma; 684 pp.
67. Penning de Vries, FWT, AHM Brunsting, and HH van Laar (1974): Products, requirements and efficiency of biosynthesis. A quantitative approach. J. Theor. Biol. 45: 339–377.
68. Phillips, DE (1980): Efficiency of symbiotic nitrogen fixation in legumes. Ann. Rev. Plant Physiol. 31: 29–49.
69. Poehlman, JM, Freda-Fu-Mei Yu-Jen (1972): Bibliography of Mungbean Research Special Report 184: 15; Univ. Missouri Agric. Exper. Station.
70. Rachie, KO, and LM Roberts (1974): Grain legumes of low land tropics. Adv. Agron. 26: 1–132.
71. Ramanujam, S (1976): Chickpea. In: Simmonds, NW (ed.): Evolution of Crop Plants: 157–159; Longmans, London.
72. Reddy, LJ, JM Green, U Singh, SS Bisen, and R Jambunathan (1979): Seed protein studies on *Cajanus cajan*, *Atylosia* spp. and some hybrid derivatives. Seed Protein Improvement in Cereals and Grain Legumes II: 105–117; IAEA Vienna.
73. Royes, MV (1976): Pigeonpea (*Cajanus cajan*). In: Simmonds, NW (ed.): Evolution of Crop Plants: 154–156; Longmans, London.
74. Ryle, GJA, CE Powell, and AJ Gordon (1979): The respiratory cost of nitrogen fixation in soyabean, cowpea and white clover II. Comparison of the cost of nitrogen fixation and the utilization of combined nitrogen. J. Exper. Bot. 30: 145–153.
75. Samson, CS, and MS Tsou, Hsu (1978): The potential roles of mungbean as a diet component in Asia. Proc. First Intern. Mungbean Symp.: 40–45; AVRDC, Shanhua, Taiwan.
76. Savoy, CF (1976): Peanut (*Arachis hypogea* L.) seed protein characterization and genotype sample classification using polyacrylamide gel electrophoresis. Biochem. Biophys. Res. Commun. 68: 886–893.
77. Sharma, TR, VP Gupta, S Gassi, and AK Kaul (1973): Estimation of genetic parameters of some

quality characters in wheat (*Triticum aestivum* L.) and Bengal gram (Cicer *arietinum* L.). Nuclear Techniques for Seed Protein Improvement: 273–289; IAEA Vienna.

78. Siegel, A, and B Fawcett, (1976): Food Legumes Processing and Utilization. IDRC-TS-1, IDRC, Ottawa; 88 pp.
79. Sinclair, TR, and CT de Wit (1975): Photosynthate and nitrogen requirements for seed production by various crops. Science 189: 565–567.
80. Sinclair, TR, and CT de Wit (1976): Analysis of carbon and nitrogen limitations to soybean yield. Agron. J. 68: 319–324.
81. Singh, KB, and LJG van der Maesen. Chickpea Bibliography 1930–1974. ICRISAT, Hyderabad, India; 223 pp.
82. Singh, U, R Jambunathan, and Narayanan (1980): Biochemical changes in developing seeds of pigeonpea (*Cajanus cajan*). Phytochemistry 19: 1291–1295.
83. Sinha, SK (1977): Food Legumes: distribution, adaptability and biology of yield. Plant Production and Protection Paper 3, FAO Rome; 124 pp.
84. Sinha, SK, S Ramanujam, and MS Swaminathan (1975): Yields of main species of grain legumes experimentally and in farm practice in different agro-ecological regions of the world. The apparent technical reasons for differences between research and practice. Rep. TAC Working Group on the Biology of Yield of Grain Legumes; TAC Section; FAO Rome; 25 pp.
85. Smartt, J (1976): Tropical Pulses. Longman Group, London; 348 pp.
86. Spector, WS (1956): Handbook of Biological Data. Saunders, Philadelphia; 87 pp.
87. St. Angelo, J, and E Mann (1973): Peanut proteins. In: Wilson, CT (ed.): Peanuts – Culture and Use: 559–592; Am. Peanut Res. Educ. Assoc., Stillwater, Oklahoma.
88. Steele, WM (1976): Cowpeas. In: Simmonds, NW (ed.): Evolution of Crop Plants: 183–185; Longmans, London.
89. Subramanian, N (1980): Technology of vegetable protein foods. J. Fd. Sci. Tech. 17: 71–77.
90. Sukhatme, PV (1973): The calorie gap. Indian J. Nutr. Dietet. 10: 198–207.
91. Swaminathan, MS, and HK Jain (1973): Food legumes in Indian agriculture. In: Milner, M (ed.): Nutritional Improvement of Food Legumes by Breeding: 69–82; UN New York.
92. Tai, YP, and CT Young (1975): Genetic studies of peanut proteins and oils. J. Amer. Oil Chem. Soc. 52: 377–385.
93. Thomson, JA, and H Doll (1979): Genetics and evolution of seed storage proteins. Seed Protein Improvement in Cereals and Grain Legumes I: 109–124; IAEA Vienna.
94. Thomson, JA, A Millerd, and HE Schroeder (1979): Genotype-dependent patterns of accumulation of seed storage proteins in *Pisum*. Seed Protein Improvement in Cereals and Grain Legumes I: 231–240; IAEA Vienna.
95. Tropical Legumes: Resource for the Future (1979): National Academy of Sciences, Washington, DC; 331 pp.
96. Verdacourt, B (1970): Studies in the *Leguminosae-Papilionoideae* for the 'Flora of Tropical East Africa' III Kew Bull, 24 (Lectotype of subgenus: *Vigna radiata* (L.) Wilczek (= *Phaseolus radiatus* L.): 556–559.
97. Waterlow, JC, and PR Payne (1975): The protein gap. Nature 258: 113–117.
98. Weber, E, and D Neumann (1980): Protein bodies, storage organelles in plant seeds. Biochem. Physiol. Pflanzen 175: 279–306.
99. Whitehouse, RNH (1970): The prospects of breeding barley, wheat and oats to meet special requirements in human and animal nutrition. Proc. Nutr. Soc. 29: 31–39.
100. Yohe, JM, and JM Poehlman (1972): Genetic variability in the mungbean *Vigna radiata* (L) Wilczek. Crop Sci. 12: 461–464.
101. Zohary, D, and M Hopf (1973): Domestication of pulses in the old world. Science 182: 887–894.

16. Study of Evolutionanary Problems by Means of Seed Protein Electrophoresis

G. LADIZINSKY

INTRODUCTION

The composition and role of seed proteins were subjected to numerous studies resulting in several review articles and books. Osborne was the first to characterize seed proteins according to their solubility potential. Later, amino acid analysis, ultracentrifugation, immunology and electrophoretic techniques were employed for a more precise determination of the various fractions of seed proteins. Characteristically various plant groups contain specific seed proteins such as prolamins in cereals and globulins in legumes. During the process of germination the various fractions of seed proteins are degraded [10, 51] releasing small peptides and free amino acids, many of which are transferred to the embryo axis [4].

If seed proteins are merely a temporary arrangement of amino acids that are degraded at germination, they might tolerate a considerable number of mutations resulting in conspicuous variation among plants of the same species and parallel variation between species. Variations of this kind, however, is very rare. For example, different cereals have specific prolamin fractions, hordein in barley, avenin in oats, zein in corn etc. Likewise different globulins are typical of different legumes [4].

Adoption of electrophoretic techniques for identifying seed proteins has indicated that the seed protein profile is highly stable and species specific [7, 28]. This profile is hardly affected by growing conditions or seasonal fluctuations [1, 14, 16, 46]. Furthermore, changes in chromosome structures, i.e. translocations and inversions, or even doubling the chromosome number do not affect the seed protein profile [23, 35, 52].

Proteins are direct products of the genes and can be considered as markers of these genes, and do even characterize the genome [40]. As such, proteins can be considered as an additional means for characterizing systematic categories. However, it is imperative to have some idea about the genetic control of the various bands in the seed protein profile. In recent years, such information has been accumulated on an increasing number of bands. Some of these bands were identified according to the protein they belong to, such as albumins and globulins; others according to their specific biological properties such as proteinase inhibitors, lectins etc. The role of these proteins in the seed is not yet understood but due to their presence in large quantities, they are assumed to be primarily storage proteins [4]. Variation in the seed protein profile caused by the presense or absence of a particular band was found to be controlled by a single gene in soybeans [45] and peas [5]. Similarly soybean seed lectin was found to be governed by a single gene [55]. A great deal of information was obtained on the genetic control of the Kunitz trypsin inhibitor of soybean. Four electrophoretic variants of this inhibitor were detected in the soybean germ plasm, three of which were controlled by different alleles of the same locus, while in the fourth one the trypsin inhibitor band was totally missing in the profile [21, 53, 54].

Some information on the genetic control of trypsin and chymotrypsin inhibitor was obtained in wild fenugreek (*Trigonella brythea*). A protein with trypsin and chymotrypsin inhibitory activity forms two prominent bands in the seed protein profile, extracted at pH 8.1 and run in anodic system (+ at the bottom of the gel). These two bands are also the fastest moving bands in the profile. Among the collections examined, one particular line of *T. berythea* did not contain the faster of the two bands as well as another two bands in the profile (Figure 1). Examination of the profile of the hybrids between these two variants and of the F_2 populations revealed that the faster moving band with trypsin and chymotrypsin inhibitory activity is controlled by a single gene.

Additional information on the genetics of the various bands indicates simple inheritance for an appreciable number. If this reflects a general state for seed protein profiles, it can be surmised that diversity at the seed protein level is caused by a relatively small number of mutations. It will be extremely interesting to know more about the evolution of the seed protein profiles and particularly the rate of change among conservative and polymorphic proteins.

The actual meaning of the stability of the seed protein profile of a species is that lines from different geographical areas, or those exhibiting morphological and genetical differences still show the same or nearly the same seed protein profile. The variation most often detected is of a quantitative nature, i.e. darkness or faintness in various bands. From a genetic point of view a cultivated plant and its wild progenitor are members of the same species. Therefore it is expected that a cultigen and its spontaneous race will share the same seed protein profile. This principle has been employed in several studies aimed at obtaining primary ideas regarding the origin of the cultivated plant, or to support evidence obtained by comparative morphology, cytology and breeding experiments. To test the potential of this approach in specific crop plants, it is necessary to examine first the rate of variation in the profile of the cultigen. A stable profile could indicate a monophyletic origin

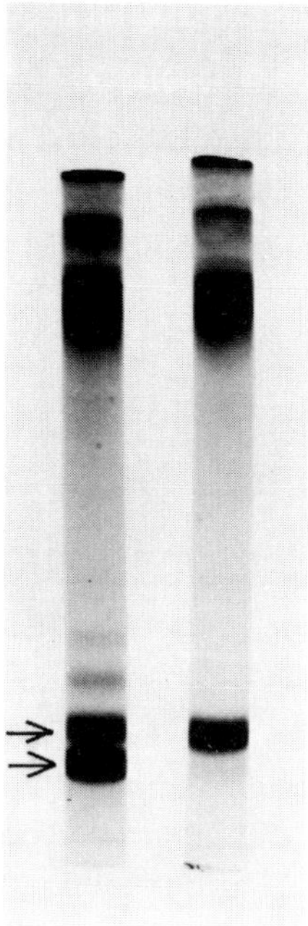

Figure 1. Variation in trypsin and chymotrypsin inhibitor in *Trigonella berythea*, arrows indicate the trypsin and chymotrypsin bands.

and in a diploid plant it could be expected that the wild progenitor would also possess the same profile. On the other hand, variations in the banding pattern can be interpreted in several ways and likewise the conclusions concerning the wild parent or parents of the crop.

THE SIGNIFICANCE OF SEED PROTEIN ELECTROPHORESIS IN CLARIFYING THE ORIGIN OF LEGUME CROPS

Soybean

The genus *Glycine* is presently divided into three subgenera [20], one subgenus *soja* is comprised of the cultivated form *G. max*, its immediate wild form *G. soja* and a weedy type *G. gracilis*. This latter type apparently evolved from a natural cross between *G. max* and *G. soja* [19].

The seed protein of soybean extracted at pH 8.1 and fractionated on 10% polycrylamide gel, contained on the lower part of the profile a band with trypsin inhibitory activity. The Rf value of this band was similar in *G. max*, *G. soja* and *G.*

gracilis but differed in the species of other subgenera. Furthermore, the taxa of the subgenera *soja* shared bands with no trypsin inhibitory activity [49]. Thus, the information from seed protein electrophoresis supports further that *G. max*, *G. soja* and *G. gracilis* are closely related and that the latter two had an important role in the evolution of the soybean. By contrast the species of other subgenera not only have different profiles, but are also cross-incompatible with the cultigen.

Chickpea

Chickpea, *Cicer arietinum*, is an old world crop and perhaps the most important field legume in the Middle East, North Africa, Ethiopia, Iran, Pakistan and India. Until recently the origin of this important legume was unknown and botanically the genus as a whole was inadequately recognized. Some authors suggested *C. judaicum* was the wild ancestor of the cultivated chickpea but this idea has not been proved.

Examination of the water soluble seed protein profile of more than 100 lines, land races and varieties of the cultigen, revealed that despite morphological variation and differences in chromosomal architecture, i.e. translocations and inversions, all the accessions examined shared the same basic profile. Differences between lines were due mainly to the intensity of various bands and rarely to an extra band. These differences, however, could not be related to geographical variation or correlated with specific morphological and agronomic types. The uniform profile thus was taken as an indication for a monophyletic origin of chickpea and it was postulated that a similar profile can be expected in the wild progenitor. Examination of the available wild annual chickpea species including *C. judaicum* revealed that none of them had a profile similar to that of the cultigen. Following a field trip to Turkey in 1974, a new type not previously described was collected in the southeast part of the country. Due to its unique morphology, it was described as a new species *C. reticulatum*. The seed protein profile of this new taxon was identical to that of the cultigen and the difference was in the intensity of two bands. Accordingly, *C. reticulatum* was proposed as the wild ancestor of the cultivated chickpea [36]. This conclusion was supported later by breeding experiments. Hybrids between these two taxa were easily obtained. The F_1 plants developed normally, at meiosis the chromosomes paired regularly in bivalents and the fertility was as good as that of the parental lines. No breakdown was noticed in the F_2 and segregation for some of the traits differentiating the parents could be detected [37].

Lentil

On morphological ground the cultivated lentil, *Lens culinaris*, is closely related to two wild taxa *L. orientalis* and *L. nigricans*. The main difference between these wild taxa is the stipule shape. It is lanceolate elliptic, entire in *L. orientalis* and semi-hustate, dentate in *L. nigricans*. This difference was taken as a sign that *L. nigricans*

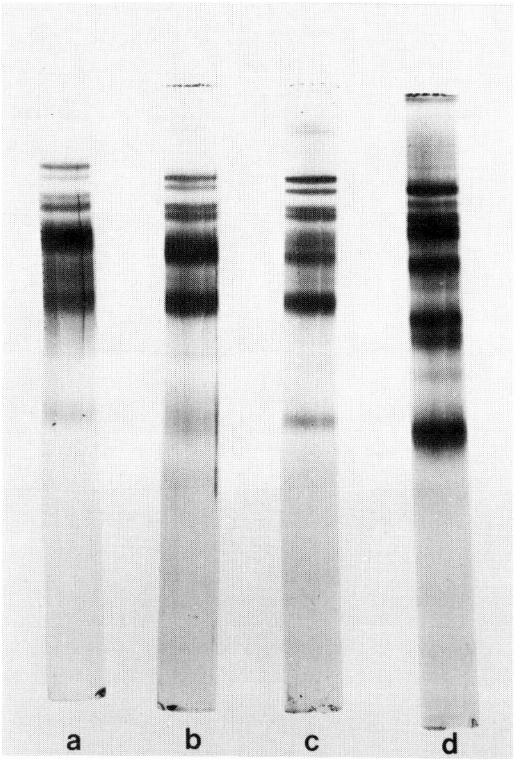

Figure 2. Seed protein profiles of *Lens* species, a. *L. culinaris*, b. *L. orientalis*, c. *L. nigricans*, d. *L. ervoides*.

is well separated from *L. orientalis* and *L. culinaris* [68]. By contrast Hegi, in Flora of Central Europe, disregarded the difference in stipule shape and treated *L. nigricans* as a subspecies of *L. culinaris*.

The seed protein profile was examined in 50 accessions of the cultivated lentil. They represented the main growing area of the crop from Afghanistan to the Iberian peninsula and from Ethiopia to Russia. They included types with small and large seeds and exhibited variation in traits such as seed and seedling color, growth habit and flower color. All the accessions had practically the same profile. Interestingly, the same profile was found also in 10 accessions of *L. orientalis* whose origin was Israel, Iran and Afghanistan, and 3 accessions of *L. nigricans* from Israel and Turkey (Figure 2). The similar profile in these three taxa clearly indicates a close genetic relationship between them. Only the wild species *L. ervoides* deviated from this profile by the migration rate of 5 bands [34].

The interrelationships among the various taxa of the genus *Lens*, as indicated by seed protein electrophoresis, were fully supported by breeding experiments. *Lens culinaris*, *L. orientalis* and *L. nigricans* are cross-compatible and hybrid seeds are easily obtained. On the other hand, *L. ervoides* is apparently cross-incompatible with these three taxa. Following hybridization the youg F_1 embryos grow for 10–14 days and then collapse. The F_1 hybrids between the cultivated lentil, and *L. orientalis* and *L. nigricans* developed normally. However, the fertility was reduced due to one chromosomal translocation differentiating *L. culinaris* and *L. orientalis* and three translocations with *L. nigricans* [33]. Thus, despite chromosomal translocations, these three taxa share the same profile. These chromosomal differences are not great enough to form reproductive barriers among the various lentil types, since in the F_2 rapid restoration of fertility was found.

Our present information on the distribution of the wild relatives of lentil indicates that *L. orientalis* is distributed in west and central Asia and *L. nigricans* in the Middle East and southern Europe [33]. Both species form small disjunct populations but almost exclusively in herbaceous formations on shallow soil and in stony, relatively dry habitats. Possibly the similarity of the seed protein profile of these two species reflects their similar ecological preferences. By contrast, ecologically, *L. ervoides* deviates from these two wild species by growing in more shady habitats, such as under or near the edge of tree canopies or in association with dwarf shrubs.

Pea

Like *Lens*, *Pisum* is a smal genus and similarly there is no full agreement among experts regarding species boundaries. Besides the cultivated pea, *P. sativum*, Boisser [6] recognized three wild species, *P. fulvum*, *P. elatius* and *P. humile*. Only two species, *P. sativum* and *P. fulvum*, were considered by Davis [11], and an even more extreme view regarded all the taxa as members of one species [44]. This latter suggestion stems from the cross compatibility among the various taxa and the full or partial fertility of the F_1 hybrids.

The cytogenetic affinities among the *Pisum* taxa are well reflected at the seed protein level. The variation observed in the banding pattern of the albumin and the globulin fractions of the seed protein could account for genotypic, rather than taxonomic differences [65]. Thus, the evidence from seed protein electrophoresis cannot support any specific taxon from which the cultivated pea was derived, including the previously suggested *P. humile* from the northern part of the Fertile Crescent [3]. It is possible that in both pea and lentil, the cultivated types evolved from several domestication events which took place at different places at different times utilizing different wild genotypes, rather than any specific taxon.

Common Vetch

Variation in the seed protein profile due to genotypic rather than taxonomic differences, is even more profound in the common vetch, known also as the *Vicia sativa* aggregate. The taxonomic boundaries in the aggregate are notoriously confused. Mettin and Hanelt [48] recognized six species *V. amphicarpa*, *V. angustifolia*, *V. cordata*, *V. macrocarpa*, *V. pilosa* and *V. sativa*. Davis and Plitmann [12] considered only five subspecies of a single species. To these five subspecies, two more were added later [70].

Not only is the taxonomic status of the *Vicia sativa* aggregate obscured but also the chromosome. number. Three different chromosome numbers have been recorded; 2n = 10, 2n = 12 and 2n = 14. While Mettin and Hanelt [48] related specific chromosome number or karyotype to their taxonomic categories, no such relationships could be confirmed in wild populations of *V. sativa* in Israel [32]. Besides the three chromosome numbers already mentioned, two different karyotypes were found in plants with 2n = 10 and 2n = 12 in these populations. In some populations, plants having different chromosome numbers or karyotypes were found side by side. Occasionally, the same chromosome number or karyotype was found in two taxonomic types, and plants which exhibited the same diagnostic morphological traits had a different chromosome number or karyotype. The complexity of the *V. sativa* aggregate is further confused by the fact that types varying in chromosome number or karyotype are cross compatible, and produce F_1 hybrids which are partially fertile [43].

Examination of the seed protein profile of representative lines of the three chromosome numbers and five karyotypes revealed remarkable variation in the banding pattern, though no band or bands could be related to a specific chromosome number of karyotype. To measure the degree of similarity among the profiles of any two accessions, the similarity index was calculated as follows:
$\text{S.I.} = \dfrac{\text{No. of common bands}}{\text{No. of common bands} + \text{No. of uncommon bands}} \times 100.$ The lower values of the similarity index within a karyotype ranged from 40.0 to 92.3 and the higher values between karyotypes ranged from 79.9 to 100 (Table 1). Thus, according to the seed protein profile it is impossible to identify *V. sativa* plants having different chromosome numbers. The remarkable variation observed is apparently of a genotypic nature and in no way represents genomic differentiation. Furthermore, variation cannot be correlated with morphological types, reaffirming the inadequacy of pure morphology for establishing natural units in the *V. sativa* aggregate. The variation in the seed protein profile and at the morphological level is probably a consequence of incomplete reproductive isolation among the various karyotypes [43]. The potential gene flow in this aggregate is not only theoretical, since mixed stands of types having different chromosome number and karyotype were detected in natural populations of *V. sativa* [32].

The great variability of the seed protein profile in *V. sativa*, compared to other legumes can probably be explained by the adaptive radiation of this aggregate [32]. It includes types, mostly with 2n = 12, that grow in primary habitats, in maquis and in dwarf shrub formations of the Mediterranean vegetation. Others have marked

Table 1. Range of similarity indices between accessions of *V. sativa* aggregates having the same or different chromosome numbers or karyotypes. A and B are karyomorphs. The number of accessions are in brackets.

	2n = 10 A (8)	2n = 10 B (8)	2n = 12 A (6)	2n = 12 B (10)	2n = 14 (10)
10 A	100-54.5				
10 B	100-47.8	100-53.8			
12 A	92.0-67.9	88.2-43.3	100-92.3		
12 B	89.2-41.8	79.3-66.6	100-48.0	100-40.0	
14	100-55.5	93.8-57.6	95.2-58.3	86.1-66.6	100-41.1

weedy tendencies and grow in man made habitats such as road sides, abandoned fields and ruins. The chromosome number of these types is usually 2n = 10, and a similar 2n number was found in many cultivated varieties. Still another type is adapted to xeric niches in steppes and in stony habitats in the Mediterranean region. This type, 2n = 14, has subterranean and aerial fruits. It is tempting to assume that the variation in the seed protein profile of *V. sativa* reflects primarily adaptation to different ecological conditions. A particular profile can be maintained by self-fertilization which is the rule in *V. sativa.* Since a certain amount of outcrossing also occurs and the F_1 hybrids are partially fertile, rigid profiles have not yet been established in the *V. sativa* aggregate.

Pigeon pea

Pigeon pea, *Cajanus cajan* is a tropical legume grown mainly in India. The 30 species recorded for *Cajanus* are considered as one polymorphic species [53]. No wild species of *Cajanus* is known and the few reports on wild pigeon pea apparently refer to types escaped from cultivation. Morphologically, the genus *Atylosia* is closely related to *Cajanus*. Further, the species *A. lineata*, *A. sericea* and *A. scarabaeoides* are cross-compatible with *C. cajan* and the F_1 hybrids of the first two combinations are even partially fertile [13].

Examination of the seed protein profile of 90 accessions of pigeon pea revealed that 86 of them share the same profile, while the other four represented three electrophoretic variants [39]. Similarity indices between the four profiles of pigeon pea indicated that the common type (A) is closer to any of the other types (B,C,D)

Table 2. Similarity indices among the profiles of *Cajanus* and *Atylosia.*

	Cajanus				*Atylosia*		
	A	B	C	D	*A. lineata*	*A. cajanifolia*	*A. scarabaeoides*
B	47.2						
C	72.2	37.5					
D	60.0	30.0	45.0				
A. lineata	57.1	33.3	42.1	38.8			
A. cajanifolia	61.5	50.0	44.4	33.3	50.0		
A. scarabaeoides	33.3	18.7	25.0	20.0	14.2	20.0	
A. platycerpa	41.1	25.0	45.0	40.0	33.3	42.8	28.5

than they are among themselves (Table 2). It is therefore probable that the three electrophoretic variants evolved independently from the common type, rather than from each other. Comparisons between the profiles of *C. cajan* and four *Atylosia* species, *A. lineata*, *A. cajanifolia*, *A. scarabaeoides* and *A. platycarpa*, showed that any band in the profiles of the wild species has a homologue in the common profile or in one of the electrophoretic variants of *C. cajan* (Table 3). Therefore it can be assumed that pigeon pea has a rather polyphyletic origin from several *Atylosia* species, or that it evolved through continuous introgression from various *Atylosia* species. This latter possibility is supported by the existence of a few bands in the *Cajanus* electrophoretic variants that are found also in one or two *Atylosia* species. For example, band no. 4 in Table 3 was detected in variant C of *C. cajan* and in *A. platycarpa;* band no. 12 in variant B and in *A. scarabaeoides;* and band no. 16 in variant D and in *A. platycarpa.* By contrast all the *A. lineata* bands were represented in the common profile of *Cajanus.*

SEED PROTEIN PROFILE OF LEGUMES WITH UNKNOWN ORIGIN

Broad Bean

The broad bean, *Vicia faba*, is apparently the only major old world legume of which the origin is still unknown. Botanically, *V. faba* belongs to section *Faba* together with several wild species. Most of these wild species are members of the narbonensis groups, namely, *V. narbonensis*, *V. galilaea*, *V. hyaeniscyamus*, *V. johunis* and *V. serratifolia*. Plitmann [57] postulated that *V. galilaea* is closely related

Table 3. R_f values of various bands in the profiles of *Cajanus* and *Atylosia*

Band No.	*Cajanus* A	B	C	D	*Atylosia* *A. lineata*	*A. cajanifolia*	*A. scarabaeoides*	*A. platycarpa*
1		.06						
2		.11			.11			
3	.15	.15	.15	.15				
4			.17					.17
5	.21		.21	.21	.21			.21
6			.23				.23	.23
7	.26	.26	.26	.26	.26			.26
8	.28	.28	.28	.28	.28	.28	.28	.28
9	.32	.32	.32	.32	.32	.32		
10	.36	.36	.36	.36	.36	.36	.36	.36
11		.42						
12		.45					.45	
13	.48	.48	.48			.48		.48
14	.50	.50	.50	.50	.50	.50		
15							.52	
16				.54				.54
17		.61		.61				
18	.66	.66	.66			.66	.66	.66
19			.68					
20	.71	.71	.71					
21	.75	.75	.75		.75	.75		
22			.77					
23			.82					
24	.89	.89	.89	.89	.89	.89		.89
25	.94		.94	.94			.94	
26		.95				.95		.95
27	1.00	1.00	1.00	1.00	1.00	1.00	1.00	1.00

to *V. faba*. Later it was suggested that the narbonensis group of species is the general stock from which the cultivated broad bean was evolved [69]. However, cytological comparisons indicated that the relationships between the broad bean and the narbonensis group of species must be very remote [28]. While the chromosome number of *V. faba* is 2n = 12, it is 2n = 14 in all the narbonensis group of species. Further, the amount of DNA in the broad bean is twice as much as *V. narbonensis* [9]. Chromosome shape also rules out any immediate relationship between *V. faba* and any of the narbonensis group of species. The karyotype of *V. faba* is extremely asymmetrical but most of the chromosomes of the wild species of section *Faba* are submetacentric [31, 61].

The results of the seed protein electrophoresis of section *Faba* are in full agree-

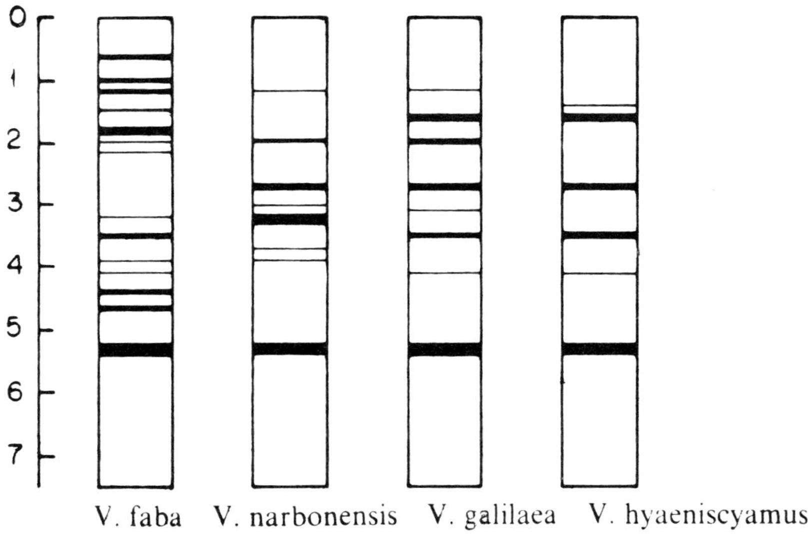

Figure 3. Seed protein profiles of wild and cultivated species of section *Faba* of *Vicia*.

ment with the morphological and the chromosomal evidence [30]. Fourteen bands were recorded in the water soluble seed protein profile of the broad bean. This profile was found in 90 accessions of *V. faba* representing the distributional range of this crop and the main taxonomical and agronomical types. Only 8–6 bands were found in the profiles of *V. narbonensis*, *V. galilaea* and *V. hyaeniscyamus* (Figure 3). Of the fourteen *V. faba* bands only seven had homologues in the profiles of the three wild species, and not more than four bands in a single species. This clearly indicates that the species examined had a negligible role in the origin and the evolution of the broad bean.

Profound differences exist between *V. faba* and the wild species of section *Faba*, including cross incompatibility. While *V. narbonensis*, *V. galilaea* and *V. hyaeniscyamus* are cross-compatible and F_1 hybrids between them occasionally can be seen in nature, they are absolutely cross-incompatible with *V. faba*. [31]. Thus, comparative morphology, cytology, cross-incompatibility and profound differences in the seed protein profile, emphasize the very remote relationships, if any, between the broad bean and the wild species of section *Faba*.

Fenugreek

The legume, *trigonella foenum graecum*, is grown for fodder and also for direct consumption by man. The seeds are used as an ingredient in seasoning and also for medicinal purposes. There have been few studies of fenugreek, and no serious efforts were made to clarify its origin. Botanically it belongs to section *Foenum graecum* together with 6 wild species [11]. On morphological grounds it was suggested that *T. berythea* is the putative ancestor of the fenugreek [67]. Examination of the seed protein profiles of the cultivated species and three wild species of section *Foenum graecum*, *T. gladiata*, *T. berythea*, and *T. macrorrhyncha*, revealed that none of them had a profile similar to that of the cultigen [35]. However, the similarity index between the cultivated fenugreek and *T. berythea* was 76.47 but 33.33 with any of the other two species. The greater genetic similarity between the fenugreek and *T. berythea*, indicated by the seed protein profiles, is further supported by breeding experiments. Numerous attempts to cross the cultivated species with either *T. gladiata* or *T. macrorrhyncha* produced not even a single hybrid seed. By contrast, many seeds were obtained in the cross with *T. berythea*, and the reciprocal. Unfortunately, all the seedlings from this cross were albinos and died a few days after germination [42].

SEED PROTEIN ELECTROPHORESIS AS A TOOL IN TRACING CEREAL EVOLUTION

In contrast to legumes, the origin and the evolution of cereals has been studied intensively by conventional methods. The information obtained by seed protein electrophoresis further supported existing ideas. A few examples will now be mentioned.

It is generally accepted that *Hordeum spontaneum* is the wild race and the progenitor of the cultivated barley. Indeed, the similarity index between the profiles of wild and cultivated barley is in the range of similarity indices among cultivated barley [47].

New evidence regarding the debatable origin of maize was obtained by seed protein electrophoresis [56]. No difference was found in the number of bands and their mobility between maize and teosinte, the wild progenitor of maize, according to various authors [2, 15, 66]. By contrast, significant differences were found in the mobility of *Tripsacum* albumin bands. Further, the pattern of the zein fraction of *Tripsacum* deviated considerably from those of maize and teosinte in the number of bands and their mobility. Thus, the elctrophoretic evidence clearly favors teosinte as the wild parent of maize.

The seed protein profile was used to estimate the degree of divergence among the *Sorghum bicolor* races. The cultivated sorghum has been divided into five races according to spikelet morphology and head type [17]. To some extent, each of these races is distributed in different parts of Africa. Study of the water soluble seed proteins has shown that the similarity indices between the profiles of the races *durra*, *candatum* and *guinea* were relatively high (58.5–95.0) but much lower (21.7–30.4) when compared with the race *kafir* [62, 63]. This deviation is quite surprising since the race *kafir* is assumed to be of fairly recent origin [17]. Therefore, it is necessary to examine more accessions of the race *kafir*. If the observed pattern is typical to *kafir*, it will be interesting to know if it resulted from accumulation of mutations or by introgression from wild sorghum native to the area where *kafir* is presently grown.

THE ADDITIVE NATURE OF THE SEED PROTEIN PROFILE AND ITS USEFULNESS IN STUDYING EVOLUTION OF POLYPLOID CEREALS

Besides the remarkable stability of the seed protein profile in many plant species, the banding pattern was found to be of an additive nature. Namely, when proteins of two electrophoretic variants are mechanically mixed and exposed to gel electrophoresis, the common bands of the two variants will merge but the uncommon bands will persist in the gel. In experiments of this type, the uncommon bands will be paler in comparison to the common bands since they are presented in half dose. Of more significance is the coexistence of the uncommon bands in the F_1 hybrid between the two variants [53].

Allopolyploids are permanent interspecific hybrids that became fertile following the doubling of their chromosome number. Their nucleus contains two or more genomes that do not intermix and which manifest themselves in various ways. Indeed protein profiles of synthetic allopolyploids were found to be the exact summation of the bands in the profile of the diploid parents [8, 18, 50]. In allopolyploids, therefore, seed protein electrophoresis can be a useful tool in obtaining quick indications regarding the putative ancestors and the course of evolution. In this respect seed protein electrophoresis is similar to extrapolation methods of comparative morphology and cytological genome analysis. The suitability of a diploid as a putative ancestor of the polyploid is determined by the possession of a specific band or bands that are present in the polyploid, but not in other diploids. Later a 1:1 mechanical mixture of the suspected diploids is expected to produce a profile which is similar or even identical to that of the natural polyploid. A word of caution should be added here. Although the additive nature of the seed protein profile has been demonstrated in several synthetic allopolyploids, one should be aware of the possiblity of intergenomic gene interactions in the allopolyploid that might shift the banding pattern by the elimination of a specific band or bands of one parent or the creation of new bands. If this is suspected to be the case, it is recommended to examine the profile of the mechanically mixed seed protein of the suspected parents and the profile of their synthetic amphidiploid.

Aegilops cylindrica is an allotetraploid, whose genomic constitution CCDD originated from the diploids *Ae. caudata* (CC) and *Ae. squarrosa* (DD). In several morphological characters the tetraploid represents an intermediate state of its parents and its range of distribution overlaps those of the diploids. Some variation was found in the profile of the alcohol soluble seed proteins of the two diploids. By contrast the profiles of the examined accessions of *Ae. cylindrica* were remarkably uniform [22]. This was taken as an indication of a monophyletic origin of *Ae.*

cylindrica. The profile of the tetraploid species was simulated by a protein mixture of specific genotypes, i.e. electrophoretic variants from *Ae. caudata* and *Ae. squarrosa*. Therefore seed protein electrophoresis not only reconfirmed the diploid progenitors of *Ae. cylindrica*, but even identified the parents at the genotypic level. It is very difficult to imagine such identification by morphological or cytological approaches alone.

The wheat group was subjected to numerous studies aimed at elucidating the evolution of this polyploid species. Seed protein electrophoresis supported some of these ideas but threw serious doubt on others.

Cytogenetically the tetraploid wheats are divided into two groups: *dicoccoides-durum* (AABB) and *araraticum-timopheevi* (AAGG). These two groups are distinct electrophoretically [23]. By contrast, the hexaploid wheats (AABBDD) that were separated in the past into many taxa are cytogenetically uniform and share the same seed protein profile [25]. This profile can be simulated by protein mixtures of *T. dicoccum* and *Ae. squarrosa*, the accepted parents of the hexaploid wheats.

According to morphological [60] and chromosomal [58] comparisons, the diploid species *Ae. speltoides* was proposed as the donor of the B genome of wheat. Comparisons of the seed protein profile of various tetraploid wheats (AABB) and of *Ae. speltoides* have shown two fast moving bands in *Ae. speltoides* that are totally missing in the tetraploid wheats [24]. Hence, according to the seed protein profiles, *Ae. speltoides* cannot be considered as a diploid parent which participated in the formation of the tetraploid wheats. A similar conclusion was reached by cytogenetic evidence [29]. It is interesting to note that all other species of section *Sitopsis* of *Aegilops*, to which *Ae. speltoides* belongs, are characterized by fast moving bands and they are not likely to contain the donor of the B genome of wheat [24].

The diploid wheats *T. boeoticum* and *T. monococcum* are considered by many wheat experts as the source of the A genome in the polyploid wheats. These diploids are characterized by the absence on the electrophoretic gel of the albumin fraction of the alcohol soluble seed protein, or by the presence of a single band [26]. These diploid wheats therefore cannot account for the five bands observed in the profile of the tetraploid wheats. The recently studied diploid wheat *T. urartu* on the other hand possesses two bands in that zone and the mobility of one of them is similar to that of the fastest moving band of the tetraploid wheats [26]. *Triticum urartu* is cross-compatible with *T. boeoticum* only when the latter serves as the pistillate parent. Despite regular chromosome pairing in meiosis the F_1 hybrids are completely self sterile. The amphidiploid, however, is partially fertile and chromosome pairing is highly diploidized. Morphologically this amphidiploid is closer to the tetraploid wheat than any of the amphidiploids between *T. boeoticum* and species of section *Sitopsis* of *Aegilops*. According to this evidence the suggestion has been made that *T. urartu* contributed the B genome to wheat [27]. More information is needed before that suggestion will be widely accepted. It is necessary to know the pattern of chromosome pairing in the hybrid between the *T. boeoticum* × *T. urartu* amphidiploid and the tetraploid wheat. However, at this stage already it is apparent that what was previously supposed to be a uniform diploid wheat is in fact a variable species or even a mixture of sibling species. It is possible that this complex contains additional types which are morphologically alike. Discovery of such types might shed new light on the origin of the B genome of wheat.

Characteristically, the seed protein profiles of polyploid plants are much more uniform compared with their diploid relatives [22, 25, 41]. The reduced variation in the polyploids can be interpreted as if the polyploid species evolved from a few polyploidization events. Another possibility is that for some reason specific genotypic combination of the interspecific diploid hybrid was more successful at the polyploid level. In any case, it is apparent that natural introgression and gene flow had a limited role in enhancing variability at the seed protein level and perhaps in other traits of the polyploid species. The highly uniform seed protein profile in many allopolyploids therefore can be considered as a consequence of genetic drift, since only a small portion of the variation of the diploids was transferred to the polyploid level. The practical implication of this situation is very clear. Vast areas exist for increasing the variability of cultivated allopolyploids, wheat and oats in particular, by artificial introgression from their diploid and tetraploid progenitors. The process of introgression by interploidy hybridization in these cereals has been studied [38, 64] and the methods for gene transfer between plants of different ploidy levels are well established. The great variation of the diploids is particularly important in view of the reduced variability and genetic erosion occurring in our cultivated polyploid cereals today. The variation of the diploid and tetraploid progenitors is dynamically maintained in natural populations and its exploitation can be an important avenue in the future breeding of wheat and oats.

REFERENCES

1. Adriaanse, A, W Klop, and JE Robbers (1969): Characterization of *Phaseolus vulgaris* cultivars by their electrophoretic pattern. J. Sci. Food Agric. 20: 647–650.
2. Beadle, GW (1972): The mystery of maize. Field Mus. Nat. History Bull. 43: 1–11.
3. Ben-ze'ev, N, and D Zohary (1973): Species relationships in the genus *Pisum* L. Israel J. Bot. 22: 73–91.
4. Bewley, JD, and M Black (1978): Physiology and Biochemistry of Seeds in Relation to Germination 1. Development, Germination and Growth. Springer-Verlag, Berlin.
5. Blixt, S, J Przybylska, and Z Zimnik-Przybylska (1980): Comparative study of seed protein in the genus *Pisum* V. Genetics of the electrophoretic patterns I and IV. Genet. Polonica 21: 153–161.
6. Boisser, E (1872): Flora orientalis 2: 622–624.
7. Boulter, D, DA Thurman, and E Derbyshire (1967): A disc electrophoresis study of globulin proteins of legume seeds with reference to their systematics. New Phytologists 66: 27–36.
8. Chen, CH, and W Bushuk (1970): Nature of proteins in *Triticale* and its parental species III. Comparison of their electrophoretic pattern. Can. J. Plant Sci. 50: 25–30.
9. Chooi, WY (1970): Variation in nuclear DNA content in the genus *Vicia*. Genetics 68: 195–211.

10. Danielsson, CE (1956) Plant proteins. Ann. Rev. Plant Physiol. 7: 215–236.
11. Davis, PH (1970): Flora of Turkey 3: 370–372; Edinburgh Univ. Press.
12. Davis, PH, and U Plitmann (1970): *Vicia* L In: Davis, PH (ed.): Flora of Turkey 3: 274–325; Edinburgh Univ. Press.
13. De, DN (1974): Pigeon pea in evolutionary studies. In: Hutchinson, JB (ed.): World Crops; Cambridge.
14. Dunnhill, PM, and L Fowden (1965): The amino acids of seeds of *Cucurbitacea.* Phytochem. 4: 933–934.
15. Galinat, WC (1971): The origin of maize. Ann. Rev. Genet. 5: 447–478.
16. Gray, JR, DE Fairbrothers, and JA Quinn (1973): Biochemical and anatomical population variation in the *Danthonoa sericea* complex. Bot. Gaz. 134: 166–173.
17. Harlan, JR, and JMJ de Wet (1972): A simplified classification of cultivated sorghum. Crop Sci. 12: 172–176.
18. Hull, O, and BL Johnson (1963): Electrophoretic analysis of the amphidiploid of *Stipa viridula* x *Orysopsis hymenoides* and its parental species. Heredity 48: 530–535.
19. Hymowitz, T (1970) On the domestication of the soybean. Econ. Bot. 24: 408–421.
20. Hymowitz, T (1976): Soybeans. In: Simmonds, NW (ed.): Evolution of Crop Plants; Longman, London.
21. Hymowitz, T, and HH Hadley (1972): Inheritance of a trypsin inhibitor variant in seed protein of soybeans. Crop Sci. 12: 197–198.
22. Johnson, BL (1967): Confirmation of the genome donors of *Aegilops cylindrica.* Nature 216: 859–862.
23. Johnson, BL (1967): Tetraploid wheats: Seed protein electrophoresis pattern of the emmer and timopheevi groups. Science 158: 131–132.
24. Johnson, BL (1972): Protein electrophoretic profiles and the origin of the B genome of wheat. Proc. Nat. Acad. Sci. USA 63: 1398–1402.
25. Johnson, BL (1972): Seed protein profile and the origin of the hexaploid wheats. Am. J. Bot. 59: 952–960.
26. Johnson, BL (1975): Identification of the apparent B genome donor of wheat. Can. J. Genet. Cytol. 17: 21–39.
27. Johnson, BL, and HS Dhaliwal (1978): *Triticum urartu* and genome evolution in the tetraploid wheats. Amer. J. Bot. 65: 907–918.
28. Johnson, BL, and O Hall (1965): Analysis of phylogenetic affinities in Triticinae by protein electrophoresis. Am. J. Bot. 52: 506–513.
29. Kimber, G, and RS Athwal (1972): A reassessment of the course of evolution of wheat. Proc. Nat. Acad. Sci. USA 69: 912–915.
30. Ladizinsky, G (1975): Seed protein electrophoresis of the wild and cultivated species of section *Faba* of *Vicia.* Euphytica 24: 785–788.
31. Ladizinsky, G (1975): On the origin of the broad bean *Vicia faba* L. Israel J. Bot. 24: 80–88.
32. Ladizinsky, G (1978): Chromosomal polymorphism in wild populations of *Vicia sativa* L. Caryologia 31: 233–241.
33. Ladizinsky, G (1979): The origin of lentil and its wild gene pool. Euphytica 28: 179–187.
34. Ladizinsky, G (1979): Species relationships in the genus *Lens* as indicated by seed protein electrophoresis. Bot. Gaz. 140: 449–451.
35. Ladizinsky, G (1979): Seed protein electrophoresis in section *Foenum-graecum* of *Trigonella* (*Fabaceae*) Pl. Syst. Evol. 133: 87–94.
36. Ladizinsky, G, and A Adler (1975): The origin of chickpea *Cicer arietinum* as indicated by seed protein electrophoresis. Israel J. Bot. 24: 183–187.
37. Ladizinsky, G, and A Adler (1975): The origin of chickpea *Cicer arietinum.* Euphytica 25: 211–217.
38. Ladizinsky, G, and R Fainstein (1977): Introgression between the cultivated hexaploid oat *A. sativa* and tetraploid wild *A. magna* and *A. murphyi.* Can. J. Genet. Cytol. 19: 59–66.

39. Ladizinsky, G, and A Hamel (1980): Seed protein profiles of pigeon pea (*Cajanus cajan*) and some *Atylosia* species. Euphytica 20: 313–317.
40. Ladizinsky, G and T Hymowitz (1979): Seed protein electrophoresis in evolutionary and taxonomic studies. Theor. Appl. Genet. 54: 145–151.
41. Ladizinsky, G, and BL Johnson (1972): Seed protein homologies and the evolution of polyploidy in Avena. Can. J. Genet. Cytol. 14: 875–888.
42. Ladizinsky, G, and N Porath (1977): On the origin of Fenugreek *Trigonella foenum graecum* L. Legume Res. 1: 38–42.
43. Ladizinsky, G, and R Tamkin (1978): The cytogenetic structure of *Vicia sativa* aggregate. Theor. Appl. Genet. 53: 33–42.
44. Lamprecht, H (1961): *Pisum fulvum* Sibth. and Sm. Genanalytische Studien zur Artberechtigung. Agri. Hort. Genet. 19: 269–297.
45. Larsen, AL, and BE Caldwell (1968): Inheritance of certain proteins in soybean seed. Crop. Sci. 8: 474–476.
46. Lee, JW, and JA Ronald (1967): Effect of environment on wheat gliaden. Nature 213: 844–846.
47. McDaniel, RG (1970): Electrophoretic characterization of proteins in *Hordeum*. J. Hered. 61: 243–247.
48. Mettin, D, and P Hanelt, (1973): Über Speziationsvorgänge in der Gattung *Vicia* L. Kulturpflanze 21: 25–54.
49. Mies, DW, and T Hymowitz (1973): Comparative electrophoretic studies of trypsin inhibitors in seed of the genus *Glycine*. Bot. Gaz. 134: 121–125.
50. Murray, BE, IL Craig, and T Rajhathy (1970): A protein electrophoretic study of three amphiploids and eight species in *Avena*. Can. J. Genet. Cytol. 12: 651–665.
51. Murray, DR (1978): A storage role for albumins in pea cotyledons. Plant Cell Environ. 2: 221–226.
52. Nakai, Y (1977): Variation of esterase isozymes and some soluble proteins in diploids and their autotetraploids in plants. Jap. J. Genet. 52: 171–181.
53. Orf, JH, and T Hymowitz (1977): Inheritance of a second trypsin inhibitor in seed protein of soybeans. Crop Sci. 17: 811–813.
54. Orf, JH, and T Hymowitz (1979): Inheritance of the absence of the Kunitz trypsin inhibitor in seed protein of soybeans. Crop Sci. 19: 107–109.
55. Orf, JH, T Hymowitz, SP Pull, and SG Pueppke (1978): Inheritance of a soybean seed lectin. Crop Sci. 18: 899–908.
56. Paulis, JW, and JS Wall (1977): Comparison of the protein compositions of selected corns and their wild relatives teosinte and *Tripsacum*. J. Agric. Food Chem. 25: 265–270.
57. Plitmann, A (1967): Biosystematic studies in the annual species of *Vicia* and *Lathyrus* of the Middle-East. Thesis, The Hebrew Univ. Jerusalem (In Hebrew).
58. Riley, R, J Unrau, and V Chapman (1958): Evidence on the origin of the B genome of wheat. J. Hered. 49: 90–98.
59. Royes, WV (1976): Pigeon pea. In: Simmonds, NW (ed.): Evolution of Crop Plants; Longman, London.
60. Sarkar, P, and GL Stebbins (1956): Morphological evidence concerning the origin of the B genome in wheat. Amer. J. Bot. 43: 297–304.
61. Schäfer, HI (1973): Zur Taxonomie der *Vicia narbonensis*-Gruppe. Kulturpflanze 21: 211–273.
62. Shechter, Y (1975): Biochemical systematic studies in *Sorghum bicolor*. Bull. Torr. Bot. Club 102: 334–339.
63. Shechter, Y, and JMJ de Wet (1975): Comparative electrophoresis and isozyme analysis of seed proteins from cultivated races of sorghum. Amer. J. Bot. 62: 254–261.
64. Vardi, A, and D Zohary, (1967): Introgression in wheat via triploid hybrids. Heredity 22: 541–560.
65. Waines, JG (1975): The biosystematics and domestication of peas (*Pisum* L.). Bull. Torr. Bot. Club. 102: 385–395.

66. Wilkes, HG (1972): Maize and its wild relatives. Science 177: 1071–1077.
67. Zohary, M (1972): Flora palaestina 2: 135–136. Israel Acad. Sci. Hum. Jerusalem.
68. Zohary, D (1977): The wild progenitor and the place of origin of the cultivated lentil. *Lens culinaris*. Econ. Bot. 26: 326–332.
69. Zohary, D, and N Hopf (1973): Domestication of pulses in the old world. Science 182: 887–894.
70. Zohary, D, and U Plitmann (1979): Chromosome polymorphism hybridization and colonization in the *Vicia sativa* group (Fabaceae). Pl. Syst. Evol. 131: 143–156.

17. The Nutritive Value of Seed Proteins

B.O. EGGUM and R.M. BEAMES

THE NUTRITIVE VALUE OF CEREAL PROTEIN

Introduction

Grains, including cereals, legumes and various dicotyledonous non-legumes, provide not only the major portion of the energy for many human populations throughout the world, but also most of the protein. In the technologically less advanced countries, food grains constitute nearly 80 per cent of the diet of a large majority of the people [71]. Quantitatively, cereals occupy first place as the source of energy and protein, with grain legumes next [1]. The grain legumes are important primarily because of their high content of essential amino acids such as lysine and threonine. The cereal grains, rice, wheat and corn, are staple foods in many parts of the world, while sorghum, millet, barley and oats are of importance in other regions [58, 87]. Information on grain legumes, presented by FAO (1966), shows consumption to be high in many countries. The popularity of legumes is based on many factors, including their capacity to fix nitrogen to produce a grain containing a high level of protein of a quality which complements the inadequacies of cereal protein. However, cereal-legume diets still are deficient in the sulphur-amino acids, methionine and cystine. Also, many legumes contain deleterious factors such as trypsin inhibitors, tannins, haemagglutinins, cynanogenic glucosides, saponins and goitrogens. Toxic compounds which cause diseases such a lathyrism and favism are also present in some legumes [54]. Fortunately, many of these compounds are inactivated by heat processing.

With animal proteins being scarce or prohibitively expensive for a large majority of low-income populations, it is essential that every encouragement be given to the development of sound breeding programmes to upgrade the quality of grain proteins.

Worldwide emphasis has been placed on the prevention and treatment of protein-calorie malnutrition in infants and preschool children [69, 73]. Although an inadequate intake and an insufficient energy value of the diet [37] are the basic problems in this age group, deficiencies in the quality and intake of protein aggravate the situation. An adequate energy supply in necessary for efficient utilization of protein for all age groups, but agreement on the relative importance of dietary protein and dietary energy has not yet been reached [5, 72].

Comprehensive reports on protein-enriched cereal foods for world needs [62] and amino acid fortification of foods [82] have been published. Also, the potential contribution of vegetable proteins in developing countries has been reviewed recently [68]. This latter work was limited to a discussion of the protein quality of seeds and some factors that may affect dietary quality of plant proteins.

Many factors must be considered when assessing the possible ways of increasing the nutritive value of foods, especially those of plant origin. Of paramount importance is an evaluation of the physiological factors and chemical components of plants that may influence the availability and subsequent utilization of nutrients from these foods. In order to accurately measure nutrient content, a test should, ideally, be sensitive to these factors. A lack of such a sensitivity is a common shortcoming of many chemical or *in vitro* tests. Consequently, the biological evaluation of a food is essential for the detection of antinutritients (or toxins) and the measurement of nutrient availability.

It is difficult to make recommendations for changes in chemical composition as nutritional goals for plant breeders, as opinions regarding requirements and priorities in human nutrition are constantly changing. Reliable information about nutrient requirements and how to satisfy them can be obtained only through research, both in the field and under controlled conditions.

Protein Quality

Cereal proteins are known to have a low content of certain essential amino acids, particularly lysine [50]. Lysine and threonine have been identified as the first and second limiting amino acids respectively in wheat [6], barley [39], rice [78], and triticale [83].

In Table 1 is shown the amino acid composition of flour and the corresponding breads of seven cereals [50]. The changes shown in the table occurred as a result of the baking process. Lysine, the most heat-sensitive amino acid, was reduced slightly in the case of wheat and maize bread, 6 per cent in barley bread and 7 per cent in sorghum bread. The content of threonine in various breads was also lowered, only slightly for millet, triticale and sorghum, 6–7 per cent for rice and wheat but 18 per cent for barley. Most severely affected was tryptophan, which was lowered by 8–10 per cent in the case of millet, barley, triticale and sorghum, 17 per cent in maize and 27 per cent in rice.

Protein score, based on the FAO scoring pattern (Table 2) indicates that lysine is

Table 1. Amino acid composition (g per 16 g N) of flour and bread.

Amino acids	Wheat		Maize		Rice		Barley		Millet		Triticale		Sorghum	
	flour	bread	flour	bread	flour	bread	flour	bread	flour	bread	flour	bread	flour	bread
Aspartic acid	4.6	4.5	5.5	5.8	8.3	8.1	5.1	4.4	6.7	6.6	5.9	5.9	6.5	6.5
Threonine	2.8	2.6	3.2	3.2	3.4	3.2	2.9	2.3	3.2	3.1	2.9	2.8	3.3	3.2
Serine	4.8	4.5	4.4	4.4	4.6	4.4	3.7	3.1	4.4	4.3	4.0	4.0	4.3	4.3
Glutamic acid	34.9	33.5	19.0	20.2	18.1	18.1	26.2	25.6	23.5	23.0	27.3	27.2	18.9	19.6
Proline	10.7	10.5	8.5	8.9	4.3	4.2	10.7	10.6	7.0	6.9	8.6	8.6	7.5	7.9
Glycine	3.7	3.6	3.4	3.4	4.2	4.1	3.6	3.7	3.3	3.2	4.1	4.1	3.8	3.7
Alanine	3.2	3.1	7.2	7.2	5.3	5.3	3.7	3.7	6.2	6.2	3.7	3.8	8.0	7.8
Valine	3.8	3.9	4.5	4.5	5.4	5.4	4.6	4.6	4.8	4.7	4.2	4.2	4.7	4.7
Isoleucine	3.4	3.6	3.4	3.5	4.0	3.9	3.5	3.4	3.8	3.8	3.4	3.5	3.6	3.6
Leucine	6.9	6.8	12.7	12.8	7.7	7.6	6.7	6.7	8.9	8.8	6.4	6.4	11.2	11.1
Tyrosine	3.0	3.1	4.0	4.1	4.6	4.6	3.0	3.0	3.0	3.0	2.8	2.9	3.6	3.5
Phenylalanine	4.7	4.7	4.5	4.6	4.8	4.7	4.9	4.9	4.5	4.4	4.2	4.2	4.4	4.4
Lysine	2.3	2.2	2.5	2.4	3.4	3.4	3.2	3.1	2.7	2.5	2.9	2.9	2.7	2.5
Histidine	2.1	2.1	2.7	2.7	2.1	2.1	2.0	2.0	2.1	2.0	2.2	2.2	2.2	2.0
Arginine	4.2	4.1	4.4	4.3	7.6	7.5	4.7	4.7	4.5	4.3	5.4	5.3	4.6	4.0
Methionine	1.6	1.6	2.1	2.1	2.9	3.1	1.7	1.6	2.2	2.2	1.7	1.7	2.3	2.3
Cystine	2.0	2.0	2.0	1.9	2.0	2.0	2.0	1.8	1.9	1.7	2.0	2.0	2.2	2.3
Tryptophan	1.0	1.0	0.6	0.5	1.1	0.8	1.1	1.0	1.3	1.2	1.1	1.0	1.0	1.0

Khan and Eggum (1978)

Table 2. Protein score, and the limiting amino acids of whole flour and bread samples.

Sample	Protein score[1]	Limiting amino acids		
		First	Second	Third
Wheat flour	49	Lysine	Threonine	Valine
Wheat bread	47	Lysine	Threonine	Valine
Maize flour	41	Lysine	Tryptophan	Threonine
Maize bread	39	Lysine	Tryptophan	Threonine
Rice flour	58	Lysine	Threonine	Isoleucine
Rice bread	58	Lysine	Threonine	Tryptophan
Barley flour	63	Lysine	Threonine	Isoleucine
Barley bread	62	Lysine	Threonine	Isoleucine
Millet flour	49	Lysine	Threonine	Isoleucine
Millet bread	46	Lysine	Threonine	Isoleucine
Triticale flour	61	Lysine	Threonine	Valine
Triticale bread	60	Lysine	Threonine	Valine
Sorghum flour	46	Lysine	Threonine	Isoleucine
Sorghum bread	43	Lysine	Threonine	Isoleucine

[1]Based on FAO/WHO 1973 scoring pattern
Khan and Eggum (1978)

the first limiting amino acid in all the flours and breads, with threonine the second limiting in wheat, rice, barley, millet, triticale and sorghum, but tryptophan the second limiting in maize. Isoleucine is the third limiting amino acid in barley, millet and sorghum, valine in wheat and triticale, threonine in maize and tryptophan in rice. Methionine was not among the first three limiting amino acids in any of the flour or bread samples.

Results obtained on the effect of baking on true protein digestibility (TD)[1],

[1] $$\text{TD} = \frac{\text{N intake} - (\text{faecal N} - \text{metabolic N})}{\text{N intake}} \cdot 100$$

$$\text{BV} = \frac{\text{N intake} - (\text{faecal N} - \text{metabolic N}) - (\text{urinary N} - \text{endogenous N})}{\text{N intake} - (\text{faecal N} - \text{metabolic N})} \cdot 100$$

$$\text{NPU} = \frac{\text{TD} \cdot \text{BV}}{100}$$

Table 3. Effect of baking on the protein quality of cereal breads.

	True protein digestibility (%)	Biological value (%)	Net protein utilization (%)	Net dietary protein calorie (%)
Wheat flour	96.0	55.0	53.0	5.7
Wheat bread	95.0	56.0	53.0	5.9
Maize flour	95.0	61.0	58.0	5.5
Maize bread	94.0	57.0[3]	53.0[3]	5.2
Rice flour	100.0	71.0	71.0	5.6
Rice bread	100.0	71.0	72.0	5.7
Barley flour	88.0	70.0	62.0	8.3
Barley bread	85.0[2]	68.0[1]	58.0[2]	7.9
Millet flour	93.0	60.0	56.0	6.3
Millet bread	92.0	62.0	57.0	6.4
Triticale flour	93.0	66.0	61.0	8.2
Triticale bread	91.0[1]	71.0[3]	65.0[3]	8.6
Sorghum flour	56.0	91.0	51.0	3.3
Sorghum bread	52.0[3]	96.0[3]	50.0	3.3

[1] $P < 0.05$ significantly different from corresponding value of flour
[2] $P < 0.01$ significantly different from corresponding value of flour
[3] $P < 0.001$ significantly different from corresponding value of flour
Khan and Eggum (1978)

biological value (BV)[1] and net protein utilization (NPU)[1] are summarised in Table 3. Baking significantly lowered the TD of barley, triticale and soghum flour, but had no effect on wheat, maize, rice and millet flours. The BV of maize and barley breads was significantly lower than that of the corresponding flour, while the BV of triticale and sorghum breads was significantly higher. The BV of wheat, rice and millet breads was not affected by baking. NPU of triticale flour was significantly increased by baking, whereas the NPU of maize and barley flour was significantly reduced. Baking had no effect on the NPU values of wheat, rice, millet and sorghum flour. The effect of baking on the digestible energy of cereals was tested by feeding two groups of rats on tritical flour and bread. No significant difference between the digestible energy of flour (87%) and bread (86%) was observed.

Net dietary protein calorie percentage (NDp cal%) of the different flour and bread samples were calculated [62] and are presented in Table 3. The NDp cal% of the breads lie between 3.3 and 8.6. The values of all the breads, except sorghum bread, were greater than 5 NDp cal %.

The processing conditions employed in the present work for preparing the different breads did not decrease the nutritive value to a large extent, even in the samples (rice, millet, maize, sorghum) exposed to 220–230°C for 7–10 min. However, making breads from barley and maize resulted in a 6 per cent and a 9 per cent decrease in NPU respectively, due to losses in the most limiting amino acids lysine, threonine and tryptophan. On the other hand, BV for triticale and sorghum breads were 8 and 6 per cent higher, respectively, than those of unprocessed samples. The moderate heat treatment which was employed apparently had a positive influence on the nutritive value due to destruction of antimetabolites. The very high BV of sorghum deserves further comment. The sorghum used in this experiment was a Pakistan variety of dark brown color and bitter taste. A high tannin content would have been expected. However, tannin, which was estimated by the method of Eggum and Christensen [22], indicated only 1.9 per cent in the sorghum flour. This tannin level cannot explain the very low digestibility values. A more specific method for identifying all the polyphenols in sorghum (as suggested by S.K. Arora, pers. comm.) could shed some light on the apparent anomaly. A higher tannin level would explain not only the low TD values but also the high BV values. Although tannin has a highly negative effect on protein digestibility, it reacts primarily with the non-essential amino acids proline, glutamic acid, glycine and alanine. The resultant increase in the BV of the absorbed protein, as was found in these experiments, results in a NPU which is not markedly inferior to that obtained with wheat, maize, barley and millet. Results of growth trials with pigs, using a variety of supplements for barley and sorghum, has shown the differences between these grains to be small and, if anything, to be in favour of sorghum [4].

Protein Composition and Distribution

A convenient method for partitioning grain proteins is to submit the ground grain to a succession of solvents, thus dividing the proteins into Osborne fractions [60] as proposed by Osborne [70] and described in detail by Shewry et al. [79] and Køie and Nielsen [52]. This subdivision is based on the solubility of albumins in water, globulins in salt solution, prolamins in alcohol and glutelins in dilute acid or alkali. In normal barley, albumins plus globulins form approximately 18 per cent of the crude protein, non protein nitrogen approximately 12 per cent, prolamins 42 per cent and glutelins approximately 23 per cent [2, 52, 60, 77]. Although the composition of each of the soluble fractions is by no means constant [12] with the

Table 4. Amino acid nitrogen as a percentage of protein nitrogen of a barley sample.

Component	Albumin (Water soluble)	Globulin (Salt soluble)	Hordein (Alcohol soluble)	Glutelins (Alkali soluble)
Amide	5.9	5.1	23.0	10.3
Aspartic Acid	8.0	5.6	1.2	4.3
Glutamic Acid	8.7	6.8	23.0	11.6
Proline	4.2	2.7	15.3	6.6
Glycine	6.7	10.7	1.7	5.2
Alanine	7.2	0.65	2.2	6.6
Valine	5.8	4.1	3.5	4.9
Leucine	5.7	4.5	4.6	5.8
Isoleucine	4.1	2.2	3.6	3.5
Phenylalanine	3.0	2.1	3.6	2.7
Tyrosine	2.7	1.5	1.6	1.9
Tryptophan	1.3	0.65	0.7	1.1
Serine	4.1	3.9	3.2	4.2
Threonine	3.4	2.4	1.9	3.1
Cystine+Cysteine	1.5	2.6	1.5	0.9
Lysine	7.9	6.3	0.80	4.8
Methionine	1.4	0.9	0.75	1.1
Arginine	13.0	22.0	6.0	12.0
Histidine	4.3	3.1	0.22	4.3

Folkes and Yemm (1956).

alcohol-soluble hordein of barley containing from almost zero to 2.3 per cent lysine [40], differences between 'average' values for each fraction are sufficiently large for this fractionation to be of some value in estimating amino acid content. Some values obtained on a single barley sample [35] are presented in Table 4. Values obtained on 'hiproly' and some standard barley varieties [38] are listed in Table 5.

Table 5. Lysine nitrogen as a percentage of protein nitrogen in Osborne fractions and in unfractionated samples.

Cultivars	Component	Water & Salt Soluble	Alcohol Soluble	Alkali Soluble	Unfractionated Sample
Hiproly	Whole Kernel	3.63	0.36	2.14	2.29
	Endosperm[1]	3.07	0.28	1.85	2.30
Orange Lemma	Whole Kernel	5.02	0.45	1.61	0.98
	Endosperm	6.40	0.38	1.90	1.23
OR61-2141-9	Whole Kernel	3.49	0.42	2.24	1.89
	Endosperm	3.57	0.27	2.53	1.51
Wocus	Whole Kernel	2.91	0.31	1.84	1.74
	Endosperm	4.54	0.16	1.89	1.58

[1] All seed part except embryo
Adapted from Helm (1972).

Factors affecting Protein Digestibility

Crude fibre. The protein components of plants are functional units of the structural tissues from which they originate. Therefore, it is not surprising that some proteins have inherent characteristics which protect them against the degradative attack of proteolytic enzymes. Since the nitrogenous constituents of cereals are largely contained inside the cellulosic structure, physical factors may influence the digestibility of proteins by preventing their contact with proteases. Other causes of a reduced protein digestibility are the presence of protease inhibitors and changes in structure resulting from processing. Several other anti-nutritional factors which indirectly affect protein utilization may also be present in cereals [84].

It has been shown [80] that both *in vivo* and *in vitro* digestibility of protein in wheat millfeeds decreases when the fibre content increases. A marked negative relationship between protein digestibility and crude fibre content has been found for pigs [47] and rats [16].

According to Saunders & Kohler [71] the indigestible protein of wheat can probably be classified into two types. The first type of protein actually resides in the aleurone layer and is of limited digestibility because of the thick cell wall that interferes with digestion; the second type of protein is tightly bound to the cellulosic matrix of the aleurone cells. The true digestibility of the individual amino acids is thus a more important factor in cereals such as barley and oats, which have higher crude fibre contents, than in maize, rye, wheat or triticale [17]. Protein fraction studies [85] revealed that most of the minor millets contained a large proportion of unextractable protein, possibly because of the fibrous envelope in which the proteins are enclosed [84].

Since any increase in the lysine content of grain generally occurs in the protein of the aleurone cells, which already contains a relatively high lysine content, the role of cellulosic supportive structures should not be disregarded when a maximum nutritional value of cereals is desired.

Inhibitors. It is well known that tannins reduce protein digestibility [22]. The tannins are present in a number of plant materials at high levels [76] and appear to be of significance in some of the cereals such as sorghum [45] and barley [22] as well as in rape seed meal [21].

Tannins affect palatability and feeding value. One per cent tannin fed to chickens influenced weight gain and depressed feed conversion by 6 per cent [3]. Glick & Joslyn [36] noted an increased excretion of protein in the faeces of rats fed 2 per cent or more tannic acid in the diet. The ability of tannins to lower protein digestibility is not the only factor explaining the adverse nutritional effects induced by them. Actually, tannins exhibit a high binding affinity for a number of enzymes and behave as enzyme inactivators [61].

The literature on protease inhibitors has been extensively reviewed [56]. Trypsin inhibitors have been identified in wheat, rye and triticale, but their importance in practical nutritional terms remains somewhat obscure [84]. Protease inhibitors not only reduce protein digestibility but also cause the pancreas to respond to the decrease of active proteases in the intestine so as to restore the level to normal [57]. Since pancreatic enzymes are rich in sulphur-containing amino acids, the inhibitor-stimulated secretion of pancreatic enzymes would create an increased requirement for methionine and/or cystine for the synthesis of other tissue proteins, thus accentuating the deficiency of sulphur-containing amino acids which already exists in many plant proteins [48].

Heat processing. Heat processing can have either a beneficial effect or a deleterious effect on the nutritional value of proteins [16]. Moderate heat-treatment, especially in combination with high moisture, not only enhances palatability and acceptance but also improves protein quality by inactivating enzymic inhibitors, lectins or other thermolabile toxic factors which may be present, [48] and by increasing the availability of amino acids through the exposure of a higher number of enzyme-susceptible bonds.

Severe heat processing such as toasting, puffing and gun explosion techniques have been reported to cause a drastic reduction in the nutritive value of wheat, oats, rice and other cereals proteins [20]. Amino acid analysis of acid and enzyme hydrolysates of raw and heat-processed oats revealed that lysine was the amino acid affected the most. Similarly the most important loss during bread-baking is that of lysine. This loss occurs mainly in the crust [20]. The loss of lysine has been found to be much higher in bread to which reducing sugar has been added [67].

Recent studies have shown that the protein of raw milled rice is better digested by in rats than wheat flour protein, with respective true digestibility (TD) values of 100 and 90 per cent [17]. The cooking of rice reduces protein digestibility but has no adverse effect on NPU (TD × biological value) because of a corresponding increase in biological value [26]. Amino acid digestibility of raw milled rice ranged from 98.8 per cent to 100.0 per cent while the corresponding amino acid digestibility of cooked rice ranged from 81.0 per cent to 100.0 per cent. Lysine, the first limiting amino acid in rice protein, retained its high digestibility on cooking.

Amino Acid Digestibility

Just as there are variations in the digestibility of protein from different sources, there are also variations in the digestibility of individual amino acids from any one protein source, such as grain, which consists of a mixture of structural and storage proteins. In the following discussion this is demonstrated in seven cereal grains for lysine and glutamic acid, the amino acids with the highest and lowest digestibilities, respectively. Corresponding values for digestibility of protein are given for purposes of comparison.

From Table 6 [18] it appears that there are considerable differences between cereals in protein digestibility. For uncooked rice there is almost no difference between total and digestible protein, whereas in rye this difference is more than 20 per cent. Such differences illustrate the possible errors associated with replacement of proteins on a total protein basis. Differences in amino acid digestibilities would compound such errors. With the other cereal grains the differences between total and digestible protein are smaller than for rye, but are still considerable. Table 7 [18] gives values for total lysine and true digestibility of lysine in the same cereal grains. Digestible amino acids were measured according to the method of Kuiken & Lyman [53]. Although this method is criticized because of microbial activity in the alimentary tract, it can still provide valuable information [16]. The differences between values of total and digestible lysine are much more pronounced than for protein. This is probably due to the fact that lysine is mainly deposited in the protein fractions of lowest digestibility [17]. In the highly digestible prolamin fraction almost no lysine is found, whereas the glutamic acid content is very high, which explains the high digestibility of glutamic acid, shown in Table 8 [19]. These differences in digestibility have been well discussed [16, 66, 86]. An intermediate true digestibility has been shown for the other amino acids [16, 74]. Aspartic acid, glycine and alanine have TD values in the lower part of the range, whereas the histidine, arginine and serine values are in the upper part. Lysine has been shown to be highly

Table 6. Total and true digestible protein in eight cereal grains.

Cereal grain	Total protein in DM[1] (%)	True digestible protein in DM[1] (%)	Difference (100-% true digestibility)
Barley	10.13	8.31	17.96
Oats	10.75	9.04	15.90
Wheat	12.63	11.32	10.37
Rye	9.13	7.03	23.00
Maize	10.06	8.81	12.42
Sorghum	12.54	10.63	15.23
Rice	8.96	8.90	0.66
Triticale	13.07	12.12	7.30

[1]DM, Dry Matter
Eggum (1977b)

Table 7. Total and true digestible lysine in seven cereal grains.

Lysine source	Total lysine (g/16g N)	True digestible lysine (g/16g N)	Difference (100-% true digestibility)
Barley	3.69	2.80	24.11
Oats	4.03	3.21	20.34
Wheat	2.55	2.02	20.80
Rye	3.67	2.40	26.60
Maize	2.73	2.31	15.38
Sorghum	1.83	1.33	27.30
Rice	3.54	3.51	0.85

Eggum (1977b)

correlated with the three amino acids, aspartic acid, glycine and alanine, and to have a high negative correlation with glutamic acid, which is one of the main amino acids of storage protein [74].

Table 8. Total and true digestible glutamic acid in cereal grains.

Grain	Total glutamic acid in DM (g/16g N)	True digestible glutamic acid in DM (g/16g N)	Difference (100-% true digestibility)
Barley	25.06	22.90	8.62
Oats	22.10	19.71	10.81
Wheat	35.77	35.41	1.01
Rye	23.62	21.52	8.89
Maize	17.46	16.05	8.08
Sorghum	21.24	19.12	9.98
Rice	17.18	17.18	0.00

Eggum (1977b)

The Effect of N-fertilization on Protein Quality

When barley, rye, wheat, maize and sorghum are fertilized with increasing amounts of nitrogen, more protein is deposited in their prolamin fraction. As this is a poor source of lysine [74] but highly digestible [17], more digestible protein but of a lower biological value is obtained [15, 31]. For oats and rice the situation is different, as glutelin (relatively rich in lysine) is the main storage protein in these cereal grains [74]. A negative correlation has been observed between lysine content of protein and protein content of brown and milled rice only in samples with protein below 10 per cent [46]. In further work with rice [23], TD and net protein utilization (NPU) were found to be positively correlated with lysine content of rice protein (3.17–4.07 g/16 g N). In this report, utilizable protein ranged from 5.19 to 11.12 g/100 g DM and was mainly determined by N content of the milled rice ($r = 0.988$). Digestibility of amino acids of milled rice protein determined by faecal analysis ranged from 94.1 to 100.0 per cent in samples with 1.38 to 2.74 per cent N, dry basis. An increase in milled rice protein content ($N \times 6.25$) from 9.12 to 11.09 in IR 8 and from 11.09 to 14.56 in IR 480–5–9 owing to N fertilizer application had no significant effect on the lysine content of the protein and had little effect on true digestibility, biological value and net protein utilization of the protein in tests with growing rats.

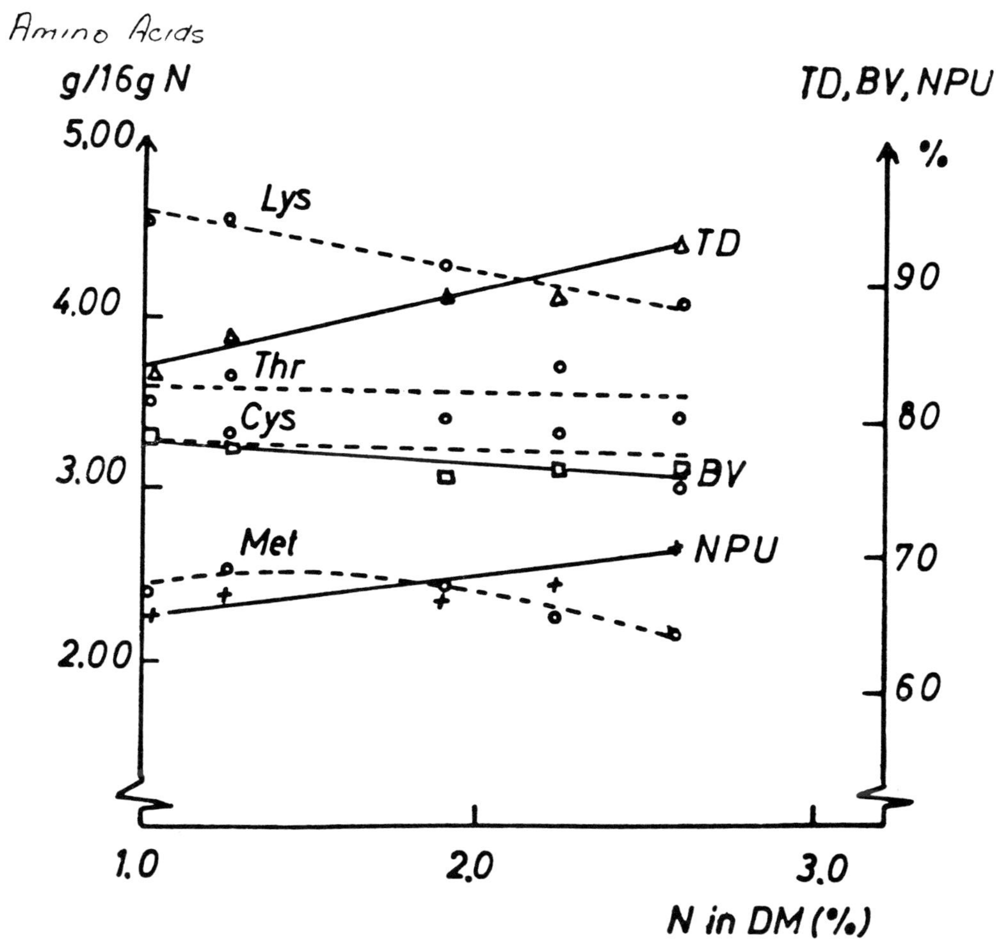

Figure 1. Relationships between true digestibility (TD), biological.

In one experiment [59], the negative correlation between protein and the lysine content of the protein appeared to become nonsignificant at higher levels of protein in wheat. However, the lysine content of the protein in genetically high-protein wheat has been found to be equal to or higher than that of normal wheats grown in the same environment. The amino acid composition of oat protein is remarkably constant over a wide range of protein content [7, 31, 81]. The relationship between protein content and protein quality in oats in response to fertilizer application [31] is illustrated in Figure 1. It is evident that the concentration of lysine, and, to a lesser extent, the concentrations of threonine, cystine and methionine, are only slightly decreased with increasing nitrogen content. This small reduction, in contrast to that

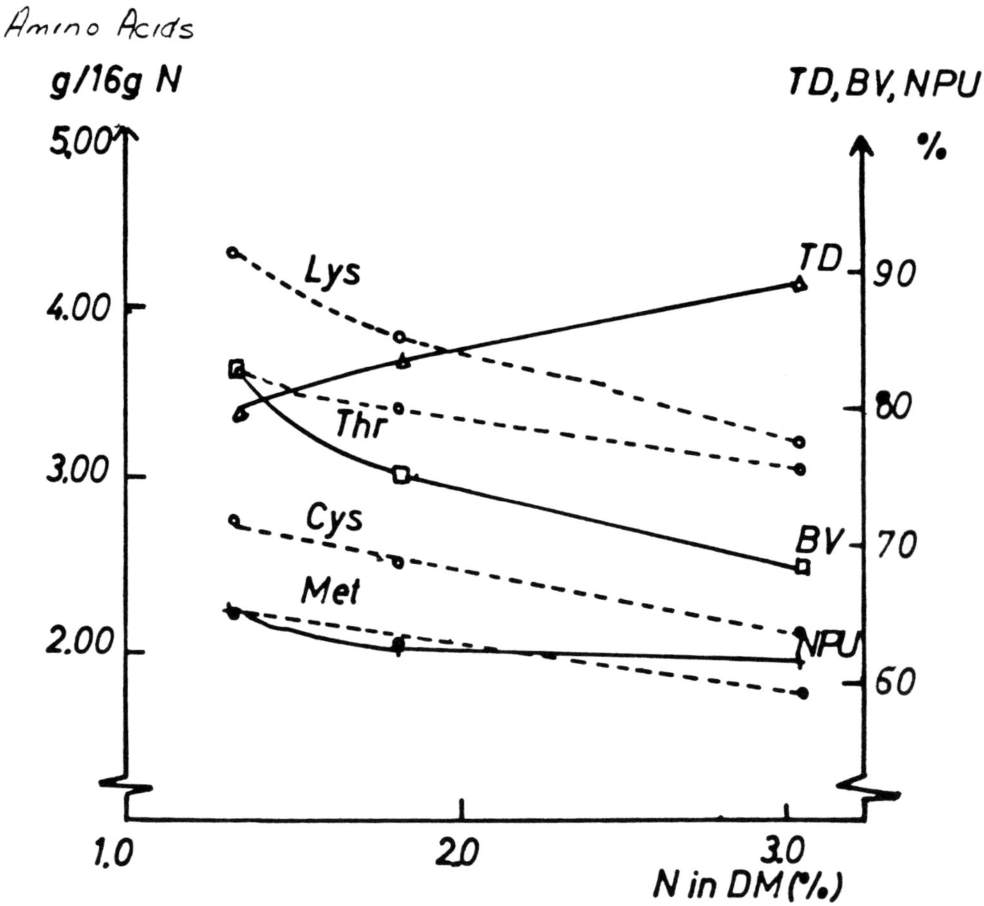

Figure 2. Relationship between true digestibility (TD), biological value (BV), net protein utilization (NPU), lysine, threonine, methionine and cystine, and the concentration of nitrogen in grain of spring rye (var. Petkus). (From Eppendorfer, 1975, with permission).

occurring in most other grains, with the exception of rice, results from the storage of the additional nitrogen as glutelin, rather than as prolamin. TD increases linearly with protein content because of the increasing proportions of the highly digestible glutelin. In combination with only a small reduction in BV, the value for NPU also increases. In Figure 2, concentration of lysine, threonine, methionine and cystine in rye are also shown as a function of N content of grain. This graphic presentation clearly shows that the decrease of the BV is very closely paralleled by decreasing concentrations of these amino acids. Figure 2 is probably representative of similar relationships between protein content and protein quality in barley, wheat, maize and sorghum, which also store most of the additional nitrogen as prolamins.

Table 9. Amino acid composition (g/16 g N), true protein digestibility (%), biological value (%), net protein utilization (%), and utilizable protein (%) of two buckwheat varieties.

Variety	Buckwheat Siva dolenjska	Buckwheat Bednja 4N
Aspartic acid	8.22	8.50
Threonine	3.15	3.30
Serine	3.88	4.16
Glutamic acid	17.65	17.77
Proline	4.30	4.22
Glycine	5.51	5.40
Alanine	4.24	4.13
Valine	4.69	4.59
Isoleucine	3.48	3.37
Leucine	6.11	5.96
Tyrosine	2.67	2.67
Phenylalanine	4.19	4.16
Lysine	5.09	5.19
Histidine	2.20	2.32
Arginine	8.55	9.00
Methionine	1.89	1.80
Cystine	2.02	2.17
Tryptophan	1.59	1.57
True digestibility	79.9 (1.7)[1]	78.8 (0.9)
Biological value	93.1 (1.4)	90.5 (1.0)
Net protein utilization	74.4 (1.5)	71.2 (1.4)
Utilizable protein	9.07 (0.2)	8.73 (0.2)

[1] Standard deviation
Eggum et al. (1981)

In grain, N applications normally increase yield, protein content, or both, and, therefore, also the production per unit area of protein and each individual amino acid, including lysine. A decrease in the nutritional value of the protein, whether expressed either as a reduction in lysine or in NPU, probably always will be more than balanced by an increased production of protein and amino acids [31].

The Protein Quality of Buckwheat (Fagopyrum spp.)

Buckwheat, although not a cereal, is usually grouped with the cereals because of its method of cultivation and utilization. Its protein is reported to be of a high quality, rich in lysine, and containing less glutamic acid and proline and more arginine and aspartic acid than cereal proteins. Due to its high lysine content, buckwheat has a higher biological value than the cereal grains, all of which have lysine as the first limiting amino acid. Results for two buckwheat samples are given in Table 9 [25]. Protein digestibility is low (TD 78.8–79.9), which is probably due to the high content of crude fibre (16–18 per cent) and tannin (1.5–1.8 per cent) [25]. In spite of this, the very high BV of the buckwheat resulted in a much higher utilizable protein value (per cent crude protein × NPU) than found in cereal grains. Based on these data it can be concluded that although protein digestibility is rather low, the protein quality of buckwheat is excellent. The high fibre content, rather than being a concern, may be found to be of some benefit in many parts of the world.

Concluding Remarks

The protein level in cereal grains is low, averaging from 9 to 12% of the dry matter for most grains, and even less for rice. The first limiting amino acid is lysine, followed by threonine, although in maize, tryptophan is probably second limiting. Lysine is not only present at low levels, but its availability is less than that of other amino acids, thus aggravating its deficiency. Also, lysine is quite heat sensitive, with both its level and availability often reduced during normal cooking procedures. These factors make the breeding of cereals with improved protein and lysine levels a matter of considerable urgency.

Compared to animal protein, the digestibility of cereal protein is rather low. As discussed in the preceding text, crude fibre in plant food has a negative effect on protein digestibility. In some of the cereal grains there are significant levels of anti-nutritional factors. The prime factor, however, seems to be tannin, which can be present in relatively high concentrations in sorghum and buckwheat, and even in barley.

In the present chapter, the topic has been discussed in relation to human nutrition. If the discussion were expanded to investigate the adequacy of grain protein for animals, particularly for pigs and poultry, the deficiencies would be even more pronounced, thereby strengthening the argument in support of research on the development of varieties with improved protein quality.

THE NUTRITIVE VALUE OF LEGUME SEED (PULSE) PROTEIN

Introduction

Experimental evidence has demonstrated conclusively that legume grain protein is the natural protein to complement that present in cereal grains. When both are ingested in the appropriate ratio, the protein quality is higher than that of the individual components [9]. Almost always, grain legumes give lower yields than the cereal grains and are thus more expensive to produce, but their production is encouraged by governments and, in particular, by international agencies in the developing countries, primarily because of their higher protein and lysine content. In recent years plant breeders have been working to select varieties of grain legumes that are not only more productive but are also of an improved nutritional quality.

Table 10. Amino acid contents (g/16 g N) of pulses as percentage of the provisional FAO/UN (1973) amino acid scoring pattern.

	Isoleucine	Leucine	Lysine	S-containing[3]	Aromatic[4]	Threonine	Tryptophan	Valine	E/T[5]
Common bean	105	108	132	54[1]	127	99	105	93[2]	2.39
Broad bean	100	101	119	44[1]	124	84[2]	90	89	2.20
Chick-pea	111	106	126	63[1]	142	94	90[2]	92	2.43
Cowpea	96	100	126	64[1]	128	90[2]	113	91	2.31
Lentil	108	108	132	49[1]	140	99[2]	100	101	2.46
Lima bean	124	116	137	64[1]	153	104[2]	105	104[2]	2.65
Lupin	110	102	97	61[1]	119	91	105	81[2]	2.18
Pea	107	97	138	58[1]	121	102	93[2]	95	2.35
Pigeon pea	78	90	141	42[1]	169	73	58[2]	73	2.25
Vetch	91	98	106	59[1]	101	84	-	80[2]	1.99
Peanut	84	90	65ℓ	68	146	65[1]	108	84	2.03
Soybean	114	110	118	74[1]	133	96[2]	133	97	2.46

[1] First limiting amino acid

[2] Second limiting amino acid

[3] Methionine + cystine

[4] Phenylalanine + tyrosine

[5] Grams of total amino acids which are essential for adult man (including cystine and tyrosine) per g of N

Burr (1975)

Table 11. Chemical score, biological value, true digestibility, and net protein utilization of various pulses.

Pulse	Chemical score	BV	TD	NPU
Bean[1]	54	58	73	42
Broad bean	44	55	87	48
Chick-pea	63	68	86	58
Cowpea	64	57	79	45
Lentil	49	45	85	38
Lima bean	64	66	78	51
Pea	58	64	88	56
Pigeon pea	42	57	78	44
Peanut	65	54	87	47
Soybean	74	73	90	66

[1] *Phaseolus* spp. excluding *P. lunatus*

Burr (1975)

In general, legume grains comprise an important part of the human diet in developing countries located in tropical and subtropical areas. There, the nutritional contribution is of paramount importance, as these populations have limited access to foods of animal origin. In Central America, beans provide 20 to 30% of the total protein intake [10]. The reason for this upper limit to legume grain consumption which is observed in most communities is not clear. A contributory factor would be the relatively low digestibility of the protein of legume grains.

Protein Quality

The quality of bean protein is quite variable, with protein efficiency ratio (PER) values as high as 2.0. In most species, and varieties within species, sulphur-containing amino acids are the most limiting. Bressani [8] drew attention to some studies suggesting that a negative relationship exists between protein content and methionine content. He recommended that this area should receive more attention, as very little is known about the direct effects of genetic composition on chemical and nutritional characteristics of food legumes or about the relationship between nutrients.

Table 12. The protein quality of some pulses expressed as amino acid content (g/16 g N), TD, BV, NPU and UP[1].

	Beans (green)	Horse-beans	Peanut meal	Soybean meal	Peas (green)
Lysine	4.73	6.34	3.07	6.00	5.39
Methionine	1.86	1.00	1.11	1.58	1.12
Cystine	1.02	1.41	1.12	1.51	0.82
Threonine	3.46	3.56	2.78	3.69	4.06
Isoleucine	3.11	4.28	3.50	4.52	3.36
Leucine	5.16	7.47	6.19	7.46	5.26
Valine	3.92	4.71	4.00	5.02	4.11
Phenylalanine	3.87	4.38	4.91	5.19	3.19
Histidine	1.86	2.60	2.14	3.37	1.47
Arginine	3.76	10.07	8.46	7.11	11.48
Protein in dry matter (%)	21.0	33.5	53.3	53.4	27.0
TD (%)	87.6	91.6	92.3	91.9	91.4
BV (%)	71.0	51.0	52.4	64.8	48.2
NPU (%)	61.6	46.7	48.4	59.6	44.1
UP[1] (%)	12.9	15.6	25.8	31.8	11.9

[1]UP = Utilisable protein (protein in dry matter x NPU/100)

Eggum (1977a)

In Table 10 are presented the average amino acid contents [33] of various pulses, expressed as percentages of the 1973 FAO/UN provisional pattern. It can be seen that in all but peanuts, the sulfur-containing amino acids are first-limiting. The role of second-limiting is shared by threonine, valine and tryptophan.

Table 11 gives average BV, TD and NPU values and chemical scores for various pulses [33]. These results suggest, that a considerable portion of pulse protein is not hydrolysed in the digestive tract for absorption.

In Table 12 are given data of some pulses frequently consumed in Scandinavia. It appears from these data that the protein quality (BV) of pulses in general is low,

Table 13. Amino acid composition (g/16 g N), true protein digestibility (TD), biological value (BV) net protein utilization (NPU) and utilizable protein (UP) of some uncooked pulses from India.

	Kasari[1]	Arhar[1]	Chick[2] pea	Pigeon[2] pea (2624)
Lysine	6.11	6.66	7.05	7.01
Methionine	0.72	1.40	1.53	1.59
Cystine	1.22	1.59	1.41	1.41
Threonine	3.38	3.61	3.48	3.73
Isoleucine	4.00	3.84	4.51	4.01
Leucine	6.54	7.26	7.85	7.57
Valine	4.38	4.65	4.57	4.77
Phenylalanine	4.06	9.97	5.50	10.10
Histidine	2.83	3.57	2.38	3.79
Arginine	8.72	6.57	9.82	6.97
Protein in dry matter (%)	39.1	22.1	24.5	22.9
TD (%)	90.6	53.3	85.2	57.0
BV (%)	52.4	79.5	63.3	70.0
NPU (%)	47.5	42.4	54.0	39.9
UP (%)	18.6	9.4	13.2	9.1

[1] Prabhu (1971)

[2] Eggum (1977a)

primarily due to the low content of sulphur-containing amino acids. In several of the pulses in Table 12 more than 50% of the nitrogen is excreted with the urine and faeces when they are fed as the only protein source. The digestibility values, however, are not as low as those given in the previous table.

In Table 13 are given values for protein quality of some pulses frequently grown in India. It should be stressed that the evaluations were performed on raw material. The lysine content of these Indian pulses is quite good whereas the concentration of sulphur-containing amino acids is very low. True protein digestibility, however, of

Table 14. Amino acid composition (g/16 gN), true protein digestibility (%), biological value (%) and net protein utilization (%) of different varieties of cooked chick peas, mash beans, mung beans and cow peas.

Amino Acid	Variety											
	Chick peas					Mash beans	Mung beans			Cow peas		
	C727	6227	6501	6560	6576	133	MgI	588	6601	Local	382	411
Aspartic acid	11.6	11.2	11.9	11.4	11.6	12.0	11.4	11.2	12.5	11.1	11.1	11.0
Threonine	3.4	3.3	3.5	3.1	3.5	3.2	3.1	3.0	2.9	3.4	3.6	3.3
Serine	4.8	4.4	4.8	4.8	4.7	5.0	4.8	4.6	4.4	4.9	4.7	4.4
Glutamic acid	17.8	17.6	18.2	17.6	17.6	21.0	18.5	18.3	18.9	18.1	18.6	18.3
Proline	4.3	4.2	4.3	4.2	4.3	4.2	4.2	4.2	4.3	4.1	4.1	4.2
Glycine	3.9	3.9	3.9	3.7	3.9	4.0	3.7	3.7	3.6	3.8	3.8	3.6
Alanine	4.1	4.1	4.2	4.0	4.2	4.3	4.2	4.1	4.3	4.1	4.2	4.1
Valine	4.3	4.3	4.4	4.1	4.3	5.3	5.0	4.9	5.1	4.7	4.8	4.8
Isoleucine	4.6	4.5	4.6	4.4	4.6	4.7	4.5	4.4	4.5	4.2	4.4	4.3
Leucine	7.6	7.5	7.7	7.2	7.6	8.3	7.7	7.7	7.9	7.6	7.7	7.8
Tyrosine	2.8	2.7	2.9	2.5	2.9	3.3	2.9	2.9	2.9	3.2	3.3	3.2
Phenylalanine	5.5	5.3	5.6	5.4	5.4	5.8	5.7	5.6	5.5	5.3	5.5	5.4
Lysine	6.1	6.0	6.3	5.8	6.1	6.1	6.4	6.3	6.2	6.1	6.3	6.2
Histidine	2.4	2.3	2.5	2.3	2.5	2.4	2.7	2.6	2.5	3.0	3.0	2.9
Arginine	9.1	9.4	9.4	11.5	8.9	6.2	6.8	6.8	6.7	8.2	7.5	7.5
Methionine	1.7	1.7	1.7	1.5	1.7	1.9	1.4	1.4	1.5	1.5	1.6	1.6
Cystine	1.6	1.5	1.4	1.2	1.5	0.7	0.6	0.7	0.6	0.7	0.8	0.8
Tryptophan	1.0	1.2	1.2	0.8	1.1	1.0	1.0	1.0	1.0	1.1	1.1	1.1
True protein digestibility	87.0 (1.3)	86.0 (1.5)	86.0 (0.4)	89.0 (1.6)	85.0 (0.7)	83.0 (0.9)	84.0 (1.5)	83.0 (0.8)	85.0 (0.7)	89.0 (1.1)	92.0 (1.5)	87.0 (0.7)
Biological value	69.0 (0.8)	67.0 (0.8)	65.0 (0.6)	62.0 (0.8)	66.0 (1.1)	62.0 (0.6)	55.0 (1.1)	54.0 (1.3)	56.0 (1.0)	57.0 (0.5)	55.0 (1.1)	59.0 (0.6)
Net protein utilization	60.0 (1.0)	58.0 (1.2)	56.0 (0.6)	55.0 (1.0)	56.0 (1.1)	51.0 (0.6)	46.0 (1.1)	45.0 (1.1)	48.0 (0.9)	51.0 (0.7)	50.0 (1.0)	51.0 (0.6)

Khan et al. (1979) Figures in parenthesis indicate SD values

Table 15. Protein score and limiting amino acids of the cooked legumes referred to in Table 14.

Legume	Varieties	Protein Score[1]	Limiting amino acids		
			First	Second	Third
Chick peas	C727	79	Threonine	Valine	Meth.+cyst.
Chick peas	6227	79	Threonine	Valine	Meth.+cyst.
Chick peas	6501	80	Threonine	Valine	Meth.+cyst.
Chick peas	6560	76	Meth.+cyst.	Threonine	Tryptophan
Chick peas	6576	81	Threonine	Valine	Meth.+cyst.
Mash bean	133	65	Meth.+cyst.	Threonine	Tryptophan
Mung bean	MgI	55	Meth.+cyst.	Threonine	Tryptophan
Mung bean	588	56	Meth.+cyst.	Threonine	Tryptophan
Mung bean	6601	58	Meth.+cyst.	Threonine	Tryptophan
Cow peas	local	61	Meth.+cyst.	Threonine	Valine
Cow peas	382	63	Meth.+cyst.	Threonine	Valine
Cow peas	411	63	Meth.+cyst.	Threonine	Valine

[1]Based on the FAO/WHO 1973 scoring pattern

Khan et al. (1979)

raw arhar and pigeon pea is extremely low. The samples with the lowest content of methionine + cystine also have very low BV values.

In one experiment [51], only small differences were found in the amino acid content of some varieties of chick peas (*Cicer arietinum*), mash beans (*Phaseolus mungo*), mung beans (*Phaseolus aureus*), and cow peas (*Vigna sinensis*), grown in Pakistan (Table 14). The highest lysine and methionine + cystine content was found in mung beans (MgI) and chick peas (C 727), respectively, and the lowest lysine, methionine + cystine and tryptophan levels per 16 g N were found in the sample of chick peas (6560) which had the highest protein content.

Table 15 presents the calculated protein scores for different varieties of legumes based upon the FAO scoring pattern. Methionine + cystine are the first limiting amino acids in all the varieties of cow peas, mung beans, mash beans and chick peas (6560) except chick pea strain Nos C 727, 6227, 6501 and 6576, where threonine is the first limiting amino acid. Threonine is the second limiting amino acid in the case of cow peas, mung beans, mash beans and chick peas (6560) and valine is the second-limiting amino acid in chick peas C 727, 6227, 6501 and 6576. Valine is the third-limiting amino acid in the case of cow peas whereas tryptophan is third-limiting in all varieties of mung beans, mash beans and chick peas (6560), methionine + cystine are third-limiting amino acids in chick peas C 727, 6227, 6501 and 6576.

The true protein digestibility (TD), biological value (BV) and net protein utilisation (NPU) of different varieties of legumes are presented in Table 14. The TD of chick peas, (6560) was highest (89%) and was significantly ($P < 0.001$), but only slightly, better than that of all the other varieties of chick peas, whereas the TD of C 727 was slightly superior ($P < 0.05$) to that of chick pea 6560. The protein of mung bean 6601 was significantly ($P < 0.01$) but slightly more digestible than that of mung bean 588. The TD of cow peas, local and 411 was lower ($P < 0.01$ and $P < 0.001$ respectively) than the TD of cow pea 382. The BV of chick pea 6560 was significantly ($P < 0.001$) lower than the BV of all the varieties of chick peas. The BV of mung bean 6601 was significantly ($P < 0.05$) higher than that of 588. The BV of cow peas local and 382 was lower ($P < 0.01$ and $P < 0.001$ respectively) than cow pea 411, but the differences were small. The NPU, a derived factor, ranged from 55 to 60% in chick peas, from 45 to 48% in mung beans and from 50 to 51% in cow peas. The BV of the samples was highly correlated with the total methionine + cystine content. The relationship is given in the following regression equation: BV (%) = 33.03 + 10.56 × methionine + cystine (g/16gN). The following values were obtained: $r = 0.97$; $s = 1.4$; $s\lambda = 0.89$; where s is the deviation from regression and $s\lambda$ is the deviation of the regression coefficient. The regression coefficient differed significantly from zero ($P < 0.001$). The results are illustrated in Figure 3.

It appears that cooking these varieties for 10–20 minutes at 121°C is sufficient to destroy the antinutritional factors [51] as indicated by a highly significant positive correlation ($r = 0.97$) between BV and the first-limiting amino acids, methionine + cystine (Figure 3). Consequently the protein quality of the processed legumes in this study [51], could be accurately estimated from their content of sulphur amino acids. In a comparison with the 1973 FAO/UN scoring pattern, these four varieties of legumes are limited by threonine for adult man.

The Problem of Legume Seed Protein Digestibility

Most nutritional and biochemical studies carried out with legume grains have dealt mainly with two factors that are important in determining their protein quality [10]. One consists of the antiphysiological substances present in legume grains, of which the trypsin inhibitors, amylase inhibitor, and haemagglutinins are the most important. The second is the well-documented deficiency of sulphur-containing amino acids. In uncooked grain, the antiphysiological compounds contribute to the low protein digestibility. As these substances are destroyed extensively, and, in some cases, completely, by heat treatment, the poor digestibility of many types of beans when cooked must result largely from other factors. One of these is the adverse effect

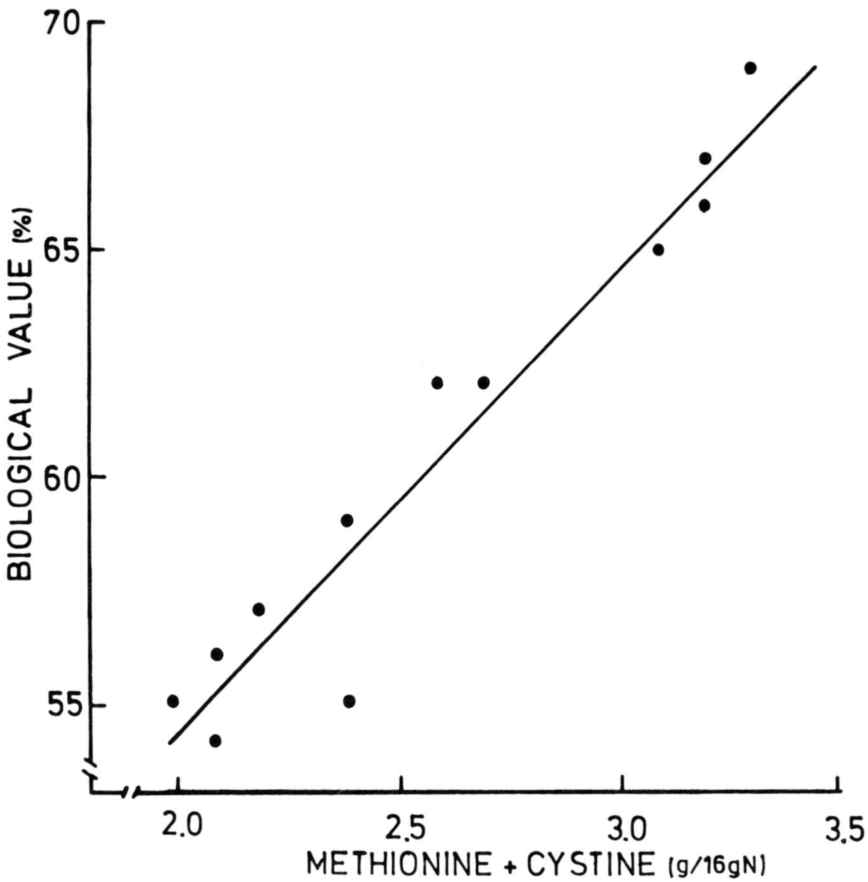

Figure 3. Relationship between total sulphur amino acids and biological value in various cooked pulses (Khan et al. 1979).

on the protein of the thermal processing *per se*. Results obtained with rats [41] show a large variation in digestibility of the protein of cooked samples within species. Large differences in protein digestibility are also evident between species.

Results from both animals and humans show that legume grain protein is, in general, of low digestibility [10]. This can be seen in Table 16, which lists results obtained with both raw and cooked samples. It is evident that raw beans of the *Phaseolus* genus are poorly digested, with values ranging from a low of 15.6% to a maximum of 56.0%. Digestion is improved dramatically by cooking. By contrast, protein digestibility of raw beans of the genus *Vigna* has been found to be relatively high (73.2–79.0%) and not modified appreciably by cooking. Digestibility figures, however, should be interpreted with caution because of variations in methods used and form of presentation. Bressani & Elias [10], in discussing the low digestibility of

Table 16. Protein digestibility of raw and cooked legume seeds.

Legume food	Scientific name genus and species	Protein Digestibility Raw (%)	Cooked (%)	Type of Test	Reference
Common beans (red)	Phaseolus vulgaris	49	71	in vitro	[43]
Common beans (red)	Phaseolus vulgaris	56	83	in vitro	[43]
Common beans (black)	Phaseolus vulgaris	55	80	in vitro	[43]
Common beans (white)	Phaseolus vulgaris	52	91	in vitro	[43]
Common beans	Phaseolus vulgaris	43.5	80.9	in vivo	[48]
Common beans (black)	Phaseolus vulgaris	15.6	71.2	in vivo	[44]
Common beans (white)	Phaseolus vulgaris	42.7	74.9	in vitro	[30]
Common beans (black)	Phaseolus vulgaris	41.1	68.1	in vitro	[30]
Common beans (red)	Phaseolus vulgaris	36.3	72.3	in vitro	[30]
Soybeans	Glycine max	70.1	85.4	in vivo	[42]
Soybeans	Glycine max	82.9	89.7	in vivo	[14]
Cowpea	Vigna sinensis	79.0	82.6	in vivo	[42]
Cowpea	Vigna sinensis	73.2	72.4	in vivo	[27]
Lima beans	Phaseolus lunatus	34.0	51.3	in vivo	[42]
Pigeon peas	Cajanus cajan	59.1	59.9	in vivo	[42]

Bressani & Elias (1977)

Table 17. Effect of various cooking processes on the digestibility of *Phaseolus vulgaris* (black) seeds.

Treatment	Digestibility (%)
Raw	15.6
Cooked in distilled water (85°C, 2h)	48.7
Cooked in 0.1% CH_3COOH (85°C, 2h)	46.6
Cooked in 0.1% $NaHCO_3$(85°C, 2h)	52.9
Autoclaved (15 psi. 30 min)	71.2
Autoclaved (15 psi. 30 min) no seed coat	73.0

Jaffé & Flores (1975)

legume protein, pondered their own interpretations, by listing similar low apparent protein digestibilities of some non-legume plant proteins, which, when corrected for basals faecal losses, gave true protein digestibility values 16–17 percentage units higher. It should be noted in Table 16 that the method of presentation does vary, e.g. one [49] is for apparent protein digestibility, while another [14] is for true digestibility.

Although it has been indicated that no correlation seems to exist between the activity of trypsin inhibitors and protein digestibility [42] there is no doubt that the presence of such activity will interfere with the process of protein breakdown in the gastro-intestinal tract. Liener [55] has discussed the stimulatory effect of trypsin inhibitor on pancreatic secretion, which is believed to be through a reduction in the depressing effect of trypsin on the release of the pancreas-stimulating hormone, cholecystokinin (CCK). As the pancreatic enzymes are particularly rich in sulphur-containing amino acids, trypsin inhibitors, by this mechanism, would aggravate the deficiency of the sulphur-containing amino acids already existing in most legume seeds. Recent findings [28], indicating the existence of heat-resistant trypsin inhibitors would suggest the desirability of a reappraisal of interpretations based on their complete inactivation by heat. Digestibility also varies with the nature of the heat treatment. This is illustrated in Table 17. The results show the importance of cooking. The low value for the raw seed is to be expected, because of its content of active antiphysiological compounds. In the cooked samples, digestibility improved

Table 18. Amino acid availability and apparent protein digestibility in *Phaseolus vulgaris* seeds.

Amino acid	Raw	Heated
Methionine	21.8	68.7
Cystine	36.6	80.6
Lysine	58.5	85.0
Leucine	47.6	85.7
Valine	46.0	84.8
Protein digestibility, %	43.5	80.9

Kakade & Evans (1966)

with increasing pressure. Optimal conditions, however, for the inactivation of antiphysiological factors by heat are not constant, but vary according to the legume grain species [29].

The application of heat may be interpreted as having a double effect [10], with each effect taking place as a consequence of the other. The first effect is to decrease or eliminate the activity of the antiphysiological factors, as is well-documented. The second is to increase amino acid availability from the protein. This is shown in Table 18. It appears that the greatest increase in availability as a result of heating was observed for the sulphur-containing amino acids, methionine and cystine, amino acids thet are present in a high concentration in the trypsin inhibitors. In fact, trypsin inhibitors may account for up to 40% of the total cystine content of some bean proteins [55].

Storage of legume grains affects protein digestibility. In one study [64], samples were stored for 0, 3 and 6 months. The samples were then soaked in water and cooked under pressure for 10, 20 and 30 min. With the 10 min cooking time, true protein digestibility fell from 79.7% with no storage to 73.7% with 3-month storage, but rose to 76.8% with 6-month storage. For all storage times, true protein disgestibility fell by approximately 1.5 percentage units for each 10 minute increment in cooking time.

It is apparent from the above discussion that there are several factors influencing

Table 19. Apparent digestibility of the water soluble, water insoluble and total nitrogen of various legume grains and casein.

	Apparent digestibility (%)		
Protein source	Soluble N	Insoluble N	Total N
Black bean	36.4	79.9	73.4
Red bean	14.2	64.3	60.1
White bean	41.2	78.7	70.3
Casein	-	94.5	89.1
Soybean	72.7	52.6	64.2
Cowpea	66.7	69.5	61.1
Pigeon pea	22.5	76.0	72.0
Casein	-	96.0	90.0

Bressani et al. (1977)

protein digestibility of legume grains. The lack of standard methods of processing, and of exact information defining or describing the conditions prevailing or applied during the various experiments would contribute considerably to the variation which has been reported in protein quality and digestibility.

In metabolic studies with dogs [11], results indicated that, with the exception of soybeans and cowpeas, apparent digestibility of the insoluble nitrogen fraction was significantly higher than that of the water-soluble nitrogen and of the total nitrogen (Table 19). These results were interpreted to mean that the protein which is more resistant to digestion is found in the water-soluble nitrogen fraction of pre-cooked bean flour. However, the authors add that this interpretation must be considered with some reservation, particularly in view of the changes occurring in these fractions in the gastro-intestinal tract. Beans contain up to 17% lignified protein [64]. If this is not digested, the level of this protein should be highly and inversely correlated with protein digestibility. Storage conditions influence the amount of protein that becomes lignified both in the seed coat and in the cotyledons (Table 20). This could explain, at least partially, any changes in digestibility with storage.

Table 20. Content of 'lignified protein' in *P. vulgaris* (black-coated) seeds.

Condition	Lignified protein (%) of total protein	
	Cotyledon	Seed coat
Stored at 4°C (9 mo)	9.2	29.5
Stored at 25°C (9 mo)	17.2	44.2

Molina et al. (1977)

In order to prevent some of the reduction in digestibility due to storage, it is recommended that the more susceptible grains be processed as soon as possible after harvest, or at a constant time after harvest. The more susceptible species belong to the genus *Phaseolus*, in particular those with red or black seed coats. Consequently, in experimental work the time between harvesting and processing should always be stated, as well as the conditions of storage.

Concluding Remarks

The full meaning of productivity is not complete if it refers merely to weight of grain per unit area. Rather, productivity must be viewed as the efficiency with which the total nutrient production meets the needs of the population, with a minimum of waste. Bressani & Elias [10], thus recommend that the basis for selection of food crops must be based on production per hectare as the first component of productivity, but for this to be modified by the nutritional quality, and, finally, by a technological index. The nutritional quality factor refers mainly to protein. The technological index refers to the attributes the food must have to be acceptable to both the consumer and the food processer. Programs in these particular research areas appear to have been impaired by the lack of a clear definition of objectives for plant breeders to strive for. If the major consideration is the improvement of the nutritional quality of the total diet, the value of legume seeds as a complement to other dietary component, rather than as a complete food, must be the overriding criterion.

The pulses remain an under-exploited source of edible protein. Greater attention needs to be given to their genetic diversity in order to improve amino acid profiles, in particular to improve the level of the sulphur-containing amino acids and to eliminate anti-nutritional factors. It is imperative that the cooking quality and consumer acceptance of the new varieties also be taken into account. Breeding for improved nutritional quality should not be undertaken at the expense of those factors contributing to an improved yield.

REFERENCES

1. Aykroyd, WR, and J Doughty (1964): Legumes in Nutrition. FAO Nutritional Studies 19: FAO Rome.
2. Balaravi, SP, HC Bansal, BO Eggum, and S Bhaskaram (1976): Characterization of induced high protein and high lysine mutants in barley. J. Sci. Fd. Agric. 27: 545–552.
3. Baelum, J, and VE Petersen (1964): Bilag til Landokonomisk Forsogslaboratoriums Efterarsmode Copenhagen, p. 311.
4. Beames, RM, and JO Sewell (1969): A comparison between barley and sorghum when combined with soybean meal or meat and bone meal in rations for growing pigs. Aust. J. Exp. Agric. Anim. Husb. 9: 482–489.
5. Beaton, GH, and LD Swiss (1974): Evaluation of the nutritional quality of food supplies; prediction of 'desirable' or 'safe' protein: calorie ratios. Am. J. Clin. Nutr. 27: 485–504.
6. Bender, AE (1968): Nutritive value of bread protein fortified with amino acids. Science N.Y. 127: 874–875.
7. Bengtsson, A, and BO Eggum (1969): Virkningen af stigende N-godskning pa havre og byg-proteinets kvalitet. Tidsskr. Plant. 73: 105–114.
8. Bressani, R (1972): A new assessment of needed research. In: Milner, M (ed.): Nutritional Improvement of Food Legumes by Breeding: 381–389; Wiley, New York.
9. Bressani, R (1975): Legumes in human diets and how they might be improved. In: Milner, M (ed.): Nutritional Improvement of Food Legumes by Breeding: 15–42; Wiley, New York.
10. Bressani, R, and LG Elias (1977): The problem of legume protein digestibility. In: Hulse, JH, KO Rachie, and LW Billingsley (eds.): Nutritional Standards and Methods of Evaluation for Food Legume Breeders: 61–73; IDRC.
11. Bressani, R, LG Elias, and MR Molina (1977): Estudios sobre la digestibilidad de la proteina de varios especies de leguminosas. Arch. Latinoam. Nutr. Quoted from Bressani and Elias, 1977.
12. Briggs, DE (1978): Barley. Chapman and Hall, London.
13. Burr, HK (1975): Pulse Protein. In: Friedmann, M (ed.): Protein Nutritional Quality of Foods and Feeds 2: 119–134; Dekker Inc., New York.
14. de Muelenaere, HJH (1964): Studies on the digestion of soybeans. J. Nutr. 82: 197–205.
15. Eggum, BO (1970): Über die Abhängigkeit der Proteinqualität vom Stickstoffgehalt der Gerste. Z. Tierphysiol., Tierernähr. Futtermittelk. 26: 65–71.
16. Eggum, BO (1973a): A study of certain factors influencing protein utilization in rats and pigs. 406. Beretn. National Institute of Animal Science, Copenhagen. 173 pp.
17. Eggum, BO (1973b): Biological availability of amino acid constituents in grain protein. Nuclear Techniques for Seed Protein Improvement: 391–408; IAEA Vienna.
18. Eggum, BO (1977a): Nutritional aspects of cereal proteins. In: Muhammed, A, R Aksel, and C von Borstel (eds.): Genetic Diversity in Plants: 349–369; Premium Press, New York, London.
19. Eggum, BO (1977b): Nutritive value of food crops and factors affecting the utilization of dietary protein and energy. Second FAO/SIDA Seminar on Field Food Crops in Africa and the Near East; Lahore, Pakistan.
20. Eggum, BO (1978): Protein quality of cereal processed in various ways. Seed Protein Improvement by Nuclear Techniques: 353–389; IAEA Vienna.
21. Eggum, BO (1980): Nutritional problems related to double low rapeseed in animal nutrition. Seminar 'Production and Utilization of Protein in Oilseed Crops'; Braunschweig.
22. Eggum, BO, and KD Christensen (1975): Influence of tannin on protein utilization in feedstuffs with special reference to barley. Breeding for Seed Protein Improvement using Nuclear Techniques: 135–143; IAEA Vienna.
23. Eggum, BO, and BO Juliano (1973): Nitrogen balance in rats fed rices differing in protein content. J. Sci. Food Agric. 24: 921–927.

24. Eggum, BO, and BO Juliano (1975): Higher protein content from nitrogen fertilizer application and nutritive value of milled rice protein. J. Sci. Food Agric. 26: 425–427.
25. Eggum, BO, I Kreft, and B Javornik (1981): Chemical composition and protein quality of buckwheat. Qual. Plant Foods Hum. Nutr. 30: 175–179.
26. Eggum, BO, AP Resureccion, and BO Juliano (1977): Effect of cooking on nutritional value of milled rice in rats. Nutr. Rep. Intern. 16: 649–655.
27. Elias, LG, and R Bressani (1977): The effect of various types of heat treatment on the protein quality of cowpea (*V. sinensis*). Quoted from Bressani and Elias (1977).
28. Elias, LG, DG de Fernandez, and R Bressani (1977): Studies of beans on the nutritive value of its protein. J. Food Sci. Quoted from Bressani and Elias (1977).
29. Elias, LG, M Hernandez, and R Bressani (1977): The nutritive value of pre-cooked legume flours processed by different methods. Quoted from Bressani and Elias (1977).
30. Elias, LG, CC Moh, and R Bressani (1973): Protein quality of a white coated mutant from irradiation of a black coated variety of beans (*P. vulgaris*). INCAP, Annual Report, 1973.
31. Eppendorfer, W (1975): Effects of fertilizer on quality and nutritional value of grain protein. 11th Kolloquium Intern. Potash Inst. Ronne, Denmark: 213–227.
32. Food and Agriculture Organization of the United Nations. (1966): Food Balance Sheets, 1960–62, Average. FAO Rome.
33. FAO/UN (1970): Amino acid content of foods and biological data on proteins. FAO Rome.
34. FAO/UN (1973): Production Year Book, 1972, Vol. 6. FAO Rome.
35. Folkes, BF, and EW Yemm (1956): The amino acid content of the proteins of barley grains. Biochem. J. 62: 4–11.
36. Glick, Z, and MA Joslyn (1970): Effect of tannic acid and related compounds on the absorption and utilization of proteins in the rat. J. Nutr. 100: 516–520.
37. Gopalan, C (1975): Protein versus calories in the treatment of protein-caloric malnutrition: Metabolic and population studies in India. In: Olson, RE (ed.): Protein-Caloric Malnutrition: 329–341; Academic Press, New York; 467 pp.
38. Helm, JH (1972): Chemical and genetic evaluation of high lysine and protein in selected barley crosses. Thesis, Oregon State University.
39. Howe, EE, GR Jansen, and EW Gilfillan (1965): Amino acid supplementation of cereal grains as related to the world food supply. Am. J. Clin. Nutr. 16: 315.
40. Ivanov, CH, B Mesrob, and Z Prusik (1968): Fractionation of hordein by preparative continuous carrier-free electrophoresis. Can. J. Biochem. 46: 1301–1307.
41. Jaffé, WG (1950a): El valor biologico comparativo de algunas leguminosas de importancia en la alimentacion venezolana. Arch. Venez. Nutr. (Venezuela) 1, 107–126.
42. Jaffé, WG (1950b): Protein digestibility and trypsin inhibitor activity of legume seeds. Proc. Soc. Exp. Biol. Med. 75: 219–220.
43. Jaffé, WG (1973): Factors affecting the nutritional value of beans. In: Milner, M (ed.): Nutritional Improvement of Food Legumes by Breeding: 43; UN New York.
44. Jaffé, WG, and ME Flores (1975): Coccion de frijoles (*Phaseolus vulgaris*). [Cooking of beans (*Phaseolus vulgaris*)]. Arch. Latinoam. Nutr. (Venezuela) 25: 79–90.
45. Jambunathan, R, and ET Mertz (1973): Relationship between tannin levels, rat growth, and distribution of proteins in sorghum. J. Agric. Fd. Chem. 21: 692–696.
46. Juliano, BO, AA Antonio, and BV Esmama (1973): Effects of protein content on the distribution and properties of rice protein. J. Sci. Fd. Agric. 24: 295–306.
47. Just Nielsen, A (1970): Alsidige foderrationers energetiske vaerdi til vaekst hos svin belyst ved forskellig metodik. 381. Beretn. National Institute of Animal Science, Copenhagen; 212 pp.
48. Kakade, ML (1974): Biochemical basis for the differences in plant protein utilization. J. Agric. Fd. Chem. 22: 550–555.
49. Kakade, ML, and RJ Evans (1966): Growth inhibition of rats fed raw navy beans (*Phaseolus vulgaris*) J. Nutr. 90: 191–198.

50. Khan, MK, and BO Eggum (1978): Effect of baking on the nutritive value of Pakistani bread. J. Sci. Fd. Agric. 29: 1069–75.
51. Khan, MK, I Jacobsen, and BO Eggum (1979): Nutrititive value of some improved varieties of legumes. J. Sci. Fd. Agric. 30: 395–400.
52. Køie, B, and G Nielsen (1977): Extraction and separation of hordeins. Techniques for the Separation of Barley and Maize Proteins. Comm. Europ. Commun. Publ. EUR 5687e: 25–35.
53. Kuiken, KA, and CM Lyman (1948): Availability of amino acids in some foods. J. Nutr. 36: 359–368.
54. Liener, IE (1962): Toxic factors in edible legumes and their elimination. Am. J. Clin. Nutr. 11: 281–298.
55. Liener, IE (1979): The nutritional significance of plant protease inhibitors. Proc. Nutr. Soc. 38: 109–113.
56. Liener, IE, and ML Kakade (1969). In: Liener, IE (ed.): Toxic Constituents in Plant Foodstuffs. Academic Press, New York.
57. Lyman, RL, BA Olds, and GM Green (1974): Chymotrypsinogen in the intestine of rats fed soybean trypsin inhibitor and its inability to suppress pancreatic enzyme secretions. J. Nutr. 104: 105–110.
58. MacKey, J (1981): Cereal production. In: Pomeranz, Y, and L Munck (eds.): Proc. Intern. Symp. on Cereals: A Renewable Resource, Theory and Practice, Copenhagen; AACC, St. Paul, MN.
59. Mattern, PJ, VA Johnson, JE Stroike, JW Schmidt, L Klepper, and RL Ulmer (1975): Status of protein quality improvement in wheat. In: High Quality Protein Maize; Proc. Symp. El Batán, Mexico: 387–397; Dowden, Hutchinson and Ross, Stroudsburg.
60. Miflin, BJ, and PR Shewry (1977): An introduction to the extraction and characterization of barley and maize prolamins. Techniques for the Separation of Barley and Maize Proteins. Comm. Europ. Commun. Publ. EUR 5687e: 13–21.
61. Milic, BL, S Stojaanovic, and N Vucurevic (1972): Lucerne tannins. II. Isolation of tannins from lucerne, their nature and influence on the digestive enzymes *in vitro*. J. Sci. Fd. Agric. 23: 1157–1162.
62. Miller, DS, and PR Payne (1961): Problems in the prediction of protein values of diets: the use of food composition tables. J. Nutr. 74: 413–419.
63. Milner, M (1969): Protein-enriched cereal foods for world needs. Am. Assoc. Cereal Chemists. St. Paul.
64. Molina, MR, MA Batten, and R Bressani (1977): Heat-treatment: A process to control the development of hard-shell in black beans (*Phaseolus vulgaris*). Quoted from Bressani & Elias, (1977).
65. Molina, MR, G de la Fuente, and R Bressani (1975): Interrelationships between storage, soaking time, cooking time, nutritive value and other characteristics of the black bean (*P. vulgaris*). J. Fd. Sci. 40: 587–591.
66. Munck, L (1964): The variation of nutritional value in barley. I. Variety and nitrogen fertilizer effects on chemical composition and laboratory feeding experiments. Hereditas 52: 1–35.
67. Munck, L (1977): Improvement of nutritional value in cereals. Hereditas 72: 1–128.
68. Oke, OL (1975): A case for vegetable proteins in developing countries. Wld. Rev. Nutr. Diet 23: 259–295.
69. Olson, RE (1975): Protein-Caloric Malnutrition. Academic Press, New York; 467 pp.
70. Osborne, TB (1909): The Vegetable Proteins. Longmans, Green & Co. London. New Impression, 1918.
71. Parpia, HAB (1972): Utilization problems in food legumes. In: Milner, M (ed.): Nutritional Improvement of Food Legumes by Breeding: 281; Wiley, New York, London, Sydney, Toronto.
72. Payne, PR (1975): Safe protein-calorie ratios in diets. The relative importance of protein and energy as causal factors in malnutrition. Am. J. Clin. Nutr. 28: 281–86.
73. Pereira, S, and A Begum (1974): The manifestations and management of severe protein-calorie malnutrition (Kwashiorkor). Wld. Rev. Nutr. Diet 19: 1–50.
74. Pomeranz, Y, SG Robbins, DM Wesenberg, EA Hockett, and JT Gilbertson (1973): Amino acid composition of two-rowed and six-rowed barleys. J. Agric. Fd. Chem. 21: 218–221.

75. Prabhu, GA (1971): The amino acid composition and nutritional evaluation of proteins in Indian feedingstuffs. Report as a FAO-Fellow. Copenhagen. 23 pp.
76. Price, ML, and LG Butler (1980): Tannins and Nutrition. Station Bulletin 272; 37 pp. Department of Biochemistry, Agricultural Experiment Station, Purdue University, West Lafayette, Indiana.
77. Rhodes, AP, and AA Gill (1980): Fractionation and amino acid analysis of the salt soluble protein fractions of normal and high lysine barleys. J. Sci. Fd. Agric. 31: 467–473.
78. Rosenberg, HR, R Culik, and RE Eckert (1959): Lysine and threonine supplementation of rice. J. Nutr. 69: 217–228.
79. Shewry, PR, HM Pratt, and BJ Miflin (1977): Extraction and separation of barley seed proteins. Techniques for the Separation of Barley and Maize Proteins. Commission of the European Communities Publ. EUR 5687e: 37–47.
80. Saunders, RM and GO Kohler (1972): In vitro determination of protein digestibility in wheat millfeeds for monogastric animals. Cereal Chem. 49: 98–103.
81. Schrickel, DJ, and WL Clark (1975): Status of protein quality improvement in oats. In: High Quality Protein Maize; Proc. Symp. El Batán, Mexico: 398–411; Dowden, Hutchinson and Ross, Stroudsburg.
82. Scrimshaw, NS, and AM Altschul (1971): Amino Acid Fortification of Protein Foods. MIT Press, Cambridge.
83. Shimada, A, and TR Cline (1974): Limiting amino acids of triticale for the growing rat and pig. J. Anim. Sci. 38: 941–946.
84. Silano, V (1977): Factors affecting digestibility and availability of proteins in cereals. Nutritional Evaluation of Cereal Mutants: 13–46; IAEA Vienna.
85. Swaminathan, MS, MS Naik, AK Kaul, and A Austin (1970): In choice of strategy for genetic upgrading of protein properties in cereals, millets and pulses. Improving Plant Protein by Nuclear Techniques: 121–183; IAEA Vienna.
86. Thomke, S (1970): Über die Veränderung des Aminosäuregehaltes der Gerste mit steigendem Stickstoffgehalt. Z. Tierphysiol., Tierernähr., Futtermittelk. 27: 23–31.
87. West, QM (1969): The qualitative role of cereals as suppliers of dietary protein. In: Milner, M (ed.): Protein Enriched Cereals for World Needs. Am. Assoc. Cereal Chemists; St. Paul.